Symbol	Meaning
LRL	Lower Reject Limit
M	Maximum allowable percent defective
Md	Median
Mo	Mode
N	Lot size (batch; population)
n	Sample size (number of observations)
np	Number nonconforming
Np_0	Standard or reference value; central line; average occurrence of an event
OC	Operating Characteristic
$P(A)$	Probability of an event
P_a, P_r	Probability of acceptance; of rejection
P_r^n	Permutation
p(F)	Proportion or fraction nonconforming (population)
p_0	Standard or reference value; central line
$\overline{P}$	Average proportion or fraction nonconforming
$100p$	Percent nonconforming
$p_{0.95}$, $p_{0.05}$	Lot or process quality related to OC curves
R	Range; reliability
R_0	Standard or reference value; central line
$\overline{R}$	Average of ranges
Q_u, Q_L	Quality indexes for U and L in variables sampling plan
q	Proportion or fraction conforming $(1 - p)$
SkSP	Skip-lot Sampling Plan
$s(s)$	Sample standard deviation (population)
$s^2(s^2)$	Sample variance population
s_0	Standard or reference value; central line
s_R	Sample standard deviation of ranges
s_p	Sample standard deviation of proportions
$S_{\overline{X}}$	Sample standard deviation of averages
$\overline{s}$	Average sample standard deviation
t	Time
USL	Upper Specification Limit
UCL	Upper Control Limit
URL	Upper Reject Limit
u	Count of nonconformities per unit
u_0	Standard or reference value; central line
$\overline{u}$	Average of nonconformities per unit
w	Weight
X_i	Observed value
$\overline{X}(\mu)$	Sample average or average (population mean)
$\overline{X}_o$	Standard or reference value; central line
$\overline{\overline{X}}$	Average of averages or grand average
X^*	Percent tolerance precontrol value
W	Standardized range value
Z	Standardized normal value; standardized individual measurement value
$\overline{Z}$	Standardized subgroup average value
α	Producer's risk; Type I error
β	Consumer's risk; Type II error; Weibull slope
λ	Failure rate
μ	See $\overline{X}$
Σ	"Sum of"
σ	See s
θ	Mean life; mean time to failure

QUALITY CONTROL

SEVENTH EDITION

Dale H. Besterfield, Ph.D., P.E.
Professor Emeritus
College of Engineering
Southern Illinois University
Principal, Besterfield & Associates

Upper Saddle River, New Jersey
Columbus, Ohio

Library of Congress Cataloging in Publication Data

Besterfield, Dale H.
Quality control/Dale H. Besterfield.—7th ed.
p. cm.
ISBN 0-13-113127-3
1. Quality control. I. Title.
TS156.B47 2004
658.5'62–dc21 2003054864

Editor in Chief: Stephen Helba
Executive Editor: Debbie Yarnell
Associate Editor: Kimberly Yehle
Production Editor: Louise N. Sette
Production Supervision: Custom Editorial Productions, Inc.
Design Coordinator: Diane Ernsberger
Cover Designer: Jim Hunter
Production Manager: Brian Fox
Marketing Manager: Jimmy Stephens

This book was set in Times Roman by Custom Editorial Productions, Inc. It was printed and bound by R. R. Donnelley & Sons Company. The cover was printed by The Lehigh Press, Inc.

Pearson Education Ltd.
Pearson Education Singapore Pte. Ltd.
Pearson Education Canada, Ltd.
Pearson Education—Japan
Pearson Education Australia Pty. Limited
Pearson Education North Asia Ltd.
Pearson Educación de Mexico, S.A. de C.V.
Pearson Education Malaysia Pte. Ltd.

10 9 8 7 6 5 4 3 2

ISBN 0-13-113127-3

PREFACE

This book provides a fundamental, yet comprehensive, coverage of quality control concepts. A practical state-of-the-art approach is stressed throughout. Sufficient theory is presented to ensure that the reader has a sound understanding of the basic principles of quality control. The use of probability and statistical techniques is reduced to simple mathematics or is developed in the form of tables and charts.

The book has served the instructional needs of technology students in technical institutes, community colleges, and universities. It has also been used by undergraduate and graduate business students. Professional organizations and industrial corporations have found the book an excellent training manual for instruction of manufacturing, quality, inspection, marketing, purchasing, and product design personnel.

Quality Control, Seventh Edition, begins with an introductory chapter about quality responsibility and two chapters that describe total quality management. These chapters are followed by chapters on fundamentals of statistics, control charts for variables, additional SPC techniques for variables, fundamentals of probability, and control charts for attributes. A subsequent group of chapters describes acceptance sampling, reliability, and management and planning tools. Also included is a CD-ROM of data files that uses Microsoft Excel.

I am indebted to the publishers and authors who have given permission to reproduce their charts, graphs, and tables. I thank Peter J. Hardro—Naval Undersea Warfare Center, New Port/University of Massachusetts, Dartmouth; Mitchel Lifland—Eastern Kentucky University; and Daryl Santos—Binghamton University, for reviewing the manuscript. Professors, practitioners, and students throughout the world have been most helpful in pointing out the need for further clarification and additional material in this seventh edition.

Dale H. Besterfield

CONTENTS

1 INTRODUCTION TO QUALITY

INTRODUCTION

Definitions

When the expression "quality" is used, we usually think in terms of an excellent product or service that fulfills or exceeds our expectations. These expectations are based on the intended use and the selling price. For example, a customer expects a different performance from a plain steel washer than from a chrome-plated steel washer because they are a different grade. When a product surpasses our expectations we consider that quality. Thus, it is somewhat of an intangible based on perception.

Quality can be quantified as follows:

$$Q = P/E$$

where Q = quality
P = performance
E = expectations

If Q is greater than 1.0, then the customer has a good feeling about the product or service. Of course, the determination of P and E will most likely be based on perception, with the organization determining performance and the customer determining expectations. Customer expectations are continually becoming more demanding.

A more definitive definition of quality is given in ISO 9000: 2000. It is defined as the degree to which a set of inherent characteristics fulfills requirements. *Degree* means that quality can be used with adjectives such as poor, good, and excellent. *Inherent* is defined as existing in something, especially as a permanent characteristic. *Characteristics* can be quantitative or qualitative. *Requirement* is a need or expectation that is stated; generally implied by the organization, its customers, and other interested parties; or obligatory.

Quality has nine different dimensions. Table 1-1 shows these nine dimensions of quality with their meanings and explanations in terms of a slide projector.

These dimensions are somewhat independent; therefore, a product can be excellent in one dimension and average or poor in another. Very few, if any, products excel in all nine dimensions. For example, the Japanese were cited for high-quality cars in the 1970s based only on the dimensions of reliability, conformance, and aesthetics. Therefore, quality products can be determined by using a few of the dimensions of quality.

Marketing has the responsibility of identifying the relative importance of each dimension of quality. These dimensions are then translated into the requirements for the development of a new product or the improvement of an existing one.

Quality control is the use of techniques and activities to achieve, sustain, and improve the quality of a product or service. It involves integrating the following related techniques and activities:

1. *Specifications* of what is needed
2. *Design* of the product or service to meet the specifications
3. *Production* or *installation* to meet the full intent of the specifications
4. *Inspection* to determine conformance to specifications
5. *Review of usage* to provide information for the revision of specifications if needed

Utilization of these activities provides the customer with the best product or service at the lowest cost. The aim should be continued quality improvement.

Statistical quality control (SQC) is a branch of Total Quality Management which is defined below. It is the collection, analysis, and interpretation of data for use in quality control activities. While much of this book emphasizes the statistical approach to quality control, this is only a part of the total picture. *Statistical process control* (SPC) and *acceptance sampling* are the two major parts of SQC.

TABLE 1-1 The Dimensions of Quality.

DIMENSION	MEANING AND EXAMPLE
Performance	Primary product characteristics, such as the brightness of the picture
Features	Secondary characteristics, added features, such as remote control
Conformance	Meeting specifications or industry standards, workmanship
Reliability	Consistency of performance over time, average time for the unit to fail
Durability	Useful life, includes repair
Service	Resolution of problems and complaints, ease of repair
Response	Human-to-human interface, such as the courtesy of the dealer
Aesthetics	Sensory characteristics, such as exterior finish
Reputation	Past performance and other intangibles, such as being ranked first

Adapted from David A. Garvin *Meanaging Quality: The Strategic and Competitive Edge* (New York: Free Press, 1988).

All the planned or systematic actions necessary to provide adequate confidence that a product or service will satisfy given requirements for quality is called *quality assurance*. It involves making sure that quality is what it should be. This includes a continuing evaluation of adequacy and effectiveness with a view to having timely corrective measures and feedback initiated where necessary.

Total Quality Management (TQM) is defined as both a philosophy and a set of guiding principles that represent the foundation of a continuously improving organization. It is the application of quantitative methods and human resources to improve all the processes within an organization and exceed customer needs now and in the future. TQM integrates fundamental management techniques, existing improvement efforts, and technical tools under a disciplined approach. It is discussed in Chapters 2 and 3.

A *process* is a set of interrelated activities that use specific inputs to produce specific outputs. The output of one process is usually the input to another. Process refers to both business and production activities. *Customer* refers to both internal and external customers, and *supplier* refers to both internal and external suppliers.

Historical Review

The history of quality control is undoubtedly as old as industry itself. During the Middle Ages, quality was to a large extent controlled by the long periods of training required by the guilds. This training instilled pride in workers for quality of a product.

The concept of specialization of labor was introduced during the Industrial Revolution. As a result, a worker no longer made the entire product, only a portion. This change brought about a decline in workmanship. Because most products manufactured during that early period were not complicated, quality was not greatly affected. In fact because productivity improved there was a decrease in cost, which resulted in lower customer expectations. As products became more complicated and jobs more specialized, it became necessary to inspect products after manufacture.

In 1924, W. A. Shewhart of Bell Telephone Laboratories developed a statistical chart for the control of product variables. This chart is considered to be the beginning of statistical quality control. Later in the same decade, H. F. Dodge and H. G. Romig, both of Bell Telephone Laboratories, developed the area of acceptance sampling as a substitute for 100% inspection. Recognition of the value of statistical quality control became apparent by 1942. Unfortunately, U.S. managers failed to recognize its value.

In 1946, the American Society for Quality was formed. This organization, through its publications, conferences, and training sessions, has promoted the use of quality control for all types of production and service.

In 1950, W. Edwards Deming, who learned statistical quality control from Shewhart, gave a series of lectures on statistical methods to Japanese engineers and on quality responsibility to the CEOs of the largest organizations in Japan. Joseph M. Juran made his first trip to Japan in 1954 and further emphasized management's responsibility to achieve quality. Using these concepts the Japanese set the quality standards for the rest of the world to follow.

In 1960, the first quality control circles were formed for the purpose of quality improvement. Simple statistical techniques were learned and applied by Japanese workers.

By the late 1970s and early 1980s, U.S. managers were making frequent trips to Japan to learn about the Japanese miracle. These trips were really not necessary—they could have read the writings of Deming and Juran. Nevertheless, a quality renaissance began to occur in U.S. products and services, and by the middle of 1980 the concepts of TQM were being publicized.

In the late 1980s the automotive industry began to emphasize statistical process control (SPC). Suppliers and their suppliers were required to use these techniques. Other industries and the Department of Defense also implemented SPC. The Malcolm Baldrige National Quality Award was established and became the means to measure TQM. Genechi Taguchi introduced his concepts of parameter and tolerance design and brought about a resurgence of design of experiments (DOE) as a valuable quality improvement tool.

Emphasis on quality continued in the auto industry in the 1990s when the Saturn automobile ranked third in customer satisfaction behind the two most expensive Japanese automobiles. In addition, ISO 9000 became the worldwide model for a quality system. The automotive industry modified ISO 9000 to place greater emphasis on customer satisfaction and added elements on production part approval process, continuous improvement, and manufacturing capabilities. ISO 14000 was approved as the worldwide model for environmental management systems.

By the year 2000, the quality focus was shifting to information technology within an organization and externally via the internet.

Metric System

In 1960, the International Committee of Weights and Measures revised the metric system. This revision is the International System of Units (SI),[1] which has the following base units:

- Length—meter (m)
- Mass—kilogram (kg)
- Time—second (s)
- Electrical current—ampere (A)
- Thermodynamic temperature—kelvin (K)
- Amount of matter—mole (mol)
- Luminous intensity—candela (cd)

These basic units are combined to form other units such as kg/m^2 (psi) and m/s (ft/sec).

This book uses the metric system of units with U.S. units given in parentheses. A detailed understanding of SI is not necessary because the concepts are independent of the units. Commonly used conversion factors are given in Table E of the appendix.

RESPONSIBILITY FOR QUALITY

Areas Responsible

Quality is not the responsibility of any one person or functional area; it is everyone's job. It includes the assembly-line worker, the typist, the purchasing agent, and the president of the company. The responsibility for quality begins when marketing determines the customer's quality requirements and continues until the product is received by a satisfied customer.

The responsibility for quality is delegated to the various areas with the authority to make quality decisions. In addition, a method of accountability, such as cost, error rate, or nonconforming units, is included with that responsibility and authority. The areas responsible for quality control are shown in Figure 1-1. They are marketing, design engineering, procurement, process design, production,

[1]Copies may be purchased from the Superintendent of Documents, Government Printing Office, Washington, D.C. 20402. (Order by SD Catalog No. C13.10: 330/3.)

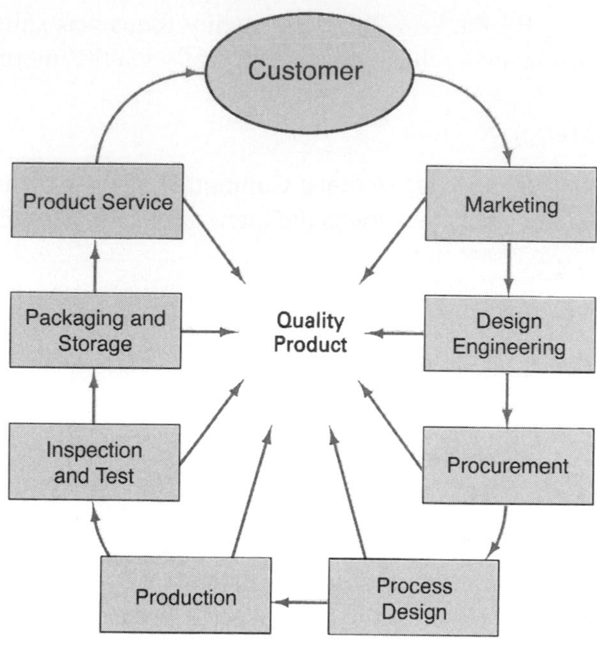

FIGURE 1-1 Areas responsible for quality.

inspection and test, packaging and storage, product service, and the customer. Figure 1-1 is a closed loop with the *customer* at the top and the areas in the proper sequence in the loop. Since the quality function does not have direct responsibility for quality, it is not included in the closed loop of the figure.

The information in this section pertains to a manufactured item; however, the concepts can be adapted to a service or any organization.

Marketing

Marketing helps to evaluate the level of product quality that the customer wants, needs, and is willing to pay for. In addition, marketing provides the product-quality data and helps to determine quality requirements.

A certain amount of marketing information is readily available to perform this function. Information concerning customer dissatisfaction is provided by customer complaints, sales representative reports, product service, and product liability cases. The comparison of sales volume with the economy as a whole is a good predictor of customer opinion of product quality. A detailed analysis of spare-part sales can locate potential quality problems. Useful market quality information is also provided by government reports on consumer product safety and independent laboratory reports on quality.

When information is not readily available, there are four methods that can be developed to obtain the desired product quality data:

1. Visit or observe the customer to determine the conditions of product use and the problems of the user.
2. Establish a realistic testing laboratory such as an automotive test track.
3. Conduct a controlled market test.
4. Organize a dealer advisory or focus group.

Marketing evaluates all the data and determines the quality requirements for the product. An information-monitoring and feedback system on a continuing basis is essential to collect data in an effective manner.

Marketing provides the company with the product brief, which translates customer requirements into a preliminary set of specifications. Among the product brief elements are

1. Performance characteristics, such as environmental, usage, and reliability considerations,
2. Sensory characteristics, such as style, color, taste, and smell,
3. Installation, configuration, or fit,
4. Applicable standards and statutory regulations,
5. Packaging, and
6. Quality verification.

Marketing is the liaison with the customer and as such is a vital link to the development of a product that surpasses customer expectations.

Design Engineering

Design engineering translates the customer's quality requirements into operating characteristics, exact specifications, and appropriate tolerances for a new product or revision of an established product. The simplest and least costly design that will meet the customer's requirements is the best design. As the complexity of the product increases, the quality and reliability decrease. Early involvement of marketing, production, quality, procurement, and the customer is essential to prevent problems before they occur. This type of involvement is called concurrent engineering.

Whenever possible, design engineering should utilize proven designs and standard components. In this regard, industry and government standards are used when applicable.

Tolerance is the permissible variation in the size of the quality characteristic, and the selection of tolerances has a dual effect on quality. As tolerances are tightened, a better product usually results; however, production and quality costs may increase. Ideally, tolerances should be determined scientifically by balancing the precision desired with the cost to achieve that precision. Since there are too many quality characteristics for scientific determination, many tolerances are set using standard dimensioning and tolerancing systems. Designed experiments are a very effective technique for determining which process and product characteristics are critical as well as their tolerances. Critical tolerances should be established in conjunction with the process capability.

The designer determines the materials to be used in the product. Material quality is based on written specifications, which include physical characteristics, reliability, acceptance criteria, and packaging.

In addition to the functional aspect, a quality product is one that can be used safely. It is also one that can be repaired or maintained easily.

Design reviews are conducted at appropriate phases in the development of the product. These reviews should identify and anticipate problem areas and inadequacies, and initiate corrective action to ensure that the final design and supporting data meet customer requirements. After the design review team approves the product for production, the final quality requirements are distributed. Quality is designed into the product before it is released to manufacturing.

No design is perfect over time; therefore, provision must be made for design-change control. Also there should be a periodic reevaluation of the product in order to ensure that the design is still valid.

Procurement

Using the quality requirements established by design engineering, procurement has the responsibility of procuring quality materials and components. Purchases fall into four categories: standard materials, such as coiled steel and angle iron; standard hardware, such as fasteners and fittings; minor components, such as gears and diodes; and major components, which perform one of the primary functions of the product. The quality requirements will vary depending on the category of the purchase.

A particular raw material or component part may have a single supplier or multiple suppliers. A single supplier as a source is usually able to provide better quality at a lower price with better service. The concept of a single supplier has been applied quite effectively in breweries, wherein the can or bottle manufacturer was located adjacent to the brewery. Multidivisional companies use the single-supplier technique and can control quality in a manner similar to the control between areas within a plant. The disadvantage of a single supplier is the potential for a material shortage that may result due to natural causes such as fire, earthquake, or flood or due to unnatural causes such as equipment breakdowns, labor problems, or financial difficulties.

To determine if a supplier is capable of providing quality materials and components, a supplier quality survey is conducted by visiting the supplier's plant. The facilities are observed, the quality control procedures studied, and pertinent data collected. From this information a reasonable decision can be made regarding the ability of the supplier to provide quality materials and components. Once a supplier is approved, other techniques of evaluation are available. Registration by means of ISO 9000 has become the best technique for approval.

There are a number of different methods used to obtain proof of conformance to quality standards. For small quantities, procurement will frequently rely on the supplier. The inspection of incoming materials and components is one of the most common methods for proof of conformance. Source inspection is identical to incoming inspection except that the inspection is conducted in the supplier's plant. Statistical evidence of quality by means of process control charts and process capability is a very effective method. Proof of conformance can also be obtained by inspection of duplicate samples that are received by procurement prior to the arrival of the shipment. Supplier surveillance is a method of controlling the quality in the supplier's plant by means of an acceptable plan and proof, such as inspection records, that the plan is followed. Any combination of these methods can be used to achieve an effective and continuing evaluation of the product.

A supplier quality rating system can be used to evaluate performance. Factors such as rejected lots, scrap and rework costs, or complaint information are used for the evaluation. In addition, delivery performance and price are included.

To improve the quality of purchased materials and components, two-way communication between the supplier and procurement is a necessity. Both positive and negative feedback should be given to the supplier. Supplier representatives may be included on design or process improvement teams.

Procurement should be concerned with the total cost and not price. For example, supplier A has a lower price than supplier B; however, the cost to utilize supplier A's material is so much greater than supplier B's that the total cost is greater.

Process Design

Process design has the responsibility of developing processes and procedures that will produce a quality product. This responsibility is achieved by specific activities, which include process selection and development, production planning, and support activities.

A process design review is conducted in order to anticipate quality problems. Quality problems are frequently related to specifications. When process capability information indicates that a tolerance is too tight for satisfactory producibility, there are five options: purchase new equipment, revise the tolerance, improve the process, revise the design, or sort out the defective product during production.

Process selection and development is concerned with cost, quality, implementation time, and efficiency. One of the basic techniques is the process capability study, which determines the ability of a process to meet specifications. Process capability information provides data for make-or-buy decisions, equipment purchases, and selection of process routes.

The sequence of operations is developed to minimize quality difficulties such as the handling of fragile products and the location of precision operations in the sequence. Methods study is used to determine the best way of performing either a production operation or an inspection operation.

Additional responsibilities include the design of equipment, the design of inspection devices, and the maintenance of the production equipment.

Production

Production has the responsibility to produce quality products. Quality cannot be inspected into a product; it must be built into the product.

The first-line supervisor is the key to the manufacture of a quality product. Since the first-line supervisor is considered by operating personnel to represent management, his ability to convey quality expectations is critical for good employee relations. A first-line supervisor who is enthusiastic in his commitment to quality can motivate the employees to build quality into each and every part and, thus, into the final unit. It is the supervisor's responsibility to provide the employee with the proper tools for the job, to provide instructions in the method of performing the job and the quality expectations of the job, and to provide feedback on performance.

In order for the operator to know what is expected, training sessions on quality should be given periodically. These training sessions reinforce management's commitment to a quality product. During the training sessions, time can be allocated to presentations by field personnel, to discussions concerning the sources of quality variations, to methods of improving quality, and so on. The primary objective of the sessions is to develop an attitude of "quality mindedness" and an environment where two-way, nonpunitive communications can flourish. Operating personnel, indeed all personnel, must not only do their jobs but search for ways to improve them.

According to Deming, only 15% of the quality problems can be attributed to operating personnel—the balance is due to the rest of the system. Statistical process control effectively controls quality and is an invaluable tool for quality improvement. Operating personnel should be trained to perform their own statistical process control.

Inspection and Test

Inspection and test has the responsibility to appraise the quality of purchased and manufactured items and to report the results. The reports are used by other departments to take corrective action when needed. Inspection and test may be an area by itself, part of production, or part of quality assurance. It might also be located in both production and quality assurance.

Although inspection is done by representatives of the inspection and test department, it does not relieve production of its responsibility to produce a quality product and make its own inspections. In fact, with automated production, workers frequently have time to perform 100% inspection before and after an operation. One of the major problems with the inspection activity is the tendency to view the inspector as a "police person" who has the quality responsibility. This attitude can lead to an ineffective inspection activity and a deterioration of quality.

In order to perform the inspection activity, accurate measuring equipment is needed. Normally, this equipment is purchased; however, it may be necessary to design and build it in cooperation with process design. In either case the equipment must be maintained in a constant state of repair and calibration.

It is necessary to continually monitor the performance of inspectors. Indications are that certain nonconformities are more difficult to find, that inspectors vary in their abilities, and that the quality level affects the number of nonconformities reported. Samples of known composition should be used to evaluate and improve the inspector's performance.

The efficiency of the appraisal activity is a function of the inspection methods and procedures (number inspected, type of sampling, and inspection location). Cooperation from process design, inspection and test, production, and quality assurance is necessary to maximize the inspector's performance.

Inspection and test should concentrate the majority of its efforts on statistical quality control, which will lead to quality improvement. Passing the conforming items and discarding the nonconforming ones is *not* quality control. Quality cannot be inspected into a product or service. Dependency on mass inspection for quality control is in most cases a waste of effort, time, and money.

Packaging and Storage

Packaging and storage has the responsibility to preserve and protect the quality of the product. Control of the product quality must extend beyond production to the distribution, installation, and use of the product. A dissatisfied customer is not concerned with where the nonconforming condition occurred.

Quality specifications are needed for the protection of the product during transit by all types of common carrier: truck, rail, boat, and air. These specifications are needed for vibration, shock, and environmental conditions such as temperature, moisture, and dust. Additional specifications are needed in regard to the handling of the product during loading, unloading, and warehousing. Occasionally, it is necessary to change the product or process design to correct quality difficulties that occur during transit. In some companies, the responsibility for the design of the package is vested in design engineering rather than packaging and storage.

Product storage, while awaiting further processing, sale, or use, presents additional quality problems. Specifications and procedures are necessary to ensure that the product is properly stored and promptly used to minimize deterioration and degradation.

Product Service

Product service has the responsibility to provide the customer with the means for fully realizing the intended function of the product during its expected life. This responsibility includes sales and distribution, installation, technical assistance, maintenance, and disposal after use. Products should be serviced quickly whenever they are improperly installed or fail during the warranty period. Prompt service can change a dissatisfied customer into a satisfied one.

Product service and marketing work closely with each other to determine the quality the customer wants, needs, and obtains.

Quality Assurance

The quality assurance or quality control (the name is not important) *does not* have direct responsibility for quality. Therefore, it is not shown in Figure 1-1. It assists or supports the other areas as they carry out their quality control responsibilities. Quality assurance *does* have the direct responsibility to continually evaluate the effectiveness of the quality system. It determines the effectiveness of the system, appraises the current quality, determines quality problem areas or potential areas, and assists in the correction or minimization of these problem areas. The overall objective is the improvement of the product quality in cooperation with the responsible departments.

CHIEF EXECUTIVE OFFICER

The chief executive officer (CEO) of a plant has responsibility for each of the areas in the closed loop of Figure 1-1 and the quality assurance area. Therefore, the CEO has the ultimate responsibility for quality. The CEO must be involved directly in the quality effort. This activity requires a knowledge of quality and direct involvement with the quality improvement program. Merely stating that quality is important is not sufficient.

Direct involvement requires the creation of a quality council and participation in meetings. It also involves being a member of a quality improvement project team, participating in recognition ceremonies, developing a mission statement, having a quarterly employee meeting, and writing a column in the monthly newsletter. Management By Walking Around (MBWA) is an excellent technique to identify quality problems.

Perhaps the best way for the CEO to be involved is to have some measure of his or her quality performance. Financial information can provide a long-term measure of quality performance. However, in the short term, it is not too difficult to make the financial data look good when in reality the product or service quality is deteriorating. Quality improvement requires a long-term financial commitment to people, programs, and equipment.

The CEO's quality performance can be effectively measured by a proportion (percent nonconforming) chart that covers the area of responsibility, whether it be a plant or a corporation. If the percent nonconforming is increasing or is constant, then, simply stated, the CEO's performance is poor. If the percent nonconforming is decreasing, the CEO's performance is good. This concept—measurement of quality performance—can be adapted for all managers, departments, and operating personnel. In conjunction with quality improvement, the proportion chart becomes a very effective technique.

Another technique is to use the Malcolm Baldrige National Quality Award criteria as a performance measure.

Each month, the CEO should review his/her appointment book to determine the percent of time spent on quality. Approximately 35% of the time should be on quality.

COMPUTERS AND QUALITY CONTROL

Computers play an essential role in the quality function. They perform very simple operations at fast speeds with an exceptionally high degree of accuracy. A computer must be programmed to execute these simple operations in the correct sequence in order to accomplish a given task. Computers can be programmed to perform complex calculations, to control a process or test, to analyze data, to write reports, and to recall information on command.

The quality function needs served by the computer are: (1) data collection, (2) data analysis and reporting, (3) statistical analysis, (4) process control, (5) test and inspection, and (6) system design. In addition, the computer serves as the platform for intranet and Internet utilization.

Data Collection

The collection, utilization, and dissemination of quality information is best accomplished when the information is incorporated into an information technology (IT) system. IT maintains relationships with other activities, such as inventory control, purchasing, design, marketing, accounting, and production control. It is essential for all the quality needs described in this chapter. Linkages are developed between the stored data records of the various activities in order to obtain additional information with a minimum of programming and to improve the storage utilization.

Computers are well suited for the collection of data. Principal benefits are faster data transmission, fewer errors, and lower collection costs. Data are transmitted to the computer by paper or magnetic tape, optical character recognition, touch telephone, wireless transmission, keyboard, voice, pointer, bar-code scan, and direct interface with a process.

The type and amount of data are the principal problems of data collection. Sources of data are process inspection stations, scrap and waste reports, product audits, testing laboratories, customer complaints, service information, process control, and incoming material inspection. From these sources a vast amount of data can be collected. The decision as to how much data to collect and analyze is based on the reports to be issued, the processes to be controlled, the records to be retained, and the nature of the quality improvement program.

A typical form for collecting data for an internal failure or deficiency is shown in Figure 1-2. In addition to the basic information concerning the internal failure or deficiency, a number of identifiers are used. Typical identifiers are part number, operator, first-line supervisor, supplier, product line, work center, and department. Identifiers are necessary for data analysis, report preparation, and record traceability. Once the disposition of the nonconforming material is determined, this particular report is sent to accounting, where the failure costs are assigned and the information is transmitted to the computer. Note that in a paperless factory the form would be viewed on a display monitor and information input directly to the computer.

Sometimes information is stored in the computer in order for it to be transmitted efficiently to remote terminals. For example, the operating instructions, specifications, drawings, tools, inspection gages, and inspection requirements for a particular job are stored in the computer. This information is then provided to the employee at the same time the work assignment is given. One of the principal advantages of this type of system is the ability to quickly update or change the information. Another advantage is the likelihood of fewer errors, since the operator is using current information rather than obsolete or hard-to-read instructions.

Periodically data are analyzed to determine what data to retain in the computer, what data to store by another method, and what data to destroy. Data can be stored on magnetic tape, CD, or a diskette and reentered into the computer if needed. Product liability requirements determine the amount and type of data to retain as well as the retention period.

Pacific Bell used handhold computers, which could be operated with one

DEFICIENCY REPORT

1. PART NUMBER | 2. ISSUE | SUF.

IDENTIFICATION

3. TELL | HUNT | 4. TELL OPER. # | 4A. RESP. OPER. # | 5. SUPERVISOR # | 6. DATE WRITTEN: MONTH | DAY | YEAR

7. PART NAME | 8. MATERIAL LOCATION | 9. DEF. OPN. | SUF. | 10. QTY. REJECTED

11. DEFICIENCY DESCRIPTION (Use Std. Abbreviations)

12. SHOP ORDER NUMBER

13. CLERK # | 14. DATE

15. DEPT. # | 16. SEC.-SH. | 17. DEFECT CODE | 18. LAST OPN. | SUF. | 19. COMPONENT S/N | 20. MODEL/S | 21. SERIAL NUMBER

22. VENDOR NAME | 23. VENDOR CODE | 24. WRITER # | REJECTED BY SIGNATURE

RESPONSIBILITY | 27. **DISPOSITION**

25. DEPT. | 26. ACCOUNT

USE AS IS ☐ | RTV ☐

REWORK ☐ | SCRAP ☐

REPAIR ☐ | HOLD FOR—SORT ☐ | DISP. ☐

28. **ROUTING** | TO 1ST | TO 2ND | TO 3RD | TO 4TH | TO LAST

FIGURE 1-2 Deficiency report. (Courtesy of Fiat-Allis Construction Machinery, Inc.)

hand, and bar codes to almost flawlessly inventory 27,000 different small metal circuit boards. The system, developed by a multifunction team, resulted in a reduction in spares from 7.5 to 2.5 per 100 in use and a savings of almost $100 million.[2]

Data Analysis, Reduction, and Reporting

While some of the quality information is merely stored in the computer for retrieval at a future time, most of the information is analyzed, reduced to a meaningful amount, and disseminated in the form of a report. These activities of analysis, reduction, and reporting are programmed to occur automatically as the data are collected or to occur on command by the computer operator.

Typical reports for scrap and rework as produced by a computer are shown in Figure 1-3. The weekly scrap and rework cost report of Figure 1-3(a) is a listing by part number of the information transmitted to the computer from the internal failure deficiency report. Identifiers reported for each transaction are a function of

[2] 1997 RIT/USA Today Quality Cup for service.

SCRAP AND REWORK COST REPORT FOR THE WEEK ENDING 11/26

PART#	CODE	TICKET	CITY	MATERIAL	LABOR	OVERHEAD	TOTAL
1194	E	2387	40000	800.00	.00	24.80	824.80
1275	E	1980	15	31.50	2.28	5.59	39.37
1276	D	2021	7	11.76	.94	2.30	15.00
1276	E	2442	10	16.80	1.34	3.28	21.42
9020	D	608	1	30.79	6.01	14.72	51.52
9600	D	2411	3	48.03	19.00	46.55	113.38
9862	D	2424	1	23.73	4.92	12.05	40.70
TOTAL				$13,627.35	2,103.65	5,153.98	21,307.41

(a) Scrap and rework Report

RECAP OF FAILURE CODES

	CODE EXPLANATION	AMOUNT	%
A	#OPERATION MISSED	5.36	
B	#BROKEN PARTS	.00	
C	#MISSING PARTS	.00	
D	#IMPROPER MACHINING	11,862.72	56
E	#DOUNDRY OR PURCHASING	8,841.79	41
F	#MECHANICAL FAILURE	.00	
G	#IMPROPER HANDLING	533.10	3
H	#OTHER	44.14	
		$21,307.41	100

(b) Summary by Failure Code

DEPARTMENT 4 MONTH OF OCTOBER

RANK	CODE	CODE DESCR.	$ SCRAP	$ RWK	TOTAL	%
01	D-T2	TURN	7,500	4,105	11,605	28.5
02	D-H1	HOB	5,810	681	6,491	16.0
03	D-G6	GRIND	4,152	1,363	5,515	13.8
04	D-D4	DRILL	793	3,178	3,971	9.8
05	D-L1	LAP	314	2,831	3,145	7.8

(c) Pareto Analysis

FIGURE 1-3 Typical scrap and rework reports: weekly cost report, weekly summary by failure code, and Pareto analysis by nonconformity code and department.

the report and the space available. For this report the identifiers are part number, operation code, and deficiency ticket number.

The basic data can be summarized in a number of different ways. Figure 1-3(b) shows a summary by failure code. Summaries are also compiled by operator, department, work center, defect, product line, part number, subassembly, vendor, and material.

A monthly Pareto analysis of the data by defect for Department 4 is shown in Figure 1-3(c). This Pareto analysis is in tabular form; however, the computer could have been programmed to present the information in graphical form, as illustrated by the Pareto analysis in Chapters 2 and 3. Pareto analyses could also have been computed for operators, work centers, departments, part numbers, and so on.

The previous paragraphs have described the reports associated with scrap and rework. Reports for inspection results, product audits, service information; customer complaints, vendor evaluation, and laboratory testing are all similar. Information of a graphical nature, such as control charts (see Chapters 5, 6, and 8), can be programmed, displayed at a terminal, and reproduced. Software programs using EXCEL are provided in the diskette located in the inside back cover.

Data can be analyzed as they are being accumulated, real time, rather than on a weekly or monthly basis. When this technique is practiced, decision rules can be employed in the program which will automatically signal the likelihood of a quality problem. In this manner, information concerning a potential problem is provided and corrective action taken in real time. For example, an operator could have a display monitor at the work station which would automatically post data to an $\overline{X}$ and R chart. The data could have been collected automatically by the equipment or by an electronic gage that would transfer the data to the monitor.

A New York Telephone multifunctional team developed a state-of-the-art fraud detection system and saved the company \$5 to \$8 million per year. The team reduced the time to detect fraud from between two to four weeks to three days. International calls that exceeded a certain amount were printed out monthly and sent through interoffice mail to a supervisor who examined it for fraud and then requested a service representative to take action. Rather than wait for the monthly printout, the team changed the program so the computer alerted the service representative for action whenever a telephone number accumulated \$200 in international calls in any three-day period.[3]

Statistical Analysis

The first and still an important use of the computer in quality control was for statistical analysis. Most of the statistical techniques discussed in this book can be easily programmed. Once programmed, considerable calculation time is saved, and the calculations are error-free. The software in the diskette inside the back cover contains many programs for statistical analysis using the EXCEL spreadsheet software.

Many statistical computer programs have been published in the *Journal of Quality Technology* and can easily be adapted to any computer or programming language. In addition, information on statistical analysis techniques has been published in *Applied Statistics*. Most of these programs have been incorporated into software packages. Additional information is available on the internet. Some major software programs such as EXCEL have very sophisticated analysis techniques, such as ANOVA, Fourier Analysis, and *t* Test.

The advantages of programmed statistical software packages are

1. Time-consuming manual calculations are eliminated.
2. Timely and accurate analyses may be performed to diagnose one-time problems or to maintain process control.
3. Many practitioners with limited knowledge in advanced statistics can perform their own statistical analyses.

[3]1993 RIT/USA Today Quality Cup for service

Once a statistical package of computer programs is developed or purchased, the quality engineer can specify a particular sequence of statistical calculations to use for a given set of conditions. The results of these calculations can provide conclusive evidence or suggest additional statistical calculations for the computer to perform. Many of these tests are too tedious to perform without the use of a computer.

Using statistical process control, the U.S. Postal Service, in Royal Oak, MI, found ways to reroute more mail to the automatic sorting machine. This improvement resulted in an annual savings of $700,000 for the facility.[4]

Process Control

The first application of computers in process control was with numerically controlled (N/C) machines. Numerically controlled machines used punched paper to transmit instructions to the computer, which then controlled the sequence of operations. Paper tape is no longer used to provide instructions to a machine. Computer Numerically Controlled (CNC) machines, robots, and Automatic Storage and Retrieval Systems (ASRS) provide the basic equipment for an automated factory. The measurement and control of critical variables to keep them on target with minimum variation and within acceptable control limits requires sophisticated equipment.

An automated process control system is illustrated by the flow diagram of Figure 1-4. While the computer is a key part of automated process control, it is not the only part. There are two major interfacing subsystems between the computer and the process.

One subsystem has a sensor that measures a process variable such as temperature, pressure, voltage, length, weight, moisture content, and so on, and sends an analog signal to the digital computer. However, the digital computer can receive information only in digital form, so the signal is converted by an analog-to-digital interface. The variable value in digital form is evaluated by the computer to determine if the value is within the prescribed limits. If such is the case, no further action is necessary; however, if the digital value is outside the limits, corrective action is required. A corrected digital value is sent to the digital-to-analog interface, which converts it to an analog signal that is acceptable to an actuator mechanism, such as a valve. Then the actuator mechanism increases or decreases the variable. Some systems are designed to contain only digital information.

The other subsystem is essentially an attribute type, which either determines if a contact is on/off or controls an on/off function. Through the contact input interface, the computer continuously scans the actual on/off status of switches, motors, pumps, and so on, and compares these to the desired contact status. The

[4]1999 RIT/USA Today Quality Cup for government.

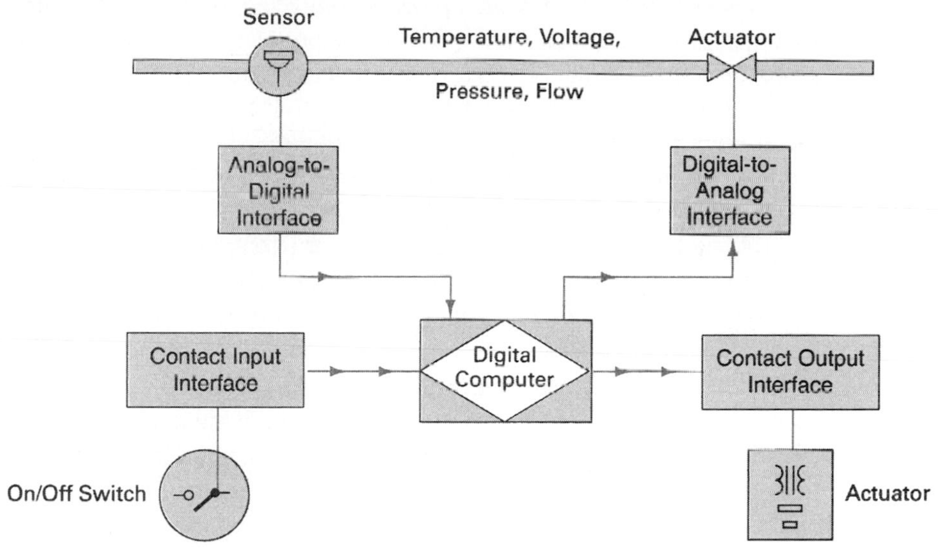

FIGURE 1-4 Automated process control system.

computer program controls the sequence of events performed during the process cycle. Operating instructions are initiated by specific process conditions or as a function of time and are sent to the contact output interface. This interface activates a solenoid, sounds an alarm, starts a pump, stops a conveyor, and so on.

The four interfaces in Figure 1-4 are capable of handling a number of signals at the same time. Also, the two subsystems can operate independently or in conjunction with each other. Since the computer operates in microseconds and the subsystems operate in milliseconds, a timing problem can occur unless the feedback loops are as tight as possible so that corrective action is immediate.[5] The benefits that are obtained from automatic process control are

1. Constant product quality, due to a reduction in process variation.
2. More uniform startup and shutdown, since the process can be monitored and controlled during these critical periods.
3. Increased productivity, because fewer people are needed to monitor the controls.
4. Safer operation for personnel and equipment, by either stopping the process or failing to start the process when an unsafe condition occurs.

[5] N. A. Poisson, "Interfaces for Process Control," *Textile Industries*, 134, No. 3 (March 1970), 61–65.

One of the first automated process-controlled installations occurred at Western Electric's North Carolina plant in 1960. The product variables were controlled by the computer using $\overline{X}$ and R control chart techniques. For example, the resistance value of deposited carbon resistors coming out of the furnace was controlled by the amount of methane in the furnace and by the speed through the furnace. Since the inspection and packaging operations were also under computer control, the entire production facility was completely automated.[6]

A nuclear generating station is another example of a fully automated system wherein the only human interaction occurs at the computer console.

An example of an automated process control for a business operation is given by the Naval Aviation Depot Operations Center at Patuxent River, MD. A multifunctional team automated the travel process of request, reservations, and reimbursement. The computer program holds profiles on individual travelers so that two-thirds of the information on the travel form is in the computer and the traveler only needs to input the itinerary. The computer makes all of the calculations for the cash advance and the reimbursement. Each week the commanding officer receives for his review and signature a one-page summary of all planned trips. The travel department can determine the year-to-date history of any traveler and whether there are any outstanding transactions. Results of the automated system are: (1) travel changes have dropped from 100 to five per month, (2) virtually 100% of travel plans are for real trips that are taken rather than 56% in the past, (3) 95% of reimbursement claims are error-free rather than 67% in the past, (4) The department has saved $42,000 in typist salaries and administrative staff has fallen from 50 to 22, and (5) a survey of travelers showed a satisfaction rating of 3.87 out of 4.00.[7]

Automated Test and Inspection

If test and inspection are considered as a process in itself or a part of a production process, then automated test and inspection is similar to the previous section on automated process control. Computer-controlled test and inspection systems offer the following advantages: improved test quality, lower operating cost, better report preparation, improved accuracy, automated calibration, and malfunction diagnostics. Their primary disadvantage is the high cost of the equipment.

Computer-controlled, automated inspection can be used for go/no-go inspection decisions or for sorting and classifying parts in selective assembly. Artificial vision is sometimes used in these processes. Automated inspection systems have the capacity and speed to be used on high-volume production lines.

---

[6]J. H. Boatwright, "Using a Computer for Quality Control of Automated Production," *Computers and Automation,* 13, No. 2 (February 1964), 10–17.

[7]1992 RIT/USA Today Quality Cup for government.

Automated test systems can be programmed to perform a complete quality audit of a product. Testing can be sequenced through the various product components and subassemblies. Parameters such as temperature, voltage, and force can be varied to simulate environmental and wear conditions. Reports are automatically prepared to reflect the performance of the product.

When automated test and inspection is applied to automated or semiautomated production, the computer can generate the inspection instructions at the same time the product is designed.

System Design

Software applications adapted to the quality function are becoming more sophisticated and comprehensive. There are numerous packages that combine many of the quality functions described previously. These software packages are user-friendly with a help provision and tutorials. Package software is much cheaper than custom software. It usually has the benefit of proven usage and technical support. Each March, *Quality Progress* publishes an updated directory of applications software particular to the quality function.

The integration of the various quality functions with other activities requires an extremely sophisticated system design. Components of a total system are available in

- CADD: Computer-Aided Drafting and Design
- CAM: Computer-Aided Manufacturing
- CAE: Computer-Aided Engineering
- MRP: Materials Requirements Planning
- MRP II: Manufacturing Resource Planning
- CAPP: Computer-Aided Process Planning
- CIM: Computer-Integrated Manufacturing
- MIS: Management Information System
- MES: Manufacturing Execution Systems
- ERP: Enterprise Resource Planning
- HRIS: Human Resource Information Systems
- TQM: Total Quality Management

Integration of these components into a total system will become commonplace in the near future. It will require the use of expert systems, relational databases, and adaptive systems.

Expert systems are computer programs that capture the knowledge of experts as a set of rules and relationships used for such applications as problem diagnosis or system performance assessment. This technology permits the thought patterns and lessons learned by experts to be consolidated and used. It provides the foundation for many of the smart systems for learning that are part of the crystal-ball system.

Relational databases are logical pointers that create linkages among different data elements to describe the relationships between them. These relationships preserve information within the system for consistent application across the entire organization.

Adaptive systems permit a system to learn from date patterns or repetitive situations. Data flow is monitored to detect, characterize, and record events that describe the actions to be taken in similar situations.[8]

When the computer is used effectively, it becomes a powerful tool to aid in the improvement of quality. However, the computer is not a device that can correct a poorly designed system. In other words, the use of computers in quality is as effective as the people who create the total system.

Bill Gates has observed: "The computer is just a tool to help in solving identified problems. It isn't, as people sometimes seem to expect, a magical panacea. The first rule of any technology used in a business is that automation applied to an efficient operation will magnify the efficiency. The second is that automation applied to an inefficient operation will magnify the inefficiency."[9]

---

[8]Gregory Watson, "Bringing Quality to the Masses: The Miracle of Loaves and Fishes," *Quality Progress* (June 1998): 29–32.

[9]Bill Gates. *The Road Ahead*, (New York, NY: Viking Penguin Books, 1995).

2 TOTAL QUALITY MANAGEMENT: PRINCIPLES AND PRACTICES

INTRODUCTION[1]

Total Quality Management (TQM) is an enhancement to the traditional way of doing business. It is a proven technique to guarantee survival in world-class competition. Only by changing the actions of management will the culture and actions of an entire organization be transformed. TQM is for the most part common sense. Analyzing the three words, we have

Total—Made up of the whole.

Quality—Degree of excellence a product or service provides.

Management—Act, art, or manner of handling, controlling, directing, etc.

---

[1]Portions of this chapter are extracted from *Total Quality Management,* 2003, by D. Besterfield, C. Besterfield-Michna, G. Besterfield, and M. Besterfield-Sacre with the permission of Prentice Hall.

Therefore, TQM is the art of managing the whole to achieve excellence. The golden rule is a simple but effective way to explain it: Do unto others as you would have them do unto you.

TQM is defined as both a philosophy and a set of guiding principles that represent the foundation of a continuously improving organization. It is the application of quantitative methods and human resources to improve all the processes within an organization and exceed customer needs now and in the future. TQM integrates fundamental management techniques, existing improvement efforts, and technical tools under a disciplined approach.

BASIC APPROACH

TQM requires six basic concepts:

1. A committed and involved management to provide long-term top-to-bottom organizational support.
2. An unwavering focus on the customer, both internally and externally.
3. Effective involvement and utilization of the entire work force.
4. Continuous improvement of the business and production processes.
5. Treating suppliers as partners.
6. Establishing performance measures for the processes.

These concepts outline an excellent way to run a business.

The purpose of TQM is to provide a quality product to customers, which will, in turn, increase productivity and lower cost. With a higher quality product and lower price, competitive position in the marketplace will be enhanced. This series of events will allow the organization to achieve the business objectives of profit and growth with greater ease. In addition, the work force will have job security, which will create a satisfying place to work. The road to business growth is shown in Figure 2-1.

As previously stated, TQM requires a cultural change. Table 2-1 compares the previous state with the new TQM state for typical quality elements. As can be seen by the table, this change is substantial and will not be accomplished in a short period of time. Small companies will be able to make the transformation in a much faster time period than large companies.

A company will not begin the transformation to TQM until it is aware that the quality of the product or service must be improved. Awareness comes about when a company loses market shares or realizes that quality and productivity go hand-in-hand. It also occurs when TQM is mandated by the customer or when

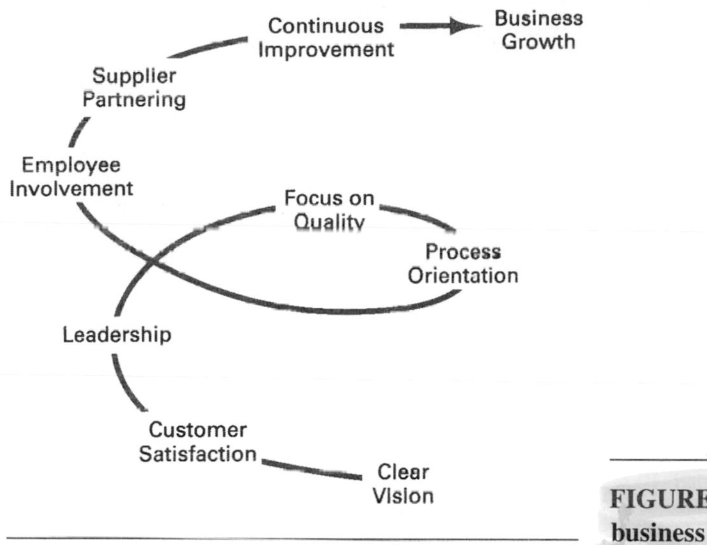

FIGURE 2-1 The road to business growth.

management realizes that TQM is a better way to run a business and compete in domestic and world markets.

Automation and other productivity enhancements will not help a corporation if it is unable to market its product or service because the quality is poor. The Japanese learned this fact from practical experience. Prior to World War II, they could sell their products only at ridiculously low prices, and even then it was difficult to secure repeat sales. Until recently, corporations have not recognized the importance of quality. Quality is first among equals of cost and service.

Quality and productivity are not mutually exclusive. Improvements in quality lead directly to increased productivity and other benefits. Table 2-2 illustrates this

TABLE 2-1 New and Old Cultures.

QUALITY ELEMENT	PREVIOUS STATE	TQM
Definition:	Product-oriented	Customer-oriented
Priorities:	Second to service and cost	First among equals of service and cost
Decisions:	Short-term	Long-term
Emphasis:	Detection	Prevention
Errors:	Operations	System
Responsibility:	Quality control	Everyone
Problem Solving:	Managers	Teams
Procurement:	Price	Life cycle costs
Manager's Role:	Plan, assign, control, and enforce	Delegate, coach, facilitate, and mentor

TABLE 2-2 Gain in Productivity with Improved Quality.

ITEM	BEFPRE IMPROVEMENT 10% NONCONFORMING	AFTER IMPROVEMENT 5% NONCONFORMING
Relative total cost for 20 units	1.00	1.00
Conforming units	18	19
Relative cost for nonconforming units	0.10	0.05
Productivity increase		$\frac{1}{18}(100) = 5.6\%$
Capacity increase		$\frac{1}{18}(100) = 5.6\%$
Profit increase		$\frac{1}{18}(100) = 5.6\%$

concept. As can be seen by the table, the improvement in quality results in a 5.6% improvement in productivity, capacity, and profit. Many quality-improvement projects are achieved with the same work force, same overhead, and no investment in new equipment.

Recent evidence suggests that more and more corporations are recognizing the importance and necessity of quality improvement if they are to survive domestic and worldwide competition. Quality improvement is not limited to the conformance of the product to specifications; it also involves the quality of the design of the product and the process. The prevention of product and process problems is a more desirable objective than taking corrective action after the product is manufactured.

TQM is not something that will occur overnight. There are no quick remedies. It takes a long time to build the appropriate emphasis and techniques into the culture. Overemphasis on short-term results and profits must be set aside so long-term planning and constancy of purpose will prevail.

The entire scope of the TQM activity is shown in Figure 2-2. There are two major headings—Principles and Practices which are briefly discussed in this chapter and Tools and Techniques. Tools and Techniques are divided into the categories of Quantitative and Non Quantitative. Under the Quantitative category the topics of SPC, Acceptance Sampling, and Reliability will be discussed in great detail in Chapters 4-11. The other topics will be very briefly discussed in Chapter 3. Under the Non Quantitative category, the topic of Management & Planning Tools is discussed in Chapter 12. In addition, ISO 9000 will be discussed in some detail and the other topics very briefly in Chapter 3.

LEADERSHIP

Senior management must recognize that the quality function is no more responsible for product quality than the finance function is responsible for profit and loss. Quality, like cost and service, is the responsibility of everyone in the corporation,

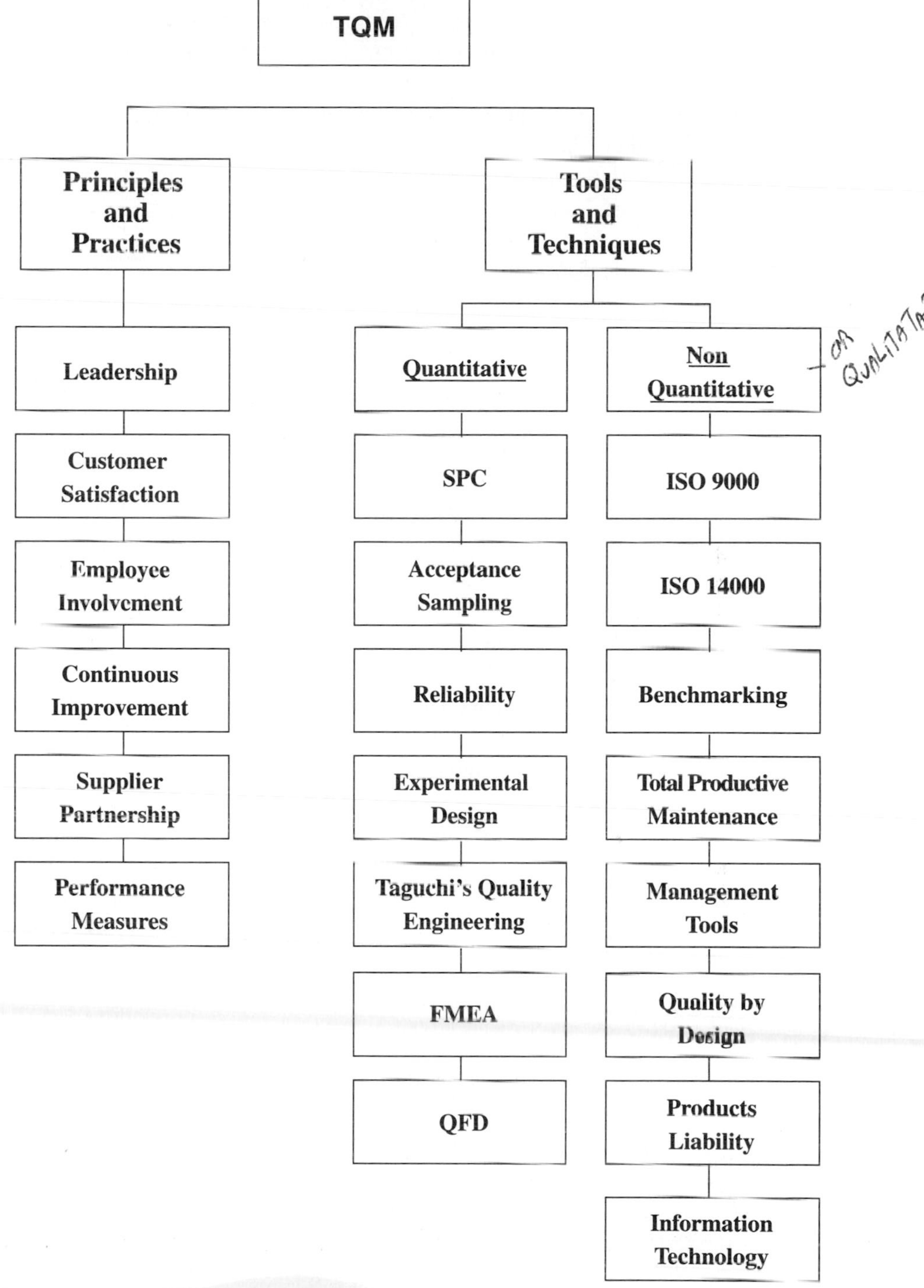

FIGURE 2-2 Scope of the TQM activity

especially the Chief Executive Officer (CEO). When a commitment to quality is made, it becomes part of the corporation's business strategy and leads to enhanced profit and an improved competitive position. To achieve never-ending quality improvement, the CEO must be directly involved in the organization and implementation of the quality improvement activity. In addition, the entire management team needs to become leaders.

There are 12 behaviors or characteristics that successful leaders demonstrate. A brief description of each is given next.[2]

1. They give priority attention to external and internal customers and their needs. Leaders place themselves in the customer's shoes and service their needs from that perspective. They continually evaluate the customers' changing requirements.

2. They empower, rather than control, subordinates. Leaders have trust and confidence in the performance of their subordinates. They provide the resources, training, and work environment to help subordinates do their jobs. However, the decision to accept responsibility lies with the individual.

3. They emphasize improvement rather than maintenance. Leaders use the phrase "If it isn't perfect, improve it" rather than "If it ain't broke, don't fix it." There is always room for improvement, even if the improvement is small. Major breakthroughs sometimes happen, but it's the little ones that keep the continuous process improvement on a positive track.

4. They emphasize prevention. "An ounce of prevention is worth a pound of cure" is certainly true. It is also true that perfection can be the enemy of creativity. There must be a balance between preventing problems and developing better processes.

5. They encourage collaboration rather than competition. When functional areas, departments, or work groups are in competition, they may find subtle ways of working against each other or withholding information. Instead, there must be collaboration among and within units.

6. They train and coach, rather than direct and supervise. Leaders know that the development of the human resource is a necessity. As coaches they help their subordinates learn to do a better job.

7. They learn from problems. When a problem exists, it is treated as an opportunity rather than something to be minimized or covered up. "What caused it?" and "How can we prevent it in the future?" are the questions asked by leaders.

---

[2]Adapted from Warren H. Schmidt and Jerome P. Finnigan, *The Race Without a finish Line* (San Francisco: Josey-Bass Publishers, 1992).

8. They continually try to improve communications. Leaders continually disseminate information about the TQM effort. They make it evident that TQM is not just a slogan. Communication is two way—ideas will be generated by people when leaders encourage them and act upon them. Communication is the glue that holds a TQM organization together.

9. They continually demonstrate their commitment to quality. Leaders walk their talk—their actions, rather than their words, communicate their level of commitment. They let the quality statements be their decision-making guide.

10. They choose suppliers on the basis of quality, not price. Suppliers are encouraged to participate on project teams and become involved. Leaders know that quality begins with quality materials and the true measure is the life-cycle cost.

11. They establish organizational systems to support the quality effort. At the senior management level a quality council is provided, and at the first-line supervisor level, work groups and project teams are organized to improve the process.

12. They encourage and recognize team effort. They encourage, provide recognition, and reward individuals and teams. Leaders know that people like to know that their contributions are important. This action is one of the leader's most powerful tools.

Implementation

The TQM implementation process begins with senior management's and, most important, the CEO's commitment. The importance of the senior management role cannot be overstated. Leadership is essential during every phase of the implementation process and particularly at the start. In fact, indifference and lack of involvement by senior management are frequently cited as the principal reasons for the failure of quality improvement efforts. Delegation and rhetoric are insufficient—involvement is required.

If senior management has not been educated in the TQM concepts, that should be accomplished next. In addition to formal education, managers should visit successful TQM organizations, read selected articles and books, and attend seminars and conferences.

Timing of the implementation process can be very important. Is the organization ready to embark on the total quality journey? There may be some foreseeable problems, such as a reorganization, change in senior management personnel, interpersonal conflicts, a current crisis, or a time-consuming activity. These problems may postpone implementation to a more favorable time.

The next step is the formation of the quality council. The membership and duties follow. Initiation of these duties is a substantial part of the implementation of TQM. The development of core values, a vision statement, a mission statement, and a quality policy statement, with input from all personnel, should be completed first.

Quality Council

In order to build quality into the culture, a quality council is established to provide overall direction. It is the driver for the TQM engine.

In a typical organization the council is composed of the CEO; the senior managers of the functional areas, such as design, marketing, finance, production, and quality; and a coordinator or consultant. If there is a union, consideration should be given to having a representative on the council. A coordinator is necessary to assume some of the added duties that a quality improvement activity requires. The individual selected for the coordinator's position should be a bright young person with executive potential. That person will report to the CEO.

The responsibility of the coordinator is to build two-way trust, propose team needs to the council, share council expectations with the team, and brief the council on team progress. In addition, the coordinator will ensure that the teams are empowered and know their responsibilities. The coordinator's activities are to assist the team leaders, share lessons learned among teams, and have regular leaders' meetings with team leaders.

In smaller organizations where managers may be responsible for more than one functional area, the number of members will be smaller. Also, a consultant would most likely be employed rather than a coordinator.

In general the duties of the council are to

1. Develop, with input from all personnel, the core values, vision statement, mission statement, and quality policy statement.
2. Develop the strategic long-term plan with goals and the annual quality improvement program with objectives.
3. Create the total education and training plan.
4. Determine and continually monitor the cost of poor quality.
5. Determine the performance measures for the organization, approve those for the functional areas, and monitor them.
6. Continually determine those projects that improve the processes, particularly those that affect external and internal customer satisfaction.
7. Establish multifunctional project and departmental or work group teams and monitor their progress.
8. Establish or revise the recognition and reward system to account for the new way of doing business.

In large organizations quality councils are established at lower levels of the corporation. Their duties are similar but relate to that particular level in the organization. Initially these activities will require additional work by council members; however, in the long term their jobs will be easier. These councils are the instruments for perpetuating the idea of never ending quality improvement.

Once the TQM program is well established, a typical meeting agenda might have the following items:

- Progress report on teams
- Customer satisfaction report
- Progress on meeting goals
- New project teams
- Recognition dinner
- Benchmarking report

Eventually, within 3 to 5 years, the quality council activities will become so ingrained in the culture of the organization that they will become a regular part of the executive meetings. When this state is achieved, a separate quality council is no longer needed. Quality becomes the first item on the executive meeting agenda or the executive meeting becomes part of the quality council.

Core Values

Core values and concepts foster TQM behavior and define the culture. Each organization will need to develop its own values. Listed here are the core values for the Malcolm Baldrige National Quality Award. They can be used as a starting point for any organization as it develops its own.

1. Customer-driven Excellence
2. Visionary leadership
3. Organizational and personal learning
4. Valuing employees and partners
5. Agility
6. Management for innovation
7. Focus on the future
8. Management by fact
9. Systems perspective
10. Social responsibility
11. Focus on results and creating value

The core values are part of the quality statements given in the next section. They should be simplified for publication within and outside the organization.

Quality Statements

In addition to the core values, the quality statements include the vision statement, mission statement, and quality policy statement. Once developed, they are only occasionally reviewed and updated. They are part of the strategic planning process, which includes goals and objectives.

The utilization of the four statements varies considerably from organization to organization. In fact, small organizations may use only the quality policy statement. Additionally, there may be considerable overlap among the statements.

The quality statements or a portion thereof may be included on employee badges. They should be developed with input from all personnel.

An example of a statement that includes vision, mission, quality policy, and core values follows.

> Geon has a clear corporate vision . . . To be the benchmark company in the polymers industry through superior performance, demonstrated by:
>
> - Living up to its established principles of excellence in environmental protection, health, and safety
> - Fully satisfying the expectations of its customers
> - Developing and commercializing innovative polymer technology
> - Utilizing all resources productively
> - Continually improving processes and products
> - Generating sustained value for customers, employees, suppliers, and investors
> - Creating an environment of Trust, Respect, Openness, and Integrity
>
> *The Geon Company*

Seven Steps to Strategic Planning[3]

There are seven basic steps to strategic quality planning. The process starts with the principle that quality and customer satisfaction are the center of an organization's future. It brings together all the key stakeholders.

1. *Customer Needs*. The first step is to discover the future needs of the customers. Who will they be? Will your customer base change? What will they want? How will the organization meet and exceed expectations?

[3]Adapted, with permission, from John R. Dew, "Seven Steps To Strategic Planning," *Quality Digest* (June 1994): 34–37.

2. *Customer Positioning*. Next, the planners determine where the organization wants to be in relation to the customers. Do they want to retain, reduce, or expand the customer base? Products or services with poor quality performance should be targeted for breakthrough or eliminated. The organization needs to concentrate its efforts on areas of excellence.

3. *Predict the Future*. Next, the planners must look into their crystal balls to predict future conditions that will affect their product or service. Demographics, economic forecasts, and technical assessments or projections are tools that help predict the future. More than one organization's product or service has become obsolete because it failed to foresee the changing technology. Note that the rate of change is continually increasing.

4. *Gap Analysis*. This step requires the planners to identify the gaps between the current state and the future state of the organization. An analysis of the core values, given earlier in the chapter, is an excellent technique for pinpointing gaps.

5. *Closing the Gap*. The plan can now be developed to close the gap by establishing goals and responsibilities. All stakeholders should be included in the development of the plan.

6. *Alignment*. As the plan is developed, it must be aligned with the mission, vision, and core values of the organization. Without this alignment, the plan will have little chance of success.

7. *Implementation*. This last step is frequently the most difficult. Resources must be allocated to collecting data, designing changes, and overcoming resistance to change. Also part of this step is the monitoring activity to ensure that progress is being made. The planning group should meet at least once per year to assess progress and take any corrective action.

Strategic planning can be performed by any organization. It can be highly effective, allowing organizations to do the right thing at the right time, every time.

Annual Quality Improvement Program

An annual program is developed along with a long-term strategic plan. Some of the strategic items will eventually become part of the annual plan, which will include new short-term items.

In addition to creating the items, the program should develop among all managers, specialists, and operating personnel

- A sense of responsibility for active participation in making improvements.
- The skills needed to make improvements.
- The habit of annual improvements so that each year the organization's quality is significantly better than the previous year.

The program is developed from the departmental level with operating personnel involvement, through the functional areas, to the organization-wide level.

Quality objectives must be stated in measurable terms, such as the following:

1. All billing clerks will receive training in error avoidance.
2. A preventive maintenance procedure will be developed and implemented for the milling department.
3. A project team will reduce field failures by 25%.
4. The wiring-harness assembly department will reduce nonconformities by 30%.

Operating personnel should be encouraged to set objectives or quality goals for themselves as indicated by item 4. Management should support these goals by training, projects, resources, and so forth.

Most likely there will be more quality objectives than can be accomplished with the available resources. Therefore, those that have the greatest opportunity for improvement will be used. Many objectives will require a project team. Some companies have well-structured annual quality-improvement programs. In companies that lack such structured programs, any improvements must come from the initiatives of middle managers and specialists. It takes a good deal of determination by these people to secure results, since they lack the legitimacy and support that comes from an official, structured program designed by the quality council.

CEO Commitment

The most important aspect of management commitment is the involvement by the CEO. This involvement can be achieved by

- Chairing or participating in the quality council
- Chairing or participating in the ISO 9000 team
- Coaching project teams
- MBWA (Management By Walking Around)
- Presiding at recognition ceremonies
- Writing a column in the newsletter
- Spending $\frac{1}{3}$ of their time on quality
- Periodically meeting with all employees

The benefits of TQM will not be obtained unless the CEO is involved.

CUSTOMER SATISFACTION

Introduction

There is an old saying that the customer is always right. That saying is as true today as it was when it was first coined. While the customer is king, education and diplomacy are sometimes necessary. In Chapter 1, we defined quality as a function of customer satisfaction. A recent study by *Quality in Manufacturing* showed that 83.6% of the respondents stated that their primary measure of quality was customer satisfaction. It is also important to recognize that expectations are constantly changing. What was acceptable 10 years ago is rejected today.

Satisfaction is a function of the entire experience with the company—not just the purchased unit. For example, a company orders 75 units from a supplier to be delivered on October 10. They receive 72 units delivered on October 15 and are billed for 74 units. The units were perfect; however, we have a dissatisfied customer.

Companies should strive to maintain customers for life. On the average, it takes five times as much money to win a new customer as to retain an existing one.

TQM implies an organizational obsession with meeting or exceeding customer expectations, to the point that customers are delighted. Understanding the customer's needs and expectations is essential to winning new business and keeping existing business. An organization must give its customers a quality product or service that meets their needs: a reasonable price, on-time delivery, and outstanding service. To attain this level, the organization continually needs to examine its quality system to see if it is responsive to ever-changing customer requirements and expectations.

Who Is the Customer?

This question is more difficult than it seems. For example, the manufacturer of the nozzle for a gasoline pump at a self-service station has the following external customers: the oil company, the service station owner, and you, the user. An auto insurance company has brokers, customer service representatives, and the insured as external customers. Internal customers are those in the next operational sequence. Some examples are: Salesperson to Order Entry Clerk, Sand Muller Operator to Mold Machine Operator, and Shipping Supervisor to Billing Clerk. The quality of everyone's performance is a function of their internal supplier.

The following checklist can be used to improve the satisfaction of both internal and external customers:

1. Who are my customers?
2. What do they need?

3. What are their measures and expectations?
4. What is my product or service?
5. Does my product exceed expectations?
6. How do I satisfy their needs?
7. What corrective action is necessary?
8. Are customers included on teams?

Each individual or group must identify and satisfy its customers while fostering a team effort where all people help the organization rather than look after personal objectives. To aid in this goal, performance measures and objectives are established for each operating unit or subunit.

Customer Feedback

In order to focus on the customer, an effective feedback program is necessary. The objectives of this program are to

1. Discover customer dissatisfaction
2. Discover relative priorities of quality with other attributes like price and delivery
3. Compare performance with the competition
4. Identify customer needs
5. Determine opportunities for improvement

Data collection will vary depending on the product and whether the customer is the end-user or not. It can be accomplished by an individual, team, or department. A brief description of various data collection tools is given in the information that follows.

For end-users, warranty cards or questionnaires that are included with the product provide an inexpensive method; however, some incentive to complete the information is needed for the customer to fill out the form. Telephone surveys collect the required information but are time-consuming and costly and may alienate the customer. Mail-out surveys for end-users are usually not completed.

Mail-out surveys work very well for commercial customers. A very good technique is to request the customer to grade (A, B, C, D, etc.) the supplier in terms of quality, delivery, and service. Another very good technique is to visit the customer, observe the product in action, and assist in any necessary problem solving. To increase public relations, bring along a few operating people.

For end-users, focus groups of current or potential customers work very well. The meetings are guided by a professional and explore the favorable and unfavorable aspects of current and future products. Feedback is also obtained by comparative evaluation of a competitor's product, packaging, and follow-up information.

Many companies rely on their service centers to provide feedback on customer satisfaction. These service centers may be organization-owned or may be operated by contract. In this regard, a Pareto analysis of repair parts can indicate the location and frequency of quality problems.

Customer complaints provide the best information; however, this information occurs in the worst situation. Complaints are discussed in the next section.

Customer Complaints

A recent survey of retail customers by the American Society for Quality (ASQ) showed that dissatisfied customers rarely complain. Products included in the study were: automobiles, mail order, radio/TV, groceries, furniture, clothing, home repair, appliances, and auto repair. The study found that, on the average, only about 1% complain to management, 18% complain to front-line people, and 81% do not complain. About 25% of the dissatisfied customers will not buy that brand again and the producer has lost sales without an explanation. While this study was conducted on retail customers, there is some evidence that the principle is applicable also to commercial customers. Therefore, when a complaint is received, it represents not only the tip of the iceberg but an opportunity for quality improvement.

Every company should have a procedure for using customer complaints. A suggested one is

1. Accept complaints—don't fight them—for they are a measure of your quality.
2. Feed back complaint information to all people.
3. Analyze complaints by doing the detective work.
4. If possible, eliminate the root cause. More inspection is not corrective action.
5. Report results of all investigations and solutions to everyone involved.

Service After the Sale

An essential characteristic of customer satisfaction occurs after the sale. An organization can create a market advantage by being the best—beyond performance, delivery, and price. Service quality is a product; therefore, it can be improved and controlled. The basic elements of service quality are

Organization

1. Identify each market
2. Establish the requirements and communicate them

Customer

3. Obtain the customer's point of view
4. Meet the customer's expectations by delivering what is promised
5. Make the customer feel valued
6. Respond to all complaints
7. Over-respond to the customer

Communication

8. Optimize the trade-off between time and personal attention
9. Minimize the number of contact points
10. Write documents in customer-friendly language

Front-Line People

11. Be sure employees are adequately trained and people-friendly
12. Serve them as internal customers
13. Give them the authority to solve problems
14. Challenge them to develop new methods
15. Establish performance measures and recognize and reward performance

Leadership

16. Lead by example
17. Listen to front-line people
18. Strive for continuous quality improvement

How these basic elements are addressed will vary among different organizations and their product or service.

Final Comments

Good experiences are repeated to six people and bad experiences are repeated to 15 people. Know your customers, listen to them, and when necessary, educate them, for they determine their needs and ultimately judge your ability to satisfy those needs.

Every company should establish a partnership relationship with its customers. The quality of U.S. products was outstanding during World War II because there was the ultimate partnership with the customer—our sons, daughters, spouses, relatives, friends, and neighbors.

EMPLOYEE INVOLVEMENT

Introduction

No resource is more valuable to an organization than its people. While the above expression is an old clichÈ, it is certainly a true one and very applicable to quality. Many companies view quality problems in terms of operating personnel. The frequent response is to develop motivation programs with goals and slogans. These programs result in an immediate "hype"; management actions (quality council) and deeds (project team success), rather than lip service, will do more to motivate people than short-lived programs.

Consider the situation where an operator produces parts that have marginal quality. If management decides to sort or scrap the parts, then concrete evidence is given that management cares about quality. However, if the decision is to take a chance and ship the parts to the customer, then all the slogans and motivational programs will have little meaning or will even be counterproductive.

Actually, management makes a serious mistake when it assumes that the quality problem is due to unconcerned operating personnel. Dr. Deming has estimated that only 15% of the quality problems of an organization are due to local faults (operators and first-line supervisors). The rest (85%) are due to the system (management).

Point 8 of Dr. Deming's 14 points is:

> Reduce fear throughout the organization by encouraging open, two-way, nonpunitive communication. The economic loss resulting from fear to ask questions or report trouble is appalling.

He further states that quick results in quality improvement can be obtained by achieving this goal. Powerful economic results are obtainable within 2 or 3 years when the employment climate is changed.

Involving people in a quality-improvement program is an effective technique to improve the quality. Management commitment, annual quality improvement, education and training, project teams, and so forth, are all effective in utilizing the human resources of an organization. People must come to work not only to do their jobs, but also to think about how to improve their jobs. People must be empowered to perform processes in an optimum manner at the lowest possible level.

Project Teams

The Japanese have had excellent success with their quality control circles. They can be used at all levels of an organization. However, the quality control circle approach is not a panacea. It is estimated that, at best, only 10% of the Japanese miracle in quality can be attributed to this approach. The reason for this low value is that the vital few (85%) quality problems are inherently due to the system (management).

For the most part, the quality circle movement outside of Japan was not successful because of a lack of management support, little first-line supervisor involvement, insufficient training, and poor projects.

In spite of the problems with quality control circles, the team concept has emerged as the key method to improve quality and productivity while breaking down barriers and enhancing morale. The project and teams of seven to ten members are authorized by the quality council with a member of the council serving as the coach or mentor. They may be departmental or multi-functional and standing or ad hoc. The supervisor will usually be the team leader for departmental teams while multi-functional teams might have a senior manager from the most affected functional area. Whenever possible, it is a good idea to have an internal or external customer and an internal or external supplier on the team.

Multi-functional project teams are used for both the development of new processes and the improvement of existing ones. They may be standing or ad hoc.

New processes are developed from research and development or from unfulfilled needs determined by functional areas. No matter what the source, it is important that there be early involvement of all appropriate personnel by placing them on project teams. Typical members will be from marketing, quality, materials management, service, finance, manufacturing, and, of course, design. The project leader will most likely be from the design area and is described as first among equals.

The basic idea is to build quality into the design at the beginning rather than fix it afterwards. Each of the team members has a role to play. For example, marketing will continue to assess the customer's requirements in regard to the design. Or, manufacturing will provide advice about the ability of the process to meet a particular specification. This approach prevents problems from occurring rather than detecting them at a later time when the cost to correct the problems may be prohibitive. Other teams might be composed of representatives from many functional areas. Occasionally, it is desirable from a management commitment and involvement viewpoint to include the CEO or senior manager on a project team. This practice is a concrete illustration of leading by example.

Department or operational areas have standing or ad hoc teams. Standing teams might have weekly meetings to discuss control chart patterns, complaints, downtime, on-time delivery, etc. The composition of ad hoc teams will be a function of the desired improvement. Some teams might be composed of simply the operator, supervisor, internal customer, and a member of the quality department.

Each team should have the proper education in the basic tools. Some initial improvement projects are order entry/invoice, customer satisfaction, supplier management, production problems, design problems, and recognition and reward.

Education and Training

The cost of education and training for all personnel is enormous and the time to achieve it is lengthy. The Japanese trained hundreds of thousands of managers and supervisors at all organizational levels plus millions of nonsupervisors. As far as quality is concerned, this massive training program has made their managers, specialists, and workers the best trained on earth. This training took more than 10 years to achieve.

To a great extent, education has been limited to the quality department. The entire company payroll must be educated in the new philosophy and the quality sciences as is appropriate to their positions. Top management will need a different education than operating personnel. While some education and training occurs concurrently, senior management will need to be first, followed by middle management and specialists, and finally by first-line supervisors and operating personnel.

Some education such as an attitude change, a rudimentary knowledge of statistical techniques, the quality-improvement sequence of events, and the concept of prevention is common to all levels. However, the educational needs of different functional areas, departments, and jobs will vary considerably. For example, purchasing personnel will need a knowledge of vendor survey, qualification, and rating; concept of price versus cost; number of suppliers; statistical process control; and acceptance sampling.

In addition to education related to quality, a rigorous program of retraining people in new skills to keep up with changes in materials, methods, product designs, and machinery is desirable. To a large extent, this retraining will be provided to the specialists within a company. For example, the product engineers of an appliance manufacturer must be aware of the impact of composite material technology on that company's products.

The education and training task is so formidable that the quality council may wish to establish a project team to do the planning on a company-wide basis. The mission of this project team is to

1. Identify the subject matter for each job category
2. Identify possible sources of training materials and leaders
3. Estimate the investment required in money, facilities, and personnel
4. Recommend a program including trainees, leaders, and a time schedule

Upper managers should become trainees in the program. Their training will be partly "by the book" and partly by the extent to which they participate in the management of the quality function.

A suggested start-up program for small companies is that

- All managers and supervisors receive 7 hours of TQM training
- All managers and supervisors receive 7 hours of SPC training
- All operating personnel receive 7 hours of TQM/SPC training

While concepts are important, the primary thrust of the training should be practical in nature.

Suggestion System

Once the proper environment is established, a suggestion system can be developed that will provide another approach for quality improvement. In order to be effective, action by management is necessary on each suggestion. This action can be quite an increase in management's workload; however, it is the only way a suggestion system will provide the maximum benefits. A few CEOs answer each suggestion with a personal letter stating why the suggestion was or was not a good one and whether it will or will not be implemented. Monetary and/or recognition reward is also an essential part of a suggestion system, and the recognition part may be the most important.

Most suggestion systems require identification of a problem and a solution. Another approach is to provide a form whereby people need only to state a problem. The appropriate functional area or department will develop the solution.

Typical problems are as follows:

1. This tool is not long enough for all the parts.
2. The sales department makes too many errors on their order entry forms.
3. We make a lot of changes in response to telephone calls, and many changes have to be redone.

Once people know that their problems will be heard and answered, communication is opened up, and the potential for quality improvement is enhanced.

Final Comments

Employee involvement in the decision-making process improves quality and increases productivity. Better decisions occur because the work force is more knowledgeable about the process. They are more likely to implement and support decisions they had a part in making. In addition, they are better able to spot and pinpoint areas for quality improvement and better able to take corrective action when a process is out of control. Employee involvement reduces the labor-management hassle and increases morale.

CONTINUOUS PROCESS IMPROVEMENT

The goal is to achieve perfection by continuously improving the business and production processes. Of course, perfection is an elusive goal; however, we must continually strive for its attainment.

We continuously improve by

- Viewing all work as a process, whether it is associated with production or business activities.
- Making all our processes effective, efficient, and adaptable.
- Anticipating changing customer needs.
- Controlling in-process performance using measures such as scrap reduction, cycle time, control charts, etc.
- Maintaining constructive dissatisfaction with the present level of performance.
- Eliminating waste and rework wherever it occurs.
- Investigating activities that do not add value to the product or service, with the aim of eliminating those activities.
- Eliminating nonconformities in all phases of everyone's work, even if the increment of improvement is small.
- Using benchmarking to improve competitive advantage.
- Innovating to achieve breakthroughs.
- Holding gains so there is no regression.
- Incorporating lessons learned into future activities.
- Using technical tools such as statistical process control (SPC), experimental design, benchmarking, quality function deployment (QFD), etc.

Process

Process refers to business and production activities of all organizations. Business processes such as purchasing, engineering, accounting, and marketing are areas where nonconformance can represent an opportunity for substantial improvement. Figure 2-3 shows a process with its inputs and outputs. Inputs may be materials, money, information, data, etc. Outputs may be information, data, products, services, etc. Actually, the output of one process can be the input to another process. Outputs usually require performance measures. They are designed to achieve certain outcomes such as customer satisfaction. Feedback is necessary to improve the process.

Process definition begins with defining the internal and/or external customers. The customer defines the purpose of the organization and every process

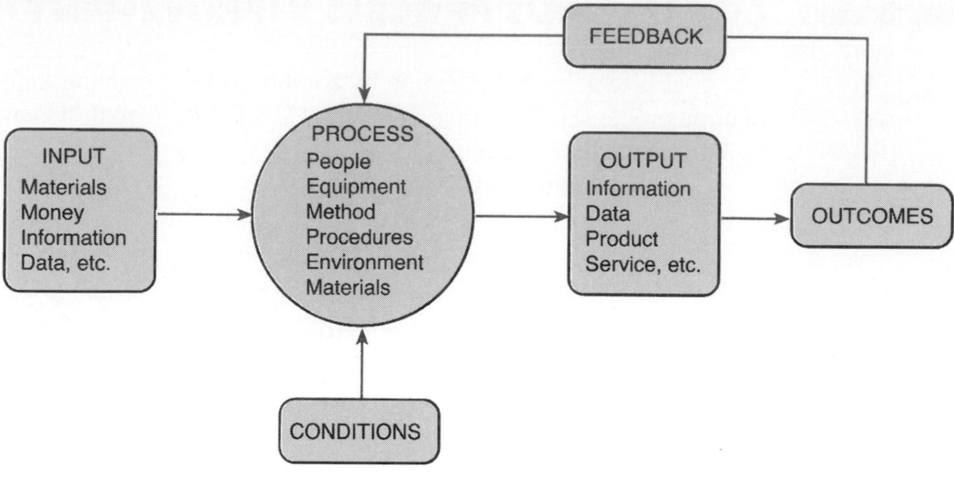

FIGURE 2-3 Input/output process model.

within it. Because the organization exists to serve the customer, process improvements must be defined in terms of increased customer satisfaction as a result of higher quality products and services.

The process is the interaction of some combination of people, materials, equipment, method, measurement, and the environment to produce an outcome such as a product, service, or an input to another process. In addition to having measurable input and output, a process must have value-added activities and repeatability. It must be effective, efficient, under control, and adaptable. In addition, it must adhere to certain conditions imposed by policies and constraints or regulations.

All processes must have an owner. In some cases, the owner is obvious, because there is only one person performing the activity. Frequently the process will cross multiple organizational boundaries, and supporting subprocesses will be owned by individuals within each of the organizations. Thus, ownership must be part of the process improvement initiatives.

At this point it is important to define an improvement. There are five basic ways: (1) reduce resources, (2) reduce errors, (3) meet or exceed expectations of downstream customers, (4) make the process safer, and (5) make the process more satisfying to the person doing it.

A process that uses more resources than necessary is wasteful. Reports that are distributed to more people than necessary waste copying and distribution time, material, user read time, and, eventually, file space.

For the most part, errors are a sign of poor workmanship. Typing errors that are detected after the computer printout require opening the file, making the correction, and printing the revised document.

By meeting or exceeding expectations of downstream customers, the process is improved. The better the oxyacetylene weld is made, the less grinding is required, making the appearance of the finish paint more pleasing.

The fourth way a process can be improved is by making it safer. A safer workplace is a more productive one with fewer lost-time accidents and less worker's compensation claims.

The fifth way to improve a process is to increase the satisfaction of the individual performing the process. Although it is difficult to quantify, the evidence suggests that a happy, satisfied employee is a more productive one. Sometimes a little change, such as a better chair, can make a substantial change in a person's attitude toward their work.

Problem-Solving Method

The project team achieves the optimal results when it operates within the framework of the problem-solving method. In the initial stages of a process-improvement program, quick results are often obtained because the solutions are obvious or someone has a brilliant idea. However, in the long term, a systematic approach will yield the greatest benefits.

The problem-solving method (also called the scientific method) as applied to process improvement has seven phases:

1. Identify the opportunity.
2. Analyze the current process.
3. Develop the optimal solution(s).
4. Implement changes.
5. Study the results.
6. Standardize the solution.
7. Plan for the future.

These steps are not totally independent; they are sometimes interrelated. In fact, some techniques such as the control chart can be effectively utilized in more than one step. Process improvement is the goal, and the problem-solving process is a framework to achieve that goal.

Phase 1 Identify the Opportunity

The objective of this phase is to identify and prioritize opportunities for improvement. It has two parts: identify the problem and form the team.

Problem identification answers the question, What are the problems? The answer leads to those problems that have the greatest potential for improvement and have the greatest need for solution. Problems can be identified from a variety of inputs, such as the following:

- Pareto analysis of repetitive external alarm signals, such as field failures, complaints, returns, and others (see Chapter 8).
- Pareto analysis of repetitive internal alarm signals (for example, scrap, re-work, sorting, and the 100% test).
- Proposals from key insiders (managers, supervisors, professionals, and union stewards).
- Proposals from suggestion schemes.
- Field study of users' needs.
- Data on performance of products versus competitors (from users and from laboratory tests).
- Comments of key people outside the organization (customers, suppliers, journalists, and critics).
- Findings and comments of government regulators and independent laboratories.
- Customer surveys.
- Employee surveys.
- Brainstorming by work groups.

Problems are not bad or good; they provide opportunities for improvement. For a condition to qualify as a problem, it must meet the following three criteria:

1. Performance varies from an established standard.
2. Deviation from the perception and the facts.
3. The cause is unknown; if we know the cause, there is no problem.

Finding problems is not too difficult, because there are many more than can be analyzed. The quality council or work group must prioritize them using the following selection criteria:

1. Is the problem important and not superficial and why?
2. Will problem solution contribute to the attainment of goals?
3. Can the problem be defined clearly using numbers?

A work group that needs to select its initial problem should find one that gives the maximum benefit for the minimum amount of effort.

The second part of Phase 1 is to form a team. If the team is a natural work group, then this part is complete. If the problem is of a multifunctional nature, as are most, then the team is selected and tasked by the quality council to address the improvement of a specific process. The team leader is selected and becomes the owner of the process. Goals and deadlines are determined.

The third part of Phase 1 is to define the scope. Failure in problem solving is frequently caused by poor definition of the problem. A problem well stated is half solved. Criteria for a good problem statement are as follows:

- It clearly describes the problem and is easily understood.
- It states the effect—what is wrong, when it happens, and where it is occurring, not why it is wrong or who is responsible.
- It focuses on what is known, what is unknown, and what needs to be done.
- It uses facts and is free of judgment.
- It emphasizes the impact on the customer.

An example of a well-written problem statement is:

> As a result of a customer satisfaction survey, a sample of 150 billing invoices showed that 18 had errors that required 1 hour to correct.

In addition to the problem statement, this phase requires a comprehensive charter for the team. The charter specifies

1. *Authority*. Who authorized the team?
2. *Objective and Scope*. What are the expected outputs and specific areas to be improved?
3. *Composition*. Who are the team members and process and subprocess owners?
4. *Direction and Control*. What are the guidelines for the internal operation of the team?
5. *General*. What are the methods to be used, the resources, and the specific milestones?

Phase 2 Analyze the Current Process

The objective of this phase is to understand the process and how it is currently performed. Key activities are to determine the measurements needed to analyze the process; gather data; define the process boundaries, outputs and customers, inputs and suppliers, and process flow; identify root causes; and determine levels of customer satisfaction.

The first step is for the team to develop a process flow diagram. A flow diagram translates complex work into an easily understood graphic description. This activity is an "eye-opening" experience for the team, because it is rare that all members of the team understand the entire process.

Next, the target performance measures are defined. Measurement is fundamental to meaningful process improvements. If something cannot be measured, it cannot be improved. There is an old saying that what gets measured gets done. The team will determine if the measurements needed to understand and improve the process are presently being used; if new ones are needed, the team will

- Establish performance measures with respect to customer requirements.
- Determine data needed to manage the process.
- Establish regular feedback with customers and suppliers.
- Establish measures for quality/cost/timelines of inputs and outputs.

Once the target performance measures are established, the team can collect all available data and information. If these data are not enough, then additional new information is obtained. Gathering data (1) helps confirm that a problem exists, (2) enables the team to work with facts, (3) makes it possible to establish measurement criteria for baseline, and (4) enables the team to measure the effectiveness of an implemented solution. It is important to collect only needed data and to get the right data for the problem. The team should develop a plan that includes input from internal and external customers and ensures the plan answers the following questions:

1. What problem or operation do we wish to learn about?
2. What are the data used for?
3. How many data are needed?
4. What conclusions can be drawn from the collected data?
5. What action should be taken as a result of the conclusion?

Data can be collected by using check sheets, by computers with their application software, by data-collection devices such as hand-held gauges, and by an on-line system.

The team will identify the customers and their expectations as well as their inputs, outputs, and interfaces of the process. Also, they will systematically review the procedures currently being used.

Common items of data and information are

- Design information, such as specifications, drawings, function, bills of materials, costs design reviews, field data, service, and maintainability
- Process information, such as routing, equipment, operators, raw material, and component parts and supplies

- Statistical information, such as average, median, range, standard deviation, skewness, kurtosis, and frequency distribution
- Quality information, such as Pareto diagrams, cause-and-effect diagrams, check sheets, scatter diagrams, control charts, histograms, process capability, acceptance sampling, run charts, life testing, and operator and equipment matrix analysis.
- Supplier information such as process variation, on-time delivery, and technical competency.

The cause-and-effect diagram is particularly effective in this phase. Determining all of the causes requires experience, brainstorming, and a thorough knowledge of the process. It is an excellent starting point for the project team. One word of caution—the object is to seek causes, not solutions. Therefore, only possible causes, no matter how trivial, should be listed.

It is important to identify the root, or most likely, cause. This activity can sometimes be determined by voting. It is a good idea to verify the most likely cause, because a mistake here can lead to the unnecessary waste of time and money. Some verification techniques are

1. Examine the most likely cause against the problem statement.
2. Recheck all data that support the most likely cause.
3. Check the process when it is performing satisfactorily against when it is not by using the who, where, when, how, what, and why approach.
4. Utilize an outside authority who plays "devil's advocate" with the data, information, and reasoning.
5. Use experimental design, Taguchi's quality engineering, and other advanced techniques to determine the critical factors and their levels.

Once the root, or most likely, cause is determined, the next phase can begin.

Phase 3 Develop the Optimal Solution(s)

This phase has the objective of establishing problem solutions and recommending the optimal solution to improve the process. Once all the information is available, the project team begins its search for possible solutions. More than one solution is frequently required to remedy a situation. Sometimes the solutions are quite evident from a cursory analysis of the data.

There are three types of creativity: (1) create new processes, (2) combine different processes, or (3) modify the existing process. The first type is innovation in its highest form, such as the invention of the transistor. Combining two or more processes is a synthesis activity to create a better process. It is a unique combination of what already exists. This type of creativity relies heavily on benchmarking. Modification involves altering a process that already exists so that it does a better job. It succeeds when managers utilize the experience, education, and energy of empowered work groups or project teams. There is not a distinct line between the three types—they overlap.[4]

In this phase, creativity plays the major role, and brainstorming is the principal technique. Other group dynamics that can be considered for this phase are the Delphi method and the nominal group technique.

Areas for possible change are the number and length of delays, number of steps, timing and number of inspections, rework, and materials handling.

Once possible solutions have been determined, evaluation or testing of the solutions comes next. As mentioned, more than one solution can contribute to the situation. Evaluation and/or testing determines which of the possible solutions have the greatest potential for success and the advantages and disadvantages of these solutions. Criteria for judging the possible solutions include such things as cost, feasibility, effect, resistance to change, consequences, and training. Solutions may be categorized as short range and long range.

One of the features of control charts is the ability to evaluate possible solutions. Whether the idea is good, poor, or has no effect is evident from the chart.

Phase 4 Implement Changes

Once the optimal solution is selected, it can be implemented. This phase has the objective of preparing the implementation plan, obtaining approval, implementing the process improvements, and studying the results.

Although the project team usually has some authority to institute remedial action, more often than not the approval of the quality council or other appropriate authority is required. If such is the case, a written and/or oral report is given.

The contents of the implementation plan report must fully describe

- Why will it be done?
- How will it be done?
- When will it be done?
- Who will do it?
- Where will it be done?

Answers to these questions will designate required actions, assign responsibility, and establish implementation milestones. The length of the report is determined by the complexity of the change. Simple changes may require only an oral report, whereas others require a detailed written report.

[4]Paul Mallette, "Improving Through Creativity," *Quality Digest* (May 1993): 81–85.

TABLE 2-3 Combination Map of Dimensions for Process Control

WHAT'S INSPECTED	TYPE OF DATA	TIMING	BY WHOM?	TYPE OF RECORD	ACTION	BY WHOM?
Process variable continuous	Variable	During run: on-line	Device	Electronic control chart	Process improved	Automated equipment
Process variable sample				Paper control chart		
		During run: off-line	Process Operator	Electronic trend chart	Process adjusted	Operator
Product sample	Attribute			Paper trend chart	Lot sorted	
		After lot: complete	Inspector	Electronic list	Sample repaired or discarded	Inspector or mechanic
100% of product				Paper list		

Reproduced, with permission, from Peter E. Pylipow, "Understanding the Hierarchy of Process Control: Using a Combination Map to Formulate an Action Plan," *Quality Progress* (October 2000): 63–66.

After approval by the quality council, it is desirable to obtain the advice and consent of departments, functional areas, teams, and individuals who may be affected by the change. A presentation to these groups will help gain support from those involved in the process and provide an opportunity for feedback with improvement suggestions.

Measurement tools such as run charts, control charts, Pareto diagrams, histograms, check sheets, and questionnaires are used to monitor and evaluate the process change.

Pylipow provides a combination map to help formulate an action plan to help measure the results of an improvement. The map, shown in Table 2-3, provides the dimensions of what is being inspected, the type of data, timing of data collection, by whom, how the results will be recorded, the necessary action that needs to be taken based on the results, and who is to take the action.

Phase 5 Study the Results

In order to study the results, measurements must be taken. Ownership of the measurement activity must be assigned. Measurement tools such as run charts, control charts, Pareto diagrams, histograms, check sheets, and questionnaires are used to monitor and evaluate the process change.

The team should meet periodically during this phase to evaluate the results to see that the problem has been solved or if fine-tuning is required. In addition, they will wish to see if any unforeseen problems have developed as a result of the changes. If the team is not satisfied then some of the phases will need to be repeated.

Phase 6 Standardize the Solution

Once the team is satisfied with the change, it must be institutionalized by positive control of the process, process certification, and operator certification. Positrol (positive control) ensures that important variables are kept under control. It specifies the what, who, how, where, and when of the process and is an updating of the combination map.

In addition, the quality peripherals—the system, environment, and supervision—must be certified. A checklist for the quality system would include items such as preventative maintenance, alarm signals, and shut-down authority. An environment checklist would include items such as electrostatic discharge, temperature control, and air purity. The supervision checklist would include items such as suggestion system, results feed back, and clear instructions. These check lists provide the means to initially evaluate the peripherals and periodically audit them to ensure the process will meet or exceed customer requirements for the product or service.

Finally, operators must be certified to know what to do and how to do it for a particular process. Also needed is cross training in other jobs within the process to ensure next-customer knowledge and job rotation. Total product knowledge is also desirable. Operator certification is an on-going process that must periodically occur.

Phase 7 Plan for the Future

This phase has the objective of achieving improved levels of process performance. Regardless of how successful initial improvement efforts are, the improvement process must continue. It is important to remember that Total Quality Management (TQM) addresses the quality of management as well as the management of quality. Everyone in the organization is involved in a systematic long-term endeavor to develop processes that are customer oriented, flexible, and responsive, and to constantly improve quality.

A key activity is to conduct regularly scheduled reviews of progress by the quality council and/or work group. Management must establish the systems to identify areas for future improvement and to track performance with respect to internal and external customers.

Continuous improvement means not being satisfied with doing a good job or process but striving to improve that job or process. It is accomplished by incorporating process measurement and team problem solving in all work activities. TQM tools and techniques are used to improve quality, delivery, and cost. We must continuously strive for excellence by reducing complexity, variation, and out-of-control processes.

Lessons learned in problem solving, communications, and group dynamics as well as technical know-how must be transferred to appropriate activities within the organization.

Although the problem-solving method is no guarantee of success, experience has indicated that an orderly approach will yield the highest probability of success. Problem solving concentrates on improvement rather than control.

SUPPLIER PARTNERSHIP

Introduction

On the average, 40% of production cost is due to purchased material; therefore, supplier management is extremely important. It follows that a substantial portion of quality problems will be due to the supplier. In order for both parties to succeed and their business to grow, a partnership is required. The supplier should be treated as an extension of the production process.

They will need to work together to achieve quality improvement. The supplier should make a positive contribution to design, production, and cost reduction. Emphasis should be placed on the total material cost, which includes both price and quality cost. The supplier should be given a long-term relationship and purchase contract. In fact, single sourcing with a large contract will create better quality at a lower cost, but delivery disruption is possible.

In order to reduce inventories, many companies are using Just-In-Time (JIT). For JIT to be effective, the supplier quality must be excellent, and the supplier must reduce the set-up time.

Supplier management activities include

1. Define the product and program requirements
2. Evaluate potential suppliers and select the best
3. Conduct joint quality planning and execution
4. Require statistical evidence of quality
5. Certify suppliers or require ISO 9000 registration
6. Conduct joint quality improvement programs
7. Create and utilize supplier ratings

Supplier Selection Criteria

Effective selection requires the supplier to be knowledgeable of the purchaser's quality philosophy and requirements. It also requires satisfaction by other customers. The supplier must demonstrate technical capability and capacity to provide quality products. Of particular importance is the credibility of the supplier—are the purchaser's secrets secure? The primary criterion is the supplier's ability to provide quality products as evidenced by the quality system and improvement program. Another criterion is the supplier's control of its suppliers. Finally, the supplier's accessibility may be critical. The criteria are applicable whether the selection is done by survey or by visit of a third party. A well-designed checklist with weights for different criteria will aid in evaluation and selection.

Supplier Certification

Incoming material is evaluated at the supplier's plant by a third party, at the purchaser's plant by 100% inspection, by acceptance sampling, or by identifying checks and statistical evidence of quality by a certified supplier. A certified supplier is one that can supply quality materials on a long-term basis. The certification process follows supplier selection and an excellent track record over time. Part of this process can include ISO 9000 registration.

Certification allows a supplier to ship to stock with only an identity check and statistical evidence of quality. The purchaser no longer needs to conduct receiving inspection, and a purchaser/supplier partnership is created. Periodic audits are conducted by the purchaser to ensure conformance. Certification reduces to a manageable level the number of suppliers.

Supplier Quality Ratings

Ratings are based on certain measures and are weighted. A typical rating system is

Percent Nonconforming	45%
Price and Quality Costs	35%
Delivery and Service	20%

Supplier quality ratings provide an objective measure of a supplier's performance. This measure can lead to a supplier review and allocation of business and identify areas for quality improvement.

In order for supplier partnering, there must be long-term commitment, trust, and shared vision.

PERFORMANCE MEASURES

The sixth and final concept of Total Quality Measurement (TQM) is performance measures. One of the Malcolm Baldrige National Quality Award core values is managing by fact rather than by gut feeling. Managing an organization without performance measures is like a captain of an ocean-going ship navigating without any instrumentation. The captain would most likely end up traveling in circles, as would an organization.

Effective management requires information obtained from the measurement of activities. Performance measures are necessary: as a baseline, to identify potential projects; to justify project resource allocation; and to assess the improvement results. Production activities use measures such as defects per million, inventory turns, and on-time delivery. Service activities use measures such as billing errors, sales per square feet, engineering changes, and activity time. There are many methods, and each has its place in the organization.

Cost of Poor Quality[5]

In the final analysis, the value of quality control must be based on its ability to contribute to profits; therefore, the most effective performance measure is the cost of poor quality. In our profit-oriented society, decisions are between alternatives and the effect each alternative will have on the expense and income of the business entity.

The efficiency of any business is measured in terms of dollars. Therefore, as with costs of maintenance, production, design, inspection, sales, and other activities, the cost of poor quality must be known. This cost is no different than other costs. It can be programmed, budgeted, measured, and analyzed to attain the objectives of better quality and customer satisfaction at less cost. A reduction in the cost of poor quality leads to increased profit.

The cost of poor quality crosses department lines by involving all activities of the company—marketing, purchasing, design, manufacturing, and service, to name a few. Some costs such as inspector salaries and rework are readily identifiable; other costs such as prevention costs associated with marketing, design, and purchasing are more difficult to identify and allocate. There are failure costs associated with lost sales and customer goodwill, which may be impossible to measure and must be estimated.

The cost of poor quality is defined as those costs associated with the nonachievement of product or service quality as defined by the requirements established by the organization and its contracts with customers and society. Simply stated, it is the cost of poor products or services.

The cost of poor quality is used by management in its pursuit of quality improvement, customer satisfaction, market share, and profit enhancement. It is the economic common denominator which forms the basic data for Total Quality Management (TQM). When the cost of poor quality is too great, it is a sign of management

[5]This section is extracted from *Guide for Reducing Quality Costs, 2 ed.*, 1987 and *Principles of Quality Costs*, 1986, by the Quality Cost Committee, with the permission of the American Society for Quality Control.

ineffectiveness, which can affect the company's competitive position. A cost program provides warning against oncoming, dangerous financial situations.

A cost of poor quality program quantifies the magnitude of the quality problem in the language that management knows best—dollars. The cost of poor quality can exceed 20% of the sales dollar in manufacturing companies and 35% of the sales dollar in service companies. In addition, the program may show quality problem areas that were not known to exist.

The cost of poor quality identifies opportunities for quality improvement and establishes funding priorities by means of Pareto analysis. This analysis allows the quality improvement program to concentrate on the vital few quality problem areas. Once corrective action has been completed, the cost of poor quality will measure the effectiveness of that action in terms of dollars.

A cost of poor quality program lends credence to management's commitment to quality. Arguments for quality improvement are stronger when the costs show a need. The program also provides cost justification for corrective action. All costs associated with poor quality and its correction are integrated into one system in order to enhance the quality management function. Quality improvement is synonymous with a reduction in the cost of poor quality. Every dollar of cost saved has a positive effect on profits.

One of the principal advantages of the program is the identification of hidden and buried costs in *all* functional areas. Costs in marketing, purchasing, and design are brought to the forefront by the system. When senior management has all the facts on hidden and buried costs, they demand a cost of poor quality program.

The program is a comprehensive system and should not be perceived as merely a "fire-fighting" technique. For example, one response to a customer's problem could be to increase inspection. Although this action might eliminate the problem, the cost of poor quality would increase. Real quality improvement occurs when the root cause of the problem is found and corrected.

Cost of Poor Quality Categories and Elements

There are four categories—prevention, appraisal, internal failure, and external failure. Each category contains elements and subelements.

The *Prevention* category is defined as the experience gained from the identification and elimination of specific causes of failure cost to prevent the recurrence of the same or similar failures in other products or services. Prevention is achieved by examining the total of such experience and developing specific activities for incorporation into the basic management system that will make it difficult or impossible for the same errors or failures to occur again. The prevention costs of quality have been defined to include the cost of all activities specifically designed for this purpose. Each activity may involve personnel from one or many departments. No attempt is made to define appropriate departments, since each organization is organized differently. These costs occur in activities associated with marketing/customer/user interface, product/service/design development, purchasing, operations, and quality administration.

The *appraisal* cost category is the assurance that the product or service is acceptable as delivered to customers. This is the responsbility for evaluating a product or service at sequential stages, from design to first delivery and through out the production process, to determine its acceptability for continuation in the production or life cycle. The frequency and spacing of these evaluations are based on a trade-off between the cost benefits of early discovery of nonconformities and the cost of the evaluations (inspections and tests) themselves. Unless perfect control can be achieved, some appraisal cost will always exist. An organization would never want the customer to be the only inspector. Thus, the appraisal costs of poor quality have been defined to include all costs incurred in the planned conduct of product or service appraisals to determine compliance to requirements. Typical appraisal costs are incoming inspection, source inspection, operations inspections, measurement equipment, and field setup.

Whenever quality appraisals are performed, there exists the possibility for discovery of a failure to meet requirements. When this happens, unscheduled and possibly unbudgeted expenses are automatically incurred. When a complete lot of metal parts, for example, is rejected for being oversize, the possibility for rework must first be evaluated. Then the cost of rework may be compared to the cost of scrapping the parts and completely replacing them. Finally, a disposition is made and the action is carried out. The total cost of this evaluation, disposition, and subsequent action is an integral part of internal failure costs. In attempting to cover all possibilities for failure to meet requirements within the internal product or service life cycle, *internal failure costs* have been defined to include basically all costs required to evaluate, dispose of, and either correct or replace nonconforming products or services prior to delivery to the customer and also to correct or replace incorrect or incomplete product or service description (documentation). In general, this includes all the material and labor expenses that are lost or wasted due to nonconforming or otherwise unacceptable work affecting the quality of end products or service. Corrective action that is directed toward elimination of the problem in the future may be classified as prevention. Typical costs are poor design, purchasing errors and losses, rework, repair, scrap, and reappraisal.

The *external failure cost category* includes all costs incurred due to nonconforming or suspected nonconforming product or service after delivery to the customer. These costs consist primarily of costs associated with the product or service not meeting customer or user requirements. The responsibility for these losses may lie in marketing or sales, design development, or operations. Determination of responsibility is not part of the cost of poor quality system. It can come about only through investigation and analysis of external failure cost inputs. Typical costs are complaint investigation, returned goods, retrofit, warranty costs, liability costs, penalties, goodwill, and lost sales.

Collection and Reporting

The measurement of actual costs is essentially an accounting function. However, the development of the collection system requires the close interaction of the quality and accounting departments. Since accounting cost data are established by departmental cost codes, a significant amount of information can be obtained from this source. In fact, the system should be designed using the company's present system and modifying it where appropriate. Some existing sources for reporting information are time sheets, schedules, minutes of meetings, expense reports, credit and debit memos, and so forth.

Some cost data cross departmental lines, and these types of costs are the most difficult to collect. Special forms may be required to report some costs of poor quality. For example, scrap and rework costs may require analysis by quality control personnel to determine the cause and the departments responsible.

In some cases, estimates are used to allocate the proportion of an activity that must be charged to a particular quality cost element. For example, when the marketing department engages in research, it is necessary for the departmental supervisor to estimate the proportion of the activity that pertains to customer quality needs and should be charged as a cost in poor quality. Work-sampling techniques can be a valuable tool for assisting the supervisor in making the estimate.

Insignificant costs of poor quality, such as a secretary retyping a letter, may be difficult to determine and may be overlooked. However, significant ones are frequently hidden or buried because the accounting system is not designed to handle them. The cost of poor quality is a tool that can determine opportunities for quality improvement, justify the corrective action, and measure its effectiveness. Including insignificant activities is not essential to use the tool effectively. All significant activities or major elements, however, must be captured even if they are only estimated.

The comptroller's office must be directly involved in the design of the collection system. Only this office has the ability to effectively create a new system that will integrate the cost of poor quality into the existing accounting system. An ideal system would be one where the cost of poor quality is the difference between actual cost and the cost if everyone did a 100% perfect job *or* the difference between actual revenues and revenues if there were no unhappy customers. This ideal is not necessary and would be impossible to obtain. By having a comptroller directly involved, high exposure for the cost of poor quality is provided. Also, involvement leads to teamwork with quality, which will enhance the company's ability to achieve cost reduction.

Costs should be collected by product line, projects, departments, operators, nonconformity classification, and work centers. This manner of collection is sufficient for subsequent cost analysis. Procedures are developed to ensure that the system functions correctly. Using spread sheet software like EXCEL, micro reports are prepared for functional areas and departments and macro reports for the TQM function.

The basic control instrument is the cost of poor quality report, which is usually issued by the accounting department. An example of this type of report for all functional areas is shown in Figure 2-4. Each functional area would most likely have its own report. Provision is made to report the current month for each element as well as the current and prior year year-to-date values for the four cost categories. Applicable baseline data and ratios are shown at the bottom of the report. Baseline data compares the cost of poor quality per some index such as net sales, direct labor cost, production cost, and unit. Ratios are also used to reflect management emphasis.

By comparing current costs with historical ones, a certain amount of control can be exercised. It is also possible to establish a budget for each cost element. By comparing actual cost of poor quality with budget costs, favorable and unfavorable variances can be determined.

Analysis

Analysis techniques for cost of poor quality are quite varied. The quality cost report provides the information for the most common techniques: trend analysis and Pareto analysis. The objective of these techniques is to determine opportunities for quality improvement.

Trend analysis involves simply comparing present cost levels to past levels. Trend analysis provides information for long-range planning. It also provides information for the instigation and assessment of quality improvement programs. Data for trend analysis come from the monthly cost of poor quality report and the detailed transactions that make up the elements.

Trend analysis can be accomplished by cost category, by subcategory, by product, by measurement base, by plants within a corporation, by department, by work center, and by combination thereof. The graphs of some of these are shown in Figure 2-5. Time scales for the graphs may be by month, quarter, or year, depending on the purpose of the analysis; therefore, these graphs are also referred to as a time series.

Figure 2-5(a) shows a graph of the four cost categories by quarter and by product. It is the cumulative type wherein the second line from the bottom includes the prevention and the appraisal costs; the third line includes the internal failure, appraisal, and prevention costs; and the top line includes all four cost categories.

COMPANY ______ FOR MONTH ENDING ______ PREPARED BY ______

Prevention Costs $ (000)	Current Month	Year-to-Date	
		Current	Prior Yr.
Marketing/Customer Product/Service Development Purchasing Operations Quality Administration			
Total			

Appraisal Costs $ (000)	Current Month	Year-to-Date	
		Current	Prior Yr.
Product/Service Development Purchasing Operations External Appraisal Costs			
Total			

Internal Failure Costs $ (000)	Current Month	Year-to-Date	
		Current	Prior Yr.
Product/Service Design Purchasing			
Operations (Subtotal) Material Review Rework Repair Reappraisal Extra Operations Scrap			
Total			

External Failure Costs $ (000)	Current Month	Year-to-Date	
		Current	Prior Yr.
Customer Complaints Returned Goods Retrofit Costs			
Warrenty Claims Liability Costs Penalties Customer Goodwill			

Baseline Data $ (000)	Current Month	Year-to-Date	
		Current	Prior Yr.
Net Sales Direct Labor Production Unit			

Ratios $ (000)	Current Month	Year-to-Date	
		Current	Prior Yr.
External Failure Cost/Net Sales Operations Failure Costs/Production Costs Operations Appraisal Costs/Production Costs Purchasing Costs of Poor Quality/Material Costs Design Costs of Poor Quality/Design Costs			

FIGURE 2-4 Cost of poor quality summary report.

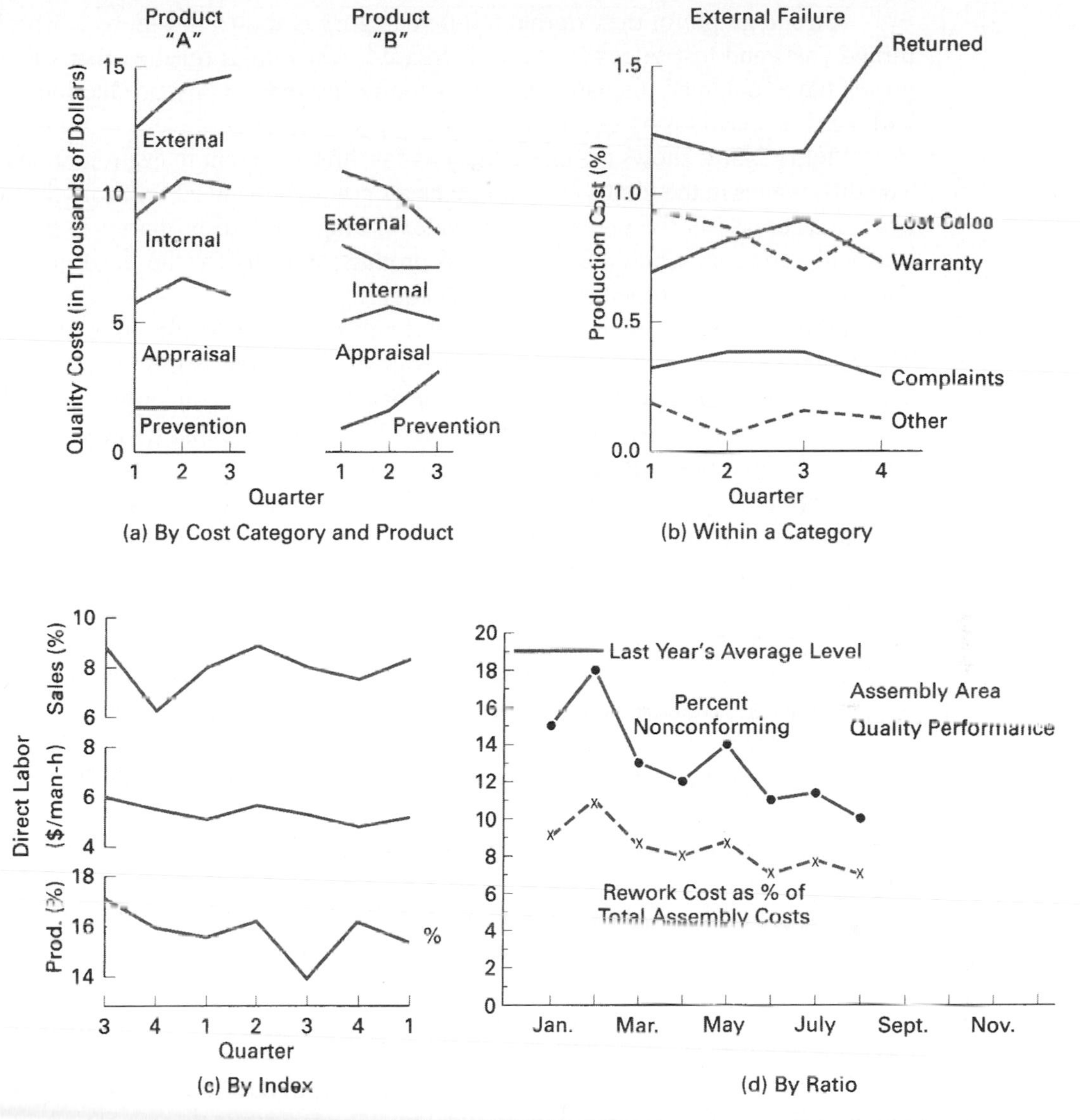

FIGURE 2-5 Typical long range trend analysis graphs.

The figure shows that the cost of poor quality for product B is better than that for product A. In fact, product B is showing a nice improvement, whereas product A's cost is increasing. An increase in prevention and appraisal costs will, it is hoped, improve the external and internal failure costs of product A. Comparisons between products and plants should be made with extreme caution.

A trend graph for the external failure category is shown in Figure 2-5(b). Returned costs and lost sales costs have increased, while costs for the other subcategories have remained unchanged. In this figure the index is by production costs, and the time period is by quarters.

Figure 2-5(c) shows the trend analysis for three different measurement bases. The differences in the trends of the three bases point up the need for more than one base. A decrease in the percent of net sales during the fourth quarter is due to a seasonal variation, while the variation in production costs for the third quarter is due to excessive overtime costs during the quarter.

Figure 2-5(d) shows a short-run trend analysis chart for the assembly area. The ratio of rework costs to total assembly costs in percent is plotted by months. This ratio is compared to the quality measure, percent nonconforming. Both curves show a decrease, which supports the basic concept that quality improvement is synonymous with reduced costs.

Trend analysis is an effective tool provided it is recognized that some period-to-period fluctuations are chance variations. These variations are similar to those that occur on an $\overline{X}$ and R chart. The important factor to observe is the cost trend. It is also important to note that there may be a time lag between the occurrence of a cost and the actual reporting of that cost.

The other effective cost-analysis tool is Pareto analysis, discussed in Chapter 3. A typical Pareto diagram for internal failures is shown in Figure 2-6(a). Items are located in descending order beginning with the largest one on the left. A Pareto diagram has a few items that represent a substantial amount of the total. These items are located on the left of the diagram and are referred to as the *vital few*. A Pareto diagram has many items that represent a small amount of the total. These items are located on the right and are referred to as the *useful many*. Pareto diagrams can be established for cost of poor quality by operator, by machine, by department, by product line, by nonconformity, by category, by element, and so forth.

Once the vital few are known, projects can be developed to reduce their costs. In other words, money is spent to reduce the vital few costs; little or no money is spent on the useful many.

Figure 2-6(b) shows a Pareto diagram by department. This Pareto diagram is actually an analysis of one of the vital few elements (operation—scrap) in the Pareto diagram for the internal failure category as shown in Figure 2-6(a). Based on the diagram, department D would be an excellent candidate for a quality-improvement program.

Optimum

In analyzing quality costs, management wants to know the optimum costs. This information is difficult to specify.

One technique is to make comparisons with other organizations. More and more organizations use net sales as an index, which makes comparison somewhat

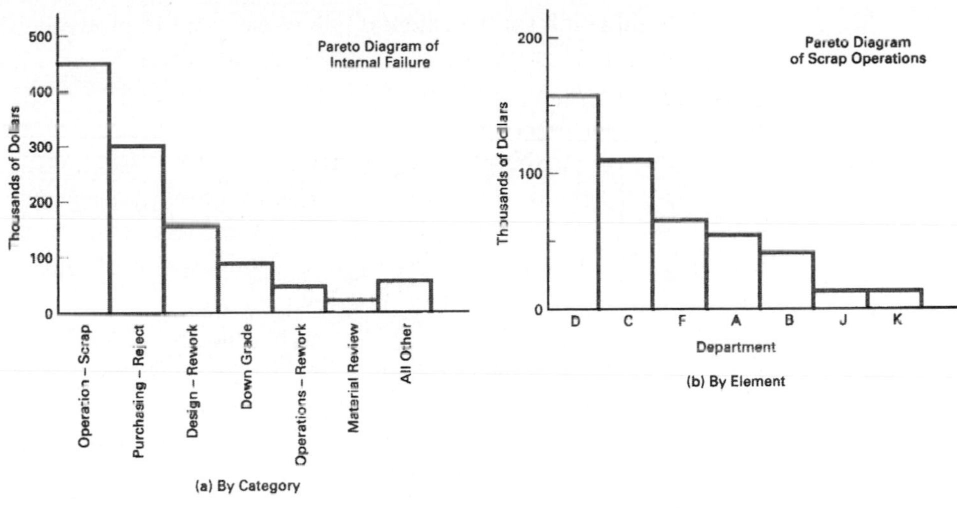

FIGURE 2-6 Pareto analysis.

easier. Difficulties arise, however, because many organizations keep their costs secret. Also, accounting systems treat the collection of costs differently. For example, overhead costs may or may not be included in a particular cost element. There are many variations in types of manufacturing and service organizations that cause quality cost to vary appreciably. Where complex, highly reliable products are involved, the cost of poor quality may be as high as 20% of sales; in industries that produce simple products with low tolerance requirements, costs of less than 5% of sales may be commonplace.

Another technique is to optimize the individual categories. Failure costs are optimized when there are no identifiable and profitable projects for reducing them. Appraisal costs are also optimized when there are no identifiable and profitable projects for reducing them. Prevention costs are optimized when most of the dollar cost is used for improvement projects, when the prevention work itself has been analyzed for improvement, and when non-project prevention work is controlled by sound budgeting.

A third technique to determine the optimum is to analyze the relationships among the cost categories. Figure 2-7 shows an economic model for the cost of poor quality. As the quality of conformance improves and approaches 100%, failure costs are reduced until they approach zero. In other words, if the product or service is perfect, there are no failure costs. To achieve a reduction in failure costs, it is necessary to increase appraisal and prevention costs. Combining the two curves gives the total cost curve. The model shows that as quality increases, the cost of poor quality decreases; however, there is a school of thought that feels it uneconomical to achieve 100% conformance. In this model, the upper two c

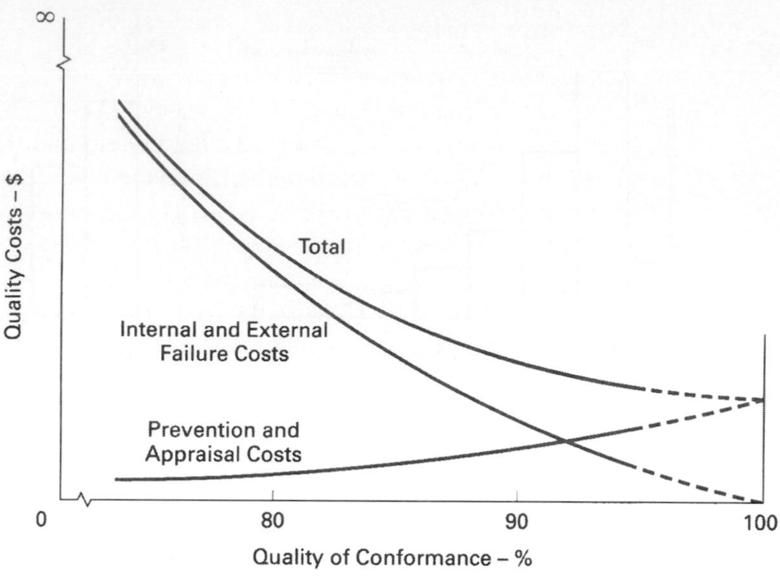

FIGURE 2-7 Optimum cost of poor quality concept.

turn upward and their costs go to infinity rather than converge, as shown by the dashed lines. However, perfection is being economically achieved as the inspection process is automated and where the customer is willing to pay for perfect quality. Also, perfection is the goal where quality has a critical impact on safety, such as in the nuclear power field, or where lost sales can lead to bankruptcy. Therefore, this theoretical model appears to represent quality costs quite well.

To further support the contention that perfection (100% conformance) is an achievable goal, 99.9% conformance for a few activities is given below:

1. 16,000 pieces of lost mail every hour
2. 500 failed surgical operations each week
3. Two unsafe airplane landings at O'Hare airport (Chicago)
4. 22,000 checks deducted from the wrong account each month
5. 2,000,000 people dead or sick from food poisoning each year
6. 45 minutes of unsafe drinking water each month

Note that the model of Figure 2-7 is for an entire system. When analyzing an individual quality characteristic, it is possible to make the quality so good that it would be uneconomical.

Quality-Improvement Strategy

Once the problem area has been determined using the analysis techniques, a project team can be established. There are two types of problems: those that a department can correct with little or no outside help and those that require coordinated action from several functional areas in the organization.

Problems of the first type do not require an elaborate system. The project team could be composed of the operating supervisor, operator, quality engineer, maintenance supervisor, and other appropriate personnel such as an internal customer or internal supplier. Usually the team has sufficient authority and resources to enact corrective action without approval of their superiors. Problems of this type usually account for about 15% of the total.

Unfortunately, about 85% of quality problems cross departmental and functional area lines. Since these problems are usually more costly and more difficult to solve, a more elaborate and structured project team is established. Members of the team would most likely be composed of personnel from operations, quality, design, marketing, purchasing, and any other area of the entity. The team receives written authority from the quality council or a similar body. Resources are allocated and a schedule of activities is prepared. Periodic reports are given to the council. A member of the quality council should act as a coach to the team.

The basic concept is that each failure has a root cause, causes are preventable, and prevention is cheaper. Based on this concept the following strategy is used:

1. Reduce failure costs by problem solving.
2. Invest in the "right" prevention activities.
3. Reduce appraisal costs where appropriate and in a statistically sound manner.
4. Continuously evaluate and redirect the prevention effort to gain further quality improvement.

1. *Reducing Failure Costs*. Most of the quality-improvement projects will be directed toward reducing failure costs. It is a fact that failures detected at the beginning of operations are less costly than failures detected at the end of operations or by the customer, and such failures are cheaper to correct. Therefore, external failures are frequently targeted for improvement because they can give the greatest return on investment, i.e., increased customer satisfaction and reduced production costs.

The project team must concentrate on finding the root cause of the problem. In this regard, it may be necessary to trace the potential cause to purchasing, design, or marketing. Care must be exercised to ensure that the basic cause has been found rather than some pseudocause. Once the cause has been determined, the project team can concentrate on developing the corrective action to control or, preferably, eliminate the problem.

Follow-up activities are conducted to ensure that the corrective action was effective in solving the problem. The team should also review similar problems to determine if a similar solution might be effective. Finally, the quality cost saving is calculated, and a final report is presented to the quality council.

2. *Prevention of Costs Due to Poor Quality*. Rather than solve problems that are costing money, it would be much better if problems could be prevented. Prevention activities are related to employee attitudes and to formal techniques to eliminate problems in the product cycle before they become costly.

Employee attitudes toward quality are determined by top management's commitment to quality and the involvement of both in the quality-improvement program. Suggestions for achieving this commitment and involvement are as follows:

1. Include both groups as members of project team
2. Establish a quality council with the CEO and functional area managers as members
3. Involve employees in the annual quality improvement program
4. Provide a system whereby employees can present quality improvement ideas
5. Communicate quality expectations to employees
6. Publish a company newsletter
7. Hold a quarterly meeting of all employees

Using formal techniques for preventing quality problems before they occur is a more desirable activity than problem solving. The following are some examples of these techniques:

1. New-product verification programs that require a comprehensive review before release for quantity production
2. Design-review programs of new or changed designs that require involvement of appropriate functional areas at the beginning of the design process (concurrent engineering)
3. Supplier selection programs that concentrate on quality rather than price
4. Reliability testing to prevent high field-failure costs
5. Thorough training and testing of employees so that their jobs are done right the first time
6. Voice of the customer, such as quality function deployment

Effective management of prevention costs will provide the greatest quality improvement potential.

3. *Reducing Appraisal Costs*. As failure costs are reduced, the need for appraisal activities will, most likely, be reduced. Programs for cost improvement can have a significant impact on the total quality costs. Periodically, a project team should review the entire appraisal activity to determine its effectiveness.

Typical questions that the project team might investigate are

1. Is 100% inspection necessary?
2. Can inspection stations be combined, relocated, or eliminated?
3. Are inspection methods the most efficient?
4. Could the inspection and test activity be automated?
5. Could data be more efficiently collected, reported, and analyzed using the computer?
6. Should statistical process control be used?
7. Should operating personnel be responsible for inspection?
8. Is appraisal being used as a substitute for prevention?

Malcolm Baldrige National Quality Award (MBNQA)[6]

A second performance measure is the Malcolm Baldrige National Quality Award criteria, an excellent performance measure for the entire organization.

The Malcolm Baldrige National Quality Award (MBNQA) is an annual award to recognize U.S. organizations for performance excellence. It was created by Public Law 100–107 on August 20, 1987. The award promotes: understanding of the requirements for performance excellence and competitiveness improvement, sharing of information on successful performance strategies, and the benefits derived from using these strategies. There are five categories: manufacturing, service, small business, health care, and education. Three awards may be given each year in each category. Competition for the awards is intense. Many organizations, that are not interested in the award, are using the categories as a technique to measure their TQM effort on an annual assessment basis.

The criteria for performance excellence are the basis for organizational self-assessment, for making awards and for giving feedback to applicants. In addition, they (1) help improve performance practices and capabilities, (2) facilitate communication and sharing of best practices information among U.S. organizations of all types, and (3) serve as a working tool for understanding and managing performance, planning, training, and assessment. The results-oriented goals are designed to

---

[6]Adapted from U.S. Department of Commerce, *Malcolm Baldrige National Quality Award 2003 Criteria*, Updated information is available at www.asq.org.

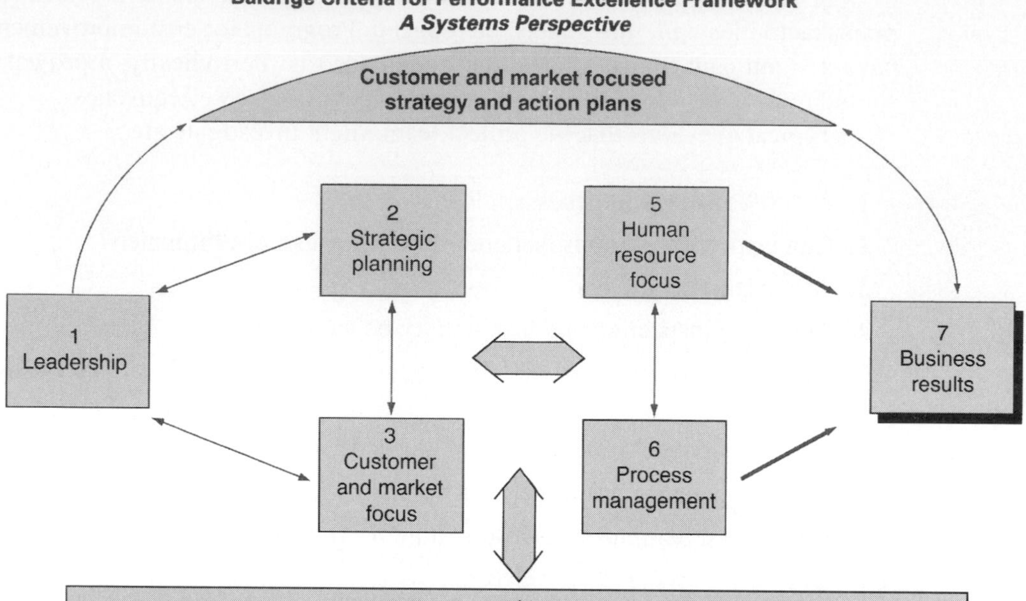

FIGURE 2-8 Award Criteria Framework

deliver ever-improving value to customers, resulting in marketplace success, and to improve overall organization performance and capability. The criteria are derived from the set of core values and concepts that were listed earlier.

The core values and concepts are embodied in seven categories, as shown in Figure 2-8. The seven categories shown in the figure are subdivided into examination items and areas to address. There are 19 examination items; item titles and point values are given later. Each examination item consists of sets of areas to address. Information is submitted by applicants in response to specific requirements of these areas.

Table 2-4 shows the seven categories and the 19 items with their point values. It is important to note that almost one-half of the total score is based on results.

The ***Leadership*** Category examines how the organization's senior leaders address values and performance expectations, as well as a focus on customers and other stakeholders, empowerment, innovation, learning, and organizational directions. Also examined is how the organization addresses its responsibilities to the public and supports its key communities.

The ***Strategic Planning*** Category examines the organization's strategy development process, including how your organization develops strategic objectives,

TABLE 2-4 Award Categories/Items and Point Values

1	**LEADERSHIP**			**120**
	1.1	Organizational Leadership	70	
	1.2	Social Responsibility	50	
2	**STRATEGIC PLANNING**			**85**
	2.1	Strategy Development	40	
	2.2	Strategy Deployment	45	
3	**CUSTOMER AND MARKET FOCUS**			**85**
	3.1	Customer and Market Knowledge	40	
	3.2	Customer Relationships and Satisfaction	45	
4	**MEASUREMENT, ANALYSIS, AND KNOWLEDGE MANAGEMENT**			**90**
	4.1	Measurement and Analysis of Organization Performance	45	
	4.2	Information and Knowledge Management	45	
5	**HUMAN RESOURCE FOCUS**			**85**
	5.1	Work systems	35	
	5.2	Employee Learning and Motivation	25	
	5.3	Employee Well-Being and Satisfaction	25	
6	**PROCESS MANAGEMENT**			**85**
	6.1	Value Creation Processes	50	
	6.2	Support Processes	35	
7	**BUSINESS RESULTS**			**450**
	7.1	Customer Focused Results	75	
	7.2	Product and Service Results	75	
	7.3	Financial and Market Results	75	
	7.4	Human Resource Results	75	
	7.5	Organizational Effectiveness Results	75	
	7.6	Governance and Social Responsibility Results	75	
		TOTAL POINTS		**1000**

action plans, and related human resource plans. Also examined are how plans are deployed and how performance is tracked.

The ***Customer and Market Focus*** Category examines how the organization determines requirements, expectations, and preferences of customers and markets. Also examined is how your organization builds relationships with customers and determines their satisfaction.

The ***Measurement, Analysis, and Knowledge Management*** Category examines the organization's performance measurement system and how the organization analyzes performance data and information.

The ***Human Resource Focus*** Category examines how the organization enables employees to develop and utilize their full potential, aligned with the organization's objectives. Also examined are the organization's efforts to build and maintain a work environment and an employee support climate conducive to performance excellence, full participation, and personal and organizational growth.

The ***Process Management*** Category examines the key aspects of the organization's process management, including customer-focused design, product and service delivery, support, and supplier and partnering processes involving all work units.

The ***Business Results*** Category examines the organization's performance and improvement in key business areas—customer satisfaction, product and service performance, financial and marketplace performance, human resource results, supplier and partner results, and operational performance. Also examined are performance levels relative to competitors.

The MBNQA provides a plan to keep improving all operations continuously and a system to measure these improvements accurately. Benchmarks are used to compare the organization's performance with the world's best and to establish stretch goals. A close partnership with suppliers and customers that feeds improvements back into the operation is required. There is a long-lasting relationship with customers, so that their wants are translated into products and services that go beyond delivery. Management from top to bottom is committed to improving quality. Preventing mistakes and looking for improvement opportunities is built into the culture. There is a major investment in human resources by means of training, motivation, and empowerment.

According to Dr. J. M. Juran, who studied the winners of the award, the gains have been stunning. The gains can be accomplished by large and small U.S. organizations and by U.S. workers. The gains include quality, productivity, and cycle time.

Competition for the award is intense, many organizations that are not interested in the award are using the categories as a technique for quality improvement.

Other Performance Measures

Cost of poor quality and the MBNQA measure the performance of the entire organization. Given below are techniques that measure a small portion of the organization.

Another performance measure is control charts, as given in Chapters 5, 6, and 8. These statistical techniques directly measure the effect of quality improvement on an existing product or process. An overall measure such as percent nonconforming is necessary for the plant or the entire corporation. This measure can also effectively evaluate the CEO's performance. Each functional area and department within a functional area should have a measure that is displayed in some type of chart so it can be viewed by all personnel. These charts create quality awareness and measure the progress of quality improvement. A few companies have improved their performance to the point where a nonconformity per million chart is more appropriate than a count of nonconformity chart.

Nonstatistical charts, such as the trend analysis (also called time series) or Pareto analysis that were previously discussed, are also used. These simple graphs effectively portray the performance of key indicators for many business and production processes.

Quality can be measured by comparing the specifications to the process capability. Dr. W. Edwards Deming has stated that we need to drive the specifications over the horizon. This statement is a figure of speech that actually means the process variability is so small around the target value and the specifications are so far away that they have the appearance of being out of sight. This concept is discussed in Chapter 5. Specifications are dynamic in that they are constantly getting smaller, which requires never-ending improvement in the process capability.

Conventional practice assumes that loss occurs only when a characteristic is outside the specifications. Dr. Genichi Taguchi states that loss to the customer and society occurs as soon as the characteristic deviates from the target value. Figure 2-9 illustrates this concept. The target value and specifications are shown on the x scale and the loss in dollars is shown on the $f(x)$ scale. The more the characteristic deviates from the target value, the greater the loss. While the actual shape of the curve may be difficult to predict, the quadratic approximation shown in the figure frequently represents the economic loss function. Where the curve crosses the specifications, the cost of repairing or discarding the product is given in dollars at D. Using this value, the equation for the curve can be determined. The concept (with a different shape curve) can be used for other situations, such as where the target value is the largest possible value or where the target value is zero. Taguchi has combined specifications, target value, minimum variation, and dollars into one package to measure quality.

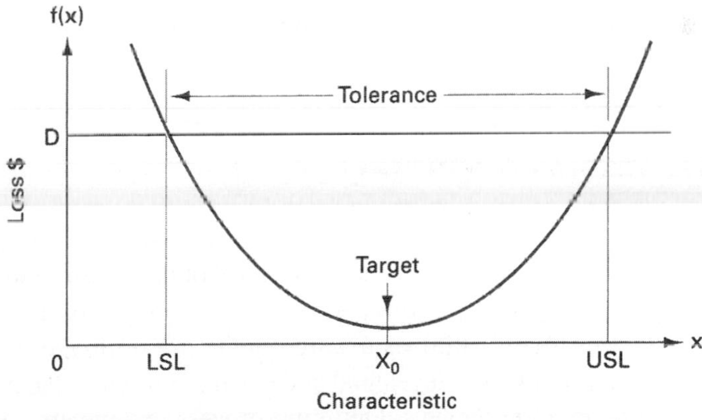

FIGURE 2-9 Taguchi's loss function.

All the measures are needed for the TQM program. They are each applied in different situations.

DEMING'S 14 POINTS

A chapter on TQM would not be complete without listing Dr. Deming's 14 obligations of senior management. They are as follows:

1. Create and publish the aims and purposes of the organization.
2. Learn the new philosophy.
3. Understand the purpose of inspection.
4. Stop awarding business based on price alone.
5. Improve constantly and forever the system.
6. Institute training.
7. Teach and institute leadership.
8. Drive out fear, create trust, and create a climate for innovation.
9. Optimize the efforts of teams, groups, and staff areas.
10. Eliminate exhortations for the work force.

11a. Eliminate numerical quotas for the work force.

11b. Eliminate management by objectives.

12. Remove barriers that rob people of pride of workmanship.
13. Encourage education and improvement.
14. Take action to accomplish the transformation.

Many of Dr. Deming's 14 points have been incorporated into the material of this chapter.

FINAL COMMENTS

Management must know that quality is first among the equals of cost and service. In this regard, there is no economic level of quality or, if there is such a level, few, if any, organizations have achieved it. The ultimate goal is exceeding customer expectations.

The old attitudes toward quality are no longer acceptable. New products and services must be developed and existing products and services modified to meet customers' requirements. Optimum process parameters need to be determined in order to achieve the smallest possible variation. The evidence shows that high-quality products and services increase productivity and provide the competitive advantage for company survival.

The effective implementation of TQM will allow companies to achieve the vision given below:

- Customers receive what they order without nonconformities, on time, in the right quantity, shipped and billed on time
- Suppliers meet our requirements
- Salespeople determine customer needs
- New products and processes are developed to agreed-upon requirements, as scheduled, and at lower costs
- People enjoy their work
- Company makes a profit

PROBLEMS

1. Working in a team of three or more people, visit one or more of the following organizations. Determine if they have a quality council or similar structure. If so describe its composition and duties.
   (a) Large bank
   (b) Health-care facility
   (c) University academic department
   (d) University nonacademic department
   (e) Large department store
   (f) Grade school
   (g) Manufacturing facility
   (h) Large grocery store
2. Design a customer satisfaction questionnaire for one of the organizations listed in Problem 1.
3. For the organizations listed in Problem 1, determine two external customers, two internal customers, and two external suppliers.
4. Using a team of three or more people design an employee opinion survey for a work unit of one of the organizations listed in Problem 1. Conduct the survey and analyze the results.
5. Working in a team of six or more people, implement the seven phases of the problem-solving method. Elect a team leader and identify a substantial customer.

6. Working in a team of three or more people develop a supplier selection plan for one of the organizations listed in Problem 1.

7. Working in a team of three or more people visit two of the organizations listed in Problem 1. Determine the performance measures used and their adequacy.

8. Write a plan to implement TQM in one of the organizations listed in Problem 1.

3 TOTAL QUALITY MANAGEMENT—TOOLS AND TECHNIQUES

Tools and techniques are divided into the categories of quantitative and non quantitative and are shown in Figure 2-2. The quantitative ones are statistical process control (SPC), acceptance sampling, reliability, experimental design, Taguchi's quality engineering, failure mode and effect analysis (FMEA), and quality function deployment (QFD). The non quantitative ones are ISO 9000, ISO 14000, benchmarking, total productive maintenance (TPM), management tools, quality by design, products liability, and information technology. A portion of SPC and all of ISO 9000 are discussed in detail in this chapter. The rest of the tools and techniques are described in summary form.

STATISTICAL PROCESS CONTROL (SPC)

SPC is comprised of seven tools: Pareto diagram, cause and effect diagram, check sheets, process flow diagram, scatter diagram, histogram, and control charts. The first five are discussed in detail and the last two in summary form with reference to the appropriate chapters.

Pareto Diagram

Alfredo Pareto (1848–1923) conducted extensive studies of the distribution of wealth in Europe. He found that there were a few people with a lot of money and many people with little money. This unequal distribution of wealth became an integral part of economic theory. Dr. Joseph Juran recognized this concept as a universal that could be applied to many fields. He coined the phrases *vital few* and *useful many*.[1]

A Pareto diagram is a graph that ranks data classifications in descending order from left to right, as shown in Figure 3-1. In this case, the data classifications are types of field failures. Other possible data classifications are problems, causes, types of nonconformities, and so forth. The vital few are on the left, and the useful many are on the right. It is sometimes necessary to combine some of the useful many into one classification called *other* and labeled *O* in the figure. When the other category is used, it is always on the far right. The vertical scale is dollars, frequency, or percent. Pareto diagrams can be distinguished from histograms (to be discussed) by the fact that the horizontal scale of a Pareto is categorical, whereas the scale for the histogram is numerical.

Sometimes a Pareto diagram has a cumulative line, as shown in Figure 3-2. This line represents the sum of the data as they are added together from left to right. Two scales are used: The one on the left is either frequency or dollars, and the one on the right is percent.

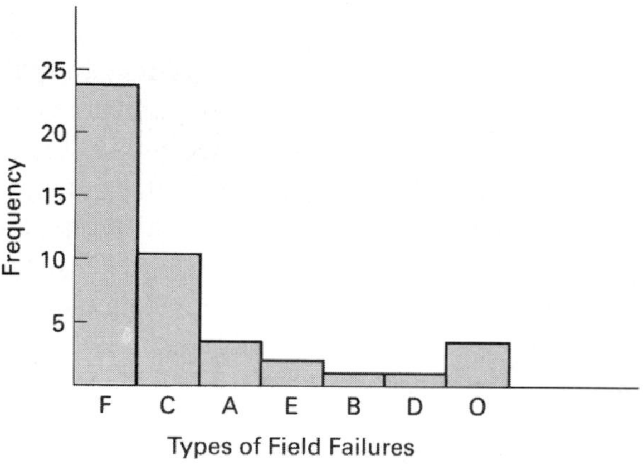

FIGURE 3-1 Pareto diagram.

[1]Dr. Juran recently changed this terminology from *trivial many* to *useful many* because there is no trivial quality problem.

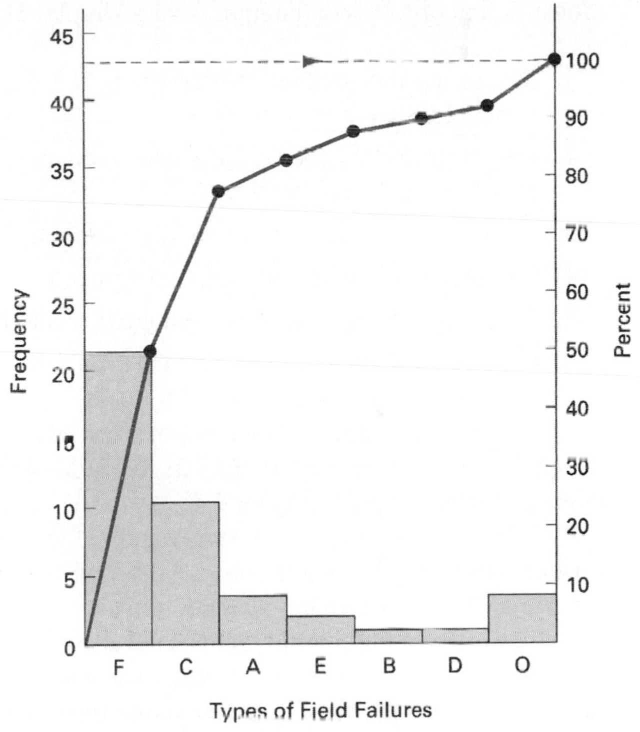

FIGURE 3-2 Cumulative line of Pareto diagram.

Pareto diagrams are used to identify the most important problems. Usually, 80% of the total results from 20% of the items. This fact is shown in Figure 3-2, where the F and C types of field failures account for almost 80% of the total. Actually, the most important items could be identified by listing the items in descending order. However, the graph has the advantage of providing a visual impact of those vital few characteristics that need attention. Resources are then directed to take the necessary corrective action. Examples of the vital few are

- A few customers account for the majority of sales.
- A few products, processes, or quality characteristics account for the bulk of the scrap or rework cost.
- A few nonconformities account for the majority of customer complaints.
- A few vendors account for the majority of rejected parts.
- A few problems account for the bulk of the process downtime.
- A few products account for the majority of the profit.
- A few items account for the bulk of the inventory cost.

Construction of a Pareto diagram is very simple. There are six steps:

1. Determine the method of classifying the data: by problem, cause, type of nonconformity, and so forth.
2. Decide if dollars (best), weighted frequency, or frequency is to be used to rank the characteristics.
3. Collect data for an appropriate time interval.
4. Summarize the data and rank order categories from largest to smallest.
5. Compute the cumulative percentage if it is to be used.
6. Construct the diagram and find the vital few.

The cumulative percentage scale, when used, must match with the dollar or frequency scale such that 100% is at the same height as the total dollars or frequency. See the arrow in Figure 3-2.

It is noted that a quality improvement of the vital few, of say 50%, is a much greater return on investment than a 50% improvement of the useful many. Also, experience has shown that it is easier to make a 50% improvement in the vital few.

The use of a Pareto diagram is a never-ending process. For example, let's assume that F is the target for correction in the improvement program. A project team is assigned to investigate and make improvements. The next time a Pareto analysis is made, another field failure, say C, becomes the target for correction, and the improvement process continues until field failures become an insignificant quality problem.

The Pareto diagram is a powerful quality-improvement tool. It is applicable to problem identification and the measurement of progress.

Cause-and-Effect Diagram

A cause-and-effect (C&E) diagram is a picture composed of lines and symbols designed to represent a meaningful relationship between an effect and its causes. It was developed by Dr. Kaoru Ishikawa in 1943 and is sometimes referred to as an Ishikawa diagram.

C&E diagrams are used to investigate either a "bad" effect and to take action to correct the causes or a "good" effect and to learn those causes responsible. For every effect, there are likely to be numerous causes. Figure 3-3 illustrates a C&E diagram with the effect on the right and causes on the left. The effect is the quality characteristic that needs improvement. Causes are usually broken down into the major causes of work methods, materials, measurement, people, and the environment. Management and maintenance are also sometimes used for the major cause. Each major cause is further subdivided into numerous minor causes. For example, under work

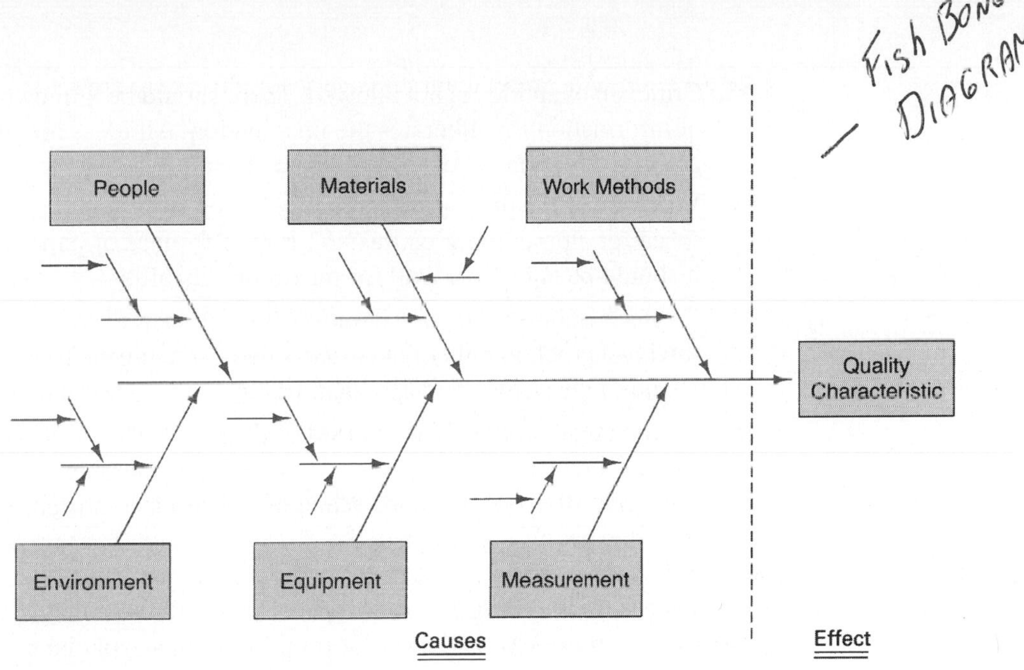

FIGURE 3-3 Cause-and-effect diagram.

methods, we might have training, knowledge, ability, physical characteristics, and so forth. C&E diagrams (frequently called "fish-bone diagrams" because of their shape) are the means of picturing all these major and minor causes.

The first step in the construction of a C&E diagram is for the project team to identify the effect or quality problem. It is placed on the right side of a large piece of paper by the team leader. Next, the major causes are identified and placed on the diagram.

Determining all the minor causes requires brainstorming by the project team. Brainstorming is an idea-generating technique that is well-suited to the C&E diagram. It uses the creative thinking capacity of the team.

Attention to a few essentials will provide a more accurate and usable result:

1. Participation by every member of the team is facilitated by each member taking a turn giving one idea at a time. If a member cannot think of a minor cause, he or she passes for that round. Another idea may occur at a later round. By following this procedure, one or two individuals do not dominate the brainstorming session.
2. Quantity of ideas, rather than quality, is encouraged. One person's idea will trigger someone else's idea, and a chain reaction occurs. Frequently, a trivial or "dumb" idea will lead to the best solution.

3. Criticism of an idea is not allowed. There should be a freewheeling exchange of information that liberates the imagination. All ideas are placed on the diagram. Evaluation of ideas occurs at a later time.
4. Visibility of the diagram is a primary factor of participation. In order to have space for all the minor causes, a 2-ft by 3-ft piece of paper is recommended. It should be taped to a wall for maximum visibility.
5. Create a solution-oriented atmosphere and not a gripe session. Focus on solving a problem rather than discussing how it began. The team leader should ask questions using the why, what, where, when, who, and how techniques.
6. Let the ideas incubate for a period of time (at least overnight), and then have another brainstorming session. Provide team members with a copy of the ideas after the first session. When no more ideas are generated, the brainstorming activity is terminated.

Once the C&E diagram is complete, it must be evaluated to determine the most likely causes. This activity is accomplished in a separate session. The procedure is to have each person vote on the minor causes. Team members may vote on more than one cause, and they do not need to vote on a cause they presented. Those causes with the most votes are circled, and the four or five most likely causes of the effect are determined.

Solutions are developed to correct the causes and improve the process. Criteria for judging the possible solutions include cost, feasibility, resistance to change, consequences, training, and so forth. Once the solutions have been agreed to by the team, testing and implementation follow.

Diagrams are posted in key locations to stimulate continued reference as similar or new problems arise. The diagrams are revised as solutions are found and improvements are made.

The cause-and-effect diagram has nearly unlimited application in research, manufacturing, marketing, office operations, and so forth. One of its strongest assets is the participation and contribution of everyone involved in the brainstorming process. The diagrams are useful in

1. *Analyzing* actual conditions for the purpose of product or service quality improvement, more efficient use of resources, and reduced costs
2. *Elimination* of conditions causing nonconforming product or service and customer complaints
3. *Standardization* of existing and proposed operations
4. *Education and training* of personnel in decision-making and corrective-action activities

The previous paragraphs have described the *cause-enumeration* type of C&E diagram, which is the most common type. There are two other types of C&E diagrams that are similar to the cause enumeration. They are the dispersion-analysis and process-analysis types. The only difference among the three methods is the organization and arrangement.

The *dispersion-analysis* type of C&E diagram looks just like the cause-enumeration type when both are complete. The difference is in the approach to constructing it. For this type, each major branch is filled in completely before starting work on any of the other branches. Also, the objective is to analyze the causes of dispersion or variability.

The *process-analysis* type of C&E diagram is the third type, and it does look different from the other two. In order to construct this diagram, it is necessary to write each step of the production process. Steps in the production process such as load, cut, bore, c'sink, chamfer, and unload become the *major causes,* as shown in Figure 3-4. Minor causes are then connected to the major ones. This C&E diagram is for elements within an operation. Other possibilities are operations within a process, an assembly process, a continuous chemical process, and so forth. The advantage of this type of C&E diagram is the ease of construction and its simplicity, since it follows the production sequence.

Check Sheets

The main purpose of check sheets is to ensure that the data are collected carefully and accurately by operating personnel for process control and problem solving. Data should be presented in such a form that it can be quickly and easily used and

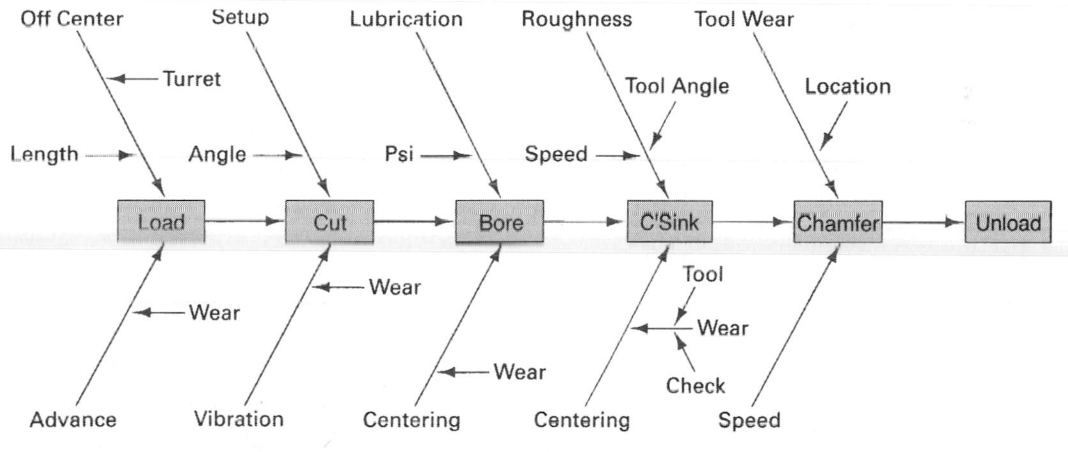

FIGURE 3-4 Process analysis C&E diagram.

CHECK SHEET

Product: Bicycle—32	**Date:** Jan. 21
Stage: Final Inspection	**ID:** Paint
Number Inspected: 2217	**Inspector/Operator:** Jane Doe

Nonconformity Type	Check	Total
Blister	𝍸 𝍸 𝍸 𝍸 |	21
Light Spray	𝍸 𝍸 𝍸 𝍸 𝍸 𝍸 𝍸 |||	38
Drips	𝍸 𝍸 𝍸 𝍸 ||	22
Overspray	𝍸 𝍸 |	11
Splatter	𝍸 |||	8
Runs	𝍸 𝍸 𝍸 𝍸 𝍸 𝍸 𝍸 𝍸 𝍸 ||	47
Others	𝍸 𝍸 ||	12
	Total	159
Number Nonconforming	𝍸 |||	113

FIGURE 3-5 Check sheet for paint nonconformities.

analyzed. The form of the check sheet is individualized for each situation and is designed by the project team. Figure 3-5 shows a check sheet for paint nonconformities for bicycles. Figure 3-6 on page 83 shows a maintenance check sheet for the swimming pool of a major motel chain. Checks are made on a daily and weekly basis, and some checks, such as temperature, are measured. This type of check sheet ensures that a check or test is made.

Figure 3-7 on page 84 shows a check sheet for temperature. The scale on the left represents the midpoint and boundaries for each temperature range. Data collection for this type of check sheet is frequently accomplished by placing an x in the appropriate square. In this case, the time has been recorded in order to provide additional information for problem solving.

Whenever possible, check sheets are also designed to show location. For example, the check sheet for bicycle paint nonconformities could have shown an outline of a bicycle with small x's indicating the location of the nonconformities.

Figure 3-8 shows a check sheet for a 9-cavity plastic mold. This check sheet clearly shows that there are quality problems at the upper corners of the mold. What additional information would the reader suggest?

Creativity plays a major role in the design of a check sheet. It should be user friendly and, whenever possible, include information on time and location.

D = Daily   A = As Needed

Hot Tub		Mon.	Tues.	Wed.	Th.	Fri.	Sat.	Sun.
Chemical Test (Add if Needed) ph/chlorine	(D)	7.4						
Temperature	(D)	81°						
Add Water (If Needed)	(D)							
Clean Deck Around Hot Tub	(D)	✓						
Pool								
Chemical Test (Add if Needed)	(D)	7.6						
Add Water (If Needed)	(D)	300 gals.						
Check Temperature	(D)	78°						
Vacuum Pool (If Needed)	(A)							
Filter Backwash (20 lb.)	(A)	✓						
Lint Filter	(D)	✓						
Sweep and Hose Off Deck	(D)	✓						
General Cleaning								
Vacuum Carpets	(D)	✓						
Vacuum and Sweep Building B	(D)	✓						
Clean Tables	(D)	✓						
Sweep and Mop Wooden Deck	(D)	✓						
Clean Outside Deck, Bring in Chairs	(D)	✓						
Take Out Trash	(D)	✓						
Empty Building B Trash Cans	(D)	✓						
Wash Windows	(D)	✓						
Bathrooms								
Scrub Sinks, Toilets, and Showers	(D)	✓						
Sweep and Mop Floors	(D)	✓						
Empty Trash and Check Lockers	(D)	✓						
Cover Hot Tub (At End of the Night)	(D)	✓						
Check Pool Filters—Be Sure It's On	(D)	✓						

List any and all deviations from this work schedule on the reverse side, date it, and initial it.

FIGURE 3-6 Check sheet for swimming pool.

TEMPERATURE CHECK SHEET

385	387.4 382.5								
380	382.4 377.5								
375	377.4 372.5	10.0							
370	372.4 367.5								
365	367.4 362.5	7.0	7.5	9.0					
360	362.4 357.5	8.0	8.5						
355	357.4 352.5	9.5							

FIGURE 3-7 **Check sheet for temperature.**

XXXX XX	X	XXXX X
	XX	
	X	X

FIGURE 3-8 **Check sheet for plastic mold nonconformities.**

Process Flow Diagram

For many products and services, it may be useful to construct a flow diagram. It is a schematic diagram that shows the flow of the product or service as it moves through the various processing stations or operations. The diagram makes it easy to visualize the entire system, identify potential trouble spots, and locate control activities.

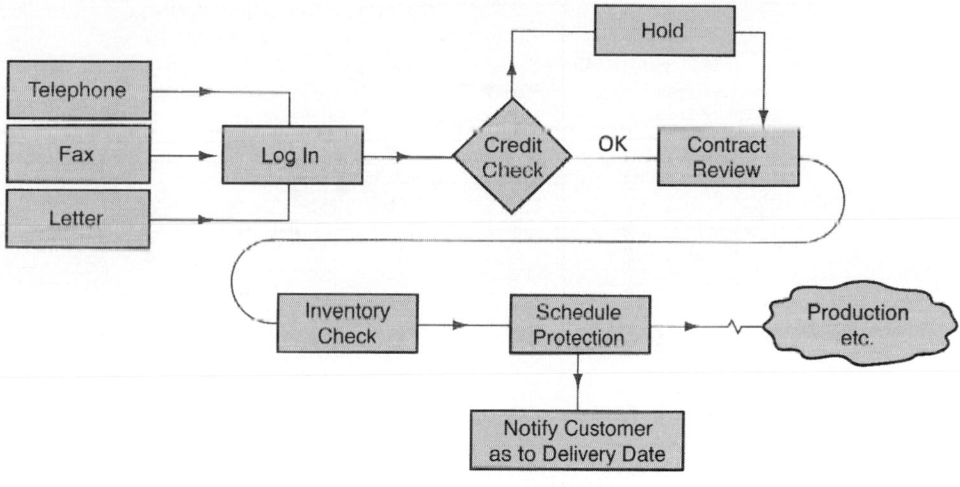

FIGURE 3-9 Flow diagram for order entry.

Standardized symbols are used by industrial engineers; however, they are not necessary for problem solving. Figure 3-9 shows a flow diagram for the order entry activity of a made-to-order company. Enhancements to the diagram are the addition of time to complete an operation and the number of people performing an operation.

The diagram shows who is the next customer in the process, thereby increasing the understanding of the process. Flow diagrams are best constructed by a team, because it is rare for one individual to understand the entire process.

Improvements to the process can be accomplished by eliminating steps, combining steps, or making frequently occurring steps more efficient.

Scatter Diagram

The simplest way to determine if a cause-and-effect relationship exists between two variables is to plot a scatter diagram. Figure 3-10 shows the relationship between automotive speed and gas mileage. The figure shows that as speed increases, gas mileage decreases. Automotive speed is plotted on the *x*-axis and is the independent variable. The independent variable is usually controllable. Gas mileage is on the *y*-axis and is the dependent, or response, variable. Other examples of relationship are as follows:

Cutting speed and tool life

Moisture content and thread elongation

Temperature and lipstick hardness

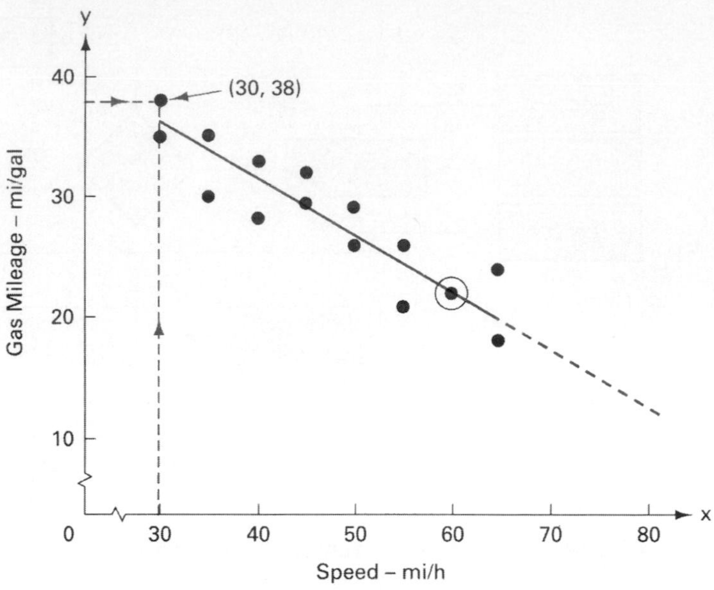

FIGURE 3-10 Scatter diagram.

Striking pressure and electrical current

Temperature and percent foam in soft drinks

Yield and concentration

Breakdowns and equipment age

There are a few simple steps in constructing a scatter diagram. Data are collected as ordered pairs (*x, y*). The automotive speed (cause) is controlled and the gas mileage (effect) is measured. Table 3–1 shows resulting *x, y* paired data.

The horizontal and vertical scales are constructed with the higher values on the right for the *x*-axis and on the top for the *y*-axis. After the scales are labeled, the data are plotted. Using dotted lines, the technique of plotting sample number 1 (30, 38) is illustrated in Figure 3–10. The *x* value is 30, and the *y* value is 38. Sample numbers 2 through 16 are plotted, and the scatter diagram is complete. If two points are identical, concentric circles can be used, as illustrated at 60 mi/h.

Once the scatter diagram is complete, the relationship or correlation between the two variables can be evaluated. Figure 3-11 shows different patterns and their interpretation. At (a), we have a positive correlation between the two variables because as *x* increases, *y* increases. At (b), there is a negative correlation between the two variables because as *x* decreases, *y* decreases. At (c), there is no correlation, and this pattern is sometimes referred to as a shotgun pattern.

TABLE 3-1 Data on Automotive Speed vs. Gas Mileage.

SAMPLE NUMBER	SPEED (MI/H)	MILEAGE (MI/GAL)	SAMPLE NUMBER	SPEED (MI/H)	MILEAGE (MI/GAL)
1	30	38	9	50	26
2	30	35	10	50	29
3	35	35	11	55	32
4	35	30	12	55	21
5	40	33	13	60	22
6	40	28	14	60	22
7	45	32	15	65	18
8	45	29	16	65	24

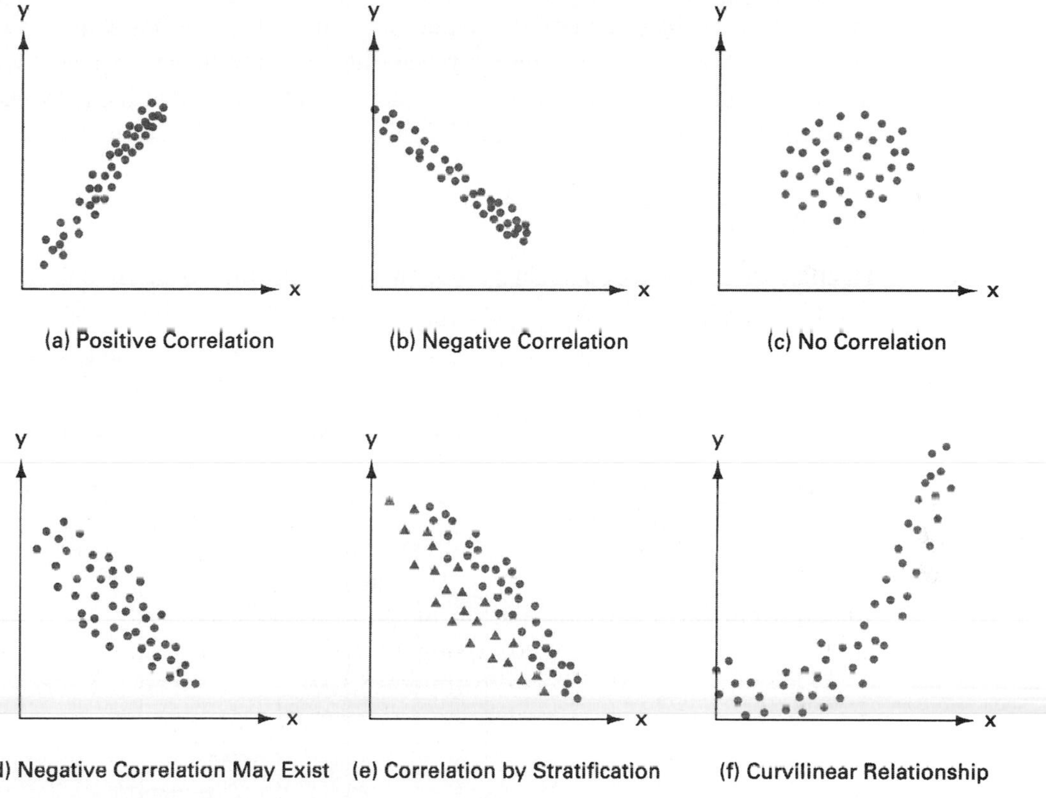

FIGURE 3-11 Different scatter diagram patterns.

The patterns described in (a), (b), and (c) are easy to understand; however, those described in (d), (e), and (f) are more difficult. At (d), there may or may not be a relationship between the two variables. There appears to be a negative relationship between x and y, but it is not too strong. Further statistical analysis is needed to evaluate this pattern. At (e), we have stratified the data to represent different causes for the same effect. Some examples are gas mileage with the wind versus against the wind, two different suppliers of material, and two different machines. One cause is plotted with a small solid circle, and the other cause is plotted with a solid triangle. When the data are separated, we see that there is a strong correlation. At (f), we have a curvilinear relationship rather than a linear one.

When all the plotted points fall on a straight line, we have a perfect correlation. Because of variations in the experiment and measurement error, this perfect situation will rarely, if ever, occur.

It is sometimes desirable to fit a straight line to the data in order to write a prediction equation. For example, we may wish to estimate the gas mileage at 43 mi/h. A line can be placed on the scatter diagram by sight or mathematically using least squares analysis. In either approach, the idea is to make the deviation of the points on each side of the line approximately equal. Where the line is extended beyond the data, a dashed line is used because there are no data in that area.

Histogram

Histograms are discussed in Chapter 4. They describe the variation in the process as illustrated by Figure 3-12. The histogram graphically shows the process capability

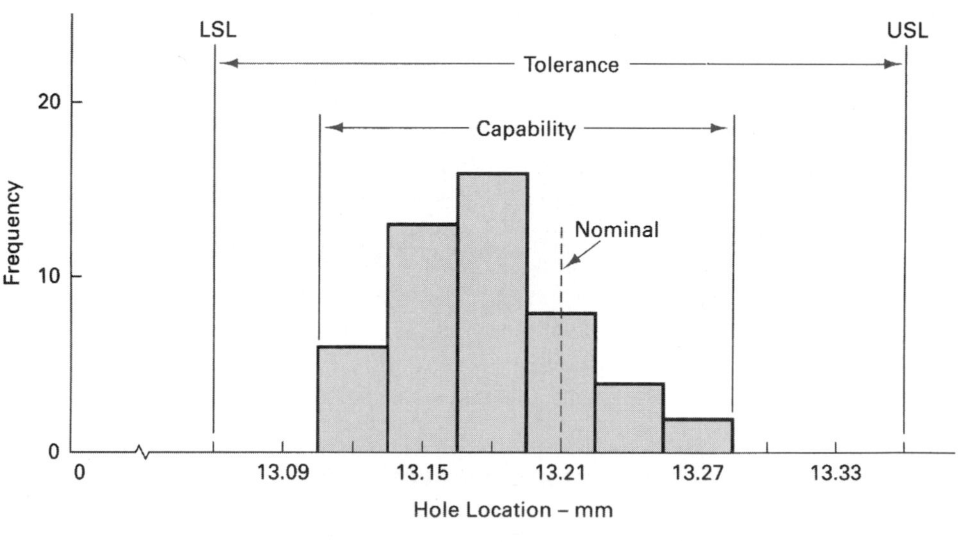

FIGURE 3-12 Histogram for hole location.

and, if desired, the relationship to the specifications and the nominal. It also suggests the shape of the population and indicates if there are any gaps in the data.

Control Charts

Control charts are discussed in Chapters 5, 6, and 8. A control chart, illustrating quality improvement, is shown in Figure 3-13. Control charts are an outstanding technique for problem solving and the resulting quality improvement.

Quality improvement occurs in two situations. When a control chart is first introduced, the process usually is unstable. As assignable causes for out-of-control conditions are identified and corrective action taken, the process becomes stable, with a resulting quality improvement.

The second situation concerns the testing or evaluation of ideas. Control charts are excellent decision makers because the pattern of the plotted points will determine if the idea is a good one, poor one, or has no effect on the process. If the idea is a good one, the pattern of plotted points of the $\overline{X}$ chart will converge on the central line, $\overline{X}_0$. In other words, the pattern will get closer to perfection, which is the central line. For the R chart and the attribute charts, the pattern will tend toward zero, which is perfection. These improvement patterns are illustrated in Figure 3-13. If the idea is a poor one, an opposite pattern will occur. Where the pattern of plotted points does not change, then the idea has no effect on the process.

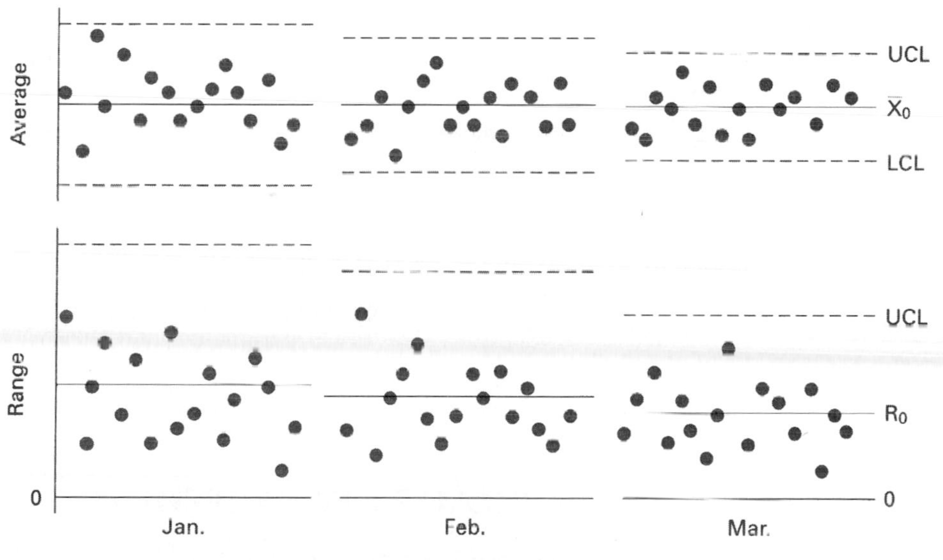

FIGURE 3-13 ***X* and *R* charts, showing quality improvement.**

While the control charts are excellent for problem solving by improving the quality, they have limitations when used to monitor or maintain a process. The pre-control technique is much better at monitoring.

ACCEPTANCE SAMPLING

Chapters 8 and 9 cover the topic of acceptance sampling. With the increased emphasis on the use of SPC as evidence of conformance to quality requirements, the need for acceptance sampling has declined. There will, however, continue to be situations where it will be necessary.

RELIABILITY

Reliability is the ability of a product to perform its intended function over a period of time. A product that works over a long period of time is a reliable one. Reliability is discussed in Chapter 11.

DESIGN OF EXPERIMENTS (DOE)

The objective of DOE is to determine those variables in a process or product that are the critical parameters and their target values. By using formal experimental techniques, the effect of many variables can be studied at one time. Changes to the process or product are introduced in a random fashion or by carefully planned, highly structured experiments.

There are three approaches to DOE: classical, Taguchi, and Shainin. The classical is based on the work of Sir Ronald Fischer in agriculture during the 1930s. Dr. Genichi Taguchi simplified the classical method and introduced additional engineering design concepts. The Dorian Shainin approach uses a variety of problem-solving methods after a product is in production. The wise practitioner will become familiar with all three approaches and develop his own methodology.

Since DOE identifies the critical parameters and their target values, its use should actually precede SPC in many circumstances. It is not unusual to find after an experiment that SPC was controlling the wrong variable or the target was incorrect.

TAGUCHI'S QUALITY ENGINEERING

Most of the body of knowledge associated with the quality sciences was developed in the United Kingdom as design of experiments and in the United States as statistical quality control. More recently. Dr. Genichi Taguchi, a mechanical engineer who has won four Deming Awards, has added to this body of knowledge. In particular, he introduced the loss function concept, which combines cost, target, and variation into one metric with specifications being of secondary importance. Furthermore, he developed the concept of robustness, which means that noise factors are taken into account to ensure that the system functions correctly. Noise factors are uncontrollable variables that can cause significant variability in the process, product, or service.

FAILURE MODE AND EFFECT ANALYSIS (FMEA)

Failure Mode and Effect Analysis (FMEA) is an analytical technique (a paper test) that combines the technology and experience of people in identifying foreseeable failure modes of a product, service, or process and planning for its elimination. In other words, FMEA can be explained as a group of activities intended to

- Recognize and evaluate the potential failure of a product, service, or process and its effects.
- Identify actions that could eliminate or reduce the chance of the potential failure occurring.
- Document the process.

FMEA is a "before the-event" action requiring a team effort to alleviate most easily and inexpensively changes in design and production. There are two types of FMEA: Design FMEA and Process FMEA.

QUALITY FUNCTION DEPLOYMENT (QFD)

QFD is a system that identifies and sets the priorities for product, service and process improvement opportunities that lead to increased customer satisfaction. It ensures the accurate deployment of the "voice of the customer" throughout the organization from product planning to field service. The multi-functional team approach to QFD improves those processes necessary to provide goods and services that meet or exceed customer expectations.

The QFD process answers the following questions:

1. What do customers want?
2. Are all wants equally important?
3. Will delivering perceived needs yield a competitive advantage?
4. How can we change the product, service, or process?
5. How does an engineering decision affect customer perception?

6. How does an engineering change affect other technical descriptors?
7. What is the relationship to parts deployment, process planning, and production planning?

QFD reduces start-up costs, reduces engineering design changes, and most important, leads to increased customer satisfaction.

ISO 9000[2]

ISO stands for International Organization for Standards. The 9000 series is a standardized Quality Management System (QMS) that has been approved by over 100 countries. It consists of three standards: (1) ISO 9000 which covers fundamentals and vocabulary, (2) ISO 9001 which is the requirements, and (3) ISO 9004 which provides guidance for performance improvement. The latest revision occurred in the year 2000 hence the designation ISO 9000:2000. Only 9001 is discussed in this book.

The requirements define the criteria for an acceptable QMS. Figure 3-14 shows the five clauses of the system and their relationship to customer requirements and customer satisfaction. The five clauses of QMS are continual improvement; management responsibility; resource management; product/service realization; and measurement, analysis, and improvement. They are identified below by the numbering system used in the standard.

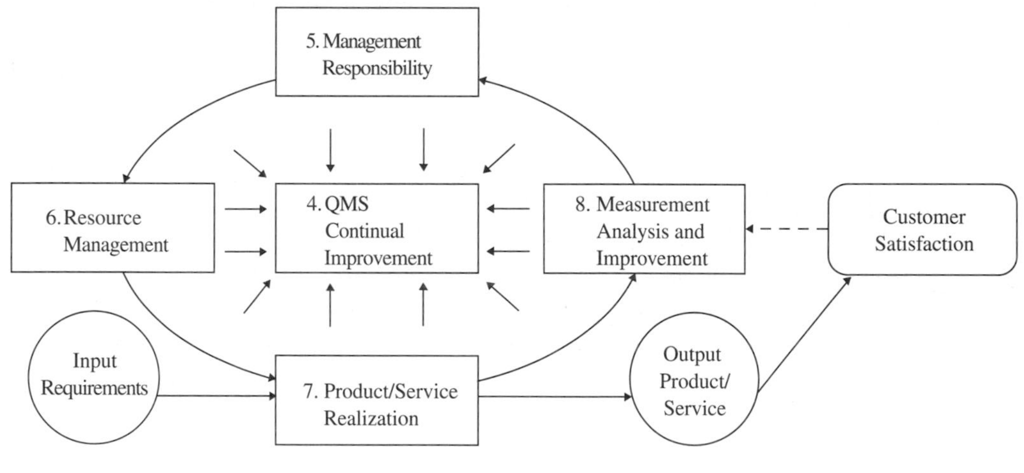

FIGURE 3-14 Model of a Process-Based Quality Management System

[2]Adapted with permission from Quality Management Systems—Requirements, ANSI/ISO/ASQ QA001: 2000, © ASQ Quality Press, (Milwaukee, WI:ASQ, 2000).

4. Quality Management System (QMS)

4.1 General Requirements

The organization shall establish, document, implement, and maintain a QMS and continually improve its effectiveness. The organization shall (a) identify needed processes such as management activities, provision of resources, product or service realization, and measurement, (b) determine their sequence and interaction, (c) determine criteria and methods for effective operation and control of these processes, (d) ensure the availability of resources and information necessary to support and monitor these processes, (e) monitor, measure, and analyse these processes, and (f) implement actions to achieve planned results and continual improvement of these processes. Out-sourced processes that affect the quality of the product shall be identified and included in the system.

4.2 Documentation

4.2.1 General. Documentation shall include (a) statements of a quality policy and quality objectives, (b) a quality manual; (c) required documented procedures; (d) needed documents to ensure effective planning, operation, and control of processes; and (e) required records. A procedure or work instruction is needed if its absence could adversely affect the product quality. The extent of the documentation will depend on the organization's size and type of activities; the complexity of the processes and their interactions; and the competency of the employees. For example: a small organization may verbally notify a manager of an upcoming meeting, whereas a large organization would need written notification. The standard should satisfy the contractual, statutory and regulatory requirements and the needs and expectations of customers and other interested parties. Documentation may be in any form or type of medium.

4.2.2 Quality Manual A quality manual shall be established and maintained that includes: (a) the scope of the QMS with details and justification for any exclusions, (b) the documented procedures or reference to them, and (c) a description of the interaction among the QMS processes.

4.2.3 Control of Documents Documents required by the QMS shall be controlled. A documented procedure shall be in place to define the controls needed to (a) approve documents prior to use, (b) review, update, and re-approve as necessary, (c) identify the current revision status, (d) ensure that current versions are available at the point of use, (e) ensure that documents are legible and readily identified, (f) identify and distribute documents of external origin, and (g) provide for the prompt removal of obsolete documents and suitably identify any that may be retained. Documented procedure means that the procedure is established, documented, implemented, and maintained. They are required in elements 4.2.3, 4.2.4, 8.2.2, 8.3, 8.5.2, and 8.5.3.

4.2.4 Control of Records Records shall be established and maintained to provide evidence of conformity to requirements and the effective operation of the QMS. They shall be legible, readily identifiable, and retrievable. A documented procedure shall be established to define the controls needed for the identification, storage, protection, retrieval, retention time, and disposition of records. Records can be used to document traceability and to provide evidence of verification, preventive action, and corrective action. They are required in elements 5.5.6, 5.6.3, 6.2.2, 7.2.2, 7.3.4, 7.3.6, 7.3.7, 7.4.1, 7.5.2, 7.6, and 8.2.4.

5. Management Responsibility

5.1 Management Commitment

Top management shall provide evidence of its commitment to the development, implementation, and continual improvement of the QMS by (a) communicating the need to meet customer, legal, and regulatory expectations, (b) establishing a quality policy, (c) ensuring that quality objectives are established, (d) conducting management reviews, and (e) ensuring the availability of resources. Top management is defined as the person or group of people who direct and control an organization.

5.2 Customer Focus

Top management shall ensure that customer requirements are determined and met in order to enhance customer satisfaction.

5.3 Quality Policy

Top management shall ensure that the quality policy (a) is aligned with the organization's purpose or mission, (b) includes a commitment to comply with requirements and continually improve the effectiveness of the QMS, (c) provides a framework to establish and review the quality objectives, (d) is communicated and understood throughout the organization, and (d) is periodically reviewed for continuing stability. The policy gives the overall intentions and direction of the organization related to quality.

5.4 Planning

5.4.1 Quality Objectives Top management shall ensure that quality objectives are established at relevant functions and levels within the organization and include product and service requirements. They shall be measurable and consistent with the quality policy. In addition, they should ensure that customer expectations are met. Quality objectives are something sought or aimed for related to quality. For example, finishing department scrap will be reduced from 5.0% to 4.3% and the first line supervisor is the person responsible.

5.4.2 Quality Management System Planning Top management shall ensure that the planning of the QMS is accomplished in order to meet the requirements of the QMS as stated in the General Requirements, Element 4.1, as well as the Quality Objectives, 5.4.1. In addition, the integrity of the QMS is maintained when changes are planned and implemented.

5.5 Responsibility, Authority, and Communication

5.5.1 Responsibility and Authority Top management shall ensure that responsibilities and authorities are defined and communicated within the organization. Responsibilities can be defined in job descriptions, procedures, and work instructions. Authorities and interrelationships can be defined in an organization chart.

5.5.2 Management Representative A member of management, shall be appointed regardless of his/her other duties, and have the responsibility and authority that includes ensuring that (a) processes needed for the QMS system are established, implemented, and maintained, (b) reports are given to top management on the performance of the QMS and any need for improvement, and (c) promoting the awareness of customer requirements throughout the organization. Appointment of a member of top management as the representative can contribute to the effectiveness of the QMS.

5.5.3 Internal Communication Top management shall ensure that appropriate communication channels are established within the organization and that communication takes place regarding the effectiveness of the QMS. Typical communication techniques are management workplace briefing, recognition of achievement, bulletin boards, e-mail, and in-house news brochures.

5.6 Management Review

5.6.1 General Top management shall periodically review the QMS to ensure its continuing suitability, adequacy, and effectiveness. This review shall include assessing opportunities for improvement and the need for changes to the QMS including the policy and objectives. Records from the reviews shall be maintained.

5.6.2 Review Input The input shall include information on (a) audit results, (b) customer feedback, (c) process, product, and service performance, (d) corrective and preventative performance, (e) follow-up actions from previous management reviews, (f) changes that could affect the QMS, and (g) improvement recommendations

5.6.3 Review Output The output shall include any decisions and actions related to (a) improvement of the effectiveness of the QMS and its processes, (b) improvement of the product and service related to customer requirements, and (c) resource needs. Top management can use the outputs as inputs to improvement opportunities.

6. Resource Management

6.1 Provision of Resources

The organization shall determine and provide the resources needed (a) to implement, maintain, and continually improve the QMS, and (b) to enhance customer satisfaction. Resources may be people, infrastructure, work environment, information, suppliers, natural resources, and financial resources. Resources can be aligned with quality objectives.

6.2 Human Resources

6.2.1 General Personnel performing work that affects product or service quality shall be competent on the basis of appropriate education, training, skills and experience.

6.2.2 Competence, Awareness and Training The organization shall (a) determine the necessary competence for personnel performing work affecting product and service quality, (b) provide training or take other actions to satisfy these competencies, (c) evaluate the effectiveness, (d) ensure that its personnel are aware of the relevance and importance of their activities and how they contribute to the achievement of the quality objectives, and (e) maintain appropriate records. Competency is defined as the demonstrated ability to apply knowledge and skills. It can be contained in the job description by function, group, or specific position. Training effectiveness can be determined by before and after tests, performance, or turnover.[3] *ISO 10015 Guidelines for Training* will help organizations comply with this standard.

6.3 Infrastructure

The organization shall determine, provide, and maintain the infrastructure needed to achieve conformity to product or service requirements. Infrastructure includes, as applicable (a) buildings, workspace, and associated utilities, (b) hardware and software process equipment, and (c) supporting services such as transport or communication.

6.4 Work Environment

The organization shall determine and manage the work environment needed to achieve conformity to product or service requirements. Creation of a suitable work environment can have a positive influence on employee motivation, satisfaction, and performance.

---

[3]Jeanne Ketola and Kathy Roberts, "Demystify ISO 9001:2000," *Quality Progress* (September 2001): 65–70.

7. Product Realization

7.1 Planning of Product Realization

The organization shall plan and develop the processes needed for product or service realization. Planning of product or service realization shall be consistent with the requirements of the other processes of QMS. In planning product or service realization, the organization shall determine the following, as appropriate: (a) quality objectives and requirements for the product or service; (b) the need to establish processes, documents, and provide resources specific to the product or service; (c) required verification, validation, monitoring, inspection, and test activities specific to the product or service and the criteria for its acceptance; and (d) records needed to verify this clause. The output of this planning shall be in a form suitable for the organization's method of operations. The organization may also apply the requirements given in 7.3 to the development of the product or service realization processes.

7.2 Customer-Related Processes

7.2.1 Determination of Requirements Related to the Product The organization shall determine the (a) requirements specified by the customer, including the requirements for delivery and post-delivery activities, (b) requirements not stated by the customer but necessary for specified or intended use, where known, (c) statutory and regulatory requirements related to the product or service, and (d) any additional requirements determined by the organization.

7.2.2 Review of Requirements Related to the Product The organization shall review the requirements related to the product or service. This review shall be conducted prior to the organization's commitment to supply a product or service to the customer (for example, submission of tenders, acceptance of contracts or orders, acceptance of changes to contracts or orders). It shall ensure that (a) product or service requirements are defined, (b) contract or order requirements differing from those previously expressed are resolved, and (c) the organization has the ability to meet the defined requirements. Records of the results and actions of the review shall be maintained. Requirements shall be confirmed by the organization before acceptance even where no stated customer requirements exist. Where product or service requirements are changed, the organization shall ensure that relevant documents are amended and that relevant personnel are made aware of the changed requirements. In many situations, a formal review is impractical for each order. Instead, the review can cover relevant product or service information such as catalogs or advertising material.

7.2.3 Customer Communication The organization shall determine and implement effective arrangements for communicating with customers in relation to (a) product or service information, (b) inquiries and documentation, and (c) customer feedback.

7.3 Design and Development

7.3.1 Design and Development Planning The organization shall plan and control the design and development of the product or service. During the planning, the organization shall determine (a) the design and development stages, (b) the review, verification and validation that are appropriate for each design and development stage, and (c) the responsibilities and authorities for the stages. The organization shall manage the interfaces between different groups involved in order to ensure effective communication and clear assignment of responsibility. Planning output shall be updated, as appropriate.

7.3.2 Design and Development Inputs Inputs relating to product or service requirements shall be determined and records maintained. These inputs shall include (a) functional and performance requirements, (b) applicable statutory and regulatory requirements, (c) information derived from previous similar designs, and (d) other essential requirements. These inputs shall be reviewed for adequacy. They shall be complete, unambiguous and not in conflict with each other.

7.3.3 Design and Development Outputs The outputs shall be provided in a form that enables verification against the input and shall be approved prior to release. Outputs shall (a) meet the input requirements, (b) provide appropriate information for purchasing, production, and maintenance, (c) contain or reference product or service acceptance criteria, and (d) specify the characteristics of the product or service that are essential for its safe and proper use.

7.3.4 Design and Development Review At suitable stages, systematic reviews shall be performed in accordance with planned arrangements to evaluate the ability of the results of design and development to meet requirements, and to identify any problems and propose necessary actions. Participants in such reviews shall include representatives of functions concerned with the stage(s) being reviewed. Records of the results of the reviews and any necessary actions shall be maintained. Risk assessment such as FMEA, reliability prediction, and simulation techniques can be undertaken to determine potential failures in products or processes.

7.3.5 Design and Development Verification Verification shall be performed in accordance with planned arrangements to ensure that the outputs have met the input requirements. Records of the results of the verification and any necessary actions shall be maintained. Verification confirms, through objective evidence, that the specified requirements have been fulfilled. Confirmation can comprise activities such as performing alternate calculations, comparing the new design specification to a similar proven design specification, undertaking tests and demonstrations, and reviewing documents prior to issue.

7.3.6 Design and Development Validation Validation shall be performed in accordance with planned arrangements to ensure that the resulting product or service is capable of meeting the requirements for the specified application or intended use, when known. Wherever practicable, validation shall be completed prior to the delivery or implementation of the product or service. Records of the results of validation and any necessary actions shall be maintained. Validation confirms, through objective evidence, that the requirements for a specific intended use have been fulfilled.

7.3.7 Control of Design and Development Changes Changes shall be identified and records maintained. They shall be reviewed, verified and validated, as appropriate, and approved before implementation. The review of design and development changes shall include evaluation of the effect of the changes on future products and product or service already delivered. Records of the results of the review of changes and any necessary actions shall be maintained.

7.4 Purchasing

7.4.1 Purchasing Process Purchased product shall conform to specified purchase requirements. The type and extent of control applied to the supplier and the purchased product or service shall be dependent upon its effect on the purchased product or service. The organization shall evaluate and select suppliers based on their ability to supply product or service in accordance with the organization's requirements. Criteria for selection, evaluation, and re-evaluation shall be established. Records of the results of evaluations and any necessary actions arising from the evaluation shall be maintained. This standard does not apply to items such as office and maintenance supplies, unless they are a product or service.

7.4.2 Purchasing Information Information shall describe the product or service, including where appropriate (a) requirements for approval of product or service, procedures, processes and equipment, (b) requirements for qualification of supplier personnel, and (c) supplier QMS requirements. The organization shall ensure the adequacy of specified requirements prior to their communication to the supplier.

7.4.3 Verification of Purchased Product The organization shall establish and implement the inspection or other activities necessary to ensure that purchased product meets requirements. Where the organization or its customer intends to perform verification at the supplier's premises, the purchasing information shall state the intended verification arrangements and method of product release.

7.5 Production and Service Provision

7.5.1 Control of Production and Service Provision The organization shall plan and carry out production and service provision under controlled conditions.

Controlled conditions shall include, as applicable (a) information that describes the characteristics of the product, (b) work instructions, as necessary, (c) the use of suitable equipment, (d) the use of monitoring and measuring devices, (e) the implementation of monitoring and measurement, and (f) the implementation of release, delivery, and post-delivery activities.

7.5.2 Validation of Processes for Production and Service Provision The organization shall validate any processes where the resulting output cannot be verified by subsequent monitoring or measurement. This includes any processes where deficiencies become apparent only after the product or service is in use. Validation shall demonstrate the ability of these processes to achieve planned results. Arrangements shall be established for these processes including, as applicable (a) defined criteria for review and approval of the processes, (b) approval of equipment and qualification of personnel, (c) use of specific methods and procedures, (d) requirements for records, and (e) revalidation.

7.5.3 Identification and Traceability Where appropriate, the organization shall identify the product or service by suitable means throughout the realization process. The status shall be identified with respect to monitoring and measurement requirements. Where traceability is a requirement, the organization shall control and record the unique identification of the product or service. In some industry sectors, configuration management is a means by which identification and traceability are maintained. Identification can frequently be accomplished with a production router or traveller.

7.5.4 Customer Property Care shall be exercised with customer property while it is under the organization's control or being used by the organization. They shall identify, verify, protect and safeguard customer property provided for use or incorporation into the product or service. If any customer property is lost, damaged, or otherwise found to be unsuitable for use, it shall be reported to the customer and records maintained. Customer property can include intellectual property.

7.5.5 Preservation of Product The conformity of product or service during internal processing and delivery to the intended destination shall be preserved. This preservation shall include identification, handling, packaging, storage, and protection. Preservation shall also apply to the constituent parts of a product or service.

7.6 Control of Monitoring and Measuring Devices

The organization shall determine the scope of the monitoring and measurement activity needed to provide evidence of conformity of product or service to requirements. Processes shall be established to ensure that monitoring and measurement are carried out in a manner that is consistent with the monitoring and measurement requirements. Where necessary to ensure valid results, measuring equipment shall (a) be calibrated or verified at specified intervals or prior to use, against measurement standards; where no such standards exist, the basis used for calibration or verification shall be recorded; (b) be adjusted or re-adjusted as necessary; (c) be identified to enable calibration status to be determined; (d) be safeguarded from adjustments that would invalidate the measurement result; and (e) be protected from damage and deterioration during handling, maintenance and storage. In addition, when the equipment is found not to conform to requirements the organization shall check and record the validity of the previous measuring results. The organization shall take appropriate action on the equipment and any product affected. Records of the results of calibration and verification shall be maintained. When computer software is used in the monitoring and measurement of specified requirements, its ability to satisfy the intended application shall be confirmed. This action shall be undertaken prior to initial use and reconfirmed as necessary. *ISO 10012-1:1992 Quality assurance requirements for measuring equipment—Part 1, ISO 10012-2:1997 Quality assurance for measuring equipment—Part 2, and ISO 17025-1999 General requirements for the competence of testing and calibration laboratories* can be used for guidance.

8. Measurement, Analysis, and Improvement

8.1 General

The organization shall plan and implement the monitoring, measurement, analysis, and improvement processes needed (a) to demonstrate conformity of the product or service, (b) to ensure conformity of the QMS, and (c) to continually improve the effectiveness of the QMS. This process shall include determination of applicable methods, including statistical techniques.

8.2 Monitoring and Measurement

8.2.1 Customer Satisfaction The organization shall monitor information relating to customer perception as to whether the organization has met customer requirements. The methods used to obtain and use this information shall be determined.

8.2.2 Internal Audit Internal audits shall be conducted at planned intervals to determine that the QMS (a) conforms to the planned processes (see 7.1), and to the requirements established by the organization, and (b) is effectively implemented and maintained. An audit program shall be planned, taking into consideration the status and importance of the processes and areas to be audited, as well as the results of previous audits. The criteria, scope, frequency and methods shall be defined. Selection of auditors and conduct of audits shall ensure objectivity and impartiality of the audit process. Auditors shall not audit their own work. The responsibilities and requirements for planning and conducting audits and for reporting results and maintaining records shall be defined in a documented procedure. Managers responsible for the area being audited shall ensure that actions are taken without undue delay to eliminate detected nonconformities and their causes. Follow-up activities shall include the verification of the actions taken and the reporting of verification results. *ISO 19011 Guidelines on quality and/or environmental management auditing* can be used for guidance.

8.2.3 Monitoring and Measurement of Processes Suitable methods for monitoring and, where applicable, measurement of the QMS processes shall be applied. These methods shall demonstrate the ability of the processes to achieve planned results. When planned results are not achieved, correction and corrective action shall be taken, as needed, to ensure conformity of the product or service.

8.2.4 Monitoring and Measurement of Product and Service The organization shall monitor and measure the characteristics of the product to verify that requirements have been met. This process shall be carried out at appropriate stages of the product or service realization process. Records shall provide evidence of conformity and indicate the person(s) authorizing release of product or service. Product release and delivery shall not proceed until the planned arrangements have been satisfactorily completed, unless otherwise approved by a relevant authority and, where applicable, by the customer.

8.3 Control of Nonconforming Product

Product or service which does not conform to requirements shall be identified and controlled to prevent its unintended use or delivery. Controls and related responsibilities and authorities for dealing with nonconforming product or service shall be defined in a document procedure. The organization shall take action in one or more of the following ways: (a) by taking action to eliminate the detected nonconformity; (b) by authorizing its use, release or acceptance under concession by a relevant authority and, where applicable, by the customer; and (c) by taking action to preclude its original intended use or application. When nonconforming product or service is corrected, it shall be subject to re-verification. In addition, when nonconforming product or service is detected after delivery or use has started, the organization shall take appropriate action. Records of the nature of nonconformities and any subsequent actions taken, including concessions obtained, shall be maintained.

8.4 Analysis of Data

The organization shall determine, collect, and analyze appropriate data to demonstrate the suitability, effectiveness, and continual improvement of the effectiveness of the QMS. This activity shall include data generated from all relevant sources. The analysis of data shall provide information relating to (a) customer satisfaction, (b) conformity to product or service requirements, (c) characteristics and trends of processes, including opportunities for preventive action, and (d) suppliers.

8.5 Improvement

8.5.1 Continual Improvement The organization shall continually improve the effectiveness of the QMS through the use of the quality policy, quality objectives, audit results, analysis of data, corrective and preventive actions, and management review.

8.5.2 Corrective Action The organization shall take action appropriate to the effects of the nonconformities in order to eliminate the cause of non-conformities and prevent recurrence. A documented procedure shall be established to define requirements to (a) review nonconformities (including customer complaints), (b) determine the causes of nonconformities, (c) evaluate the need for action to ensure that nonconformities do not recur, (d) determine and implement needed actions, (e) record the results of actions taken, and (f) review corrective actions taken.

8.5.3 Preventive Action The organization shall determine action to eliminate the causes of potential nonconformities in order to prevent their occurrence. These actions shall be appropriate to the effects of the potential problems. A documented procedure shall be established to define requirements for (a) determining potential nonconformities and their causes, (b) evaluating the need for action to prevent occurrence of nonconformities, (c) determining and implementing action needed, (d) records of results of action taken, and (e) reviewing preventive action taken. Preventive action is taken to prevent occurrence while corrective action is taken to prevent reoccurrence.

Eight total quality management principles form the basis for the QMS standards. They are customer focus, leadership, employee involvement, process approach, system approach to management, continual improvement, factual approach to decision making, and mutually-beneficial supplier relationships. These principles are similar to the core values of the Malcolm Baldrige National Quality Award.

Internal Audits

After the policies, procedures, and work instructions have been developed and implemented, checks must be made to ensure that the system is being followed and the expected results are being obtained. This activity is accomplished through the internal audit, which is one of the key elements of the ISO 9000 standard. All elements should be audited at least once per year and some more frequently, depending on need.

Objectives

There are five objectives of the internal audit. They are to:

- Determine that actual performance conforms to the documented QMS.
- Initiate corrective action activities in response to deficiencies.
- Follow up on noncompliance items from previous audits.
- Provide continued improvement in the system through feedback to management.
- Cause the auditee to think about the process, thereby encouraging possible improvements.

Auditor

Audits should be performed by qualified individuals who have received training in auditing principles and procedures. Training programs are available from ASQ and RAB. Training should include classroom information as well as practical demonstration by the trainer and a critiqued audit by the trainee. To be able to audit efficiently, an individual should possess good written and oral communication skills, be a good listener, and be good at taking notes. Other skills should include the ability to concentrate on the task at hand and not be distracted by other activities that are taking place at the same time, be observant and questioning, and be able to separate relevant facts from other information.

The auditor should be objective, honest, and impartial. Of course, the auditor should be prepared by being knowledgeable about the standards.

Techniques

During the actual audit, there are a number of techniques that the auditor should employ. The objective is to collect evidence, and there are three methods: examination of documents, observation of activities, and interviews.

The easiest method is to examine the documents. The auditor should start with the quality manual to determine that the policies cover the QMS standards, and that they are controlled and assessable. Next, the documents are examined in a systematic manner. For example, the auditor would check the purchase orders to determine whether they were accurate and followed the procedures; all appropriate attachments were present; all orders were numbered, signed, and dated; only approved suppliers were used; and so forth. Document control ensures that (1) documents are identified with a title, revision date, and responsible owner; (2) documents are readily available to users; (3) a master list by department or function for procedures, work instructions,

[4]William A. Stimson, "Internal Quality Auditing," *Quality Progress* (November 2001): 39–43.

and records is appropriately located; (4) there are no obsolete documents at workstations; and (5) changes follow a prescribed procedure.[4]

Observation of activities is also an easy method that requires an aptitude for detail. For example, to evaluate the preservation of product, the auditor would observe the identification, handling, packaging, storage, and protection of the product.

The most difficult method of collecting evidence is by interviewing the employee or auditee. However, there are ways to make the process easier. First, place the auditee in a nonthreatening environment by starting with introductions and an explanation of the purpose of the audit. This initial conversation can be followed by easy questions such as, "How long have you been working for the organization?" Humor is also very effective in placing one at ease. In addition, use basic human behavior techniques such as giving compliments, using a person's first name, encouraging suggestions, and so forth.

Second, spend as much time listening and as little time as possible talking. Encourage employees to talk about the process. Then paraphrase your interpretations of their statements so there are no misunderstandings.

Third, if and when you find deficiencies in processes and systems, separate the significant from the trivial. Reserve the major issues for your report and the minor ones for the auditee. Focus on the system and not on the auditee.

Fourth, discuss the major issues informally with the auditee first. The auditor's job is to identify problems and allow the organization to determine solutions. Be sure that the auditee understands the problem, agrees that it is a problem, and agrees that corrective action is necessary. If the auditee does not agree, there will be little or no cooperation. Sometimes the auditor, based on his experience, will have an idea that might solve the problem. It should be discussed in such a manner that the auditee believes it is his/her idea.[5]

Fifth, use the appropriate type of question. There are open questions, closed questions, clarifying questions, leading questions, and aggressive questions. Each type is discussed in the paragraphs that follow.

Examples of open questions are:

- "When are supplier reviews performed?"
- "How is the inspection status identified on this item?"
- "Where does this document come from?"

This type of question is designed to get a wide range of answers rather than a simple "yes" or "no." They are used to obtain an opinion, an explanation of a process, a

[5]Peter Hawkins, ed., "Five Steps to 'Win-Win' Audits," *Quality Management* (Issue 1915, August 10, 1996): 1–4.

person's attitudes, or the reasoning behind an action. The disadvantage of open questions is that the auditor can receive more information than desired.

Examples of closed questions are:

- "Do you have a work instruction for this operation?"
- "Does this instrument require calibration?"
- "Is this die supplied by the customer?"

This type of question can be answered with yes or no and provides evidence or facts quickly. Closed questions are used to gather specific evidence and reduce any misunderstanding. The disadvantage of closed questions is that the interview can appear to be an interrogation.

Examples of clarifying questions are:

- "Tell me more about this operation."
- "Please give me some examples."
- "What do you mean by parting line mismatch?"

This type of question is used to obtain further information. It helps to prevent misunderstanding and encourages the auditee to relax and be more open. The disadvantages are that these questions can give the impression that the auditor is not listening or that the auditor is stupid. Also, when used too often, they are time consuming.

An example of a leading question is:

> "Don't you agree that the nonconformity was caused by not understanding the purchase order?"

This type of question should be avoided, because it encourages the auditee to provide a particular answer and will bias the audit findings.

An example of an aggressive question is:

> "You don't mean to tell me that this test is the only one you perform?"

This type of question should be avoided because it is offensive and argumentative.

The auditor should primarily use open questions with an occasional closed and clarifying question, as the interview may necessitate. For effective communication, there must be mutual trust between auditor and auditee.

Procedure

Before the audit takes place, an audit plan and checklist should be prepared by the lead auditor. As much time is spent planning as doing. The contents of an audit plan should identify the activity or department to be audited; list the procedures, documents, and regulatory requirements involved; name the audit team; and list who is to be notified of the audit and who will receive audit reports. The plan should also contain a schedule, which includes audit notification, audit conducted, corrective action required, if any, and follow-up, if any.

Checklists ensure that the audit is efficient and give the auditor control of the process. It can take the form of questions to ask, the sequence in which they should be asked, and space for writing the results. Checklist questions should be based on the procedures, records, and work instructions to be audited, referencing the specific paragraphs being addressed.

The audit itself has three parts, the preaudit meeting, the audit, and a closing meeting. During the preaudit meeting, the audit process and timetable are discussed and prior audits are reviewed. Minutes of the meeting should be recorded and included with the audit documentation. A list of those attending the meeting is recorded in the minutes.

The purpose of the audit is to determine how well the quality system has been implemented and maintained. In large organizations, an escort should be provided by the area being audited. Escorts become witnesses who can provide backup to an event should a finding be challenged at a later time. The escort is usually a supervisor or key person of the audited area. The audit includes interviewing people working in the area and checking various records that back up the interviews. Often what surfaces from records of one area will lead to further questions that will have to be answered in other areas. Notes should be made to be sure that there is adequate follow-up. The audit is not only a measure of conformity to the system, it is also a measure of the system itself. It should determine if the procedure is adequate or if it is time for a change. The object of the auditing process is to provide for continuous improvement and increased customer satisfaction. The audit findings should be written out in detail from the auditors' notes and should include the conforming as well as the nonconforming items. Separate reports are prepared for each nonconformance and should include:

1. The element title and a unique identification number such as NC 7.2.3, where the NC stands for nonconformance and the other numbers give the element number.
2. Where the nonconformance was observed.
3. Objective evidence used as a basis for the nonconformance.
4. The nonconformance worded as closely as possible to the language of the requirement.

At the closing meeting, the lead auditor presents a summary of the audit findings along with the evidence that supports them. An estimate is made of when the final report will be issued. The distribution of the report is agreed upon. Again, minutes of the meeting are recorded, along with a record of attendance. The audit report will:

1. Have a cover sheet that includes the audit date, names of the audit team, areas audited, distribution list, a statement that the audit is only a sample, and a unique reference number, and it will be signed by the lead auditor.
2. List the nonconformances and copies of all nonconformance reports.
3. Outline procedures for corrective action and subsequent follow-up.

Benefits

There are various reasons for implementing a quality system that conforms to an ISO standard. The primary reason is that customers or marketing are suggesting or demanding compliance to a quality system. Other reasons are needed improvement in processes or systems and a desire for global deployment of products and services.[6] As more and more organizations become registered, they are requiring their subcontractors or suppliers to be registered, creating a snowball effect. Consequently, in order to maintain or increase market share, many organizations are finding they must be in conformance with an ISO standard.

Internal benefits that can be received from developing and implementing a well-documented quality system can far outweigh the external pressures. These are improved quality, production reliability, time performance, and cost of poor quality.

ISO 14000

ISO 14000 is the international standard for an environmental management system (EMS). It provides organizations with the EMS elements which can be integrated into other management systems to help achieve environmental and economic goals. The standard describes the requirements for registration and/or self-declaration of the organization's EMS. Demonstration of successful implementation of the system can be used to assure other parties that an appropriate EMS is in place. It was written to be applicable to all types and sizes of organizations and to accommodate diverse geographical, cultural, and social conditions. The requirements are based on the process and not on the product or service. It does, however, require commitment to the organization's EMS policy, applicable regulations, and continual improvement.

---

[6]F.C. Weston, Jr., "What Do Managers Really Think of the ISO 9000 Registration Process?" *Quality Progress* (October 1995): 67–73.

The basic approach to EMS begins with the environmental policy, which is followed by planning, implementation and operation; checking and corrective action; and management review. There is a logical sequence of events to achieve continual improvement. Many of the requirements may be developed concurrently or revisited at any time. The overall aim is to support environmental protection and prevention of pollution in balance with socioeconomic needs.

BENCHMARKING

It is the search for industry's best practices that leads to superior performance. Benchmarking is a new way of doing business that was developed by Xerox in 1979. The idea is to find another company that is doing a particular process better than your company and then, using that information, improve your process. For example, suppose a small company takes 15 hours to complete a payroll of 75 people while the local bank takes 10 hours to complete one with 80 people. Since both processes are similar, the small company should find out why the bank is more efficient in their payroll process.

Benchmarking forces constant testing of internal processes against industry's best practices. It promotes teamwork by directing attention to business practices as well as production to remain competitive. The technique is unarguable—if another company can do a particular process or practice better, why can't our company? Benchmarking allows a company to establish realistic and credible goals.

TOTAL PRODUCTIVE MAINTENANCE (TPM)

Total Productive Maintenance is a technique that utilizes the entire work force to obtain the optimum use of equipment. There is a continuous search to improve maintenance activities. Emphasis is placed on an interaction between operators and maintenance to maximize up-time. The technical skills in TPM are: daily equipment checking, machine inspection, fine-tuning machinery, lubrication, trouble-shooting, and repair.

MANAGEMENT AND PLANNING TOOLS

Most of these tools have their roots in post–World War II operations research work and in the work of leaders of the Japanese total quality control movement of the 1970s. They are affinity diagram, interrelationship diagram, tree diagram, prioritization matrices, matrix diagram, process decision program chart, and activity network diagram.

Descriptions of these tools are given in Chapter 12.

QUALITY BY DESIGN

Quality by design is the practice of using a multidisciplinary team to conduct product or service concepting, design, and production planning at one time. It is also known as simultaneous engineering or parallel engineering. The team is composed of specialists from design engineering, marketing, purchasing, quality, manufacturing engineering, finance, and the customer. Suppliers of process equipment, purchased parts, and services are included on the team at appropriate times.

In the past, the major functions would complete their task, "throw it over the wall" to the next department in the sequence, and not be concerned with any internal customer problems that may arise. Quality by design requires the major functions to be going on at the same time. This system provides for immediate feedback and prevents quality and productivity problems from occurring.

The major benefits are faster product development, shorter time to market, better quality, less work-in-process, fewer engineering change orders, and increased productivity. Design For Manufacturing and Assembly (DFMA) is an integral part of the process.

PRODUCTS LIABILITY

Consumers are initiating lawsuits in record numbers as a result of injury, death, and property damage from faulty product or service design or faulty workmanship. The number of liability lawsuits has skyrocketed since 1965. Jury verdicts in favor of the injured party have continued to rise in recent years. The size of the judgment or settlement has also increased significantly, which has caused an increase in product liability insurance. Although the larger manufacturers have been able to absorb the judgment or settlement cost and pass the cost on to the consumer, smaller manufacturers have occasionally been forced into bankruptcy. Although injured consumers must be compensated, it is also necessary to maintain viable manufacturing entities.

Reasons for injuries fall generally into three areas—the behavior or knowledge of a user, the environment where the product is used, and whether the factory has designed and constructed the product carefully using safety analysis and quality control. The safety and quality of products has been steadily improving. Manufacturers have met the challenge admirably: for instance, using safety glass where previously glass shards caused many severe injuries, placing safety guards around lawn mower blades to prevent lacerations and amputations, redesigning hot water vaporizers to reduce the risk of burns to children, and removing sharp edges on car dashboards to minimize secondary collision injuries.

Resources are limited; therefore, the perfect product or service is, in many cases, an unattainable goal. In the long term, customers pay for the cost of regulations and lawsuits. It is appropriate to mention the old cliché, "An ounce of prevention is worth a pound of cure." An adequate prevention program can substantially reduce the risk of damaging litigation.

INFORMATION TECHNOLOGY

Information Technology (IT) is a tool like the other tools presented in this textbook. And like the other tools, it helps the TQM organization achieve its goals. Over the past few decades, computers and quality management practices have evolved together and have supported each other. This interdependence will continue in the near future.

Information Technology is defined as computer technology (either hardware or software) for processing and storing information, as well as communications technology for transmitting information.[7] There are three levels of information technology;[8]

Data are alphanumeric and can be moved about without regard to meaning.

Information is the meaningful arrangement of data that creates patterns and activates meanings in a person's mind. It exists at the point of human perception.

Knowledge is the value-added content of human thought, derived from perception and intelligent manipulation of information. Therefore, it is the basis for intelligent action.

Organizations need to become proficient in converting information to knowledge. According to Alan Greenspan, Chairman of the Federal Reserve, "Our economy is benefiting from structural gains in productivity that have been driven by a remarkable wave of technological innovation. What differentiates this period from other periods in our history is the extraordinary role played by information and communication technologies."[9]

COMPUTER PROGRAM

The EXCEL software in the diskette in the inside back cover will solve the Pareto diagram. Its file name is Pareto.

---

[7]E. Wainright Martin, et. al., *Managing Information Technology* 4 ed (Upper Saddle River, NJ: Prentice-Hall, 2001).

[8]Kurt Albrecht, "Information: The Next Quality Revolution," *Quality Digest* (June 1999): 30–32.

[9]The Associated Press, "Information Technology Raises Productivity, Greenspan Says," *St. Louis Post-Dispatch*, (June 14, 2000): C2.

PROBLEMS

1. Construct a Pareto diagram for replacement parts for an electric stove. Six-months' data are: oven door, 193; timer, 53; front burners, 460; rear burners, 290; burner control, 135; drawer rollers, 46; other, 84; and oven regulators, 265.
2. A project team is studying the downtime cost of a soft-drink bottling line. Data analysis in thousands of dollars for a 3-month period are: back pressure regulator, 30; adjust feed worm, 15; jam copper head, 6; lost cooling, 52; valve replacement, 8; and other, 5. Construct a Pareto diagram.
3. Approximately two-thirds of all automobile accidents are due to improper driving. Construct a Pareto diagram without the cumulative line for the data: improper turn, 3.6%; driving too fast for conditions, 28.1%; following too closely, 8.1%; right-of-way violations, 30.1%; driving left of center, 3.3%; improper overtaking, 3.2%; and other, 23.6%.
4. A major record-of-the-month club collected data on the reasons for returned shipments during a quarter. Results are: wrong selection, 50,000; refused, 195,000; wrong address, 68,000; order canceled, 5,000; and other, 15,000. Construct a Pareto diagram.
5. Paint nonconformities for a 1-month period for a riding lawn mower manufacturer are: blister, 212; light spray, 582; drips, 227; overspray, 109; splatter, 141; bad paint, 126; runs, 434; and other, 50. Construct a Pareto diagram.
6. Construct a Pareto diagram for the analysis of internal failures for the following data:

TYPE OF COST	DOLLARS (IN THOUSANDS)
Purchasing—rejects	205
Design—scrap	120
Operations—rework	355
Purchasing—rework	25
All other	65

7. Construct a Pareto diagram for the analysis of the external failure costs for a wireless telephone manufacturer using the following data:

TYPE OF COST	DOLLARS (IN THOUSANDS)
Customer complaints	20
Returned goods	30
Retrofit costs	50
Warranty claims	90
Liability costs	10
Penalties	5
Customer goodwill	25

8. A building construction company needs a Pareto diagram for the analysis of the following design department cost of poor quality:

ELEMENT	DOLLARS (IN THOUSANDS)
Progress reviews	5
Support activities	3
Qualification tests	2
Corrective action	15
Rework	50
Scrap	25
Liaison	2

9. Construct a trend-analysis graph for the four quality costs categories and the total. Cost of poor quality data for a wheelbarrow manufacturer as a percent of net sales are as follows:

YEAR	PREVENTION	APPRAISAL	INTERNAL FAILURE	EXTERNAL FAILURE	TOTAL
1	0.2	2.6	3.7	4.7	11.2
2	0.6	2.5	3.3	3.6	10.0
3	1.2	2.8	4.0	1.8	9.8
4	1.2	1.7	3.4	1.2	7.5
5	1.0	1.3	1.8	0.9	5.0

10. Prepare a trend (time series) graph for nonconformities per unit for hospital Medicare claims and analyze the results. Data are: 1986—0.20, 1987—0.15, 1988—0.16, and 1989—0.12.

11. Form a project team of six or seven people, elect a leader, and construct a cause-and-effect diagram for bad coffee from a 22-cup appliance used in the office.

12. Form a project team of six or seven people, elect a leader, and construct a C&E diagrom for:
    (a) Dispersion analysis type for a quality characteristic.
    (b) Process analysis type for a sequence of office activities on an insurance form.
    (c) Process analysis type for a sequence of production activities on a lathe: load 25 mm dia.—80 mm long rod, rough turn 12 mm dia.—40 mm long, UNF thread—12 mm dia., thread relief, finish turn 25 mm dia.—20 mm long, cut off, and unload.

13. Design a check sheet for the maintenance of a piece of equipment such as a gas furnace, laboratory scale, or computer.

14. By means of a scatter diagram, determine if a relationship exists between product temperatures and percent foam for a soft drink. Data are

DAY	°F PRODUCT TEMPERATURE	% FOAM	FAY	°F PRODUCT TEMPERATURE	% FOAM
1	36	15	11	44	32
2	38	19	12	42	33
3	37	21	13	38	20
4	44	30	14	41	27
5	46	36	15	45	35
6	39	20	16	49	38
7	41	25	17	50	40
8	47	36	18	48	42
9	39	22	19	46	40
10	40	23	20	41	30

15. By means of a scatter diagram, determine if there is a relationship between hours of machine use and millimeters off the target. Data for 20 (x, y) pairs with hours of machine use as the x variable are (30, 1.10), (31, 1.21), (32, 1.00), (33, 1.21), (34, 1.25), (35, 1.23), (36, 1.24), (37, 1.28), (38, 1.30), (39, 1.30), (40, 1.38), (41, 1.35), (42, 1.38), (43, 1.38), (44, 1.40), (45, 1.42), (46, 1.45), (47, 1.45), (48, 1.50), and (49, 1.58). Draw a line for the data using eyesight only and estimate the number of millimeters off the target at 55 h.

16. Data on gas pressure (kg/cm^2) and its volume (liters) are as follows: (0.5, 1.62), (1.5, 0.75), (2.0, 0.62), (3.0, 0.46), (2.5, 0.52), (1.0, 1.00), (0.8, 1.35), (1.2, 0.89), (2.8, 0.48), (3.2, 0.43), (1.8, 0.71), and (0.3, 1.80). Construct a scatter diagram and determine the relationship.

17. The following data (tensile strength, hardness) are for tensile strength (100 psi) and hardness (Rockwell E) of die-cast aluminum. Construct a scatter diagram and determine the relationship: (293, 53), (349, 70), (368, 40), (301, 55), (340, 78), (308, 64), (354, 71), (313, 53), (322, 82), (334, 67), (377, 70), (247, 56), (348, 86), (298, 60), (287, 72), (292, 51), (345, 88), (380, 95), (257, 51), (258, 75).

18. Data on the amount of water applied in inches and the yield of alfalfa in tons per acre are:

Water	12	18	24	30	36	42	48	60
Yield	5.3	5.7	6.3	7.2	8.2	8.7	8.4	8.2

Prepare a scatter diagram and analyze the results.

19. Construct a flow diagram for the manufacture of a product or the providing of a service.

20. Using the EXCEL software in the diskette in the inside back cover solve Problems 1, 2, 3, 4, 5, 6, 7, and 8.

21. Working in a team of three or more people visit an organization and evaluate their quality management system.

4 FUNDAMENTALS OF STATISTICS

INTRODUCTION

Definition of Statistics

The word *statistics* has two generally accepted meanings:

1. A collection of quantitative data pertaining to any subject or group, especially when the data are systematically gathered and collated. Examples of this meaning are blood pressure statistics, statistics of a football game, employment statistics, and accident statistics, to name a few.
2. The science that deals with the collection, tabulation, analysis, interpretation, and presentation of quantitative data.

It is noted that the second meaning is broader than the first, since it, too, is concerned with collection of data. The use of statistics in quality deals with the second and broader meaning and involves the divisions of collecting, tabulating, analyzing, interpreting, and presenting the quantitative data. Each division is dependent on the accuracy and completeness of the preceding one. Data may be collected by a technician measuring the tensile strength of a plastic part or by a market researcher determining consumer color preferences. It may be tabulated by simple paper-and-pencil techniques or by the use of a computer. Analysis may involve a cursory visual examination or exhaustive calculations. The final results are interpreted and presented to assist in making decisions concerning quality.

There are two phases of statistics:

1. *Descriptive* or *deductive statistics,* which endeavor to describe and analyze a subject or group.
2. *Inductive statistics,* which endeavor to determine from a limited amount of data (sample) an important conclusion about a much larger amount of data (population). Since these conclusions or inferences cannot be stated with absolute certainty, the language of *probability* is often used.

This chapter covers the statistical fundamentals necessary to understand the subsequent quality control techniques. Fundamentals of probability are discussed in Chapter 7. An understanding of statistics is vital for an understanding of quality and, for that matter, many other disciplines.

Collection of Data

Data may be collected by direct observation or indirectly through written or verbal questions. The latter technique is used extensively by market research personnel and public opinion pollsters. Data that are collected for quality purposes are obtained by direct observation and are classified as either variables or attributes. *Variables* are those quality characteristics that are measurable, such as a weight measured in grams. *Attributes,* on the other hand, are those quality characteristics that are classified as either conforming or not conforming to specifications such as a "go/no go gage."

A variable that is capable of any degree of subdivision is referred to as *continuous*. The weight of a gray iron casting, which can be measured as 11 kg, 11.33 kg, or 11.3398 kg (25 lb), depending on the accuracy of the measuring instrument, is an example of a continuous variable. Measurements such as meters (feet), liters (gallons), and pascals (pounds per square inch) are examples of continuous data. Variables that exhibit gaps are called *discrete*. The number of nonconforming rivets in a travel trailer can be any whole number, such as 0, 3, 5, 10, 96, . . .; however, there cannot be, say, 4.65 nonconforming rivets in a particular trailer. In general, continuous data are measurable, whereas discrete data are countable.

Sometimes it is convenient for verbal or nonnumerical data to assume the nature of a variable. For example, the quality of the surface finish of a piece of furniture can be classified as poor, average, or good. The poor, average, or good classification can be replaced by the numerical values of 1, 2, or 3, respectively. In a similar manner, educational institutions assign to the letter grades of A, B, C, D, and F the numerical values of 4, 3, 2, 1, and 0, respectively, and use those discrete numerical values as discrete variables for computational purposes.

While many quality characteristics are stated in terms of variables, there are many characteristics that must be stated as attributes. Frequently, those characteristics that are judged by visual observation are classified as attributes. The wire on an electric motor is either attached to the terminal or it is not; the words on this page are correctly spelled or they are incorrectly spelled; the switch is on or it is off; and the answer is right or it is wrong. The examples given in the previous sentence show conformance to a particular specification or nonconformance to that specification.

It is sometimes desirable for variables to be classified as attributes. Factory personnel are frequently interested in knowing if the product they are producing conforms to the specifications. For example, the numerical values for the weight of a package of sugar may not be as important as the information that the weight is within the prescribed limits. Therefore, the data, which are collected on the weight of the package of sugar, are reported as conforming or not conforming to specifications.

In collecting data the number of figures is a function of the intended use of the data. For example, in collecting data on the life of light bulbs, it is acceptable to record 995.6 h; however, recording a value of 995.632 h is too accurate and unnecessary. Similarly, if a keyway specification has a lower limit of 9.52 mm (0.375 in.) and an upper limit of 9.58 mm (0.377 in.), data would be collected to the nearest 0.001 mm and rounded to the nearest 0.01 mm. In general, the more figures to the right of the decimal point, the more sophisticated the measuring instrument.

Measuring instruments may not give a true reading because of problems due to accuracy and precision. Figure 4-1 (a) shows an accurate series of repeated measurements because their average is close to the true value, which is at the center. In Figure 4-1(b) the repeated measurements in the series are precise (very close together), but are not close to the true value. Figure 4-1(c) shows the series of repeated measurements tightly compacted around the true value, and these measurements are both accurate and precise.

The rounding of data requires that certain conventions be followed. In rounding the numbers 0.9530, 0.9531, 0.9532, 0.9533, and 0.9534 to the nearest thousandth, the answer is 0.953, since all the numbers are closer to 0.953 than they are to 0.954. And in rounding the numbers 0.9535, 0.9536, 0.9537, 0.9538, and 0.9539, the answer is 0.954, since all the numbers are closer to 0.954 than to 0.953. In other words, if the last digit is 5 or greater, the number is rounded up.[1]

[1]This rounding rule is the simplest to use. Another one decides whether to round up or down based on the previous digit. Thus, a number that ends in x5 will be rounded up if x is odd and down if x is even. For example, rounding to two decimal places, the number 6.415 would equal or approximate 6.42 and the number 3.285 would equal 3.28.

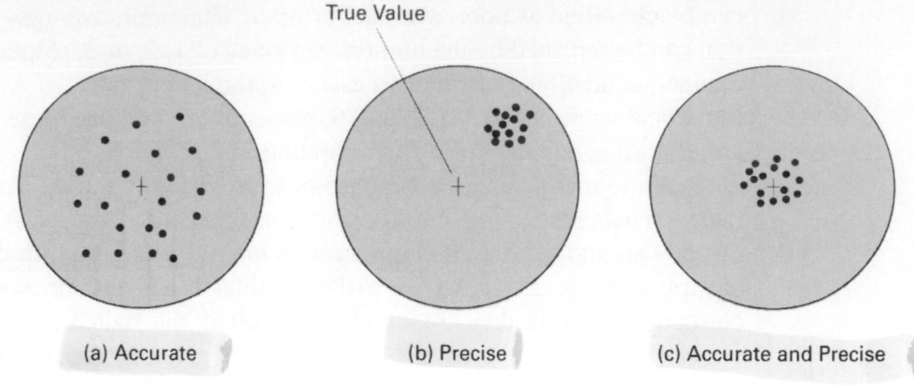

FIGURE 4-1 Difference between accuracy and precision.

Based on the above rounding rule, a rounded number is an approximation of the exact number. Thus, the rounded number 6.23 lies between 6.225 and 6.235 and is expressed as:

$$6.225 \leq 6.23 < 6.235$$

The precision is 0.010, which is the difference between 6.225 and 6.235. An associated term is greatest possible error, g.p.e., which is one-half the precision or $0.010 \div 2 = 0.005$.

Sometimes precision and g.p.e. are not adequate to describe error. For example, the numbers 8,765.4 and 3.2 have the same precision and g.p.e. of 0.10 and 0.05, respectively; however, the relative error, r.e., is much different. It equals the g.p.e. of the number divided by the number. Thus,

$$\text{r.e. of } 8{,}765.4 = 0.05 \div 8{,}765.4 = 0.000006$$

$$\text{r.e. of } 3.2 = 0.05 \div 3.2 = 0.02$$

The following examples will help to clarify these concepts:

Rounded Number with Boundaries	Precision	G.P.E.	R.E.
$5.645 \leq 5.65 < 5.655$	0.01	0.005	0.0009
$431.5 \leq 432 < 432.5$	1.0	0.5	0.001

In working with numerical data, *significant figures* are very important. The significant figures of a number are the digits exclusive of any leading zeros needed to locate the decimal point. For example, the number 3.69 has three significant figures; 36.900 has five significant figures; 2700 has four significant figures; 22.0365 has six significant figures; and 0.00270 has three significant figures. Trailing zeros are counted as being significant, while leading zeros are not. The rule gives some difficulty when working with whole numbers since the number 300 can have one, two, or three significant figures. This difficulty can be eliminated by the use of scientific notation. Therefore, 3×10^2 has one significant figure, 3.0×10^2 has two significant figures, and 3.00×10^2 has three significant figures. Numbers with leading zeros can be written as 2.70×10^{-3} for 0.00270. Numbers that are associated with counting have an unlimited number of significant figures, and the counting number 65 can be written as 65 or 65.000. . . .

The following examples illustrate the number 600 with three, two, and one significant figures, respectively:

Rounded Number with Boundaries	Precision	G.P.E.	R.E.
$599.5 \leq 6.00 \times 10^2 < 600.5$	1	0.5	0.0008
$595 \leq 6.0 \times 10^2 < 605$	10	5	0.008
$550 \leq 6 \times 10^2 < 650$	100	50	0.08

When performing the mathematical operations of multiplication, division, and exponentiation, the answer has the same number of significant figures as the number with the fewest significant figures. The following examples will help to clarify this rule:

$$\sqrt{81.9} = 9.05$$

$$6.59 \times 2.3 = 15$$

$$32.65 \div 24 = 1.4 \qquad \text{(24 is not a counting number)}$$

$$32.65 \div 24 = 1.360 \qquad \text{(24 is a counting number with a value of 24.00. . .)}$$

When performing the mathematical operations of addition and subtraction, the final answer can have no more significant figures after the decimal point than the number with the fewest significant figures after the decimal point. In cases involving numbers without decimal points, the final answer has no more significant figures than the number with the fewest significant figures. Examples to clarify this rule are as follows:

$$38.26 - 6 = 32 \qquad \text{(6 is not a counting number)}$$

$$38.26 - 6 = 32.26 \qquad \text{(6 is a counting number)}$$

$$38.26 - 6.1 = 32.2 \qquad \text{(answer was rounded from 32.16)}$$

$$8.1 \times 10^3 - 1232 = 6.9 \times 10^3 \qquad \text{(fewest significant figures are two)}$$

$$8.100 \times 10^3 - 1232 = 6868 \qquad \text{(fewest significant figures are four)}$$

Utilization of the rules above will avoid discrepancies in answers among quality personnel; however, some judgment may sometimes be required. When a series of calculations, is made, significant figure and rounding determinations can be

TABLE 4-1 Number of Daily Billing Errors

0	1	3	0	1	0	1	0
1	5	4	1	2	1	2	0
1	0	2	0	0	2	0	1
2	1	1	1	2	1	1	
0	4	1	3	1	1	1	
1	3	4	0	0	0	0	
1	3	0	1	2	2	3	

evaluated at the end of the calculations. In any case, the final answer can be no more accurate than the incoming data. For example, if the incoming data is to two decimal places (x.xx), then the final answer must be to two decimal places (x.xx).

Describing the Data

In industry, business, and government the mass of data that has been collected is voluminous. Even one item, such as the number of daily billing errors of a large organization, can represent such a mass of data that it can be more confusing than helpful. Consider the data shown in Table 4-1.

Clearly these data, in this form, are difficult to use and are not effective in describing the data's characteristics. Some means of summarizing the data are needed to show what value or values the data tend to cluster about and how the data are dispersed or spread out. Two techniques are available to accomplish this summarization of data—graphical and analytical.

The graphical technique is a plot or picture of a *frequency distribution,* which is a summarization of how the data points (observations) occur within each subdivision of observed values or groups of observed values. Analytical techniques summarize data by computing a *measure of central tendency* and a *measure of the dispersion*. Sometimes both the graphical and analytical techniques are used.

These techniques will be described in the subsequent sections of this chapter.

FREQUENCY DISTRIBUTION

Ungrouped Data

Ungrouped data comprise a listing of the observed values, while grouped data represent a lumping together of the observed values. The data can be discrete, as they are in this section, or continuous, as in the next section.

Because unorganized data are virtually meaningless, a method of processing the data is necessary. Table 4-1 will be used to illustrate the concept. An analyst

TABLE 4-2 Tally of Number of Daily Billing Errors

NUMBER NONCONFORMING	TABULATION	FREQUENCY
0	𝍸 𝍸 𝍸	15
1	𝍸 𝍸 𝍸 𝍸	20
2	𝍸 |||	8
3	𝍸	5
4	|||	3
5	|	1

reviewing the information as given in this table would have difficulty comprehending the meaning of the data. A much better understanding can be obtained by tallying the frequency of each value, as shown in Table 4-2.

The first step is to establish an *array,* which is an arrangement of raw numerical data in ascending or descending order of magnitude. An array of ascending order from 0 to 5 is shown in the first column of Table 4-2. The next step is to tabulate the frequency of each value by placing a tally mark under the tabulation column and in the appropriate row. Start with the numbers 0, 1, 1, 2, . . . of Table 4-1 and continue placing tally marks until all the data have been tabulated. The last column of Table 4-2 is the numerical value for the number of tallies and is called the *frequency*.

Analysis of Table 4-2 shows that one can visualize the distribution of the data. If the "Tabulation" column is eliminated, the resulting table is classified as a *frequency distribution,* which is an arrangement of data to show the frequency of values in each category.

The frequency distribution is a useful method of visualizing data and is a basic statistical concept. To think of a set of numbers as having some type of distribution is fundamental for solving quality problems. There are different types of frequency distributions, and the type of distribution can indicate the problem-solving approach.

Frequency distributions are presented in graphical form when greater visual clarity is desired. There are a number of different ways to present the frequency distribution.

A *histogram* consists of a set of rectangles that represent the frequency in each category. It represents graphically the frequencies of the observed values. Figure 4-2(a) is a histogram for the data in Table 4-2. Since this is a discrete variable, a vertical line in place of a rectangle would have been theoretically correct (see Figure 4-5). However, the rectangle is commonly used.

Another type of graphic representation is the relative frequency distribution. Relative, in this sense, means the proportion or fraction of the total. Relative frequency is calculated by dividing the frequency for each data value (in this case, number nonconforming) by the total, which is the sum of the frequencies. These calculations are shown in the third column of Table 4-3. Graphical representation

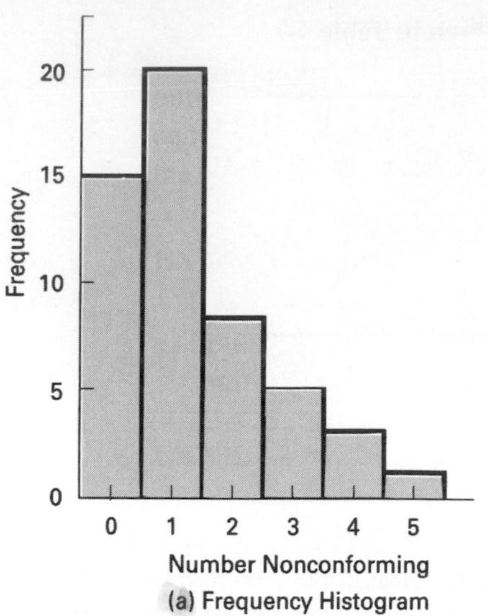

(a) Frequency Histogram

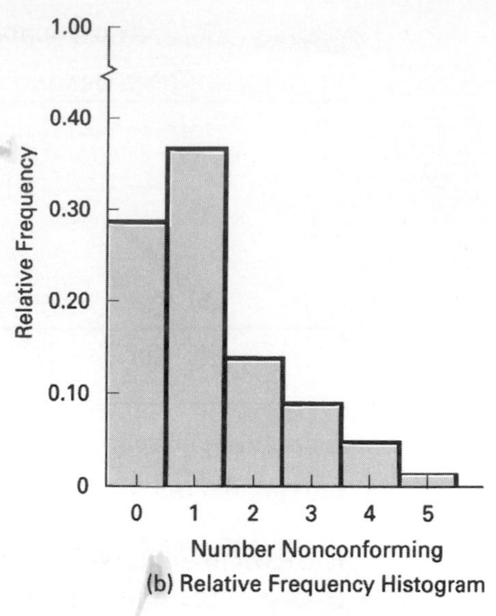

(b) Relative Frequency Histogram

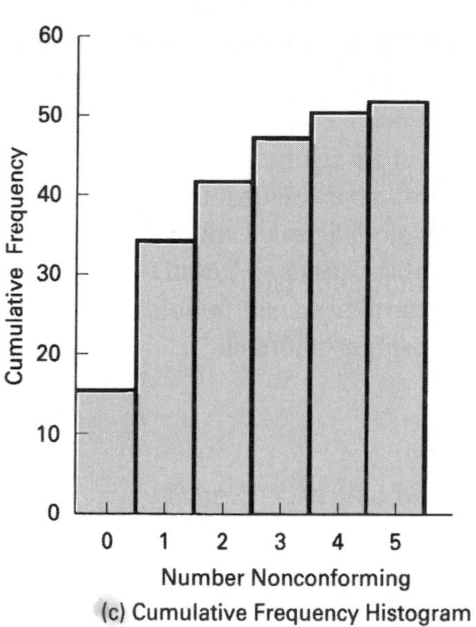

(c) Cumulative Frequency Histogram

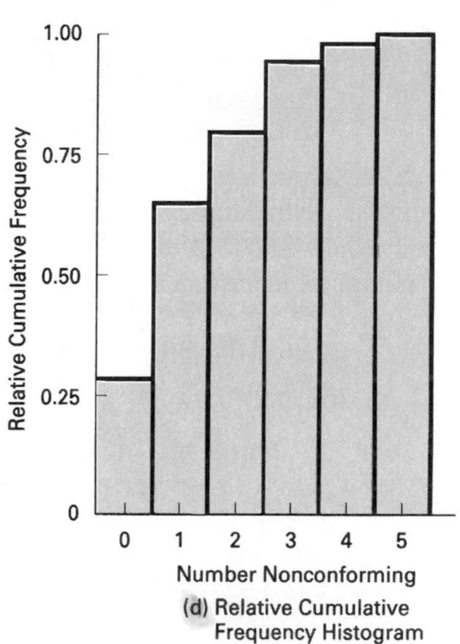

(d) Relative Cumulative Frequency Histogram

FIGURE 4-2 Graphic representation of data given in Tables 4-2 and 4-3.

TABLE 4-3 **Different Frequency Distributions of Data Given in Table 4-1**

NUMBER NONCONFORMING	FREQUENCY	RELATIVE FREQUENCY	CUMULATIVE FREQUENCY	RELATIVE CUMULATIVE FREQUENCY
0	15	15 ÷ 52 = 0.29	15	15 ÷ 52 = 0.29
1	20	20 ÷ 52 = 0.38	15 + 20 = 35	35 ÷ 52 = 0.67
2	8	8 ÷ 52 = 0.15	35 + 8 = 43	43 ÷ 52 = 0.83
3	5	5 ÷ 52 = 0.10	43 + 5 = 48	48 ÷ 52 = 0.92
4	3	3 ÷ 52 = 0.06	48 + 3 = 51	51 ÷ 52 = 0.98
5	1	1 ÷ 52 = 0.02	51 + 1 = 52	52 ÷ 52 = 1.00
Total	52	1.00		

is shown in Figure 4-2(b). Relative frequency has the advantage of a reference. For example, the proportion of 15 nonconforming units is 0.29. Some practitioners prefer to use percents for the vertical scale rather than fractions.

Cumulative frequency is calculated by adding the frequency of each data value to the sum of the frequencies for the previous data values. As shown in the fourth column of Table 4-3, the cumulative frequency for 0 nonconforming units is 15; for 1 nonconforming unit, 15 + 20 = 35; for 2 nonconforming units, 35 + 8; and so on. Cumulative frequency is the number of data points equal to or less than a data value. For example, this value for 2 or less nonconforming units is 43. Graphic representation is shown in Figure 4-2(c).

Relative cumulative frequency is calculated by dividing the cumulative frequency for each data value by the total. These calculations are shown in the fifth column of Table 4-3, and the graphical representation is shown in Figure 4-2(d). The graph shows that the proportion of the billing errors that have 2 or fewer nonconforming units is 0.83 or 83%.

The foregoing example is limited to a discrete variable with six values. Although this example is sufficient for a basic introduction to the frequency distribution concept, it does not provide a thorough knowledge of the subject. Most data are continuous rather than discrete and require grouping.

Grouped Data

The construction of a frequency distribution for grouped data is more complicated because there is usually a larger number of categories. An example problem using a continuous variable illustrates the concept.

1. *Collect data and construct a tally sheet.* Data collected on the weights of 110 steel shafts are shown in Table 4-4. The first step is to make a tally of the

TABLE 4-4 Steel Shaft Weight (kilograms)

2.559	2.556	2.566	2.546	2.561
2.570	2.546	2.565	2.543	2.538
2.560	2.560	2.545	2.551	2.568
2.546	2.555	2.551	2.554	2.574
2.568	2.572	2.550	2.556	2.551
2.561	2.560	2.564	2.567	2.560
2.551	2.562	2.542	2.549	2.561
2.556	2.550	2.561	2.558	2.556
2.559	2.557	2.532	2.575	2.551
2.550	2.559	2.565	2.552	2.560
2.534	2.547	2.569	2.559	2.549
2.544	2.550	2.552	2.536	2.570
2.564	2.553	2.558	2.538	2.564
2.552	2.543	2.562	2.571	2.553
2.539	2.569	2.552	2.536	2.537
2.532	2.552	2.575(h)	2.545	2.551
2.547	2.537	2.547	2.533	2.538
2.571	2.545	2.545	2.556	2.543
2.551	2.569	2.559	2.534	2.561
2.567	2.572	2.558	2.542	2.574
2.570	2.542	2.552	2.551	2.553
2.546	2.531(l)	2.563	2.554	2.544

values, as shown in Table 4-5. In order to be more efficient, the weights are coded from 2.500 kg, which is a technique used to simplify data. Therefore, a weight with a value of 31 is equivalent to 2.531 kg (2.500 + 0.031). Analysis of Table 4-5 shows that more information is conveyed to the analyst than from the data of Table 4-4; however, the general picture is still somewhat blurred.

In this problem there are 45 categories, which are too many and must be reduced by grouping into cells.[2] A cell is a grouping within specified boundaries of observed values along the abscissa (horizontal axis) of the histogram. The grouping of data by cells simplifies the presentation of the distribution; however, some of the detail is lost. When the number of cells is large, the true picture of the distribution is distorted by cells having an insufficient number of items or none at all. Or, when the number of cells is small, too many items are concentrated in a few cells and the distribution is also distorted.

The number of cells or groups in a frequency distribution is largely a matter of judgment by the analyst. This judgment is based on the number of observations

[2]The word *class* is sometimes used in place of the word *cell*.

TABLE 4-5 Tally Sheet Shaft Weight (Coded from 2.500 kg)

WEIGHT	TABULATION	WEIGHT	TABULATION	WEIGHT	TABULATION
31	ǀ	46	ǀǀǀǀ	61	𝍸
32	ǀǀ	47	ǀǀǀ	62	ǀǀ
33	ǀ	48		63	ǀ
34	ǀǀ	49	ǀǀ	64	ǀǀǀ
35		50	ǀǀǀǀ	65	ǀǀ
36	ǀǀ	51	𝍸 ǀǀǀ	66	ǀ
37	ǀǀ	52	𝍸 ǀ	67	ǀǀ
38	ǀǀǀ	53	ǀǀǀ	68	ǀǀ
39	ǀ	54	ǀǀ	69	ǀǀǀ
40		55	ǀ	70	ǀǀǀ
41		56	𝍸	71	ǀǀ
42	ǀǀǀ	57	ǀ	72	ǀǀ
43	ǀǀǀ	58	ǀǀǀ	73	
44	ǀǀ	59	𝍸	74	ǀǀ
45	ǀǀǀǀ	60	𝍸	75	ǀǀ

and can require trial and error to determine the optimum number of cells. In general, the number of cells should be between 5 and 20. Broad guidelines are as follows: use 5 to 9 cells when the number of observations is less than 100; use 8 to 17 cells when the number of observations is between 100 and 500; and use 15 to 20 cells when the number of observations is greater than 500. To provide flexibility, the number of cells in the guidelines are overlapping. It is emphasized that these guidelines are not rigid and can be adjusted when necessary to present an acceptable frequency distribution.

2. *Determine the range.* It is the difference between the highest observed value and the lowest observed value as shown by the formula

$$R = X_h - X_l$$

where R = range

X_h = highest number

X_l = lowest number

From Table 4-4 or Table 4-5 the highest number is 2.575 and the lowest number is 2.531. Thus

$$\begin{aligned} R &= X_h - X_l \\ &= 2.575 - 2.531 \\ &= 0.044 \end{aligned}$$

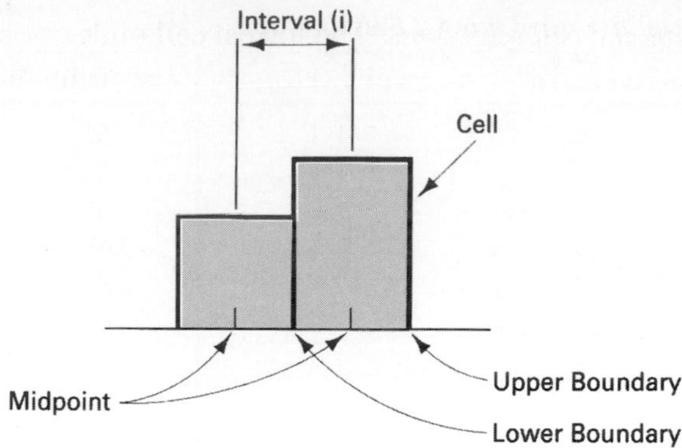

FIGURE 4-3 Cell nomenclature.

3. *Determine the cell interval*. The *cell interval* is the distance between adjacent cell midpoints as shown in Figure 4-3. Whenever possible, an odd interval such as 0.001, 0.07, 0.5, or 3 is recommended so that the midpoint values will be to the same number of decimal places as the data values. The simplest technique is to use Sturgis' rule, which is

$$i = \frac{R}{1 + 3.322 \log n}$$

For the example problem the answer is

$$i = \frac{R}{1 + 3.322 \log n} = \frac{0.044}{1 + 3.322(2.041)} = 0.0057$$

and the closest odd interval for the data is 0.005.

Another technique uses trial and error. The cell interval (i) and the number of cells (h) are interrelated by the formula, $h = R/i$. Since h and i are both unknown, a trial-and-error approach is used to find the interval that will meet the guidelines.

$$\text{Assume that } i = 0.003\text{; then } h = \frac{R}{i} = \frac{0.044}{0.003} = 15$$

$$\text{Assume that } i = 0.005\text{; then } h = \frac{R}{i} = \frac{0.044}{0.005} = 9$$

$$\text{Assume that } i = 0.007\text{; then } h = \frac{R}{i} = \frac{0.044}{0.007} = 6$$

A cell interval of 0.005 with 9 cells will give the best presentation of the data based on the guidelines for the number of cells given in step 1.

Both techniques give similar answers.

4. *Determine the cell midpoints*. The lowest cell midpoint must be located to include the lowest data value in its cell. The simplest technique is to select the lowest data point (2.531) as the midpoint value for the first cell. A better technique is to use the formula

$$MP_l = X_l + \frac{i}{2} \quad \text{(Do not round answer)}$$

where MP_l = midpoint for lowest cell

For the example problem, the answer is

$$MP_l = X_l + \frac{i}{2} = 2.531 + \frac{0.005}{2} = 2.533$$

The answer cannot be rounded using this formula. Since the interval is 0.005, there are 5 data values in each cell; therefore, a midpoint value of 2.533 can be used for the first cell. This value will have the lowest data value (2.531) in the first cell, which will have data values of 2.531, 2.532, 2.533, 2.534, and 2.535.

Midpoint selection is a matter of judgment, and in this case a midpoint of 2.533 was selected and the number of cells is 9. Selection of any other midpoint, although not incorrect, would have given 10 cells in the frequency distribution. Selection of different midpoint values will produce different frequency distributions—5 are possible. The midpoints for the other 8 cells are obtained by adding the cell interval to the previous midpoint: 2.533 + 0.005 = 2.538, 2.538 + 0.005 = 2.543, 2.543 + 0.005 = 2.548, . . ., and 2.568 + 0.005 = 2.573. These midpoints are shown in Table 4-6.

The midpoint value is the most representative value within a cell provided that the number of observations in a cell is large and the difference in boundaries

TABLE 4-6 Frequency Distribution of Steel Shaft Weight (kilograms)

CELL BOUNDARIES	CELL MIDPOINT	FREQUENCY
2.531–2.535	2.533	6
2.536–2.540	2.538	8
2.541–2.545	2.543	12
2.546–2.550	2.548	13
2.551–2.555	2.553	20
2.556–2.560	2.558	19
2.561–2.565	2.563	13
2.566–2.570	2.568	11
2.571–2.575	2.573	8
Total		110

is not too great. Even if this condition is not met, the number of observations above and below the midpoint of a cell will frequently be equal. And even if the number of observations above and below a cell midpoint is unbalanced in one direction, it will probably be offset by an unbalance in the opposite direction of another cell. Midpoint values should be to the same degree of accuracy as the original observations.

5. *Determine the cell boundaries. Cell boundaries* are the extreme or limit values of a cell, referred to as the upper boundary and the lower boundary. All the observations that fall between the upper and lower boundaries are classified into that particular cell. Boundaries are established so there is no question as to the location of an observation. Therefore, the boundary values are an extra decimal place or significant figure in accuracy than the observed values. Since the interval is odd, there will be an equal number of data values on each side of the midpoint. For the first cell with a midpoint of 2.533 and an interval of 0.005, there will be two values on each side. Therefore, that cell will contain the values 2.531, 2.532, 2.533, 2.534, and 2.535. To prevent any gaps, the true boundaries are extended about halfway to the next number, which gives values of 2.5305 and 2.5355. The following number line illustrates this principle:

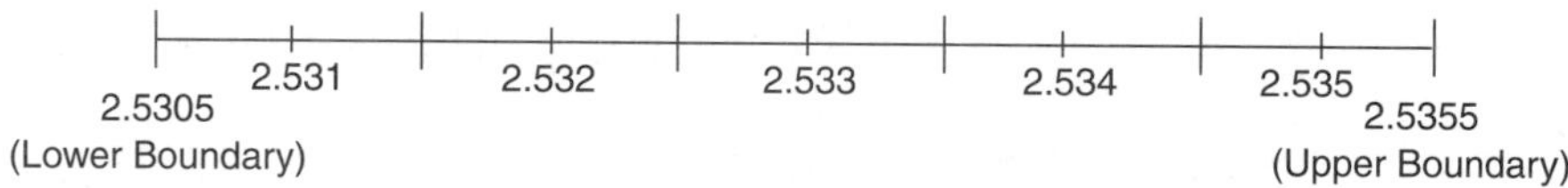

Some analysts prefer to leave the boundaries at the same number of decimal places as the data. No difficulty is encountered with this practice as long as the cell interval is odd and it is understood that the true boundaries are extended halfway to the next number. This practice is followed in this book and in EXCEL. It is shown in Table 4-6. Therefore, the lower boundary for the first cell is 2.531.

Once the boundaries are established for one cell, the boundaries for the other cells are obtained by successive additions of the cell interval. Therefore, the lower boundaries are 2.531 + 0.005 = 2.536, 2.536 + 0.005 = 2.541, . . ., 2.566 + 0.005 = 2.571. The upper boundaries are obtained in a similar manner and are shown in the first column of Table 4-6.

6. *Post the cell frequency*. The amount of numbers in each cell is posted to the frequency column of Table 4-6. An analysis of Table 4-5 shows that for the lowest cell there are: one 2.531, two 2.532, one 2.533, two 2.534, and zero 2.535. Therefore, there is a total of six values in the lowest cell, and the cell with a midpoint of 2.533 has a frequency of 6. The amounts are determined for the other cells in a similar manner.

The completed frequency distribution is shown in Table 4-6. This frequency distribution gives a better conception of the central value and how the data are

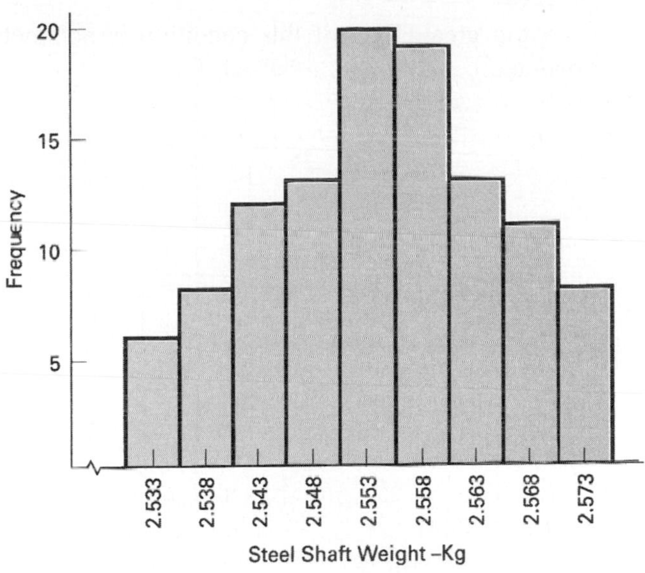

FIGURE 4-4 Histogram of data given in Table 4-6.

dispersed about that value than the unorganized data or a tally sheet. The histogram is shown in Figure 4-4.

Information on the construction of the relative frequency, cumulative frequency, and relative cumulative frequency histograms for grouped data is the same as for ungrouped data, but with one exception. With the two cumulative frequency histograms, the true upper cell boundary is the value labeled on the abscissa. Construction of these histograms for the example problem is left to the reader as an exercise.

The histogram describes the variation in the process. It is used to

1. Solve problems,
2. Determine the process capability,
3. Compare with specifications,
4. Suggest the shape of the population, and
5. Indicate discrepancies in data such as gaps.

Other Types of Frequency Distribution Graphs

The bar graph can also represent frequency distributions, as shown in Figure 4-5(a) using the data of Table 4-1. As mentioned previously, the bar graph is theoretically correct for discrete data but is not commonly used.

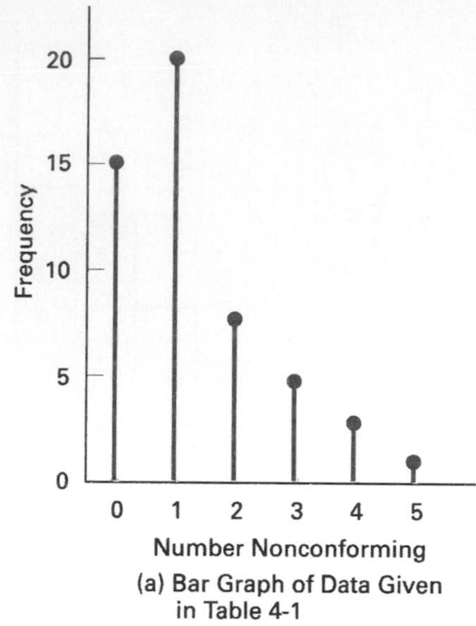

(a) Bar Graph of Data Given in Table 4-1

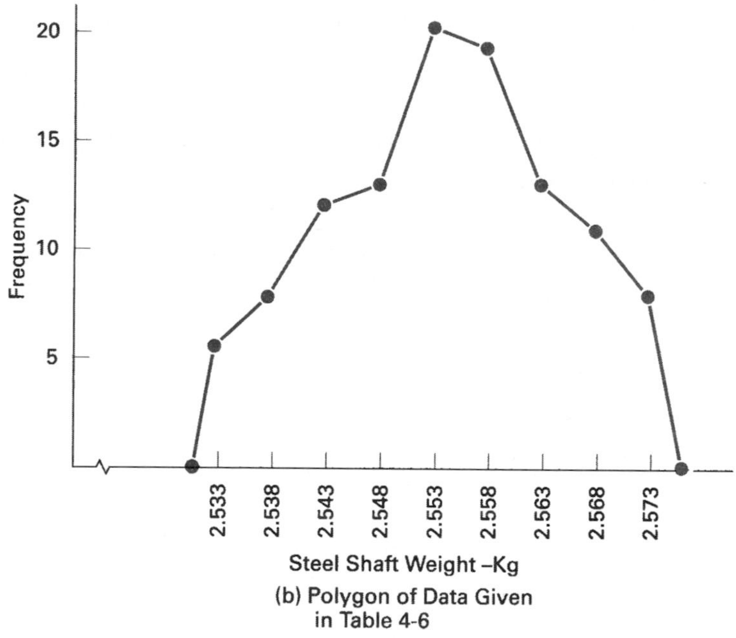

(b) Polygon of Data Given in Table 4-6

FIGURE 4-5 Other types of frequency distribution graphs.

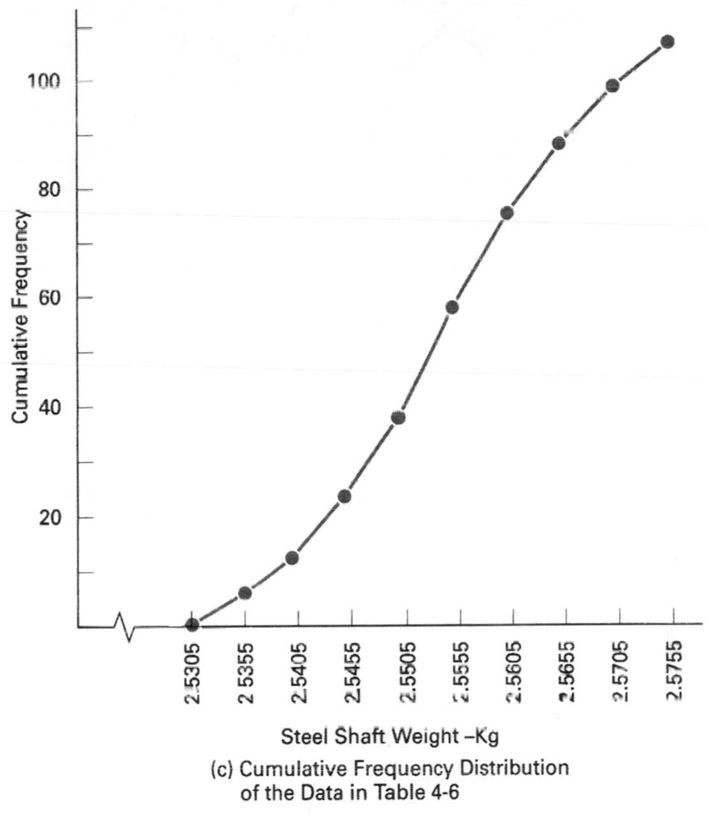

(c) Cumulative Frequency Distribution of the Data in Table 4-6

FIGURE 4-5 (Continued)

The *polygon* or *frequency polygon* is another graphic way of presenting frequency distributions and is illustrated in Figure 4-5(b) using the data of Table 4-6. It is constructed by placing a dot over each cell midpoint at the height indicated for each frequency. The curve is extended at each end in order for the figure to be enclosed. Since the histogram shows the area in each cell, it is considered to present a better graphical picture than the polygon and is the one most commonly used.

The graph that is used to present the frequency of all values less than the upper cell boundary of a given cell is called a *cumulative frequency,* or *ogive.* Figure 4-5(c) shows a cumulative frequency distribution curve for the data in Table 4-6. The cumulative value for each cell is plotted on the graph and joined by a straight line. The true upper cell boundary is labeled on the abscissa except for the first cell, which also has the true lower boundary.

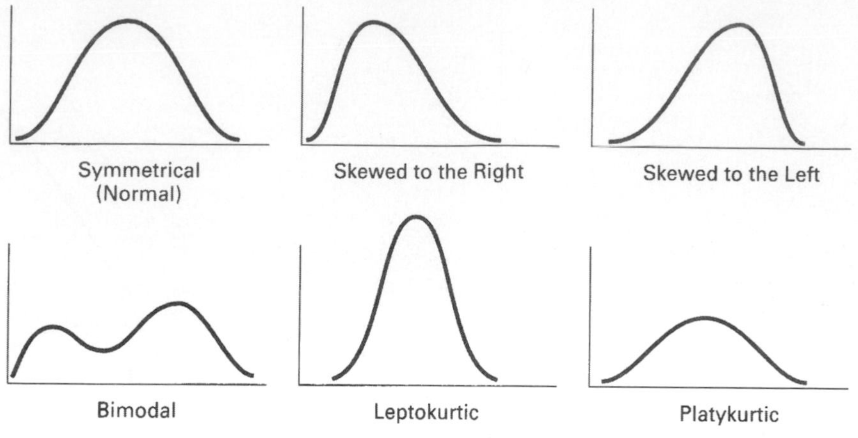

FIGURE 4-6 Characteristics of frequency distributions.

Characteristics of Frequency Distribution Graphs

The graphs of Figure 4-6 use smooth curves rather than the rectangular shapes associated with the histogram. A smooth curve represents a population frequency distribution, whereas the histogram represents a sample frequency distribution. The difference between a population and a sample is discussed in a later section of this chapter.

Frequency distribution curves have certain identifiable characteristics. One characteristic of the distribution concerns the symmetry or lack of symmetry of the data. Are the data equally distributed on each side of the central value, or are the data skewed to the right or to the left? Another characteristic concerns the number of modes or peaks to the data. There can be one mode, two modes (bimodal), or multiple modes. A final characteristic concerns the peakedness of the data. When the curve is quite peaked, it is referred to as *leptokurtic,* and when it is flatter, it is referred to as *platykurtic*.

Frequency distributions can give sufficient information about a quality control problem to provide a basis for decision making without further analysis. Distributions can also be compared in regard to location, spread, and shape as illustrated in Figure 4-7.

Analysis of Histograms

Analysis of a histogram can provide information concerning specifications, the shape of the population frequency distribution, and a particular quality problem.

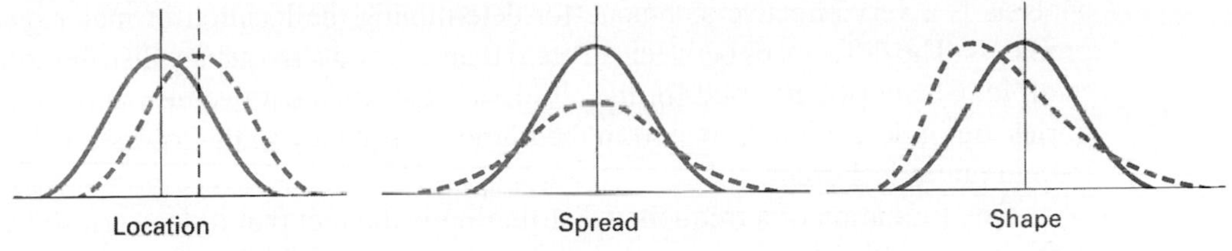

FIGURE 4-7 Differences due to location. spread, and shape.

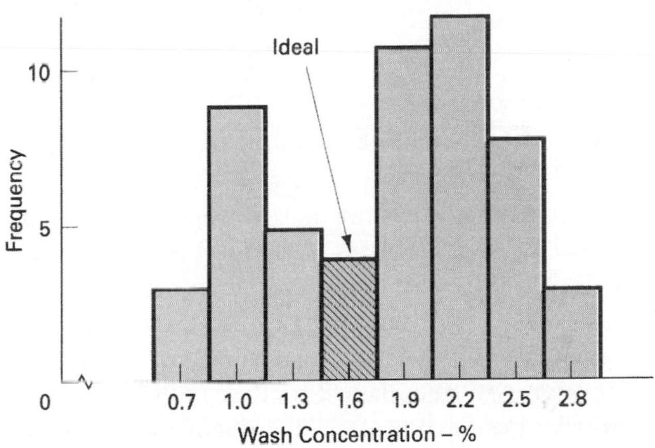

FIGURE 4-8 Histogram of wash concentration.

Figure 4-8 shows a histogram for the percentage of wash concentration in a steel tube cleaning operation prior to painting. The ideal concentration is between 1.45 and 1.74%, as shown by the crosshatched rectangle. Concentrations less than 1.45% produce poor quality; concentrations greater than 1.75%, while producing more than adequate quality, are costly and therefore reduce productivity. No complex statistics are needed to show that corrective measures are needed to bring the spread of the distribution closer to the ideal value of 1.6%.

By adding specifications to the graph, additional problem-solving information is created. Since the spread of the distribution is a good indication of the process capability, the graph will show the degree of capability.

Final Comments

Another type of distribution that is similar to the histogram is the Pareto diagram. The reader is referred to Chapter 3 for a discussion of this type of distribution. A Pareto analysis is a very effective technique for determining the location of major quality problems. The differences between a Pareto diagram and a frequency distribution are twofold. Categories are used for the abscissa rather than data values, and the categories are in descending order from the highest frequency to the lowest one rather than in numerical order.

One limitation of a frequency distribution is the fact that it does not show the order in which the data were produced. In other words, initial data could all be located on one side and later data on the other side. When this situation occurs, the interpretation of the frequency distribution will be different. A run chart, discussed in Chapter 5, shows the order in which the data were produced and can aid in the analysis.

MEASURES OF CENTRAL TENDENCY

A frequency distribution is sufficient for many quality problems. However, with a broad range of problems a graphical technique is either undesirable or needs the additional information provided by analytical techniques. Analytical methods of describing a collection of data have the advantage of occupying less space than a graph. They also have the advantage of allowing for comparisons between collections of data. And, they also allow for additional calculations and inferences. There are two principal analytical methods of describing a collection of data—measures of central tendency and measures of dispersion. The latter measure is described in the next section, while this section covers measures of central tendency.

A *measure of central tendency* of a distribution is a numerical value that describes the central position of the data or how the data tend to build up in the center. There are three measures in common use: (1) the average, (2) the median, and (3) the mode.

Average

The average is the sum of the observations divided by the number of observations. It is the most common measure of central tendency. There are three different techniques available for calculating the average: (1) ungrouped data, (2) grouped data, and (3) weighted average.

1. *Ungrouped data*. This technique is used when the data are unorganized. The average is represented by the notation $\overline{X}$, which is read as "X bar" and is given by the formula

$$\overline{X} = \frac{\sum_{i=1}^{n} X_i}{n} = \frac{X_1 + X_2 + \cdots + X_n}{n}$$

where $\overline{X}$ = average

n = number of observed values

$X_1, X_2, \ldots, X_n$ = observed value identified by the subscript 1, 2, . . ., n or general subscript i

Σ = symbol meaning "sum of"

The first expression is a simplified method of writing the formula whereby $\Sigma_{i=1}^{n} X_i$ is read as "summation from 1 to n of X sub i" and means to add together the values of the observations.

EXAMPLE PROBLEM

A technician checks the resistance value of 5 coils and records the values in ohms (Ω): $X_1 = 3.35$, $X_2 = 3.37$, $X_3 = 3.28$, $X_4 = 3.34$, and $X_5 = 3.30$. Determine the average.

Know

$$\overline{X} = \frac{\sum_{i=1}^{n} X_i}{n}$$

$$= \frac{3.35 + 3.37 + 3.28 + 3.34 + 3.30}{5}$$

$$= 3.33\Omega$$

Most electronic hand calculators have the capability of automatically calculating the average after the data are entered.

2. *Grouped data.* When the data have been grouped into a frequency distribution, the following technique is applicable. The formula for the average of grouped data is

$$\overline{X} = \frac{\sum_{i=1}^{h} f_i X_i}{n} = \frac{f_1 X_1 + f_2 X_2 + \cdots + f_h X_h}{f_1 + f_2 + \cdots + f_h}$$

where n = sum of the frequencies

f_i = frequency in a cell or frequency of an observed value

X_i = cell midpoint or an observed value

h = number of cells or number of observed values

The formula is applicable when the grouping is by cells with more than one observed value per cell, as illustrated by the steel shaft problem (Table 4-6). It is also applicable when each observed value, X_i, has its own frequency, f_i, as illustrated by the billing error problem (Table 4-1). In this situation, h is the number of observed values.

In other words, if the frequency distribution has been grouped into cells, X_i is the cell midpoint and f_i is the number of observations in that cell. If the frequency distribution has been grouped by individual observed values, X_i is the observed value and f_i is the number of times that observed value occurs in the data. This practice holds for both discrete and continuous variables.

Each cell midpoint is used as the representative value of that cell. The midpoint is multiplied by its cell frequency; the products are summed; and they are divided by the total number of observations. In the example problem that follows, the first three columns are those of a typical frequency distribution. The fourth column is derived from the product of the second column (midpoint) and third column (frequency) and is labeled "f_iX_i."

EXAMPLE PROBLEM

Given the frequency distribution of the life of 320 automotive tires in 1000 km (621.37 mi) as shown in Table 4-7, determine the average.

$$\overline{X} = \frac{\sum_{i=1}^{h} f_i X_i}{n}$$

$$= \frac{11{,}549}{320}$$

$$= 36.1 \quad \text{(which is in 1000 km)}$$

Therefore, $\overline{X} = 36.1 \times 10^3$ km.

TABLE 4-7 Frequency Distributions of the Life of 320 Tires in 1000 km

BOUNDARIES	MIDPOINT X_i	FREQUENCY f_i	COMPUTATION f_iX_i
23.6–26.5	25.0	4	100
26.6–29.5	28.0	36	1,008
29.6–32.5	31.0	51	1,581
32.6–35.5	34.0	63	2,142
35.6–38.5	37.0	58	2,146
38.6–41.5	40.0	52	2,080
41.6–44.5	43.0	34	1,462
44.6–47.5	46.0	16	736
47.6–50.5	49.0	6	294
Total		$n = 320$	$\Sigma f_iX_i = 11{,}549$

When comparing an average calculated from this technique with one calculated using the ungrouped technique, there can be a slight difference. This difference is caused by the observations in each cell being unevenly distributed in the cell. In actual practice the difference will not be of sufficient magnitude to affect the accuracy of the problem.

3. *Weighted average.* When a number of averages are combined with different frequencies, a *weighted average* is computed. The formula for the weighted average is given by

Know

$$\overline{X}_w = \frac{\sum_{i=1}^{n} w_i \overline{X}_i}{\sum_{i=1}^{n} w_i}$$

where $\overline{X}_w$ = weighted average

w_i = weight of the ith average

EXAMPLE PROBLEM

Tensile tests on aluminum alloy rods are conducted at three different times, which results in three different average values in megapascals (MPa). On the first occasion, 5 tests are conducted with an average of 207 MPa (30,000 psi); on the second occasion, 6 tests, with an average of 203 MPa; and on the last occasion, 3 tests, with an average of 206 MPa. Determine the weighted average.

$$\overline{X}_w = \frac{\sum_{i=1}^{n} w_i \overline{X}_i}{\sum_{i=1}^{n} w_i}$$

$$= \frac{(5)(207) + (6)(203) + (3)(206)}{5 + 6 + 3}$$

$$= 205 \text{ MPa}$$

The weighted average technique is a special case of the grouped data technique wherein the data are not organized into a frequency distribution. In the example above, the weights are whole numbers. Another method of solving the same problem is to use proportions. Thus,

$$w_1 = \frac{5}{5 + 6 + 3} = 0.36$$

$$w_2 = \frac{6}{5 + 6 + 3} = 0.43$$

$$w_3 = \frac{3}{5 + 6 + 3} = 0.21$$

and the sum of the weights equals 1.00. The latter technique would be necessary when the weights are given in percent or the decimal equivalent.

Unless otherwise noted, $\overline{X}$ stands for the average of observed values, $\overline{X}_x$. The same equation is used to find

$\overline{X}_{\bar{x}}$ or $\overline{\overline{X}}$—average of averages

$\overline{R}$—average of ranges

$\overline{c}$—average of count of nonconformities

$\overline{s}$—average of sample standard deviations, etc.

A bar on top of any variable indicates that it is an average.

Median

Another measure of central tendency is the *median,* which is defined as the value which divides a series of ordered observations so that the number of items above it is equal to the number below it.

1. *Ungrouped technique.* Two situations are possible in determining the median of a series of ungrouped data—when the number in the series is odd and when the number in the series is even. When the number in the series is odd, the median is the midpoint of the values. Thus, the ordered set of numbers 3, 4, 5, 6, 8, 8, and 10 has a median of 6, and the ordered set of numbers 22, 24, 24, 24, and 30 has a median of 24. When the number in the series is even, the median is the average of the two middle numbers. Thus, the ordered set of numbers 3, 4, 5, 6, 8, and 8 has a median that is the average of 5 and 6, which is $(5 + 6)/2 = 5.5$. If both middle numbers are the same, as in the ordered set of numbers 22, 24, 24, 24, 30, and 30, it is still computed as the average of the two middle numbers, since $(24 + 24)/2 = 24$. The reader is cautioned to be sure the numbers are ordered before computing the median.

2. *Grouped technique.* When data are grouped into a frequency distribution, the median is obtained by finding the cell that has the middle number and then interpolating within the cell. The interpolation formula for computing the median is given by

$$\text{Md} = L_m + \left(\frac{\frac{n}{2} - cf_m}{f_m} \right) i$$

where Md = median

L_m = lower boundary of the cell with the median

n = total number of observations

cf_m = cumulative frequency of all cells below L_m

f_m = frequency of median cell

i = cell interval

To illustrate the use of the formula, data from Table 4-7 will be used. By counting up from the lowest cell (midpoint 25.0), the halfway point (320/2 = 160) is reached in the cell with a midpoint value of 37.0 and a lower limit of 35.6. The cumulative frequency (cf_m) is 154, the cell interval is 3, and the frequency of the median cell is 58.

$$\text{Md} = L_m + \left(\frac{\frac{n}{2} - cf_m}{f_m}\right)i$$

$$= 35.6 + \left(\frac{\frac{320}{2} - 154}{58}\right)3$$

$$= 35.9 \qquad \text{(which is in 1000 km)}$$

If the counting is begun at the top of the distribution, the cumulative frequency is counted to the cell upper limit and the interpolated quantity is subtracted from the upper limit. However, it is more common to start counting at the bottom of the distribution.

The median of grouped data is not used too frequently.

Know

Mode

The *mode* (Mo) of a set of numbers is the value that occurs with the greatest frequency. It is possible for the mode to be nonexistent in a series of numbers or to have more than one value. To illustrate, the series of numbers 3, 3, 4, 5, 5, 5, and 7 has a mode of 5; the series of numbers 22, 23, 25, 30, 32, and 36 does not have a mode; and the series of numbers 105, 105, 105, 107, 108, 109, 109, 109, 110, and 112 has two modes, 105 and 109. A series of numbers is referred to as *unimodal* if it has one mode, *bimodal* if it has two modes, and *multimodal* if there are more than two modes.

When data are grouped into a frequency distribution, the midpoint of the cell with the highest frequency is the mode, since this point represents the highest point (greatest frequency) of the histogram. It is possible to obtain a better estimate of the mode by interpolating in a manner similar to that used for the median. However, this is not necessary, since the mode is employed primarily as an inspection method for determining the central tendency, and greater accuracy than the cell midpoint is not required.

Relationship Among the Measures of Central Tendency

Differences among the three measures of central tendency are shown in the smooth polygons of Figure 4-9. When the distribution is symmetrical, the values for the average, median, and mode are identical; when the distribution is skewed, the values are different.

The average is the most commonly used measure of central tendency. It is used when the distribution is symmetrical or not appreciably skewed to the right or left; when additional statistics, such as measures of dispersion, control charts, and so on, are to be computed based on the average; and when a stable value is needed for inductive statistics.

The median becomes an effective measure of the central tendency when the distribution is positively (to the right) or negatively (to the left) skewed. It is used when an exact midpoint of a distribution is desired. When a distribution has extreme values, the average will be adversely affected while the median will remain unchanged. Thus, in a series of numbers such as 12, 13, 14, 15, 16, the median and average are identical and equal to 14. However, if the first value is changed to a 2, the median remains at 14, but the average becomes 12. A control chart based on the median is user-friendly and excellent for monitoring quality.

The mode is used when a quick and approximate measure of the central tendency is desired. Thus, the mode of a histogram is easily found by a visual examination. In addition, the mode is used to describe the most typical value of a distribution, such as the modal age of a particular group.

Other measures of central tendency are the geometric mean, harmonic mean, and quadratic mean. These measures are not used in quality.

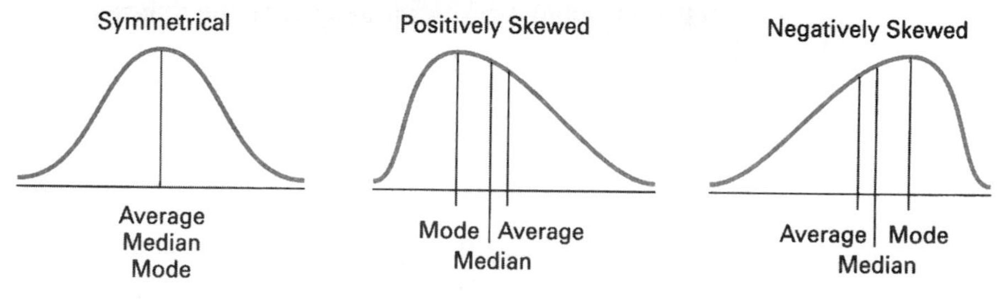

FIGURE 4-9 Relationship among average, median, and mode.

MEASURES OF DISPERSION

Introduction

In the preceding section, techniques for describing the central tendency of data were discussed. A second tool of statistics is composed of the *measures of dispersion,* which describe how the data are spread out or scattered on each side of the central value. Measures of dispersion and measures of central tendency are both needed to describe a collection of data. To illustrate, the employees of the plating and the assembly departments of a factory have identical average weekly wages of \$225.36; however, the plating department has a high of \$230.72 and a low of \$219.43, while the assembly department has a high of \$280.79 and a low of \$173.54. The data for the assembly department are spread out or dispersed farther from the average than are those of the plating department.

Measures of dispersion discussed in this section are range, standard deviation, and variance. Other measures such as mean deviation and quartile deviation are not used in quality.

Range

The *range* of a series of numbers is the difference between the largest and smallest values or observations. Symbolically, it is given by the formula

$$R = X_h - X_l$$

where R = range

X_h = highest observation in a series

X_l = lowest observation in a series

EXAMPLE PROBLEM

If the highest weekly wage in the assembly department is \$280.79 and the lowest weekly wage is \$173.54, determine the range.

$$\begin{aligned} R &= X_h - X_l \\ &= \$280.79 - \$173.54 \\ &= \$107.25 \end{aligned}$$

The range is the simplest and easiest to calculate of the measures of dispersion. A related measure, which is occasionally used, is the *midrange,* which is the range divided by 2, $R/2$.

Standard Deviation

The *standard deviation* is a numerical value in the units of the observed values that measures the spreading tendency of the data. A large standard deviation shows greater variability of the data than does a small standard deviation. In symbolic terms it is given by the formula

$$s = \sqrt{\frac{\sum_{i=1}^{n}(X_i - \overline{X})^2}{n - 1}}$$

where s = sample standard deviation

X_i = observed value

$\overline{X}$ = average

n = number of observed values

Table 4-8 will be used to explain the standard deviation concept. The first column (X_i) gives six observed values in kilograms, and from these values the average, $\overline{X} = 3.0$, is obtained. The second column $(X_i - \overline{X})$ is the deviation of the individual observed values from the average. If we sum the deviations, the answer will be 0, which is always the case, but it will not lead to a measure of dispersion. However, if the deviations are squared, they will all be positive and their sum will be greater than zero. Calculations are shown in the third column, $(X_i - \overline{X})^2$, with a resultant sum of 0.08, which will vary depending on the observed values. The average of the squared deviations can be found by dividing by n; however, for theoretical reasons we divide by $n - 1$.[3] Thus,

$$\frac{\sum(X_i - \overline{X})^2}{n - 1} = \frac{0.08}{6 - 1} = 0.016 \text{ kg}^2$$

TABLE 4-8 Standard Deviation Analysis

X_i	$X_i - \overline{X}$	$(X_i - \overline{X})^2$
3.2	+0.2	0.04
2.9	−0.1	0.01
3.0	0.0	0.00
2.9	−0.1	0.01
3.1	+0.1	0.01
2.9	−0.1	0.01
$\overline{X} = 3.0$	$\Sigma = 0$	$\Sigma = 0.08$

[3]The reason for using $n - 1$ is because one degree of freedom is lost due to the use of the sample statistic, $\overline{X}$, rather than the population parameter, μ.

which gives an answer that has the units squared. This result is not acceptable as a measure of the dispersion but is valuable as a measure of variability for advanced statistics. It is called the *variance* and is given the symbol s^2. If we take the square root, the answer will be in the same units as the observed values. Calculations are

$$s = \sqrt{\frac{\sum (X_i - \overline{X})^2}{n - 1}} = \sqrt{\frac{0.08}{6 - 1}} = 0.13 \text{ kg}$$

This formula is for explanation rather than for the purpose of calculation. Because the form of the data can be either grouped or ungrouped, there are different computing techniques.

1. *Ungrouped technique*. The formula used in the definition of standard deviation can be used for ungrouped data. However, an alternative formula is more convenient for computation purposes:

$$s = \sqrt{\frac{n \sum_{i=1}^{n} X_i^2 - \left(\sum_{i=1}^{n} X_i\right)^2}{n(n - 1)}}$$

EXAMPLE PROBLEM

Determine the standard deviation of the moisture content of a roll of kraft paper. The results of six readings across the paper web are 6.7, 6.0, 6.4, 6.4, 5.9, and 5.8%.

$$s = \sqrt{\frac{n \sum_{i=1}^{n} X_i^2 - \left(\sum_{i=1}^{n} X_i\right)^2}{n(n - 1)}}$$

$$= \sqrt{\frac{6(231.26) - (37.2)^2}{6(6 - 1)}}$$

$$= 0.35\%$$

After entry of the data, many hand calculators compute the standard deviation on command.

2. *Grouped technique*. When the data have been grouped into a frequency distribution, the following technique is applicable. The formula for the standard deviation of grouped data is

$$s = \sqrt{\frac{n \sum_{i=1}^{h} (f_i X_i^2) - \left(\sum_{i=1}^{h} f_i X_i\right)^2}{n(n - 1)}}$$

TABLE 4-9 Passenger Car Speeds (in km/h) During a 15-Minute Interval on I-57 at Location 236

BOUNDARIES	MIDPOINT X_i	FREQUENCY f_i	COMPUTATIONS f_iX_i	$f_iX_i^2$
72.6–81.5	77.0	5	385	29,645
81.6–90.5	86.0	19	1634	140,524
90.6–99.5	95.0	31	2945	279,775
99.6–108.5	104.0	27	2808	292,032
108.6–117.5	113.0	14	1582	178,766
Total		$n = 96$	$\Sigma fX = 9354$	$\Sigma fX^2 = 920{,}742$

where the symbols f_i, X_i, n, and h have the same meaning as given for the average of grouped data.

To use this technique, two additional columns are added to the frequency distribution. These additional columns are labeled "f_iX_i" and "$f_iX_i^2$," as shown in Table 4-9. It will be recalled that the "f_iX_i" column is needed for the average computations; therefore, only one additional column is required to compute the standard deviation. The technique is shown by the following example problem.

Do not round ΣfX or ΣfX^2, as this action will affect accuracy. Most hand calculators have the capability to enter grouped data and calculate s on command.

Unless otherwise noted, s stands for s_x, the sample standard deviation of observed values. The same formula is used to find

$s_{\bar{x}}$—sample standard deviation of averages

s_p—sample standard deviation of proportions

s_R—sample standard deviation of ranges

s_s—sample standard deviation of standard deviations, etc.

EXAMPLE PROBLEM

Given the frequency distribution of Table 4-9 for passenger car speeds during a 15-minute interval on 1-57, determine the average and standard deviation.

$$\bar{x} = \frac{\sum_{i=1}^{h} f_iX_i}{n} = \frac{9354}{96} = 97.4 \text{ km/h}$$

$$s = \sqrt{\frac{n\sum_{i=1}^{h}(f_iX_i^2) - \left(\sum_{i=1}^{h} f_iX_i\right)^2}{n(n-1)}} = \sqrt{\frac{96(920{,}742) - (9354)^2}{96(96-1)}} = 9.9 \text{ km/h}$$

The standard deviation is a reference value that measures the dispersion in the data. It is best viewed as an index that is defined by the formula. The smaller the value of the standard deviation, the better the quality, since the distribution is more closely compacted around the central value. Also, the standard deviation helps to define populations.

Relationship Between the Measures of Dispersion

In quality the range is a very common measure of the dispersion; it is used in one of the principal control charts. The primary advantages of the range are in providing a knowledge of the total spread of the data and in its simplicity. It is also valuable when the amount of data is too small or too scattered to justify the calculation of a more precise measure of dispersion. The range is not a function of a measure of central tendency. As the number of observations increases, the accuracy of the range decreases, since it becomes easier for extremely high or low readings to occur. It is suggested that the use of the range be limited to a maximum of 10 observations.

The standard deviation is used when a more precise measure is desired. Figure 4-10 shows two distributions with the same average, $\overline{X}$, and range R; however, the distribution on the bottom is much better. The sample standard deviation is much smaller on the bottom distribution, indicating that the data are more compact

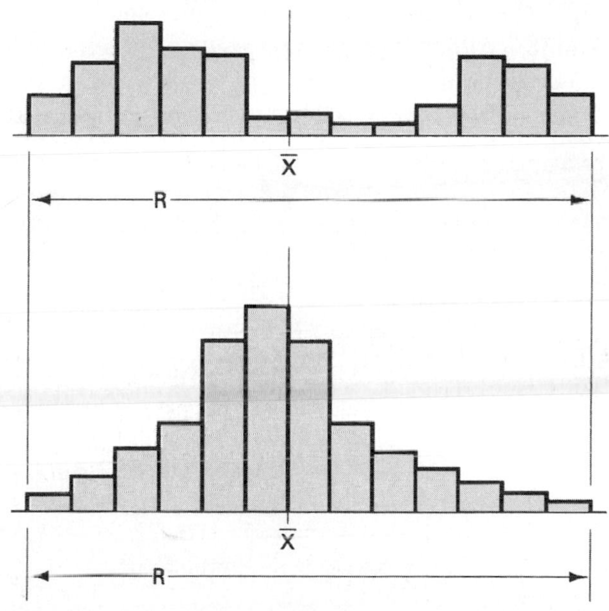

FIGURE 4-10 Comparison of two distributions with equal average and range.

around the central value, $\overline{X}$. As the sample standard deviation gets smaller, the quality gets better. It is also the most common measure of the dispersion and is used when subsequent statistics are to be calculated. When the data have an extreme value for the high or the low, the standard deviation is more desirable than the range.

OTHER MEASURES

There are two other measures that are frequently used to analyze a collection of data—skewness and kurtosis.

Know

Skewness

As indicated previously, *skewness* is a lack of symmetry of the data. The formula is given by[4]

$$a_3 = \frac{\sum_{i=1}^{h} f_i(X_i - \overline{X})^3/n}{s^3}$$

where a_3 represents skewness.

Skewness is a number whose size tells us the extent of the departure from symmetry. If the value of a_3 is 0, the data are symmetrical; if greater than 0 (positive), the data are skewed to the right, which means that the long tail is to the right; and if less than 0 (negative), the data are skewed to the left, which means that the long tail is to the left. See Figure 4-11 for a graphical representation of skewness. Values of $+1$ or -1 imply a strong unsymmetrical distribution.

EXAMPLE PROBLEM

Determine the skewness of the frequency distribution of Table 4-10. The average and sample standard deviation are calculated and are 7.0 and 2.30, respectively.

$$a_3 = \frac{\sum_{i=1}^{h} f_i(X_i - \overline{X})^3/n}{s^3}$$

$$= \frac{-648/124}{2.30^3}$$

$$= -0.43$$

[4]This formula is an approximation that is good enough for most purposes.

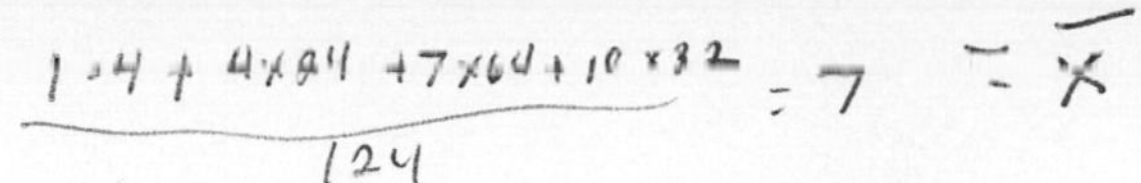

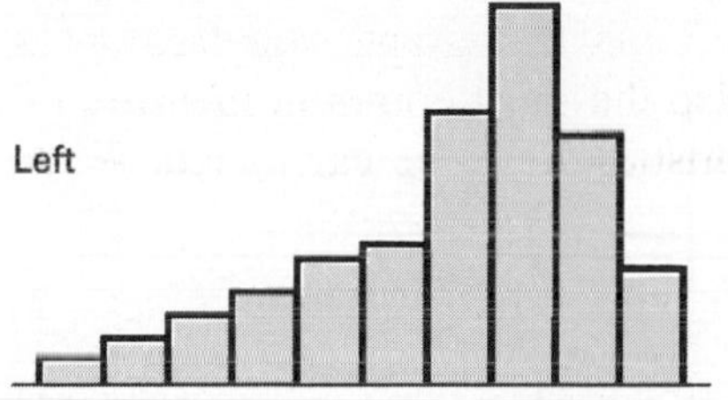

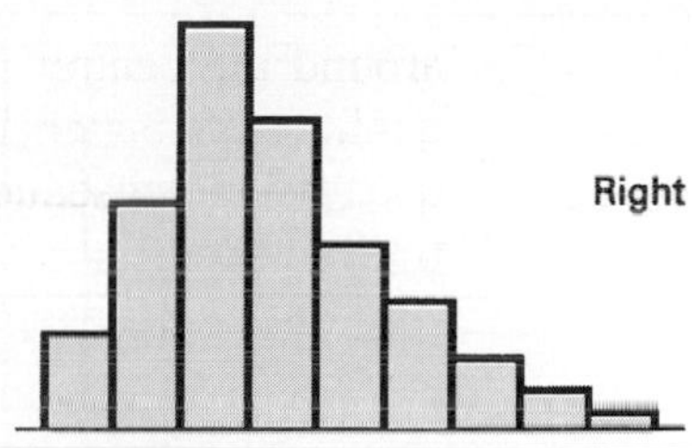

FIGURE 4-11 Left (negative) and right (positive) skewness distributions.

TABLE 4-10 Data for Skewness and Kurtosis Example Problems

X_i	f_i	$X_i - \overline{X}$	$f_i(X_i - \overline{X})^3$	$f_i(X_i - \overline{X})^4$
1	4	$(1 - 7) = -6$	$4(-6)^3 = -864$	$4(-6)^4 = 5184$
4	24	$(4 - 7) = -3$	$24(-3)^3 = -648$	$24(-3)^4 = 1944$
7	64	$(7 - 7) = \ \ 0$	$64(0)^3 = \ \ 0$	$64(0)^4 = \ \ 0$
10	32	$(10 - 7) = +3$	$32(+3)^3 = +864$	$32(+3)^4 = 2592$
	$\Sigma = 124$		$\Sigma = -648$	$\Sigma = 9720$

The skewness value of -0.43 tells us that the data are skewed to the left. Visual examination of the X and f columns or a histogram would have indicated the same information.

In order to determine skewness, the value of n must be large, say, at least 100. Also, the distribution must be unimodal. The skewness value provides information concerning the shape of the population distribution. For example, a normal distribution has a skewness value of zero, $a_3 = 0$.

Kurtosis

As indicated previously, *kurtosis* is the peakedness of the data. The formula is given by[5]

$$a_4 = \frac{\sum_{i=1}^{h} f_i(X_i - \overline{X})^4/n}{s^4}$$

where a_4 represents kurtosis.

Kurtosis is a dimensionless value that is used as a measure of the height of the peak in a distribution. Figure 4-12 shows a leptokurtic (more peaked) distribu-

[5]This formula is an approximation that is good enough for most purposes.

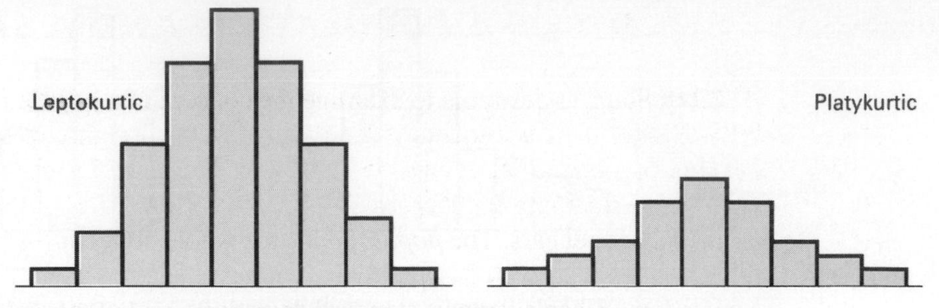

FIGURE 4-12 Leptokurtic and platykurtic distributions.

tion and a platykurtic (flatter) distribution. Between these two distributions is one referred to as mesokurtic, which is the normal distribution.

EXAMPLE PROBLEM

Determine the kurtosis of the frequency distribution of Table 4-10, which has $\overline{X} = 7.0$ and $s = 2.30$.

$$a_4 = \frac{\sum_{i=1}^{h} f_i(X_i - \overline{X})^4/n}{s^4}$$

$$= \frac{9720/124}{2.30^4}$$

$$= 2.80$$

The kurtosis value of 2.80 does not provide any information by itself—it must be compared to another distribution. Use of the kurtosis value is the same as skewness—large sample size, n, and unimodal distribution. It provides information concerning the shape of the population distribution. For example, a normal distribution, mesokurtic, has a kurtosis value of 3, $a_4 = 3$. If $a_4 > 3$, then the height of the distribution is more peaked than normal, leptokurtic, and if $a_4 < 3$, the height of the distribution is less peaked than normal, platykurtic. Some software packages such as EXCEL normalize the data to zero by subtracting 3 from the answer.

The concepts of skewness and kurtosis are useful in that they provide some information about the shape of the distribution. Calculations are best made by a computer program.

CONCEPT OF A POPULATION AND A SAMPLE

At this point, it is desirable to examine the concept of a population and a sample. In order to construct a frequency distribution of the weights of steel shafts, a small portion, or *sample*, is selected to represent all the steel shafts. Similarly, the data collected concerning the passenger car speeds represented only a small portion of all the passenger cars. The *population* is the whole collection of measurements, and in the examples above, the population would be all the steel shafts and all the passenger cars. When averages, standard deviations, and other measures are computed from samples, they are referred to as *statistics*. Since the composition of samples will fluctuate, the computed statistics will be larger or smaller than their true population values, or *parameters*. Parameters are considered to be fixed reference (standard) values or the best estimate of these values available at a particular time.

The population may have a finite number of items, such as a day's production of steel shafts. It may be infinite or almost infinite, such as the number of rivets in a year's production of jet airplanes. The population may be defined differently depending on the particular situation. Thus, a study of a product could involve the population of an hour's production, a week's production, 5000 pieces, and so on.

Since it is rarely possible to measure all of the population, a sample is selected. Sampling is necessary when it may be impossible to measure the entire population; when the expense to observe all the data is prohibitive; when the required inspection destroys the product; or when a test of the entire population may be too dangerous, as would be the case with a new medical drug. Actually, an analysis of the entire population may not be as accurate as sampling. It has been shown that 100% manual inspection is not as accurate as sampling when the percent nonconforming is very small. This is probably due to the fact that boredom and fatigue cause inspectors to prejudge each inspected item as being acceptable.

When designating a population, the corresponding Greek letter is used. Thus, the sample average has the symbol $\overline{X}$, and the population mean the symbol μ (mu).

Note that the word *average* changes to *mean* when used for the population. The symbol $\overline{X}_0$ is the standard or reference value. Mathematical concepts are based on μ, which is the true value—$\overline{X}_0$ represents a practical equivalent in order to use the concepts. The sample standard deviation has the symbol s, and the population standard deviation the symbol σ (sigma). The symbol s_0 is the standard or reference value and has the same relationship to σ that $\overline{X}_0$ has to μ. The true population value may never be known; therefore, the symbol $\hat{\mu}$ and $\hat{\sigma}$ are sometimes used to indicate "estimate of." A comparison of sample and population is given in Table 4-11. A sample frequency distribution is represented by a histogram, while a population frequency distribution is represented by a smooth curve. Additional comparisons will be given as they occur.

TABLE 4-11 Comparison of Sample and Population

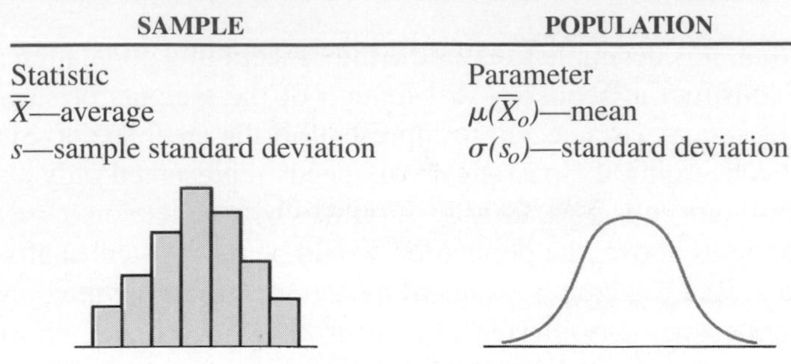

SAMPLE	POPULATION
Statistic	Parameter
$\overline{X}$—average	$\mu(\overline{X}_o)$—mean
s—sample standard deviation	$\sigma(s_o)$—standard deviation

The primary objective in selecting a sample is to learn something about the population that will aid in making some type of decision. The sample selected must be of such a nature that it tends to resemble or represent the population. How successfully the sample represents the population is a function of the size of the sample, chance, sampling method, and whether the conditions change or not.

Table 4-12 shows the results of an experiment that illustrates the relationship between samples and the population. A container holds 800 blue and 200 green spheres 5 mm (approximately 3/16 in.) in diameter. The 1000 spheres are considered to be the population, with 20% being green. Samples of size 10 are selected and posted to the table and then replaced in the container. The table illustrates the differences between the sample results and what should be expected from the known population. Only in samples 2 and 7 are the sample statistics equal to the population parameter. There definitely is a chance factor that determines the composition of the sample. When the eight individual samples are combined into one

TABLE 4-12 Results of Eight Samples of Blue and Green Spheres from a Known Population

SAMPLE NUMBER	SAMPLE SIZE	NUMBER OF BLUE SPHERES	NUMBER OF GREEN SPHERES	PERCENTAGE OF GREEN SPHERES
1	10	9	1	10
2	10	8	2	20
3	10	5	5	50
4	10	9	1	10
5	10	7	3	30
6	10	10	0	0
7	10	8	2	20
8	10	9	1	10
Total	80	65	15	18.8

large one, the percentage of green spheres is 18.8, which is close to the population value of 20%.

While inferences are made about the population from samples, it is equally true that a knowledge of the population provides information for analysis of the sample. Thus, it is possible to determine whether a sample came from a particular population. This concept is necessary to understand control chart theory. A more detailed discussion is delayed until Chapter 5.

THE NORMAL CURVE

Description

Although there are as many different populations as there are conditions, they can be described by a few general types. One type of population that is quite common is called the *normal curve*, or *Gaussian distribution*. The normal curve is a symmetrical, unimodal, bell-shaped distribution with the mean, median, and mode having the same value.

A population curve or distribution is developed from a frequency histogram. As the sample size of a histogram gets larger and larger, the cell interval gets smaller and smaller. When the sample size is quite large and the cell interval is very small, the histogram will take on the appearance of a smooth polygon or a curve representing the population. A curve of the normal population of 1000 observations of the resistance in ohms of an electrical device with population mean, μ, of 90 Ω and population standard deviation, σ, of 2 Ω is shown in Figure 4-13. The interval between dotted lines is equal to one standard deviation, σ.

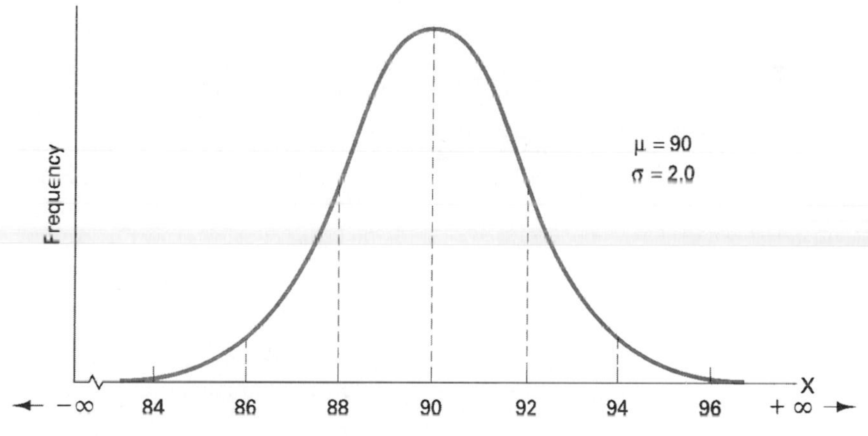

FIGURE 4-13 Normal distribution for resistance of an electrical device with μ = 90 ohms and σ = 2.0 ohms.

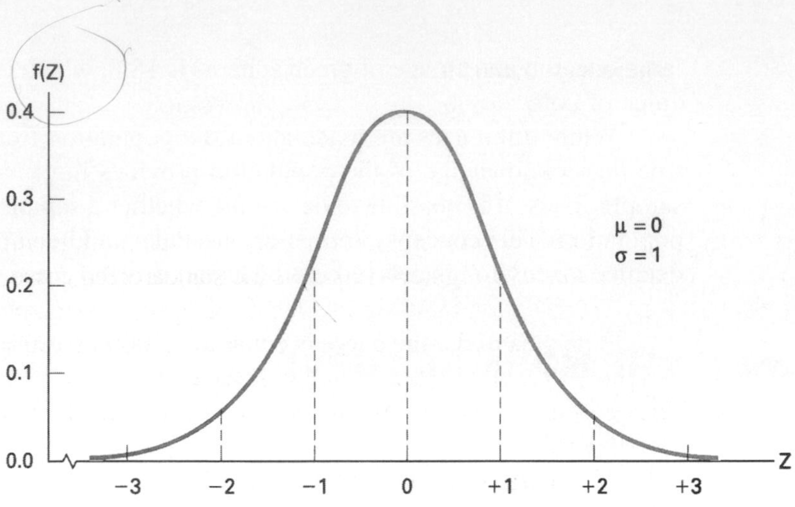

FIGURE 4-14 Standardized normal distribution with $\mu = 0$ and $\sigma = 1$.

Much of the variation in nature and in industry follows the frequency distribution of the normal curves. Thus, the variations in the weight of elephants, the speed of antelopes, and the height of human beings will follow a normal curve. Also, the variations found in industry, such as the weight of gray iron castings, the life of 60-W light bulbs, and the dimensions of a steel piston ring, will be expected to follow the normal curve. When considering the heights of human beings, we can expect a small percentage of them to be extremely tall and a small percentage to be extremely short, with the majority of human heights clustering about the average value. The normal curve is such a good description of the variations that occur to most quality characteristics in industry that it is the basis for many techniques.

All normal distributions of continuous variables can be converted to the standardized normal distribution (see Figure 4-14) by using the *standardized normal value,* Z. For example, consider the value of 92 Ω in Figure 4-14, which is one standard deviation above the mean ($\mu + 1\sigma = 90 + 1(2) = 92$). Conversion to the Z value is

$$Z = \frac{X_i - \mu}{\sigma} = \frac{92 - 90}{2} = +1$$

which is also 1σ above μ on the Z scale of Figure 4-14.

The formula for the standardized normal curve is

$$f(Z) = \frac{1}{\sqrt{2\pi}} e^{-Z^2} 2 = 0.3989 e^{-Z^2} 2$$

where $\pi = 3.14159$

$$e = 2.71828$$

$$Z = \frac{X_i - \mu}{\sigma}$$

A table is provided in the Appendix (Table A); therefore, it is not necessary to use the formula. Figure 4-14 shows the standardized curve with its mean of zero and standard deviation of 1. It is noted that the curve is asymptotic at $Z = -3$ and $Z = +3$.

The area under the curve is equal to 1.0000 or 100% and therefore can easily be used for probability calculations. Since the area under the curve between various points is a very useful statistic, a normal area table is provided as Table A in the appendix.

The normal distribution can be referred to as a normal probability distribution. While it is the most important population distribution, there are a number of other ones for continuous variables. There are also a number of probability distributions for discrete variables. These distributions are discussed in Chapter 7.

Relationship to the Mean and Standard Deviation

As seen by the formula for the standardized normal curve, there is a definite relationship among the mean, the standard deviation, and the normal curve. Figure 4-15 shows three normal curves with different mean values; it is noted that the only change is in the location. Figure 4-16 shows three normal curves with the same mean but different standard deviations. The figure illustrates the principle that the larger the standard deviation, the flatter the curve (data are widely dispersed), and the smaller the standard deviation, the more peaked the curve (data are narrowly dispersed). If the standard deviation is zero, all values are identical to the mean and there is no curve.

The normal distribution is fully defined by the population mean and population standard deviation. Also, as seen by Figures 4-15 and 4-16, these two parameters are independent. In other words, a change in one has no effect on the other.

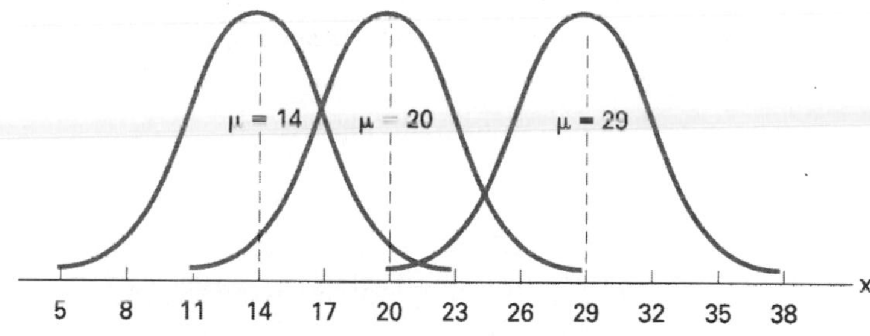

FIGURE 4-15 Normal curve with different means but identical standard deviations.

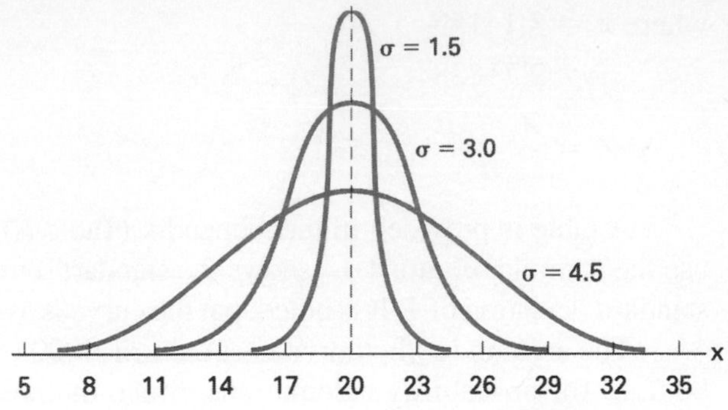

FIGURE 4-16 Normal curve with different standard deviations but identical means.

A relationship exists between the standard deviation and the area under the normal curve as shown in Figure 4-17. The figure shows that in a normal distribution 68.26% of the items are included between the limits of $\mu + 1\sigma$ and $\mu - 1$, 95.46% of the items are included between the limits $\mu + 2$ and $\mu - 2\sigma$, and 99.73% of the items are included between $\mu + 3\sigma$ and $\mu - 3\sigma$. One hundred percent of the items are included between the limits $+\infty$ and $-\infty$. These percentages hold true regardless of the shape of the normal curve. The fact that 99.73% of the items are included between $\pm 3\sigma$ is the basis for control charts that are discussed in Chapter 5.

Applications

The percentage of items included between any two values can be determined by calculus. However, this is not necessary, since the areas under the curve for various Z values are given in Table A in the Appendix. Table A, "Areas Under the Normal Curve," is a left-reading table,[6] which means that the given areas are for that portion of the curve from $-\infty$ to a particular value, X_i.

The first step is to determine the Z value using the formula

$$Z = \frac{X_i - \mu}{\sigma}$$

where Z = standard normal value

X_i = individual value

μ = mean

σ = population standard deviation

[6]In some texts the table for the areas under the normal curve is arranged in a different manner.

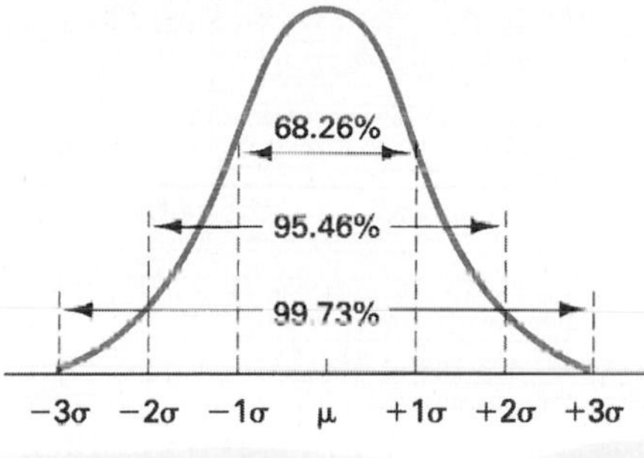

FIGURE 4-17 Percent of items included between certain values of the standard deviation.

Next, using the calculated Z value, the area under the curve to the left of X_i is found in Table A. Thus, if a calculated Z value is -1.76, the value for the area is 0.0392. Since the total area under the curve is 1.0000, the 0.0392 value for the area can be changed to a percent of the items under the curve by moving the decimal point two places to the right. Therefore, 3.92% of the items are less than the particular X_i value.

Assuming that the data are normally distributed, it is possible to find the percent of the items in the data that are less than a particular value, greater than a particular value, or between two values. When the values are upper and/or lower specifications, a powerful statistical tool is available. The following example problems will illustrate the technique.

EXAMPLE PROBLEM

The mean value of the weight of a particular brand of cereal for the past year is 0.297 Kg (10.5 oz) with a standard deviation of 0.024 Kg. Assuming normal distribution, find the percent of the data that falls below the lower specification limit of 0.274 Kg. (*Note:* Since the mean and standard deviation were determined from a large number of tests during the year, they are considered to be valid estimates of the population values.)

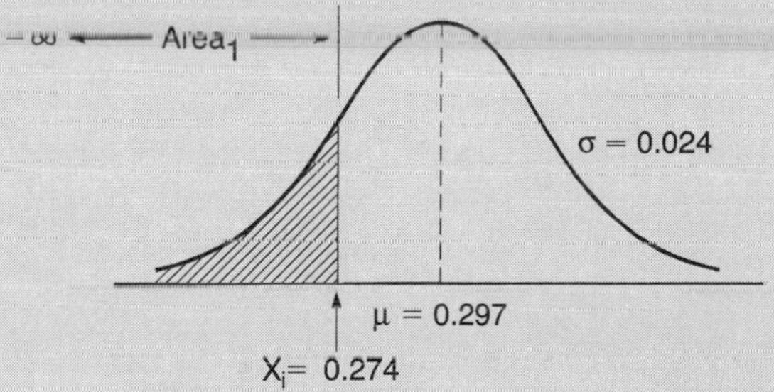

$$Z = \frac{X_i - \mu}{\sigma}$$

$$= \frac{0.274 - 0.297}{0.024}$$

$$= -0.96$$

From Table A it is found that for $Z = -0.96$,

$$\text{Area}_1 = 0.1685 \text{ or } 16.85\%$$

Thus, 16.85% of the data are less than 0.274 Kg.

EXAMPLE PROBLEM

Using the data from the preceding problem, determine the percentage of the data that fall above 0.347 kg.

Since Table A is a left-reading table, the solution to this problem requires the use of the relationship: $\text{Area}_1 + \text{Area}_2 = \text{Area}_T = 1.0000$. Therefore, Area_2 is determined and subtracted from 1.0000 to obtain Area_1.

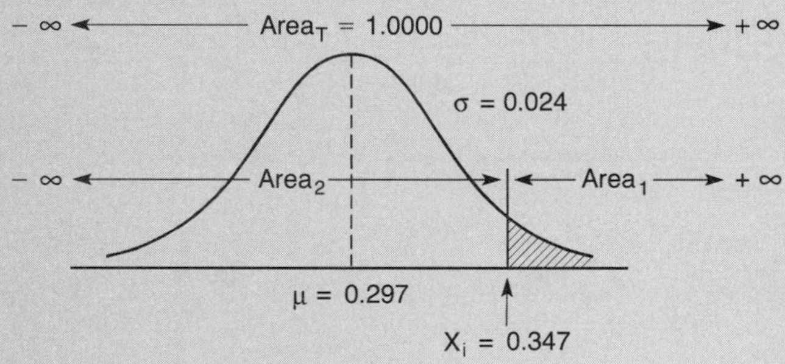

$$Z_2 = \frac{X_i - \mu}{\sigma}$$

$$= \frac{0.347 - 0.297}{0.024}$$

$$= +2.08$$

From Table A it is found that for $Z_2 = +2.08$,

$$\text{Area}_2 = 0.9812$$

$$\text{Area}_1 = \text{Area}_T - \text{Area}_2$$

$$= 1.0000 - 0.9812$$

$$= 0.0188 \text{ or } 1.88\%$$

Thus, 1.88% of the data are above 0.347 Kg.

EXAMPLE PROBLEM

A large number of tests of line voltage to home residences show a mean of 118.5 V and a population standard deviation of 1.20 V. Determine the percentage of data between 116 and 120 V.

Since Table A is a left-reading table, the solution requires that the area to the left of 116 V be subtracted from the area to the left of 120 V. The graph and calculations show the technique.

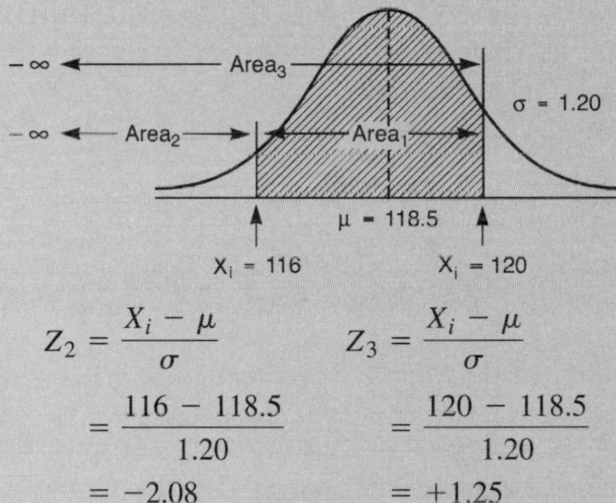

$$Z_2 = \frac{X_i - \mu}{\sigma} \qquad Z_3 = \frac{X_i - \mu}{\sigma}$$

$$= \frac{116 - 118.5}{1.20} \qquad = \frac{120 - 118.5}{1.20}$$

$$= -2.08 \qquad = +1.25$$

From Table A it is found that for $Z_2 = -2.08$, $Area_2 = 0.0188$; for $Z_3 = +1.25$, $Area_3 = 0.8944$.

$$Area_1 = Area_3 - Area_2$$

$$= 0.8944 - 0.0188$$

$$= 0.8756 \text{ or } 87.56\%$$

Thus, 87.56% of the data are between 116 and 120 V.

EXAMPLE PROBLEM

If it is desired to have 12.1% of the line voltage below 115 V, how should the mean voltage be adjusted? The dispersion is $\sigma = 1.20$ V.

The solution to this type problem is the reverse of the other problems. First 12.1%, or 0.1210, is found in the body of Table A. This gives a Z value and using the formula for Z, we can solve for the mean voltage. From Table A with $\text{Area}_1 = 0.1210$, the Z value of -1.17 is obtained.

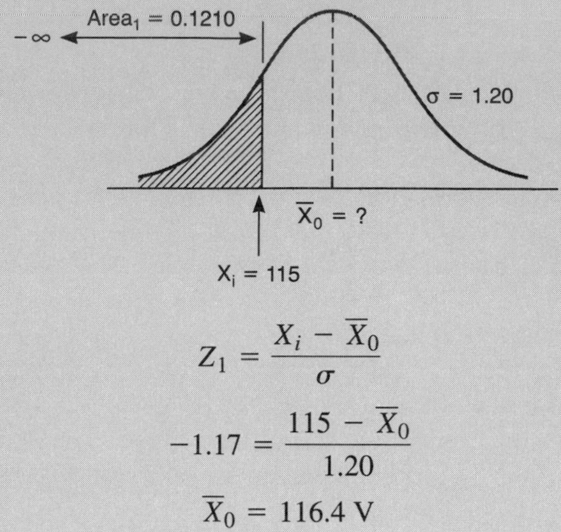

$$Z_1 = \frac{X_i - \overline{X}_0}{\sigma}$$

$$-1.17 = \frac{115 - \overline{X}_0}{1.20}$$

$$\overline{X}_0 = 116.4\text{ V}$$

Thus, the mean voltage should be centered at 116.4 V for 12.1% of the values to be less than 115 V.

Note that $\overline{X}_0$ has been substituted in the equation. The normal curve concept is based on the values of μ and σ; however, $\overline{X}_0$ and s_o can substitute provided there is some evidence that the distribution is normal. The last example problem illustrates the independence of μ and σ. A small change in the centering of the process does not affect the dispersion.

TESTS FOR NORMALITY

Because of the importance of the normal distribution, it is frequently necessary to determine if the data are normal. In using these techniques the reader is cautioned that none are 100% certain. The techniques of histogram, skewness and kurtosis, probability plots, and chi-square test are also applicable with some modification to other population distributions.

Histogram

Visual examination of a histogram developed from a large amount of data will give an indication of the underlying population distribution. If a histogram is unimodal, is symmetrical, and tapers off at the tails, normality is a definite possibility and may be sufficient information in many practical situations. The histogram of Figure 4-4 of steel shaft weight is unimodal, tapers off at the tails, and is somewhat symmetrical except for the upper tail. If a sorting operation had discarded shafts with weights above 2.575, this would explain the upper tail cutoff.

The larger the sample size, the better the judgment of normality. A minimum sample size of 50 is recommended.

Skewness and Kurtosis

Skewness and kurtosis measurements are another test of normality. From the steel shaft data of Table 4-6, we find that $a_3 = -0.11$ and $a_4 = 2.19$. These values indicate that the data are moderately skewed to the left but are close to the normal value of 0 and that the data are not as peaked as the normal distribution, which would have an a_4 value of 3.0.

These measurements tend to give the same information as the histogram. As with the histogram, the larger the sample size, the better the judgment of normality. A minimum sample size of 100 is recommended.

Probability Plots

Another test of normality is the plotting of the data on normal probability paper. This type of paper is shown in Figure 4-18. Different probability papers are used for different distributions. To illustrate the procedure we will again use the steel shaft data in its coded form. The step-by-step procedure follows.

1. *Order the data.* The data from the first column of Table 4-4 are used to illustrate the concept. Each observation is recorded as shown in Table 4-13 from the smallest to the largest. Duplicate observations are recorded as shown by the value 46.

2. *Rank the observations.* Starting at 1 for the lowest observation, 2 for the next lowest observation, and so on, rank the observations. The ranks are shown in column 2 of Table 4-13.

3. *Calculate the plotting position.* This step is accomplished using the formula

$$PP = \frac{100(i - 0.5)}{n}$$

where i = rank

PP = plotting position in percent

n = sample size

The first plotting position is $100(1 - 0.5)/22$, which is 2.3%; the others are calculated similarly and posted to Table 4-13.

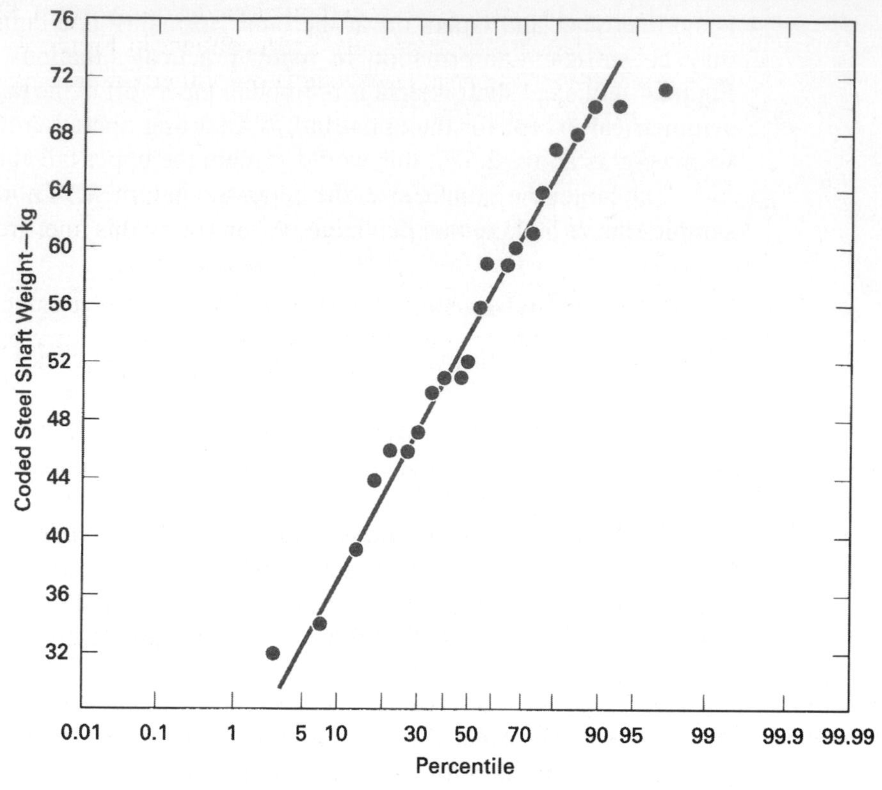

FIGURE 4-18 Probability plots of data from Table 4-13.

TABLE 4-13 Data on Steel Shaft Weight for Probability Plotting

OBSERVATION x_i	RANK i	PLOTTING POSITION	OBSERVATION x_i	RANK i	PLOTTING POSITION
32	1	2.3	56	12	52.3
34	2	6.8	59	13	56.8
39	3	11.4	59	14	61.4
44	4	15.9	60	15	65.9
46	5	20.5	61	16	70.5
46	6	25.0	64	17	75.0
47	7	29.5	67	18	79.5
50	8	34.1	68	19	84.1
51	9	38.6	70	20	88.6
51	10	43.2	70	21	93.2
52	11	47.7	71	22	97.7

4. *Label the data scale.* The coded values range from 32 to 71, so the vertical scale is labeled appropriately and is shown in Figure 4–18. The horizontal scale represents the normal curve and is preprinted on the paper.

5. *Plot the points.* The plotting position and the observation are plotted on the normal probability paper.

6. *Attempt to fit by eye a "best" line.* A clear plastic straightedge will be most helpful in making this judgment. When fitting this line, greater weight should be given to the center values than to the extreme ones.

7. *Determine normality.* This decision is one of judgment as to how close the points are to the straight line. If we disregard the extreme points at each end of the line, we can reasonably assume that the data are normally distributed.

If normality appears reasonable, additional information can be obtained from the graph. The mean is located at the 50th percentile, which gives a value of approximately 55. Standard deviation is two-fifths the difference between the 90th percentile and the 10th percentile, which would be approximately 14[(2/5)(72−38)]. We can also use the graph to determine the percent of data below, above, or between data values. For example, the percent less than 48 is approximately 31%. Even though the example problem used 22 data points with good results, a minimum sample size of 30 is recommended.

The "eye-balling" technique is a judgement decision and for the same data different people will have different slopes to their straight line. An analytical technique using the Weibull distribution effectively overcomes this limitation. It is similiar to normal probability plots and can use spread sheet software such as EXCEL.[7] The Weibull distribution is discussed in the chapter on reliability wherein normality is determined if the slope parameter, β, is approximately equal to 3.4.

Chi-Square Goodness of Fit Test

The chi-square (χ^2) test is another technique of determining if the sample data fits a normal distribution or other distribution.

This test uses the equation

$$\chi^2 = \sum_{i=1}^{k} \frac{(O_i - E_i)^2}{E_i}$$

where χ^2 = Chi−squared

O_i = Observed value in a cell

E_i = Expected value for a cell

[7]EXCEL has a normal probability plot available under regression analysis; however the X-axis is linear and the straight line interpretation is not valid.

The expected value is determined from the normal distribution or from any distribution. After χ^2 is determined it is compared to the chi-squared distribution to determine whether the observed data is from the expected distribution. An example is given in the diskette. While the χ^2 test is the best method of determined normalcy, it does require a minimum sample size of 125.

It is important for the analyst to understand that none of these techniques prove that the data are normally distributed. We can only conclude that there is no evidence that the data cannot be treated as if they were normally distributed.

COMPUTER PROGRAM

The EXCEL software in the diskette inside the back cover will solve for the histogram with descriptive statistics and the chi-squared test. Their file names are *histogram* and *chi-squared*. It should be noted that the histogram is an add-in under data analysis for the tools menu.

PROBLEMS

1. Round the following numbers to two decimal places.
   (a) 0.862
   (b) 0.625
   (c) 0.149
   (d) 0.475

2. Find the g.p.e. of the following numbers.
   (a) 8.24
   (b) 522
   (c) 6.3×10^2
   (d) 0.02

3. Find the relative error of the numbers in Problem 2.

4. Perform the operation indicated and leave the answer in the correct number of significant figures.
   (a) (34.6)(8.20)
   (b) (0.035)(635)
   (c) 3.8735/6.1
   (d) 5.362/6 (6 is a counting number)
   (e) 5.362/6 (6 is not a counting number)

5. Perform the operation indicated and leave the answer in the correct number of significant figures.
   (a) $64.3 + 2.05$
   (b) $381.0 - 1.95$
   (c) $8.652 - 4$ (4 is not a counting number)
   (d) $8.652 - 4$ (4 is a counting number)
   (e) $6.4 \times 10^7 + 24.32$

6. In his last 70 games a professional basketball player made the following scores:

10	17	9	17	18	20	16
7	17	19	13	15	14	13
12	13	15	14	13	10	14
11	15	14	11	15	15	16
9	18	15	12	14	13	14
13	14	16	15	16	15	15
14	15	15	16	13	12	16
10	16	14	13	16	14	15
6	15	13	16	15	16	16
12	14	16	15	16	13	15

   (a) Make a tally sheet in ascending order.
   (b) Using the data above, construct a histogram.

7. A company that fills bottles of shampoo tries to maintain a specific weight of the product. The table gives the weight of 110 bottles that were checked at random intervals. Make a tally of these weights and construct a frequency histogram. (Weight is in kilograms.)

6.00	5.98	6.01	6.01	5.97	5.99	5.98	6.01	5.99	5.98	5.96
5.98	5.99	5.99	6.03	5.99	6.01	5.98	5.99	5.97	6.01	5.98
5.97	6.01	6.00	5.96	6.00	5.97	5.95	5.99	5.99	6.01	6.00
6.01	6.03	6.01	5.99	5.99	6.02	6.00	5.98	6.01	5.98	5.99
6.00	5.98	6.05	6.00	6.00	5.98	5.99	6.00	5.97	6.00	6.00
6.00	5.98	6.00	5.94	5.99	6.02	6.00	5.98	6.02	6.01	6.00
5.97	6.01	6.04	6.02	6.01	5.97	5.99	6.02	5.99	6.02	5.99
6.02	5.99	6.01	5.98	5.99	6.00	6.02	5.99	6.02	5.95	6.02
5.96	5.99	6.00	6.00	6.01	5.99	5.96	6.01	6.00	6.01	5.98
6.00	5.99	5.98	5.99	6.03	5.99	6.02	5.98	6.02	6.02	5.97

8. Listed next are 125 readings obtained in a hospital by motion-and-time-study analyst who took 5 readings each day for 25 days. Construct a tally sheet. Prepare a table showing cell midpoints, cell boundaries, and observed frequencies. Plot a frequency histogram.

DAY	DURATION OF OPERATION TIME (MIN)				
1	1.90	1.93	1.95	2.05	2.20
2	1.76	1.81	1.81	1.83	2.01
3	1.80	1.87	1.95	1.97	2.07
4	1.77	1.83	1.87	1.90	1.93
5	1.93	1.95	2.03	2.05	2.14
6	1.76	1.88	1.95	1.97	2.00
7	1.87	2.00	2.00	2.03	2.10
8	1.91	1.92	1.94	1.97	2.05
9	1.90	1.91	1.95	2.01	2.05
10	1.79	1.91	1.93	1.94	2.10
11	1.90	1.97	2.00	2.06	2.28
12	1.80	1.82	1.89	1.91	1.99
13	1.75	1.83	1.92	1.95	2.04
14	1.87	1.90	1.98	2.00	2.08
15	1.90	1.95	1.95	1.97	2.03
16	1.82	1.99	2.01	2.06	2.06
17	1.90	1.95	1.95	2.00	2.10
18	1.81	1.90	1.94	1.97	1.99
19	1.87	1.89	1.98	2.01	2.15
20	1.72	1.78	1.96	2.00	2.05
21	1.87	1.89	1.91	1.91	2.00
22	1.76	1.80	1.91	2.06	2.12
23	1.95	1.96	1.97	2.00	2.00
24	1.92	1.94	1.97	1.99	2.00
25	1.85	1.90	1.90	1.92	1.92

9. The relative strength of 150 silver solder welds are tested, and the results are given in the following table. Tally these figures and arrange them in a frequency distribution. Determine the cell interval and the approximate number of cells. Make a table showing cell midpoints, cell boundaries, and observed frequencies. Plot a frequency histogram.

1.5	1.2	3.1	1.3	0.7	1.3
0.1	2.9	1.0	1.3	2.6	1.7
0.3	0.7	2.4	1.5	0.7	2.1
3.5	1.1	0.7	0.5	1.6	1.4
1.7	3.2	3.0	1.7	2.8	2.2
1.8	2.3	3.3	3.1	3.3	2.9
2.2	1.2	1.3	1.4	2.3	2.5
3.1	2.1	3.5	1.4	2.8	2.8
1.5	1.9	2.0	3.0	0.9	3.1
1.9	1.7	1.5	3.0	2.6	1.0

2.9	1.8	1.4	1.4	3.3	2.4
1.8	2.1	1.6	0.9	2.1	1.5
0.9	2.9	2.5	1.6	1.2	2.4
3.4	1.3	1.7	2.6	1.1	0.8
1.0	1.5	2.2	3.0	2.0	1.8
2.9	2.5	2.0	3.0	1.5	1.3
2.2	1.0	1.7	3.1	2.7	2.3
0.6	2.0	1.4	3.3	2.2	2.9
1.6	2.3	3.3	2.0	1.6	2.7
1.9	2.1	3.4	1.5	0.8	2.2
1.8	2.4	1.2	3.7	1.3	2.1
2.9	3.0	2.1	1.8	1.1	1.4
2.8	1.8	1.8	2.4	2.3	2.2
2.1	1.2	1.4	1.6	2.4	2.1
2.0	1.1	3.8	1.3	1.3	1.0

10. Using the data of Problem 6, construct:
(a) A relative frequency histogram
(b) A cumulative frequency histogram
(c) A relative cumulative frequency histogram

11. Using the data of Problem 7, construct:
(a) A relative frequency histogram
(b) A cumulative frequency histogram
(c) A relative cumulative frequency histogram

12. Using the data of Problem 8, construct:
(a) A relative frequency histogram
(b) A cumulative frequency histogram
(c) A relative cumulative frequency histogram

13. Using the data of Problem 9, construct:
(a) A relative frequency histogram
(b) A cumulative frequency histogram
(c) A relative cumulative frequency histogram

14. Construct a bar graph of the data in:
(a) Problem 6
(b) Problem 7

15. Using the data of Problem 8, construct:
(a) A polygon
(b) An ogive

16. Using the data of Problem 9, construct:
(a) A polygon
(b) An ogive

17. An electrician testing the incoming line voltage for a residential house obtains 5 readings: 115, 113, 121, 115, 116. What is the average?

18. An employee makes 8 trips to load a trailer. If the trip distances in meters are 25.6, 24.8, 22.6, 21.3, 19.6, 18.5, 16.2, and 15.5, what is the average?

19. Tests of noise ratings at prescribed locations throughout a large stamping mill are given in the frequency distribution below. Noise is measured in decibels. Determine the average.

CELL MIDPOINT	FREQUENCY
148	2
139	3
130	8
121	11
112	27
103	35
94	43
85	33
76	20
67	12
58	6
49	4
40	2

20. The weight of 65 castings in kilograms is distributed as follows:

CELL MIDPOINT	FREQUENCY
3.5	6
3.8	9
4.1	18
4.4	14
4.7	13
5.0	5

Determine the average.

21. Destructive tests on the life of an electronic component were conducted on 2 different occasions. On the first occasion 3 tests had a mean of 3320 h; on the second occasion 2 tests had a mean of 3180 h. What is the weighted average?

22. The average height of 24 students in Section 1 of a course in quality control is 1.75 m; the average height of 18 students in Section 2 of quality control is 1.79 m; and the average height of 29 students in Section 3 of quality control is 1.68 m. What is the average height of the students in the 3 sections of quality control?

23. Determine the median of the following numbers.
   (a) 22, 11, 15, 8, 18
   (b) 35, 28, 33, 38, 43, 36

24. Determine the median for the following:
   (a) The frequency distribution of Problem 8
   (b) The frequency distribution of Problem 9
   (c) The frequency distribution of Problem 19
   (d) The frequency distribution of Problem 20
   (e) The frequency distribution of Problem 30
   (f) The frequency distribution of Problem 32

25. Given the following series of numbers, determine the mode.
   (a) 50, 45, 55, 55, 45, 50, 55, 45, 55
   (b) 89, 87, 88, 83, 86, 82, 84
   (c) 11, 17, 14, 12, 12, 14, 14, 15, 17, 17

26. Determine the modal cell of the data in:
   (a) Problem 6
   (b) Problem 7
   (c) Problem 8
   (d) Problem 9
   (e) Problem 19
   (f) Problem 20

27. Determine the range for each set of numbers.
   (a) 16, 25, 18, 17, 16, 21, 14
   (b) 45, 39, 42, 42, 43
   (c) The data in Problem 6
   (d) The data in Problem 7

28. Frequency tests of a brass rod 145 cm long give values of 1200, 1190, 1205, 1185, and 1200 vibrations per second. What is the sample standard deviation?

29. Four readings of the thickness of the paper in this textbook are 0.076 mm, 0.082 mm, 0.073 mm, and 0.077 mm. Determine the sample standard deviation.

30. The frequency distribution given here shows the percent of organic sulfur in Illinois No. 5 coal. Determine the sample standard deviation.

CELL MIDPOINT (%)	FREQUENCY (NUMBER OF SAMPLES)
0.5	1
0.8	16
1.1	12

CELL MIDPOINT (%)	FREQUENCY (NUMBER OF SAMPLES)
1.4	10
1.7	12
2.0	18
2.3	16
2.6	3

31. Determine the sample standard deviation for the following.
(a) The data of Problem 9
(b) The data of Problem 19

32. Determine the average and sample standard deviation for the frequency distribution of the number of inspections per day as follows:

CELL MIDPOINT	FREQUENCY
1000	6
1300	13
1600	22
1900	17
2200	11
2500	8

33. Using the data of Problem 19, construct:
(a) A polygon
(b) An ogive

34. Using the data of Problem 20, construct:
(a) A polygon
(b) An ogive

35. Using the data of Problem 30, construct:
(a) A polygon
(b) An ogive

36. Using the data of Problem 32, construct:
(a) A polygon
(b) An ogive

37. Using the data of Problem 19, construct:
(a) A histogram
(b) A relative frequency histogram
(c) A cumulative frequency histogram
(d) A relative cumulative frequency histogram

38. Using the data of Problem 20, construct:
(a) A histogram

(b) A relative frequency histogram
(c) A cumulative frequency histogram
(d) A relative cumulative frequency histogram

39. Using the data of Problem 30, construct:
(a) A histogram
(b) A relative frequency histogram
(c) A cumulative frequency histogram
(d) A relative cumulative frequency histogram

40. Using the data of Problem 32, construct:
(a) A histogram
(b) A relative frequency histogram
(c) A cumulative frequency histogram
(d) A relative cumulative frequency histogram

41. Determine the skewness and kurtosis of:
(a) Problem 6
(b) Problem 7
(c) Problem 8
(d) Problem 9
(e) Problem 20
(f) Problem 32

42. If the maximum allowable noise is 134.5 db, what percent of the data of Problem 19 is above that value?

43. Evaluate the histogram of Problem 20, where the specifications are 4.25 ± 0.60 kg.

44. A utility company will not use coal with a sulfur content of more than 2.25%. Based on the histogram of Problem 30, what percent of the coal is in that category?

45. The population mean of a company's racing bicycles is 9.07 kg (20.0 lb) with a population standard deviation of 0.40 kg. If the distribution is approximately normal, determine (a) the percentage of bicycles less than 8.30 kg, (b) the percentage of bicycles greater than 10.00 kg, and (c) the percentage of bicycles between 8.00 and 10.10 kg.

46. If the mean time to clean a motel room is 16.0 min and the standard deviation is 1.5 min, what percentage of the rooms will take less than 13.0 min to complete? What percentage of the rooms will take more than 20.0 min to complete? What percentage of the rooms will take between 13.0 and 20.5 min to complete? The data are normally distributed.

47. A cold-cereal manufacturer wants 1.5% of the product to be below the weight specification of 0.567 kg (1.25 lb). If the data are normally distributed and the standard deviation of the cereal filling machine is 0.018 kg, what mean weight is required?

48. In the precision grinding of a complicated part, it is more economical to rework the part than to scrap it. Therefore, it is decided to establish the rework percentage at 12.5%. Assuming normal distribution of the data, a standard deviation of 0.01 mm, and an upper specification limit of 25.38 mm (0.99 in.), determine the process center.

49. Using the information of Problem 41, what is your judgment concerning the normality of the distribution in each of the following?
(a) Problem 6
(b) Problem 7
(c) Problem 8
(d) Problem 9
(e) Problem 20
(f) Problem 32

50. Using normal probability paper, determine (judgment) the normality of the distribution of the following.
(a) Second column of Table 4-4
(b) First three columns of Problem 7
(c) Second column of Problem 8

51. Using the software determine the descriptive statistics and histogram using the data of the following.
(a) Problem 6
(b) Problem 7
(c) Problem 8
(d) Problem 9
(e) Problem 19
(f) Problem 20
(g) Problem 30
(h) Problem 32

52. Using the software with the file name *Weibull* determine the normality of Problem 50(a), 50(b), and 50(c).

53. Using the chi-squared software determine the normality of the distribution of the following.
(a) Problem 19
(b) Problem 20
(c) Problem 30
(d) Problem 32

54. Obtain data from production or the laboratory and determine the descriptive statistics and histogram.

5 CONTROL CHARTS FOR VARIABLES[1]

INTRODUCTION

Variation

One of the axioms or truisms of manufacturing is that no two objects are ever made exactly alike. In fact, the variation concept is a law of nature in that no two natural items in any category are the same. The variation may be quite large and easily noticeable, such as the height of human beings, or the variation may be very small, such as the weight of fiber-tipped pens or the shape of snowflakes. When variations are very small, it may appear that items are identical; however, precision instruments will show differences. If two items appear to have the same measurement, it is due to the limits of our measuring instruments. As measuring instruments have become more refined, variation has continued to exist; only the increment of variation has changed. The ability to measure variation is necessary before it can be controlled.

[1]The information in this chapter is based on ANSI/ASQC B1-B3—1996.

There are three categories of variations in piece part production:

1. *Within-piece variation.* This type of variation is illustrated by the surface roughness of a piece wherein one portion of the surface is rougher than another portion, or the width of one end of a keyway varies from the other end.

2. *Piece-to-piece variation.* This type of variation occurs among pieces produced at the same time. Thus, the light intensity of four consecutive light bulbs produced from a machine will be different.

3. *Time-to-time variation.* This type of variation is illustrated by the difference in product produced at different times of the day. Thus, a service given early in the morning would be different from that given later in the day, or as a cutting tool wears, the cutting characteristics change.

Categories of variation for other types of processes such as a continuous chemical process will not be exactly the same; however, the concept will be similar.

Variation is present in every process due to a combination of the equipment, materials, environment, and operator. The first source of variation is the *equipment.* This source includes tool wear, machine vibration, workholding-device positioning, and hydraulic and electrical fluctuations. When all these variations are put together, there is a certain capability or precision within which the equipment operates. Even supposedly identical machines will have different capabilities, and this fact becomes a very important consideration when scheduling the manufacture of critical parts.

The second source of variation is the *material.* Since variation occurs in the finished product, it must also occur in the raw material (which was someone else's finished product). Such quality characteristics as tensile strength, ductility, thickness, porosity, and moisture content can be expected to contribute to the overall variation in the final product.

A third source of variation is the *environment.* Temperature, light, radiation, electrostatic discharge, particle size, pressure, and humidity can all contribute to variation in the product. In order to control this source, products are sometimes manufactured in white rooms. Experiments are conducted in outer space to learn more about the effect of the environment on product variation.

A fourth source is the *operator.* This source of variation includes the method by which the operator performs the operation. The operator's physical and emotional well-being also contribute to the variation. A cut finger, a twisted ankle, a personal problem, or a headache can make an operator's quality performance very. An operator's lack of understanding of equipment and material variations due to lack of training may lead to frequent machine adjustments, thereby compounding the variability. As our equipment has become more automated, the operator's effect on variation has lessened.

The above four sources account for the true variation. There is also a reported variation, which is due to the *inspection* activity. Faulty inspection equipment, the incorrect application of a quality standard, or too heavy a pressure on a micrometer can be the cause of the incorrect reporting of variation. In general, variation due to inspection should be one-tenth of the four other sources of variations. It should be noted that three of these sources are present in the inspection activity—an inspector, inspection equipment, and the environment.

As long as these sources of variation fluctuate in a natural or expected manner, a stable pattern of many *chance causes* (random causes) of variation develops. Chance causes of variation are inevitable. Because they are numerous and individually of relatively small importance, they are difficult to detect or identify. Those causes of variation that are large in magnitude, and therefore readily identified, are classified as *assignable causes*.[2] When only chance causes are present in a process, the process is considered to be in a state of statistical control. It is stable and predictable. However, when an assignable cause of variation is also present, the variation will be excessive, and the process is classified as out of control or beyond the expected natural variation.

The Control Chart Method

In order to indicate when observed variations in quality are greater than could be left to chance, the control chart method of analysis and presentation of data is used. The control chart method for variables is a means of visualizing the variations that occur in the central tendency and dispersion of a set of observations. It is a graphical record of the quality of a particular characteristic. It shows whether or not the process is in a stable state.

An example of a control chart is given in Figure 5-1. This particular chart is referred to as an $\overline{X}$ chart and is used to record the variation in the average value of samples. Another chart, such as the *R* chart (range), would have also served for explanation purposes. The horizontal axis is labeled "Subgroup Number," which identifies a particular sample consisting of a fixed number of observations. These subgroups are in order, with the first one inspected being 1 and the last one inspected being 14. The vertical axis of the graph is the variable, which in this particular case is weight measured in kilograms.

Each small solid circle represents the average value within a subgroup. Thus, subgroup number 5 consists of, say, four observations. 3.46, 3.49, 3.45, and 3.44, and their average is 3.46 kg. This value is the one posted on the chart for subgroup number 5. Averages are used on control charts rather than individual observations because average values will indicate a change in variation much faster.[3] Also, with

[2]Dr. W. Edwards Deming uses the words *common* and *special* for chance and assignable.

[3]For a proof of this statement, see J. M. Juran, ed., *Quality Control Handbook*, 4th ed. (New York: McGraw-Hill Book Company, 1988), Sec. 24, p. 10.

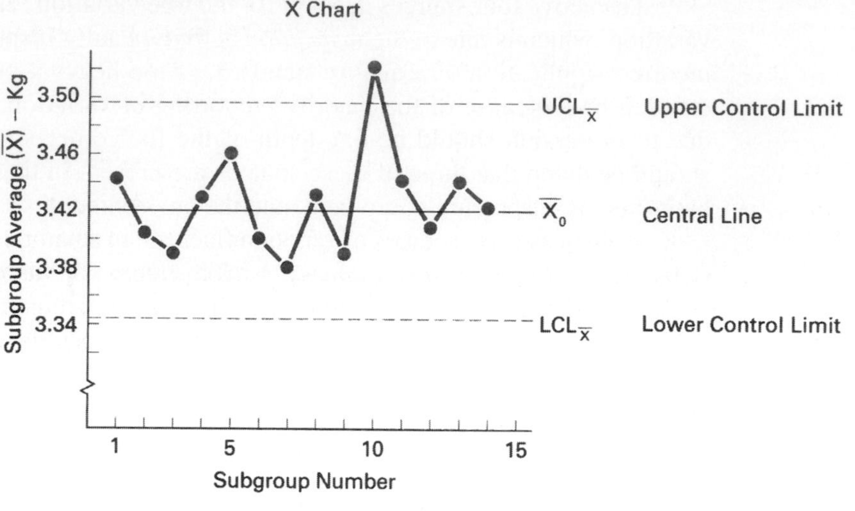

FIGURE 5-1 Example of a control chart.

two or more observations in a sample, a measure of the dispersion can be obtained for a particular subgroup.

The solid line in the center of the chart can have three different interpretations depending on the available data. First, it can be the average of the plotted points, which in the case of an $\overline{X}$ chart is the average of the averages or "X-double bar," $\overline{\overline{X}}$. Second, it can be a standard or reference value, $\overline{X}_0$, based on representative prior data, an economic value based on production costs or service needs, or an aimed-at value based on specifications. Third, it can be the population mean, μ, if that value is known.

The two dashed outer lines are the upper and lower control limits. These limits are established to assist in judging the significance of the variation in the quality of the product. Control limits are frequently confused with *specification limits,* which are the permissible limits of a quality characteristic of each *individual* unit of a product. However, *control limits* are used to evaluate the variations in quality from subgroup to subgroup. Therefore, for the $\overline{X}$ chart, the control limits are a function of the subgroup averages. A frequency distribution of the subgroup averages can be determined with its corresponding average and standard deviation. The control limits are usually, established at ± 3 standard deviations from the central line. One recalls, from the discussion of the normal curve, that the number of items between $+3\sigma$ and -3σ equals 99.73%. Therefore, it is expected that over 997 times out of 1000, the subgroup values will fall between the upper and lower limits, and when this occurs, the process is considered to be in control. When a subgroup value falls outside the limits, the process is considered to be out of control and an assignable cause for the variation is present. Subgroup number 10 in Figure 5-1 is beyond the upper control limit; therefore, there has been a change in the stable nature of the process, causing the out-of-control point.

In practice, control charts are posted at individual machines or work centers to control a particular quality characteristic. Usually, an $\overline{X}$ chart for the central tendency and an R chart for the dispersion are used together. An example of this dual charting is illustrated in Figure 5-2, which shows a method of charting and reporting inspection results for rubber durometers. At work center number 365-2 at 8:30 A.M., the operator selects four items for testing, and records the observations of 55, 52, 51, and 53 in the rows marked X_1, X_2, X_3, and X_4, respectively. A subgroup average value of 52.8 is obtained by summing the observation and dividing by 4, and the range value of 4 is obtained by subtracting the low value, 51, from the high value, 55. The operator places a small solid circle at 52.8 on the $\overline{X}$ chart and a small solid circle at 4 on the R chart and then proceeds with his other duties.

The frequency with which the operator inspects a product at a particular machine or work center is determined by the quality of the product. When the process is in control and no difficulties are being encountered, fewer inspections may be required, and, conversely, when the process is out of control or during start-up, more inspections may be needed. The inspection frequency at a machine or work center can also be determined by the amount of time that must be spent on non-inspection activities. In the example problem, the inspection frequency appears to be every 60 or 65 minutes.

At 9:30 A.M. the operator performs the activities for subgroup 2 in the same manner as for subgroup 1. It is noted that the range value of 7 falls right on the upper control limit. Whether to consider this in control or out of control would be a matter of company policy. It is suggested that it be classified as in control and a cursory examination for an assignable cause be conducted by the operator. A plotted point that falls on the control limit will be a rare occurrence.

The inspection results for subgroup 2 show that the third observation, X_3, has a value of 57, which exceeds the upper control limit. The reader is cautioned to remember the earlier discussion on control limits and specifications. In other words, the 57 value is an individual observation and does not relate to the control limits. Therefore, the fact that an individual observation is greater than or less than a control limit is meaningless.

Subgroup 4 has an average value of 44, which is less than the lower control limit of 45. Therefore, subgroup 4 is out of control, and the operator will report this fact to the departmental supervisor. The operator and supervisor will then look for an assignable cause and, if possible, take corrective action. Whatever corrective action is taken will be noted by the operator on the $\overline{X}$ and R chart or on a separate

$\overline{X}$ AND R CHART

Work Center Number 365–2
Quality Characteristic Durometer Date 3/6

Time	8:30 AM	9:30 AM	10:40 AM	11:50 AM	1:30 PM									
Subgroup	1	2	3	4	5	6	7	8	9	10	11	12	13	14
X_1	55	51	48	45	53									
X_2	52	52	49	43	50									
X_3	51	57	50	45	48									
X_4	53	50	49	43	50									
Sum	211	210	196	176	201									
$\overline{X}$	52.8	52.5	49	44	50.2									
R	4	7	2	2	5									

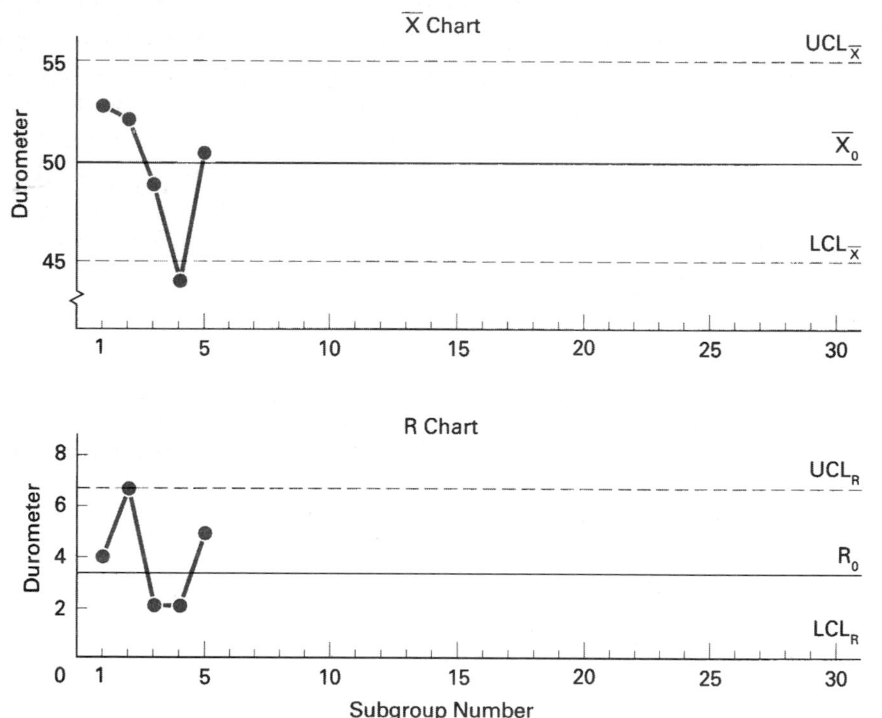

FIGURE 5-2 Example of a method of reporting inspection results.

form. The control chart indicates when and where trouble has occurred; the identification and elimination of the difficulty is a production problem. Ideally, the control chart should be maintained by the operator provided time is available and proper training has been given. When the operator cannot maintain the chart, then it is maintained by quality control.

A control chart is a statistical tool that distinguishes between natural and unnatural variation as shown in Figure 5-3. Unnatural variation is the result of assignable causes. It usually, but not always, requires corrective action by people close to the process such as operators, technicians, clerks, maintenance workers, and first-line supervisors.

Natural variation is the result of chance causes. It requires management intervention to achieve quality improvement. In this regard, between 80% and 85% of the quality problems are due to management or the system, and 15% to 20% are due to operations.

The control chart is used to keep a continuing record of a particular quality characteristic. It is a picture of the process over time. When the chart is completed, it is replaced by a fresh chart, and the completed chart is stored in an office file. The chart is used to improve the process quality, to determine the process capability, to

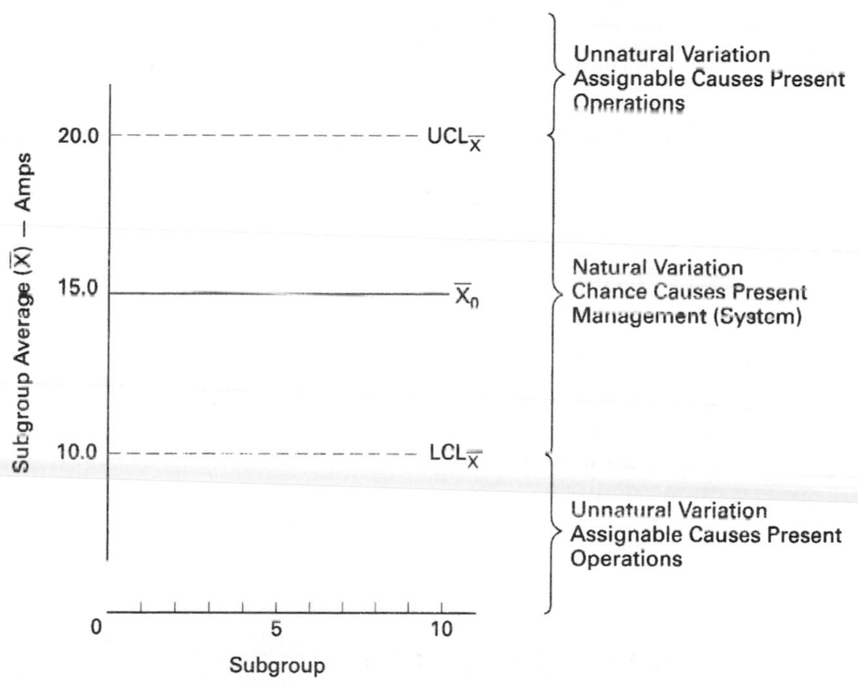

FIGURE 5-3 Natural and unnatural causes of variation.

help determine effective specifications, to determine when to leave the process alone and when to make adjustments, and to investigate causes of unacceptable or marginal quality.

Objectives of Variable Control Charts

Variable control charts provide information:

1. *For quality improvement.* Having a variable control chart merely because it indicates that there is a quality control program is missing the point. A variable control chart is an excellent technique for achieving quality improvement.

2. *To determine the process capability.* The true process capability can be achieved only after substantial quality improvement has been achieved. During the quality improvement cycle, the control chart will indicate that no further improvement is possible without a large dollar expenditure. At that point the true process capability is obtained.

3. *For decisions in regard to product specifications.* Once the true process capability is obtained, effective specifications can be determined. If the process capability is ± 0.003, then specifications of ± 0.004, are realistically obtainable by operating personnel.

4. *For current decisions in regard to the production process.* First a decision is needed to judge whether control exists. If not, the control chart is used to attain control. Once control is obtained, the control chart is used to maintain control. Thus, the control chart is used to decide when a natural pattern of variation occurs and the process should be left alone, and when an unnatural pattern of variation is occurring, which requires action to find and eliminate the assignable causes.

   In this regard, operating personnel are giving a quality performance as long as the plotted points are within the control limits. If this performance is not satisfactory, the solution is the responsibility of the system rather than the operator.

5. *For current decisions in regard to recently produced items.* Thus, the control chart is used as one source of information to help decide whether an item or items should be released to the next phase of the sequence or some alternative disposition made, such as sorting and repairing.

These purposes are frequently dependent on each other. For example, quality improvement is needed prior to determining the true process capability, which is needed prior to determining effective specifications. Control charts for variables should be established to achieve a particular purpose. Their use should be discontinued when the purpose has been achieved or their use continued with the inspection substantially reduced.

CONTROL CHART TECHNIQUES

Introduction

In order to establish a pair of control charts for the average ($\overline{X}$) and the range (R), it is desirable to follow a set procedure. The steps in this procedure are as follows:

1. Select the quality characteristic.
2. Choose the rational subgroup.
3. Collect the data.
4. Determine the trial central line and control limits.
5. Establish the revised central line and control limits.
6. Achieve the objective.

The procedure presented in this section relates to an $\overline{X}$ and R chart. Information on an s chart is also presented.

Select the Quality Characteristic

The variable that is chosen for an $\overline{X}$ and R chart must be a quality characteristic that is measurable and can be expressed in numbers. Quality characteristics that can be expressed in terms of the seven basic units—length, mass, time, electrical current, temperature, substance, or luminous intensity—are appropriate, as well as any of the derived units, such as power, velocity, force, energy, density, and pressure.

Those quality characteristics affecting the performance of the product would normally be given first attention. These may be a function of the raw materials, component parts, subassemblies, or finished parts. In other words, give high priority to the selection of those characteristics that are giving difficulty in terms of production problems and/or cost. An excellent opportunity for cost savings is frequently selected where spoilage and rework costs are high. A Pareto analysis is also useful to establish priorities. Another possibility occurs where destructive testing is used to inspect a product.

In any manufacturing plant there are a large number of variables that make up a product. It is, therefore, impossible to place $\overline{X}$ and R charts on all variables, and a judicious selection of those quality characteristics is required. Since all variables can be treated as attributes, an attribute control chart (see Chapter 8) can also be used to achieve quality improvement.

Choose the Rational Subgroup

As previously mentioned, the data that are plotted on the control chart consist of groups of items that are called rational subgroups. It is important to understand that data collected in a random manner do *not* qualify as rational. A rational subgroup is one in which the variation within the group is due only to chance causes. This within-subgroup variation is used to determine the control limits. Variation between subgroups is used to evaluate long-term stability. There are two schemes for selecting the subgroup samples:

1. The first scheme is to select the subgroup samples from product produced at one instant of time or as close to that instant as possible. Four consecutive parts from a machine or four parts from a tray of recently produced parts would be an example of this subgrouping technique. The next subgroup sample would be similar but for product produced at a later time—say, 1 h later. This scheme is called the instant-time method.

2. The second scheme is to select product produced over a period of time so that it is representative of all the product. For example, an inspector makes a visit to a circuit breaker assembling process once every hour. The subgroup sample of, say, four is randomly selected from all the circuit breakers produced in the previous hour. On his next visit, the subgroup is selected from the product produced between visits, and so forth. This scheme is called the period-of-time method.

In comparing the two schemes, the instant-time method will have a minimum variation *within* a subgroup and a maximum variation *among* subgroups. The period-of-time method will have a maximum variation *within* a subgroup and a minimum variation *among* subgroups. Some numerical values may help to illustrate this difference. Thus, for the instant-time method, subgroup average values ($\overline{X}$'s) could be from, say, 26 to 34 with subgroup range values (R's) from 0 to 4; whereas for the period-of-time method, the subgroup average values ($\overline{X}$'s) would vary from 28 to 32 with the subgroup range values (R's) from 0 to 8.

The instant-time method is the one most commonly used since it provides a particular time reference for determining assignable causes. It also provides a more sensitive measure of changes in the process average. Since all the values are close together, the variation will most likely be due to chance causes and thereby meet the rational subgroup criteria.

The advantage of the period-of-time method is that it provides better overall results, and, therefore, quality reports will present a more accurate picture of the quality. It is also true that because of process limitations this method may be the only practical method of obtaining the subgroup samples. Assignable causes of variation *may* be present in the subgroup, which will make it difficult to ensure that a rational subgroup is present.

In rare situations, it may be desirable to use both subgrouping methods. When this occurs, two charts with different control limits are required.

Regardless of the scheme used to obtain the subgroup, the lots from which the subgroups are chosen must be homogeneous. By "homogeneous" is meant that the pieces in the lot are as alike as possible—same machine, same operator, same mold cavity, and so on. Similarly, a fixed quantity of material, such as that produced by one tool until it wears out and is replaced or resharpened, should be a homogeneous lot. Homogeneous lots can also be designated by equal time intervals, since this technique is easy to organize and administer. No matter how the lots are designated, the items in any one subgroup should have been produced under essentially the same conditions.

Decisions on the size of the sample or subgroup require a certain amount of empirical judgment; however, some helpful guidelines can be given:

1. As the subgroup size increases, the control limits become closer to the central value, which makes the control chart more sensitive to small variations in the process average.
2. As the subgroup size increases, the inspection cost per subgroup increases. Does the increased cost of larger subgroups justify the greater sensitivity?
3. When destructive testing is used and the item is expensive, a small subgroup size of 2 or 3 is necessary, since it will minimize the destruction of expensive product.
4. Because of the ease of computation a sample size of 5 is quite common in industry; however, when inexpensive electronic hand calculators are used this reason is no longer valid.
5. From a statistical basis a distribution of subgroup averages, $\overline{X}$'s, are nearly normal for subgroups of 4 or more even when the samples are taken from a nonnormal population. Proof of this statement is made later in the chapter.
6. When the subgroup size exceeds 10, the s chart should be used instead of the R chart for the control of the dispersion.

There is no rule for the frequency of taking subgroups, but the frequency should be often enough to detect process changes. The inconveniences of the factory or office layout and the cost of taking subgroups must be balanced with the value of the data obtained. In general, it is best to sample quite often at the beginning and reduce the sampling frequency when the data permit.

The use of Table 5-1, which was obtained from ANSI/ASQ Z1.9—1993, can be a valuable aid in making judgments on the amount of sampling required. If a process is expected to produce 4000 pieces per day, then 75 total inspections are suggested. Therefore, with a subgroup size of four, 19 subgroups would be a good starting point.

The precontrol rule (see Chapter 6) for the frequency of sampling could also be used. It is based on how often the process is adjusted. If the process is adjusted every hour, then sampling should occur every 10 minutes; if the process is adjusted every 2 hours, then sampling should occur every 20 minutes; if the process is adjusted every 3 hours, then sampling should occur every 30 minutes; and so forth.

TABLE 5-1 Sample Sizes (From ANSI/ASQ Z1.9—1993, Normal Inspection, Level II)

LOT SIZE	SAMPLE SIZE
91–150	10
151–280	15
281–400	20
401–500	25
501–1,200	35
1,201–3,200	50
3,201–10,000	75
10,001–35,000	100
35,001–150,000	150

The frequency of taking a subgroup is expressed in terms of the percent of items produced or in terms of a time interval. In summary, the selection of the rational subgroup is made in such a manner that only chance causes are present in the subgroup.

Collect the Data

The next step is to collect the data. This step can be accomplished using the type of form shown in Figure 5-2, wherein the data are recorded in a vertical fashion. By recording the measurements one below the other, the summing operation for each subgroup is somewhat easier. An alternative method of recording the data is shown in Table 5-2, wherein the data are recorded in a horizontal fashion. The particular method makes no difference when an electronic hand calculator is available. For illustrative purposes, the latter method will be used.

Assuming that the quality characteristic and the plan for the rational subgroup have been selected, a technician can be assigned the task of collecting the data as part of his or her normal duties. The first-line supervisor and the operator should be informed of the technician's activities; however, no charts or data are posted at the work center at this time.

Because of difficulty in the assembly of a gear hub to a shaft using a key and keyway, the project team recommends using an $\overline{X}$ and R chart. The quality characteristic is the shaft keyway depth of 6.35 mm (0.250 in.). Using a rational subgroup of four, a technician obtains five subgroups per day for 5 days. The samples are measured, the subgroup average $(\overline{X})$ and range R are calculated, and the results are recorded on the form. Additional recorded information includes the date, time, and any comments pertaining to the process. For simplicity, individual measurements are coded from 6.00 mm. Thus, the first measurement of 6.35 is recorded as 35.

TABLE 5-2 Data on the Depth of the Keyway (millimeters)[a]

SUBGROUP NUMBER	DATE	TIME	MEASUREMENTS X_1	X_2	X_3	X_4	AVERAGE $\overline{X}$	RANGE R	COMMENT
1	12/26	8:50	35	40	32	37	6.36	0.08	
2		11:30	46	37	36	41	6.40	0.10	
3		1:45	34	40	34	36	6.36	0.06	
4		3:45	69	64	68	59	6.65	0.10	New, temporary
5		4:20	38	34	44	40	6.39	0.10	operator
6	12/27	8:35	42	41	43	34	6.40	0.09	
7		9:00	44	41	41	46	6.43	0.05	
8		9:40	33	41	38	36	6.37	0.08	
9		1:30	48	44	47	45	6.46	0.04	
10		2:50	47	43	36	42	6.42	0.11	
11	12/28	8:30	38	41	39	38	6.39	0.03	
12		1:35	37	37	41	37	6.38	0.04	
13		2:25	40	38	47	35	6.40	0.12	
14		2:35	38	39	45	42	6.41	0.07	
15		3:55	50	42	43	45	6.45	0.08	
16	12/29	8:25	33	35	29	39	6.34	0.10	
17		9:25	41	40	29	34	6.36	0.12	
18		11:00	38	44	28	58	6.42	0.30	Damaged oil line
19		2.35	35	41	37	38	6.38	0.06	
20		3:15	56	55	45	48	6.51	0.11	Bad material
21	12/30	9:35	38	40	45	37	6.40	0.08	
22		10:20	39	42	35	40	6.39	0.07	
23		11:35	42	39	39	36	6.39	0.06	
24		2:00	43	36	35	38	6.38	0.08	
25		4:25	39	38	43	44	6.41	0.06	
Sum							160.25	2.19	

[a]For simplicity in recording, the individual measurements are coded from 6.00 mm.

It is necessary to collect a minimum of 25 subgroups of data. A fewer number of subgroups would not provide a sufficient amount of data for the accurate computation of the central line and control limits, and a larger number of subgroups would delay the introduction of the control chart.

The data are plotted in Figure 5-4, and this chart is called a *run chart*. It does not have control limits but can be used to analyze the data, especially in the development stage of a product or prior to a state of statistical control. The data points are plotted by order of production, as shown by the figure. Plotting the data points is a very effective way of finding out about the process. This should be done as the first step in data analysis.

Statistical limits are needed to determine if the process is stable.

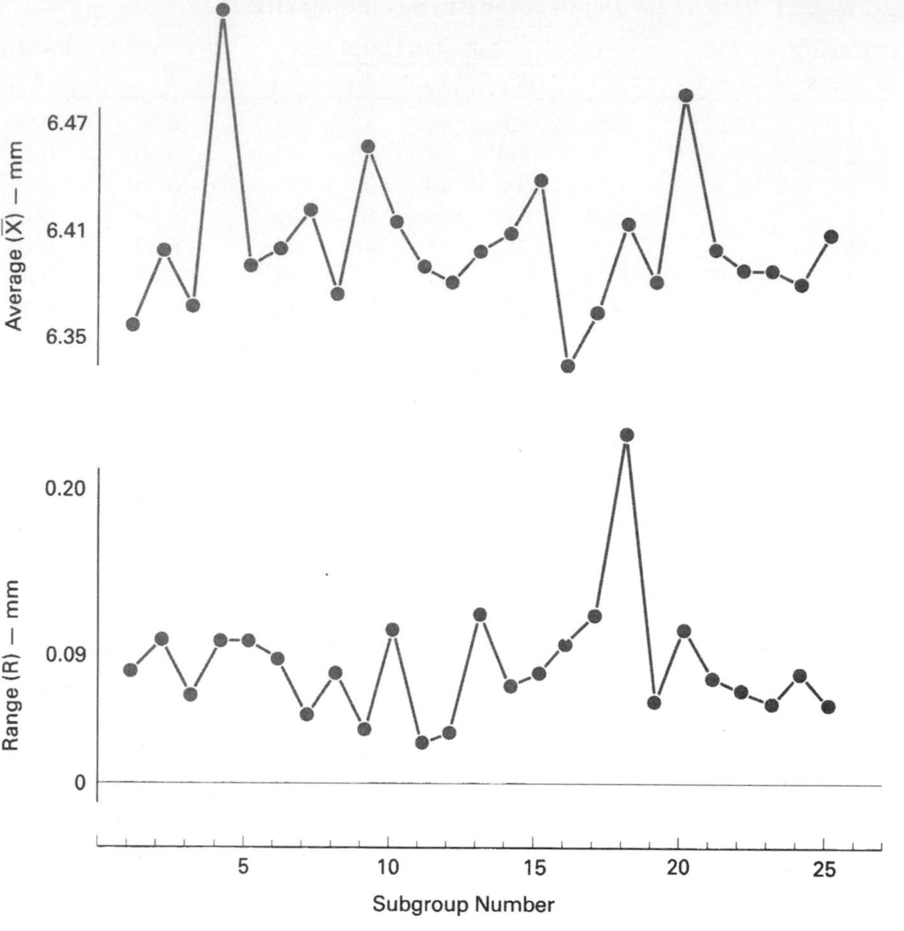

FIGURE 5-4 Run chart for data of Table 5-2.

Determine the Trial Central Line and Control Limits

The central lines for the $\overline{X}$ and R charts are obtained using the formulas

$$\overline{\overline{X}} = \frac{\sum_{i=1}^{g} \overline{X}_i}{g} \quad \text{and} \quad \overline{R} = \frac{\sum_{i=1}^{g} R_i}{g}$$

where $\overline{\overline{X}}$ = average of the subgroup averages (read "X double bar")

$\overline{X}_i$ = average of the ith subgroup

g = number of subgroups

$\overline{R}$ = average of the subgroup ranges

R_i = range of the ith subgroup

Trial control limits for the charts are established at ±3 standard deviations from the central value, as shown by the formulas

$$\text{UCL}_{\overline{X}} = \overline{\overline{X}} + 3\sigma_{\overline{X}} \quad \text{UCL}_R = \overline{R} + 3\sigma_R$$
$$\text{LCL}_{\overline{X}} = \overline{\overline{X}} - 3\sigma_{\overline{X}} \quad \text{LCL}_R = \overline{R} - 3\sigma_R$$

where UCL = upper control limit

LCL = lower control limit

$\sigma_{\overline{X}}$ = population standard deviation of the subgroup averages ($\overline{X}$'s)

σ_R = population standard deviation of the range

In practice, the calculations are simplified by using the product of the range ($\overline{R}$) and a factor (A_2) to replace the three standard deviations ($A_2\overline{R} = 3\,\sigma_{\overline{X}}$)[4] in the formulas for the $\overline{X}$ chart. For the R chart, the range $\overline{R}$ is used to estimate the standard deviation of the range (σ_R).[5] Therefore, the derived formulas are

$$\text{UCL}_{\overline{X}} = \overline{\overline{X}} + A_2\overline{R} \quad \text{UCL}_R = D_4\overline{R}$$
$$\text{LCL}_{\overline{X}} = \overline{\overline{X}} - A_2\overline{R} \quad \text{LCL}_R = D_3\overline{R}$$

where A_2, D_3, and D_4 are factors that vary with the subgroup size and are found in Table B of the Appendix. For the $\overline{X}$ chart the upper and lower control limits are symmetrical about the central line. Theoretically, the control limits for an R chart should also be symmetrical about the central line. But, for this situation to occur, with subgroup sizes of 6 or less, the lower control limit would need to have a negative value. Since a negative range is impossible, the lower control limit is located at zero by assigning to D_3 the value of zero for subgroups of 6 or less.

When the subgroup size is 7 or more, the lower control limit is greater than zero and symmetrical about the central line. However, when the R chart is posted at the work center, it may be more practical to keep the lower control limit at zero. This practice eliminates the difficulty of explaining to the operator that points below the lower control limit on the R chart are the result of exceptionally good

[4]The derivation of $3\sigma_{\overline{X}} = A_2\overline{R}$ is based on the substitution of $\sigma_{\overline{X}} = \sigma/\sqrt{n}$ and an estimate of $\sigma = \overline{R}/d_2$, where d_2 is a factor for the subgroup size.

$$3\sigma_{\bar{x}} = \frac{3\sigma}{\sqrt{n}} = \frac{3}{d_2\sqrt{n}}\overline{R}; \qquad \text{therefore, } A_2 = \frac{3}{d_2\sqrt{n}}$$

[5]The derivation of the simplified formula is based on the substitution of $d_3\sigma = \sigma_R$ and $\sigma = \overline{R}/d_2$, which gives

$$\left(1 + \frac{3d_3}{d_2}\right)\overline{R} \quad \text{and} \quad \left(1 - \frac{3d_3}{d_2}\right)\overline{R}$$

for the control limits. Thus, D_4 and D_3 are set equal to the coefficients of $\overline{R}$.

performance rather than poor performance. However, quality personnel should keep their own charts with the lower control limit in its proper location, and any out-of-control low points should be investigated to determine the reason for the exceptionally good performance. Since subgroup sizes of 7 or more are uncommon, the situation occurs infrequently.

EXAMPLE PROBLEM

In order to illustrate the calculations necessary to obtain the trial control limits and the central line, the data in Table 5-2 concerning the depth of the shaft keyway will be used. From Table 5-2, $\Sigma\overline{X} = 160.25$, $\Sigma R = 2.19$, and $g = 25$; thus, the central lines are

$$\overline{\overline{X}} = \frac{\sum_{i=1}^{g} \overline{X}_i}{g} = \frac{160.25}{25} = 6.41 \text{ mm} \qquad \overline{R} = \frac{\sum_{i=1}^{g} R_i}{g} = \frac{2.19}{25} = 0.0876 \text{ mm}$$

From Table B in the appendix, the values for the factors for a subgroup size (n) of 4 are $A_2 = 0.729$, $D_3 = 0$, and $D_4 = 2.282$. Trial control limits for the $\overline{X}$ chart are

$$\begin{aligned} \text{UCL}_{\overline{X}} &= \overline{\overline{X}} + A_2\overline{R} & \text{LCL}_{\overline{X}} &= \overline{\overline{X}} - A_2\overline{R} \\ &= 6.41 + (0.729)(0.0876) & &= 6.41 - (0.729)(0.0876) \\ &= 6.47 \text{ mm} & &= 6.35 \text{ mm} \end{aligned}$$

Trial control limits for the R chart are

$$\begin{aligned} \text{UCL}_R &= D_4\overline{R} & \text{LCL}_R &= D_3\overline{R} \\ &= (2.282)(0.0876) & &= (0)(0.0876) \\ &= 0.20 \text{ mm} & &= 0 \text{ mm} \end{aligned}$$

Figure 5-5 shows the central lines and the trial control limits for the $\overline{X}$ and R charts for the preliminary data.

Establish the Revised Central Line and Control Limits

The first step is to post the preliminary data to the chart along with the control limits and central lines. This has been accomplished and is shown in Figure 5-5.

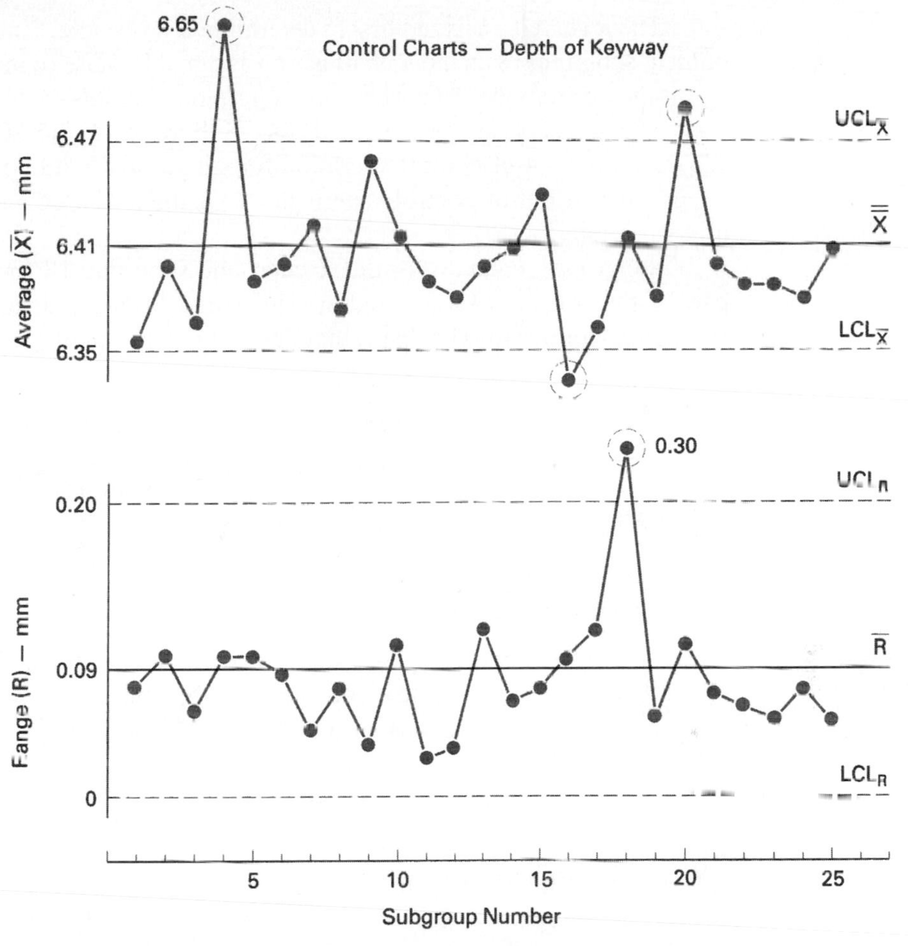

FIGURE 5-5 $\overline{X}$ **and** R **chart for preliminary data with trial control limits.**

The next step is to adopt standard values for the central lines or, more appropriately stated, the best estimate of the standard values with the available data. If an analysis of the preliminary data shows good control, then $\overline{\overline{X}}$ and $\overline{R}$ can be considered as representative of the process and these become the standard values, $\overline{X}_0$ and R_0 Good control can be briefly described as that which has no out-of-control points, no long runs on either side of the central line, and no unusual patterns of variation. More information concerning in control and out of control is provided later in the chapter.

Most processes are not in control when first analyzed. An analysis of Figure 5-5 shows that there are out-of-control points on the $\overline{X}$ chart at subgroups 4, 16, and 20 and an out-of-control point on the R chart at subgroup 18. It also appears that there are a large number of points below the central line, which is no doubt due to the influence of the high points.

The R chart is analyzed first to determine if it is stable. Since the out-of-control point at subgroup 18 on the R chart has an assignable cause (damaged oil line), it can be discarded from the data. The remaining plotted points indicate a stable process.

The $\overline{X}$ chart can now be analyzed. Subgroups 4 and 20 had an assignable cause, but the out-of-control condition for subgroup 16 did not. It is assumed that subgroup 16's out-of-control state is due to a chance cause and is part of the natural variation.

Subgroups 4 and 20 for the $\overline{X}$ chart and subgroup 18 for the R chart are not part of the natural variation and are discarded from the data, and new $\overline{\overline{X}}$ and $\overline{R}$ values are computed with the remaining data. The calculations are simplified by using the following formulas:

$$\overline{\overline{X}}_{new} = \frac{\Sigma\overline{X} - \overline{X}_d}{g - g_d} \qquad \overline{R}_{new} = \frac{\Sigma R - R_d}{g - g_d}$$

where $\overline{X}_d$ = discarded subgroup averages

g_d = number of discarded subgroups

R_d = discarded subgroup ranges

There are two techniques used to discard data. If either the $\overline{X}$ or the R value of a subgroup is out of control and has an assignable cause, both are discarded, or only the out-of-control value of a subgroup is discarded. In this book the latter technique is followed; thus, when an $\overline{X}$ value is discarded, its corresponding R value is not discarded and vice versa.

EXAMPLE PROBLEM

Calculations for a new $\overline{X}$ are based on discarding the $\overline{X}$ values of 6.65 and 6.51 for subgroups 4 and 20, respectively. Calculations for a new $\overline{R}$ are based on discarding the R value of 0.30 for subgroup 18.

$$\overline{\overline{X}}_{new} = \frac{\Sigma\overline{X} - \overline{X}_d}{g - g_d} \qquad \overline{R}_{new} = \frac{\Sigma R - R_d}{g - g_d}$$

$$= \frac{160.25 - 6.65 - 6.51}{25 - 2} \qquad = \frac{2.19 - 0.30}{25 - 1}$$

$$= 6.40 \text{ mm} \qquad = 0.079 \text{ mm}$$

These new values of $\overline{\overline{X}}$ and $\overline{R}$ are used to establish the standard values of $\overline{X}_0$, R_0, and σ_0. Thus,

$$\overline{X}_0 = \overline{\overline{X}}_{new}, \quad R_0 = \overline{R}_{new}, \quad \text{and} \quad \sigma_0 = \frac{R_0}{d_2}$$

where d_2 = a factor from Table B for estimating σ_0 from R_0. The standard or reference values can be considered to be the best estimate with the data available. As more data become available, better estimates or more confidence in the existing standard values are obtained. *Our objective is to obtain the best estimate of these population standard values*:

Using the standard values, the central lines and the 3σ control limits for actual operations are obtained using the formulas

$$\text{UCL}_{\overline{X}} = \overline{X}_0 + A\sigma_0 \qquad \text{LCL}_{\overline{X}} = \overline{X}_0 - A\sigma_0$$

$$\text{UCL}_R = D_2\sigma_0 \qquad \text{LCL}_R = D_1\sigma_0$$

where A, D_1, and D_2 are factors from Table B for obtaining the 3σ control limits from $\overline{X}_0$ and σ_0.

EXAMPLE PROBLEM

From Table B in the Appendix and for a subgroup size of 4, the factors are $A = 1.500$, $d_2 = 2.059$, $D_1 = 0$, and $D_2 = 4.698$. Calculations to determine $\overline{X}_0$, R_0, and σ_0 using the data previously given are

$$\overline{X}_0 = \overline{\overline{X}}_{\text{new}} = 6.40 \text{ mm}$$

$$R_0 = \overline{R}_{\text{new}} = 0.079 \text{ (0.08 for the chart)}$$

$$\sigma_0 = \frac{R_0}{d_2} = \frac{0.079}{2.059} = 0.038 \text{ mm}$$

Thus, the control limits are

$$\text{UCL}_{\overline{X}} = \overline{X}_0 + A\sigma_0 = 6.40 + (1.500)(0.038) = 6.46 \text{ mm} \qquad \text{LCL}_{\overline{X}} = \overline{X}_0 - A\sigma_0 = 6.40 - (1.500)(0.038) = 6.34 \text{ mm}$$

$$\text{UCL}_R = D_2\sigma_0 = (4.698)(0.038) = 0.18 \text{ mm} \qquad \text{LCL}_R = D_1\sigma_0 = (0)(0.038) = 0 \text{ mm}$$

The central lines and control limits are drawn on the $\overline{X}$ and R charts for the next period and are shown in Figure 5-6. For illustrative purposes the trial control

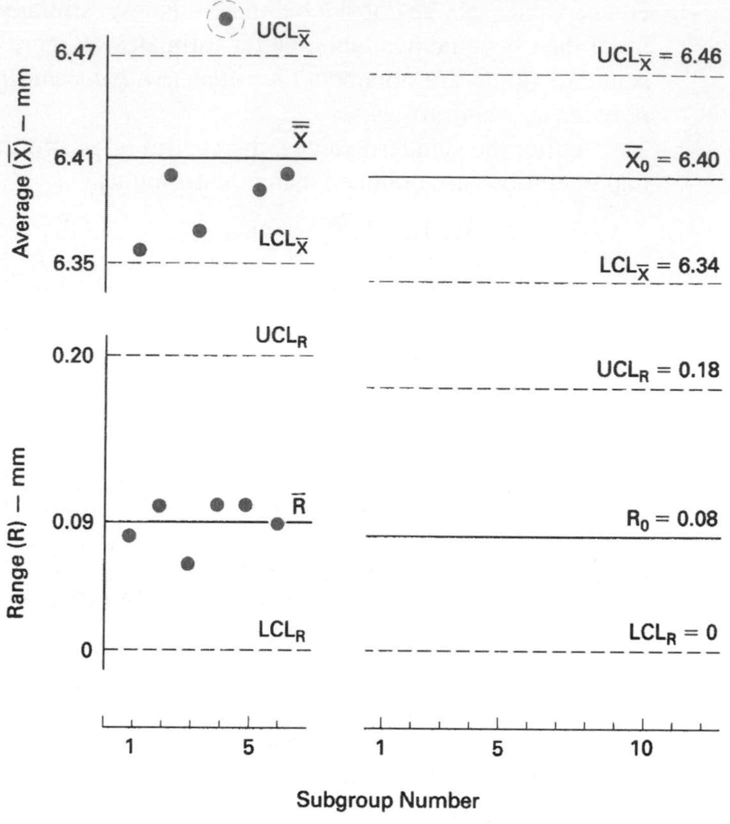

FIGURE 5-6 Trial control limits and revised control limits for $\overline{X}$ and R charts.

limits and the revised control limits are shown on the same chart. The limits for both the $\overline{X}$ and R charts became narrower, as was expected. No change occurred in LCL_R because the subgroup size is 6 or less. Figure 5-6 also illustrates a simpler charting technique in that lines are not drawn between the points.

The preliminary data for the initial 25 subgroups are not plotted with the revised control limits. These revised control limits are for reporting the results for future subgroups. To make effective use of the control chart during production, it should be displayed in a conspicuous place where it can be seen by operators and supervisors.

Before proceeding to the action step, some final comments are appropriate. First, many analysts eliminate this step in the procedure because it appears to be somewhat redundant. However, by discarding out-of-control points with assignable causes, the central line and control limits are more representative of the process.

This step may be too complicated for operating personnel. Its elimination would not affect the next step.

Second, the formula for the control limits are mathematically equal. Thus, for the upper control limit; $\overline{X}_0 + A\sigma_0 = \overline{\overline{X}}_{new} + A_2\overline{R}_{new}$. Similar equivalences are true for the lower control limit and both control limits for the R chart.

Third, the parameter σ_0 is now available to obtain the initial estimate of the process capability, which is $6\sigma_0$. The true process capability is obtained in the next step (Achieve the Objective).

Fourth, the central line $\overline{X}_0$ for the $\overline{X}$ chart is frequently based on the specifications. In such a case, the procedure is used only to obtain R_0 and σ_0. If in our example problem the nominal value of the characteristic is 6.38 mm, then $\overline{X}_0$ is set to that value and the upper and lower control limits are

$$
\begin{aligned}
UCL_{\overline{X}} &= X_0 + A\sigma_0 & LCL_{\overline{X}} &= \overline{X}_0 - A\sigma_0 \\
&= 6.38 + (1.500)(0.038) & &= 6.38 - (1.500)(0.038) \\
&= 6.44 & &= 6.32
\end{aligned}
$$

The central line and control limits for the R chart do not change. This modification can be taken only if the process is adjustable. If the process is not adjustable, then the original calculations must be used.

Fifth, it follows that adjustments to the process should be made while taking data. It is not necessary to run nonconforming material while collecting data, since we are primarily interested in obtaining R_0, which is not affected by the process setting. The independence of μ and σ provide the rationale for this concept.

Sixth, the process determines the central line and control limits. They are not established by design, manufacturing, marketing, or any other department, except for X_0 when the process is adjustable.

Finally, when population values are known (μ and σ), the central lines and control limits may be calculated immediately, saving time and work. Thus $\overline{X}_0 = \mu$; $\sigma_0 = \sigma$; and $R_0 = d_2\sigma$, and the limits are obtained using the appropriate formulas. This situation would be extremely rare.

Achieve the Objective

When control charts are first introduced at a work center, an improvement in the process performance usually occurs. This initial improvement is especially noticeable when the process is dependent on the skill of the operator. Posting a quality control chart appears to be a psychological signal to the operator to improve performance. Most workers want to produce a quality product; therefore, when management shows an interest in the quality, the operator responds.

Figure 5-7 illustrates the initial improvement that occurred after the introduction of the $\overline{X}$ and R charts in January. Owing to space limitations, only a representative

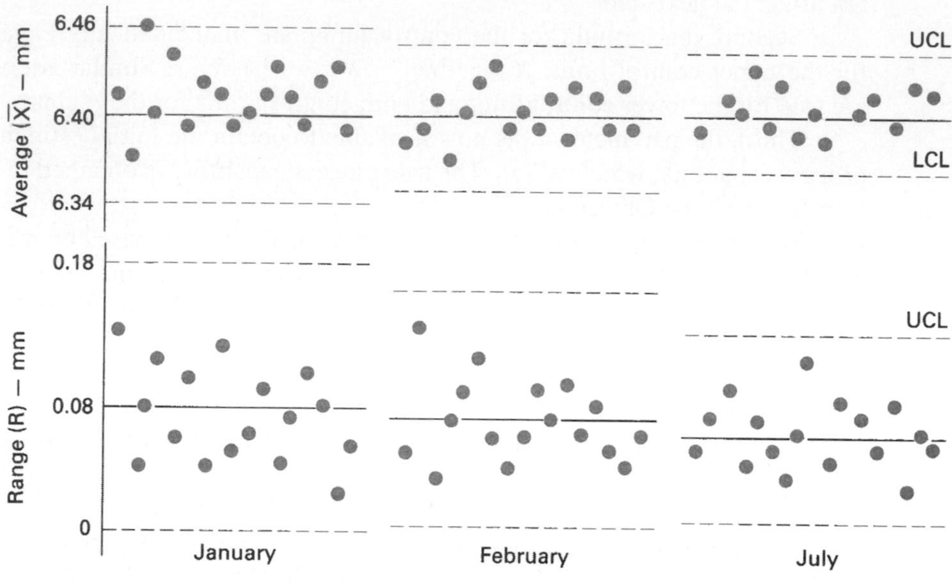

FIGURE 5-7 Continuing use of control charts, showing improved quality.

number of subgroups for each month are shown in the figure. During January the subgroup averages had less variation and tended to be centered at a slightly higher point. A reduction in the range variation occurred also.

Not all the improved performance in January was the result of operator effort. The first-line supervisor initiated a program of tool wear control, which was a contributing factor.

At the end of January new central lines and control limits were calculated using the data from subgroups obtained during the month. It is a good idea, especially when a chart is being initiated, to calculate standard values periodically to see if any changes have occurred. This reevaluation can be done for every 25 or more subgroups, and the results compared to the previous values.[6]

New control limits were established for the $\overline{X}$ and R charts and central line for the R chart for the month of February. The central line for the $\overline{X}$ chart was not changed because it is the nominal value. During the ensuing months the maintenance department replaced a pair of worn gears; purchasing changed the material supplier; and tooling modified a workholding device. All these improvements were the result of investigations that tracked down the causes for out-of-control conditions

[6]These values are usually compared without the use of formal tests. An exact evaluation can be obtained by mathematically comparing the central lines to see if they are from the same population.

or were ideas developed by a project team. The generation of ideas by many different personnel is the most essential ingredient for continuous quality improvement. Ideas by the operator, first-line supervisor, quality assurance, maintenance, manufacturing engineering, and industrial engineering should be evaluated. This evaluation or testing of an idea requires 25 or more subgroups. The control chart will tell if the idea is good, poor, or has no effect on the process. Quality improvement occurs when the plotted points of the $\overline{X}$ chart converge on the central line, or when the plotted points of the R chart trend downward, or when both actions occur. If a poor idea is tested, then the reverse occurs. Of course, if the idea is neutral it will have no affect on the plotted point pattern.

In order to speed up the testing of ideas, the taking of subgroups can be compressed in time as long as the data represents the process by accounting for any hourly or day-to-day fluctuations. Only one idea should be tested at a time; otherwise, the results will be confounded.

At the end of June, the periodic evaluation of the past performance showed the need to revise the central lines and the control limits. The performance for the month of July and subsequent months showed a natural pattern of variation and no quality improvement. At this point no further quality improvement is possible without a substantial investment in new equipment or equipment modification.

W. Edwards Deming has stated "that if he were a banker, he would not lend any money to a company unless statistical methods were used to prove that the money was necessary." This is precisely what the control chart can achieve, provided that all personnel use the chart as a method of quality improvement rather than just a maintenance function.

When the objective for initiating the charts has been achieved, its use should be discontinued or the frequency of inspection be substantially reduced to a monitoring action by the operator. Efforts should then be directed toward the improvement of some other quality characteristic. If a project team was involved it should be congratulated for its performance and disbanded.

The Sample Standard Deviation Control Chart

While the $\overline{X}$ and R charts are the most common charts for variables, some organizations prefer the sample standard deviation, s, as the measure of the subgroup dispersion. In comparing an R chart with an s chart, an R chart is easier to compute and easier to explain. On the other hand, the subgroup sample standard deviation for the s chart is calculated using all the data rather than just the high and the low value, as done for the R chart. An s chart is therefore more accurate than an R chart. When subgroup sizes are less than 10, both charts will graphically portray the same variation[7]; however, as subgroup sizes increase to 10 or more, extreme

[7]A proof of this statement can be observed by comparing the R chart of Figure 5-5 with the s chart of Figure 5-8.

values have an undue influence on the R chart. Therefore, at larger subgroup sizes the s chart must be used.

The steps necessary to obtain the $\overline{X}$ and s trial control and revised control limits are the same as those used for the $\overline{X}$ and R chart except for different formulas. In order to illustrate the method, the same data will be used. They are reproduced in Table 5-3 with the addition of an s column and the elimination of the R column. The appropriate formulas used in the computation of the trial control limits are

$$\overline{s} = \frac{\sum_{i=1}^{g} \overline{s}_i}{g} \qquad \overline{\overline{X}} = \frac{\sum_{i=1}^{g} \overline{X}_i}{g}$$

$$\text{UCL}_{\overline{X}} = \overline{\overline{X}} + A_3\overline{s} \qquad \text{UCL}_s = B_4\overline{s}$$

$$\text{LCL}_{\overline{X}} = \overline{\overline{X}} - A_3\overline{s} \qquad \text{LCL}_s = B_3\overline{s}$$

where s_i = sample standard deviation of the subgroup values

$\overline{s}$ = average of the subgroup sample standard deviations

A_3, B_3, B_4 = factors found in Table B of the Appendix for obtaining the 3σ control limits for $\overline{X}$ and s charts from $\overline{s}$

Formulas for the computation of the revised control limits using the standard values of $\overline{X}_0$ and σ_0 are

$$\overline{X}_0 = \overline{\overline{X}}_{\text{new}} = \frac{\Sigma\overline{X} - \overline{X}_d}{g - g_d}$$

$$s_0 = \overline{s}_{\text{new}} = \frac{\Sigma s - s_d}{g - g_d} \qquad \sigma_0 = \frac{s_0}{c_4}$$

$$\text{UCL}_{\overline{X}} = \overline{X}_0 + A\sigma_0 \qquad \text{UCL}_s = B_6\sigma_0$$

$$\text{LCL}_{\overline{X}} = \overline{X}_0 - A\sigma_0 \qquad \text{LCL}_s = B_5\sigma_0$$

where s_d = sample standard deviation of the discarded subgroup

c_4 = factor found in Table B for computing σ_0 from $\overline{s}$

A, B_5, B_6 = factors found in Table B for computing 3σ process control limits for $\overline{X}$ and s charts

The first step is to determine the standard deviation for each subgroup from the preliminary data. For subgroup 1, with values of 6.35, 6.40, 6.32, and 6.37, the standard deviation is

$$s = \sqrt{\frac{n\sum_{i=1}^{n} X_i^2 - \left(\sum_{i=1}^{n} X_i\right)^2}{n(n-1)}}$$

$$= \sqrt{\frac{4(6.35^2 + 6.40^2 + 6.32^2 + 6.37^2) - (6.35 + 6.40 + 6.32 + 6.37)^2}{4(4 - 1)}}$$

$$= 0.034 \text{ mm}$$

The standard deviation for subgroup 1 is posted to the s column as shown in Table 5-3, and the process is repeated for the remaining 24 subgroups. Continuation of the X and s charts is accomplished in the same manner as the $\overline{X}$ and R charts.

TABLE 5-3 Data on the Depth of the Keyway (millimeters)[a]

SUBGROUP NUMBER	DATE	TIME	MEASUREMENTS X_1	X_2	X_3	X_4	AVERAGE $\overline{X}$	SAMPLE STANDARD DEVIATION S	COMMENT
1	12/26	8:50	35	40	32	37	6.36	0.034	
2		11:30	46	37	36	41	6.40	0.045	
3		1:45	34	40	34	36	6.36	0.028	
4		3:45	69	64	68	59	6.65	0.045	New, temporary operator
5		4:20	38	34	44	40	6.39	0.042	
6	12/27	8:35	42	41	43	34	6.40	0.041	
7		9:00	44	41	41	46	6.43	0.024	
8		9:40	33	41	38	36	6.37	0.034	
9		1:30	48	44	47	45	6.46	0.018	
10		2:50	47	43	36	42	6.42	0.045	
11	12/28	8:30	38	41	39	38	6.39	0.014	
12		1:35	37	37	41	37	6.38	0.020	
13		2:25	40	38	47	35	6.40	0.051	
14		2:35	38	39	45	42	6.41	0.032	
15		3:55	50	42	43	45	6.45	0.036	
16	12/29	8:25	33	35	29	39	6.34	0.042	
17		9:25	41	40	29	34	6.36	0.056	
18		11:00	38	44	28	58	6.42	0.125	Damaged oil line
19		2:35	35	41	37	38	6.38	0.025	
20		3:15	56	55	45	48	6.51	0.054	Bad material
21	12/30	9:35	38	40	45	37	6.40	0.036	
22		10:20	39	42	35	40	6.39	0.029	
23		11:35	42	39	39	36	6.39	0.024	
24		2:00	43	36	35	38	6.38	0.036	
25		4:25	39	38	43	44	6.41	0.029	
Sum							160.25	0.965	

[a]For simplicity in recording, the individual measurements are coded from 6.00 mm.

EXAMPLE PROBLEM

Using the data of Table 5-3, determine the revised central line and control limits for $\overline{X}$ and s charts. The first step is to obtain $\bar{s}$ and $\overline{\overline{X}}$, which are computed from Σs and $\Sigma \overline{X}$, whose values are found in Table 5-3.

$$\bar{s} = \frac{\sum_{i=1}^{g} s_i}{g} = \frac{0.965}{25} = 0.039 \text{ mm} \qquad \overline{\overline{X}} = \frac{\sum_{i=1}^{g} \overline{X}_i}{g} = \frac{160.25}{25} = 6.41 \text{ mm}$$

From Table B the values of the factors—$A_3 = 1.628$, $B_3 = 0$, and $B_4 = 2.266$—are obtained, and the trial control limits are

$$\begin{aligned} UCL_{\overline{X}} &= \overline{\overline{X}} + A_3\bar{s} & LCL_{\overline{X}} &= \overline{\overline{X}} - A_3\bar{s} \\ &= 6.41 + (1.628)(0.039) & &= 6.41 - (1.628)(0.039) \\ &= 6.47 \text{ mm} & &= 6.35 \text{ mm} \\ UCL_s &= B_4\bar{s} & LCL_s &= B_3\bar{s} \\ &= (2.266)(0.039) & &= (0)(0.039) \\ &= 0.088 \text{ mm} & &= 0 \text{ mm} \end{aligned}$$

The next step is to plot the subgroup $\overline{X}$ and s on graph paper with the central lines and control limits. This step is show in Figure 5-8. Subgroups 4 and 20 are out of control on the $\overline{X}$ chart, and since they have assignable causes, they are discarded. Subgroup 18 is out of control on the s chart, and since it has an assignable cause, it is discarded. Computation to obtain the standard values of $\overline{X}_0$, s_0, and σ_0 are as follows:

$$\begin{aligned} \overline{X}_0 = \overline{\overline{X}}_{new} &= \frac{\Sigma\overline{X} - \overline{X}_d}{g - g_d} & s_0 = \bar{s}_{new} &= \frac{\Sigma s - s_d}{g - g_d} \\ &= \frac{160.25 - 6.65 - 6.51}{25 - 2} & &= \frac{0.965 - 0.125}{25 - 1} \\ &= 6.40 \text{ mm} & &= 0.035 \text{ mm} \end{aligned}$$

$$\sigma_0 = \frac{s_0}{c_4} \qquad \text{from Table B, } c_4 = 0.9213$$

$$= \frac{0.035}{0.9213}$$

$$= 0.038 \text{ mm}$$

The reader should note that the standard deviation, σ_0, is the same as the value obtained from the range in the preceding section. Using the standard values of $\overline{X}_0 = 6.40$ and $\sigma_0 = 0.038$, the revised control limits are computed.

$$UCL_{\overline{X}} = \overline{X} + A\sigma_0 \qquad LCL_{\overline{X}} = \overline{X}_0 - A\sigma_0$$

$$= 6.40 + (1.500)(0.038) \qquad = 6.40 - (1.500)(0.038)$$

$$= 6.46 \text{ mm} \qquad = 6.34 \text{ mm}$$

$$UCL_s = B_6\sigma_0 \qquad LCL_s = B_5\sigma_0$$

$$= (2.088)(0.038) \qquad = (0)(0.038)$$

$$= 0.079 \text{ mm} \qquad = 0 \text{ mm}$$

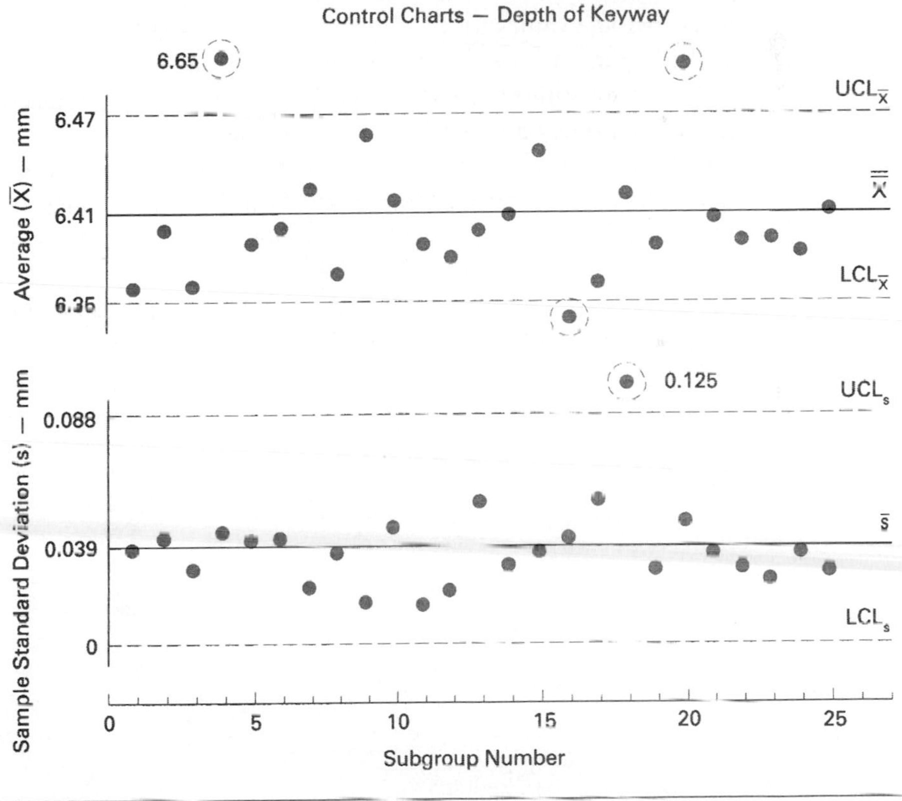

FIGURE 5-8 $\overline{X}$ **and** s **chart for preliminary data with trial control limits.**

STATE OF CONTROL

Process in Control

When the assignable causes have been eliminated from the process to the extent that the points plotted on the control chart remain within the control limits, the process is in a state of control. No higher degree of uniformity can be attained with the existing process. Greater uniformity can, however, be attained through a change in the basic process through quality improvements ideas.

When a process is in control, there occurs a natural pattern of variation, which is illustrated by the control chart in Figure 5-9. This natural pattern of variation has (1) about 34% of the plotted points in an imaginary band between one standard deviation on both sides of the central line, (2) about 13.5% of the plotted points in an imaginary band between one and two standard deviations on both sides of the central line, and (3) about 2.5% of the plotted points in an imaginary band between two and three standard deviations on both sides of the central line. The points are located back and forth across the central line in a random manner with no points beyond the control limits. The natural pattern of the points or subgroup average values forms its own frequency distribution. If all of the points were stacked up at one end they would form a normal curve (see Figure 5-11).

Control limits are usually established at three standard deviations from the central line. They are used as a basis to judge whether there is evidence of lack of control. The choice of 3σ limits is an economic one with respect to two types of errors that can occur. One error, called Type I by statisticians, occurs when looking for an assignable cause of variation when in reality a chance cause is present When the limits are set at three standard deviations, a Type I error will occur 0.27% (3 out of 1000) of the time. In other words, when a point is outside the control limits, it is assumed to be due to an assignable cause even though it would be due to a

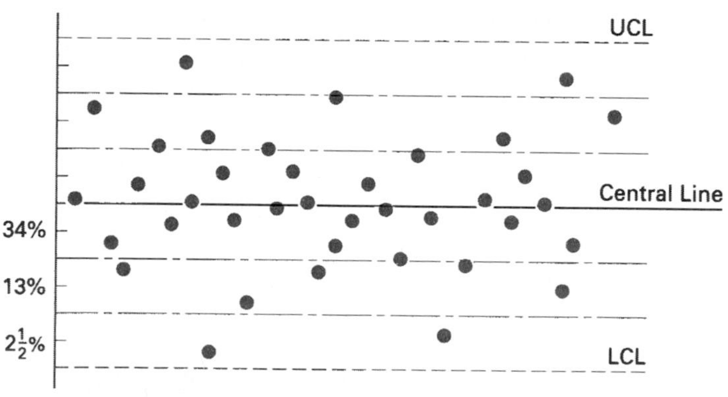

FIGURE 5-9 Natural pattern of variation of a control chart.

TABLE 5-4 Type I and Type II Errors

	PLOTTED POINT IS	
	OUTSIDE CONTROL LIMITS	INSIDE CONTROL LIMITS
Assignable Cause Present	OK	Type II Error
Chance Cause Present	Type I Error	OK

chance cause 0.27% of the time. We might think of this situation as "guilty until proven innocent." The other type error, called Type II, occurs when assuming that a chance cause of variation is present when in reality there is an assignable cause. In other words, when a point is inside the control limits, it is assumed to be due to a chance cause even though it might be due to an assignable cause. We might think of this situation as "innocent until proven guilty." Table 5-4 illustrates the difference between the Type I and Type II errors. If control limits are established at say ±2.5 standard deviations, Type I errors would increase and Type II decrease. Abundant experience since 1930 in all types of industry indicates that 3σ limits provide an economic balance between the costs resulting from the two types of errors. Unless there are strong practical reasons for doing otherwise, the ±3 standard deviation limits should be used.

When a process is in control, only chance causes of variation are present. Small variations in machine performance, operator performance, and material characteristics are expected and are considered to be part of a stable process.

When a process is in control, certain practical advantages accrue to the producer and consumer.

1. Individual units of the product will be more uniform—or, stated another way, there will be less variation and fewer rejections.

2. Since the product is more uniform, fewer samples are needed to judge the quality. Therefore, the cost of inspection can be reduced to a minimum. This advantage is extremely important when 100% conformance to specifications is not essential.

3. The process capability or spread of the process is easily attained from 6σ. With a knowledge of the process capability, a number of reliable decisions relative to specifications can be made, such as

(a) To decide the product specifications,

(b) To decide the amount of rework or scrap when there is insufficient tolerance, and

(c) To decide whether to produce the product to tight specifications and permit interchangeability of components or to produce the product to loose specifications and use selective matching of components.

4. Trouble can be anticipated before it occurs, thereby speeding up production by avoiding rejections and interruptions.

5. The percentage of product that falls within any pair of values may be predicted with the highest degree of assurance. For example, this advantage can be very important when adjusting filling machines to obtain different percentages of items below, between, or above particular values.

6. It permits the consumer to use the producer's data and, therefore, to test only a few subgroups as a check on the producer's records. The $\overline{X}$ and R charts are used as statistical evidence of process control.

7. The operator is performing satisfactorily from a quality viewpoint. Further improvement in the process can be achieved only by changing the input factors: materials, equipment, environment, and operators. These changes require action by management.

When only chance causes of variation are present, the process is stable and predictable over time, as shown in Figure 5-10(a). We know that future variation as shown by the dotted curve will be the same unless there has been a change in the process due to an assignable cause.

Process Out of Control

The term *out of control* is usually thought of as being undesirable; however, there are situations where this condition is desirable. It is best to think of the term *out of control* as a change in the process due to an assignable cause.

When a point (subgroup value) falls outside its control limits, the process is out of control. This means that an assignable cause of variation is present. Another way of viewing the out-of-control point is to think of the subgroup value as coming from a different population than the one from which the control limits were obtained.

Figure 5-11 shows a frequency distribution of plotted points that are all stacked up at one end for educational purposes to form a normal curve for averages. The data were developed from a large number of subgroups and, therefore, represent the population mean, $\mu = 450$ g, and the population standard deviation for the averages, $\sigma_{\overline{X}} = 8$ g. The frequency distribution for subgroup averages is shown by a dashed line. Future explanations will use the dashed line to represent the frequency distribution of averages and will use a solid line for the frequency distribution of individual values. The out-of-control point has a value of 483 g. This point is so far away from the 3σ limits (99.73%) that it can only be considered to have come from another population. In other words, the process that produced the subgroup average of 483 g is a different process than the stable process from which the 3σ control limits were developed. Therefore, the process has changed; some assignable cause of variation is present.

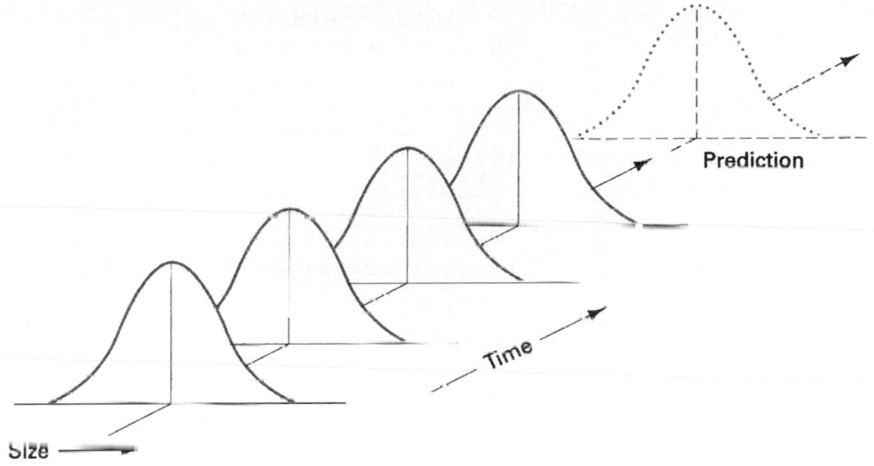

(a) Only Chance Causes of Variation Present

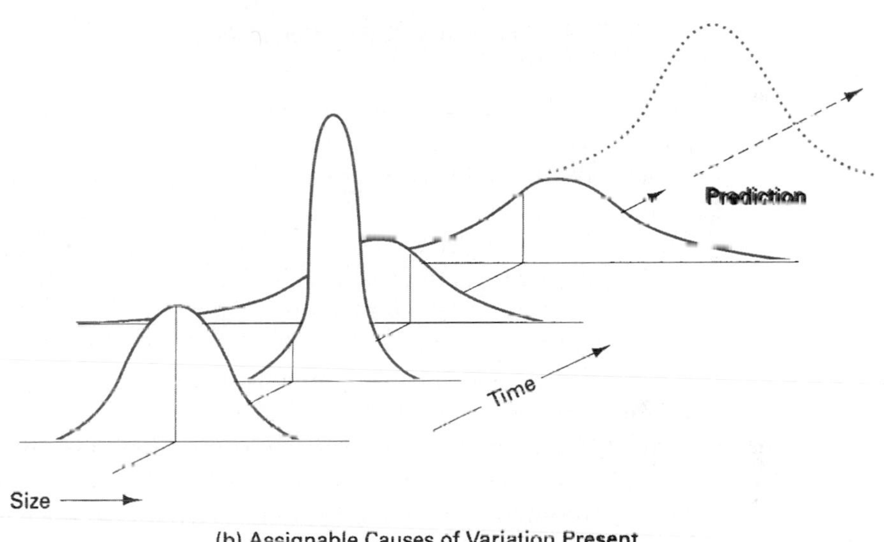

(b) Assignable Causes of Variation Present

FIGURE 5-10 Stable and unstable variation.

Figure 5-10(b) illustrates the effect of assignable causes of variation over time. The unnatural, unstable nature of the variation makes it impossible to predict future variation. The assignable causes must be found and corrected before a natural, stable process can continue.

A process can also be considered out of control even when the points fall inside the 3σ limits. This situation occurs when unnatural runs of variation are present in the process. First, let's divide the control chart into six equal standard

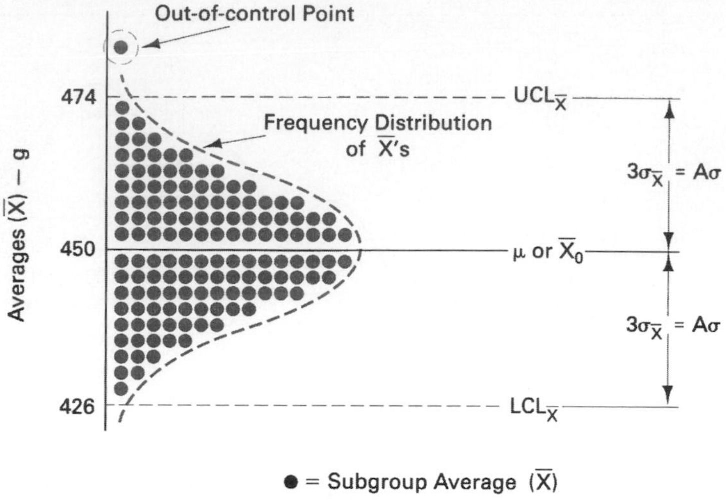

FIGURE 5-11 Frequency distribution of subgroup averages with control limits.

deviation bands in the same manner as Figure 5-9. For identification purposes the bands are labeled A, B, and C zones as shown in Figure 5-12.

It is not natural for seven or more consecutive points to be above or below the central line as shown in Figure 5-12(a). Also when 10 out of 11 points or 12 out of 14 points, etc., are located on one side of the central line, it is unnatural. Another unnatural run occurs at (b), where six points in a row are steadily increasing or decreasing. In Figure 5-12(c) we have two out of three points in a row in zone A and at (d) four out of five points in a row in zone B and beyond. There are many statistical possibilities, with the four common ones being shown in the figure. Actually, any significant divergence from the natural pattern as shown in Figure 5-9 would be unnatural and would be classified as an out-of-control condition.

Rather than divide the space into three equal zones of one standard deviation, a simplified technique would divide the space into two equal zones of 1.5 standard deviations. The process is out of control when there are two successive points at 1.5 standard deviations or beyond. The simplified rule makes for greater ease of implementation by operators without drastically sacrificing power.[8] It is shown in Figure 5-13 and replaces the information of Figures 5-12(c) and (d).

Analysis of Out-of-Control Condition

When a process is out of control, the assignable cause responsible for the condition must be found. The detective work necessary to locate the cause of the out-of-

[8]For more information, see A. M. Hurwitz and M. Mathur, "A Very Simple Set of Process Control Rules," *Quality Engineering*, Vol. 5, No. 1, 1992–93, 21–29.

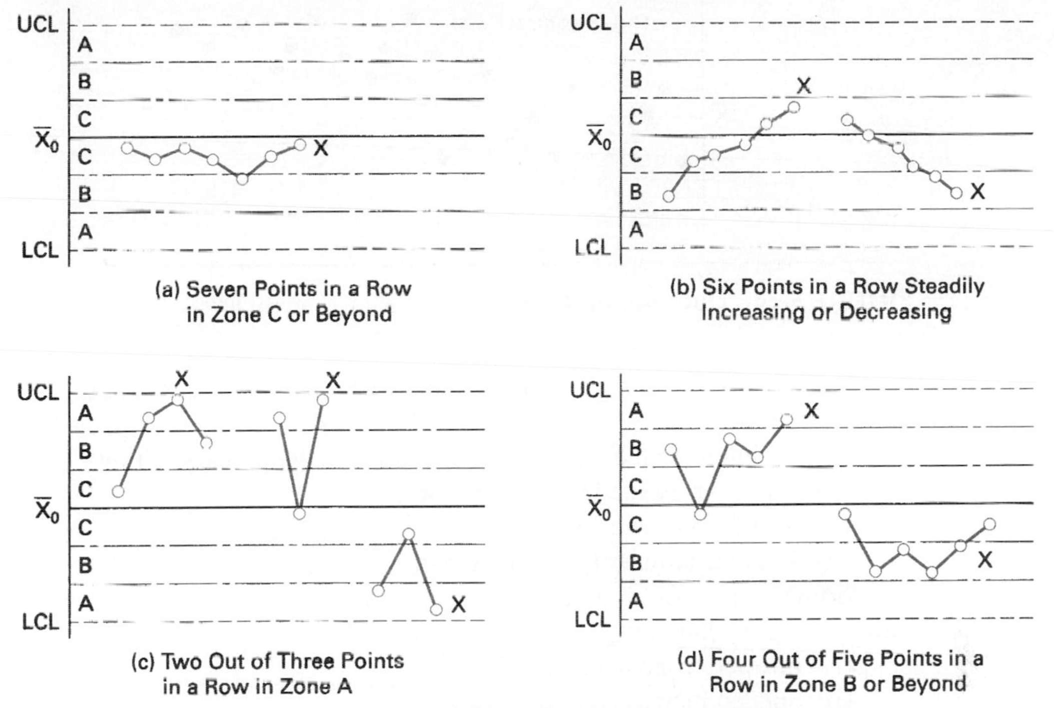

FIGURE 5-12 Some unnatural runs—process out of control.

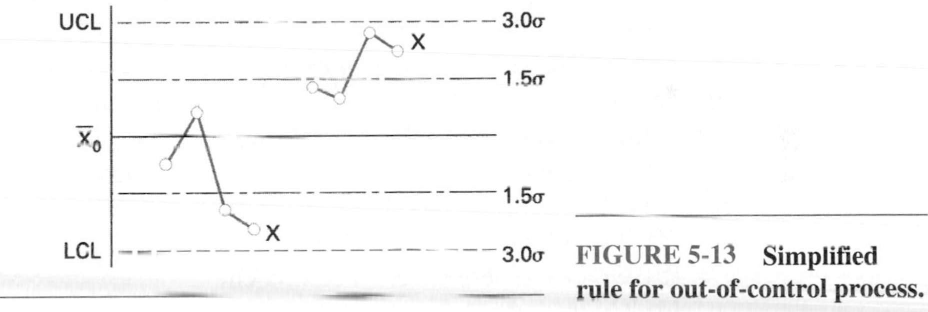

FIGURE 5-13 Simplified rule for out-of-control process.

control condition can be minimized by a knowledge of the types of out-of-control patterns and their assignable causes. Types of out-of-control $\overline{X}$ and R patterns are (1) change or jump in level, (2) trend or steady change in level, (3) recurring cycles, (4) two populations, and (5) mistakes.

1. *Change or jump in level.* This type is concerned with a sudden change in level to the $\overline{X}$ chart, to the R chart, or to both charts. Figure 5-14 illustrates

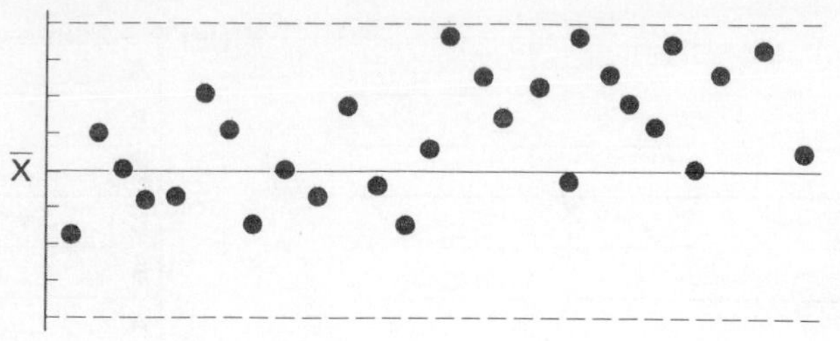

FIGURE 5-14 Out-of-control pattern: change or jump in level.

the change in level. For an $\overline{X}$ chart, the change in the process average can be due to

(a) An intentional or unintentional change in the process setting
(b) A new or inexperienced operator
(c) A different raw material
(d) A minor failure of a machine part

Some causes for a sudden change in the process spread or variability as shown on the R chart are

(a) Inexperienced operator
(b) Sudden increase in gear play
(c) Greater variation in incoming material

Sudden changes in level can occur on both the $\overline{X}$ and the R charts. This situation is common during the beginning of control chart activity prior to the attainment of a state of control. There may be more than one assignable cause, or it may be a cause that could affect both charts, such as an inexperienced operator.

2. *Trend or steady change in level*. Steady changes in control chart level are very common industrial phenomena. Figure 5-15 illustrates a trend or steady change that is occurring in the upward direction; the trend could have been illustrated in the downward direction. Some causes of steady progressive changes on an $\overline{X}$ chart are

(a) Tool or die wear
(b) Gradual deterioration of equipment
(c) Gradual change in temperature or humidity
(d) Viscosity breakdown in a chemical process
(e) Buildup of chips in a work-holding device

A steady change in level or trend on the R chart is not as common as on the $\overline{X}$ chart. It does, however, occur, and some possible causes are

(a) An improvement in worker skill (downward trend)
(b) A decrease in worker skill due to fatigue, boredom, inattention, and so on (upward trend)
(c) A gradual improvement in the homogeneity of incoming material

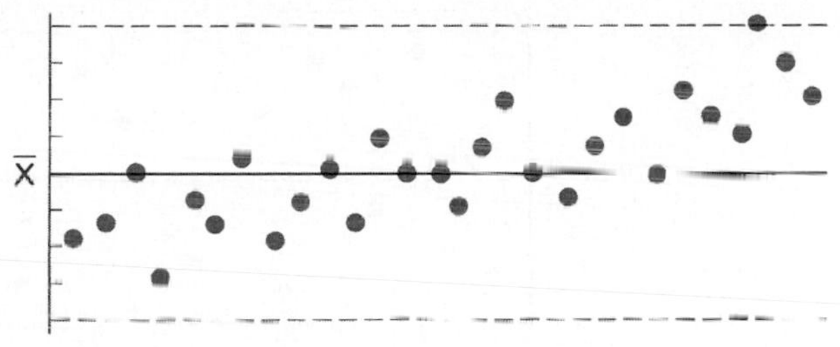

FIGURE 5-15 Out-of-control pattern: trend or steady change in level.

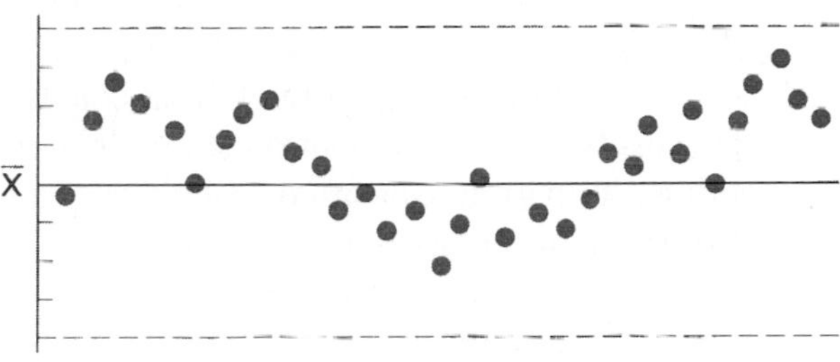

FIGURE 5-16 Out-of-control pattern: recurring cycles.

3. *Recurring cycles*. When the plotted points on an $\overline{X}$ or R chart show a wave or periodic high and low points, it is called a *cycle*. A typical recurring out-of-control pattern is shown in Figure 5-16.

For an $\overline{X}$ chart, some of the causes of recurring cycles are

(a) The seasonal effects of incoming material
(b) The recurring effects of temperature and humidity (cold morning start-up)
(c) Any daily or weekly chemical, mechanical, or psychological event
(d) The periodic rotation of operators

Periodic cycles on an R chart are not as common as for an $\overline{X}$ chart. Some affecting the R chart are due to

(a) Operator fatigue and rejuvenation resulting from morning, noon, and afternoon breaks
(b) Lubrication cycles

The out-of-control pattern of a recurring cycle sometimes goes unreported because of the inspection cycle. Thus, a cyclic pattern of a variation that occurs approximately every 2 h could coincide with the inspection frequency. Therefore, only the low points on the cycle are reported, and there is no evidence that a cyclic event is present.

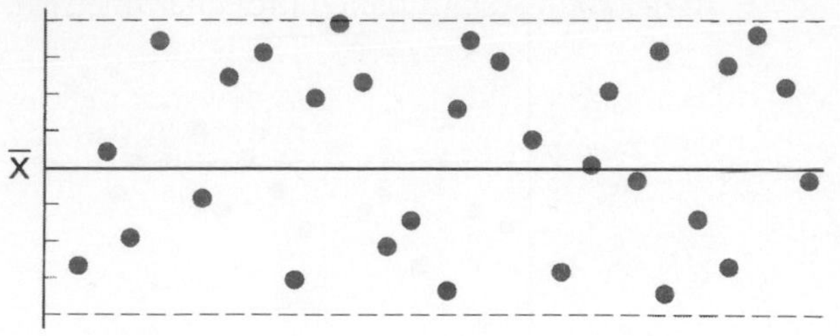

FIGURE 5-17 Out-of-control pattern: two populations.

4. *Two populations (also called mixture).* When there are a large number of points near or outside the control limits, a two-population situation may be present. This type of out-of-control pattern is illustrated in Figure 5-17.

For an $\overline{X}$ chart the out-of-control pattern can be due to

(a) Large differences in material quality
(b) Two or more machines on the same chart
(c) Large differences in test method or equipment

Some causes for an out-of-control pattern on an R chart are due to

(a) Different workers using the same chart
(b) Materials from different suppliers

5. *Mistakes.* Mistakes can be very embarrassing to quality assurance. Some causes of out-of-control patterns resulting from mistakes are

(a) Measuring equipment out of calibration
(b) Errors in calculations
(c) Errors in using test equipment
(d) Taking samples from different populations

Many of the out-of-control patterns that have been described can also be attributed to inspection error or mistakes.

The causes given for the different types of out-of-control patterns are suggested possibilities and are not meant to be all-inclusive. These causes will give production and quality personnel ideas for the solution of problems. They can be a start toward the development of an assignable cause checklist, which is applicable to their particular organization.

When out-of-control patterns occur in relation to the lower control limit of the R chart, it is the result of outstanding performance. The cause should be determined so that the outstanding performance can continue.

The preceding discussion has used the R chart as the measure of the dispersion. Information on patterns and causes also pertains to an s chart.

In the sixth step of the control chart method, it was stated that 25 subgroups were necessary to test an idea. The information given above on out of control can be used to make a decision with a fewer number of subgroups. For example, a run of six consecutive points in a downward trend on an R chart would indicate that the idea was a good one.

SPECIFICATIONS

Individual Values Compared to Averages

Before discussing specifications and their relationship with control charts, it appears desirable, at this time, to obtain a better understanding of individual values and average values. Figure 5-18 shows a tally of individual values (X's) and a tally of the subgroup averages ($\overline{X}$'s) for the data on keyway depths given in Table 5-2. The four out-of-control subgroups were not used in the two tallys; therefore, there are 84 individual values and 21 averages. It is observed that the averages are grouped much closer to the center than the individual values. When four values are

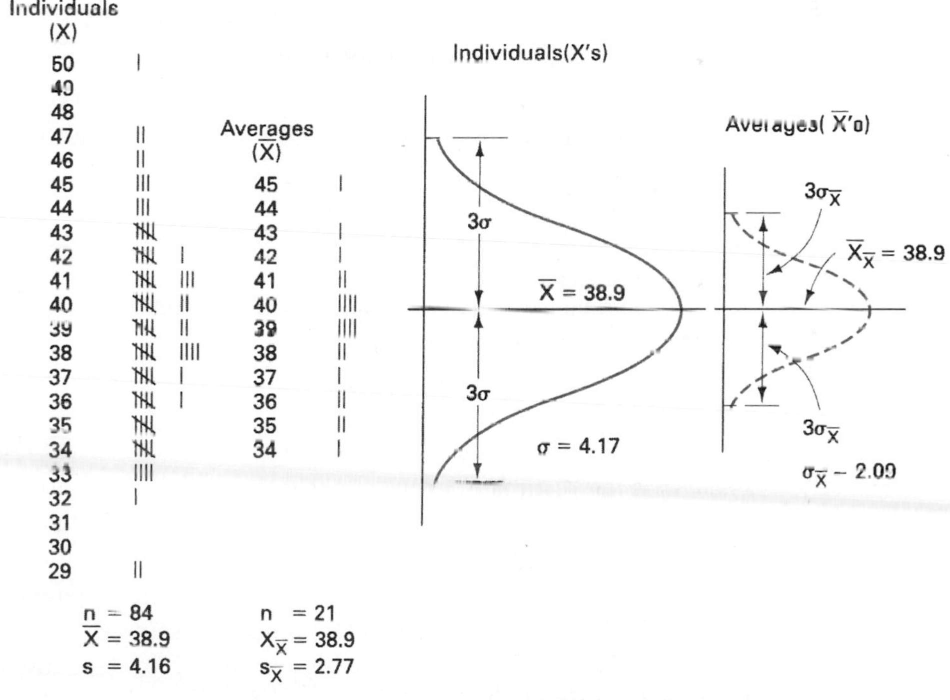

FIGURE 5-18 **Comparison of individual values and averages using the same data.**

averaged, the effect of an extreme value is minimized because the chance of four extremely high or four extremely low values in one subgroup is slight.

Calculations of the average for both the individual values and for the subgroup averages are the same, $\overline{X} = 38.9$. However, the sample standard deviation of the individual values (s) is 4.16, while the sample standard deviation of the subgroup average ($s_{\overline{X}}$) is 2.77.

If there are a large number of individual values and subgroup averages, the smooth polygons of Figure 5-18 would represent their frequency distributions if the distribution is normal. The curve for the frequency distribution of the averages has a dashed line while the curve for the frequency distribution of individual values has a solid line; this convention will be followed throughout the book. In comparing the two distributions it is observed that both distributions are normal in shape; in fact, even if the curve for individual values was not quite normal, the curve for averages would be close to a normal shape. The base of the curve for individual values is about twice as large as the base of the curve for averages. When population values are available for the standard deviation of individual values ($\hat{\sigma}$) and for the standard deviation of averages ($\sigma_{\overline{X}}$), there is a definite relationship between them, as given by the formula

$$\sigma_{\overline{X}} = \frac{\sigma}{\sqrt{n}}$$

where $\sigma_{\overline{X}}$ = population standard deviation of subgroup averages ($\overline{X}$'s)
σ = population standard deviation of individual values (X's)
n = subgroup size

Thus, for a subgroup of size 5, $\sigma_{\overline{X}} = 0.45\sigma$, and for a subgroup of size 4, $\sigma_{\overline{X}} = 0.50\sigma$.

If we assume normality (which may or may not be true), the population standard deviation can be estimated from

$$\hat{\sigma} = \frac{s}{c_4}$$

where $\hat{\sigma}$ is the "estimate" of the population standard deviation and c_4[9] is approximately equal to 0.996997 for $n = 84$. Thus, $\sigma = s/c_4 = 4.16/0.996997 = 4.17$ and $\sigma_{\overline{X}} = \sigma/\sqrt{n} = 4.17/\sqrt{4} = 2.09$. Note that $s_{\overline{X}}$, which was calculated from sample data, and $\sigma_{\overline{X}}$, which was calculated above, are different. This difference is due to sample variation or the small number of samples, which was only 21, or some combination thereof. The difference would not be caused by a nonnormal population of X's.

---

[9]Values of c_4 are given in Table B of the Appendix up to $n = 20$. For values greater than 20, $c_4 = \frac{4(n-1)}{4n-3}$.

Since the height of the curve is a function of the frequency, the curve for individual values is higher. This is easily verified by comparing the tally sheet in Figure 5-18. However, if the curves represent relative or percentage frequency distributions, then the area under the curve must be equal to 100%. Therefore, the percentage frequency distribution curve for averages, with its smaller base, would need to be much higher to enclose the same area as the percentage frequency distribution curve for individual values.

Central Limit Theorem

Now that you are aware of the difference between the frequency distribution of individual values, X's, and the frequency distribution of averages, $\overline{X}$'s, the central limit theorem can be discussed. In simple terms it is:

> If the population from which samples are taken is *not* normal, the distribution of sample averages will tend toward normality provided that the sample size, *n*, is at least 4. This tendency gets better and better as the sample size gets larger. Furthermore, the standardized normal can be used for the distribution of averages with the modification,
>
> $$Z = \frac{\overline{X} - \mu}{\sigma_{\overline{X}}} = \frac{\overline{X} - \mu}{\sigma \sqrt{n}}$$

This theorem was illustrated by Shewhart[10] for a uniform population distribution and a triangular population distribution of individual values as shown in Figure 5-19. Obviously, the distribution of X's are considerably different than a normal distribution; however, the distribution of $\overline{X}$'s is approximately normal.

The central limit theorem is one of the reasons the $\overline{X}$ chart works, in that we do not need to be concerned if the distribution of X's is not normal, provided that the sample size is 4 or more. Figure 5-20 shows the results of a dice experiment. First is a distribution of individual rolls of a six-sided die; second is a distribution of the average of rolls of two dice. The distribution of the averages ($\overline{X}$'s) is unimodal, symmetrical, and tapers off at the tails. This experiment provides practical evidence of the validity of the central limit theorem.

Control Limits and Specifications

Control limits are established as a function of the averages; in other words, control limits are for averages. Specifications, on the other hand, are the permissible variation in the size of the part and are, therefore, for individual values. The specification or tolerance limits are established by design engineers to meet a particular function. Figure 5-21 shows that the location of the specifications is optional and is not re-

[10]W. A. Shewhart, *Economic Control of Quality of Manufactured Product* (Princeton, N.J.: Van Nostrand Reinhold Company, Inc., 1931), pp. 180–186.

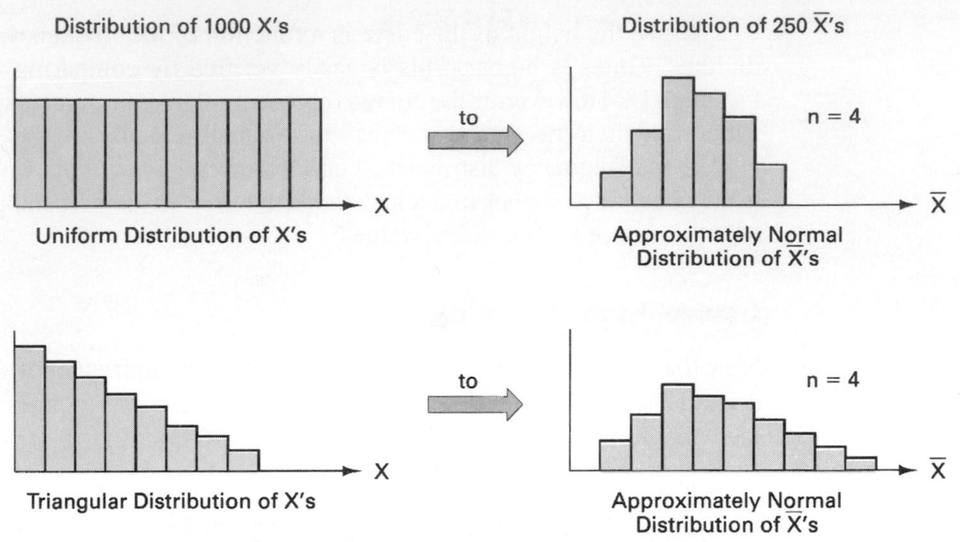

FIGURE 5-19 **Illustration of central limit theorem.**

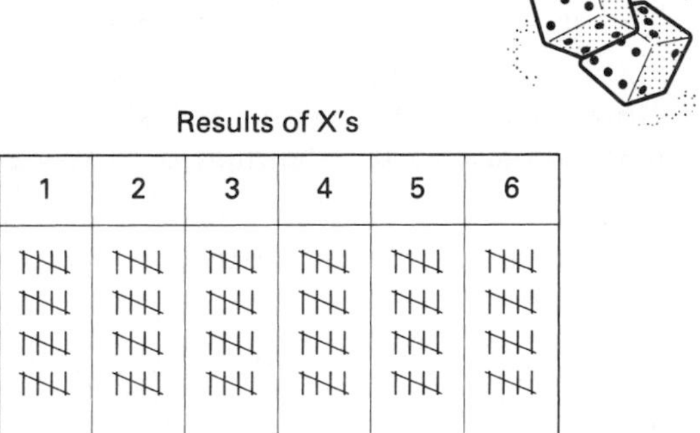

1	2	3	4	5	6
𝍸 𝍸 𝍸 𝍸	𝍸 𝍸 𝍸 𝍸	𝍸 𝍸 𝍸 𝍸	𝍸 𝍸 𝍸 𝍸	𝍸 𝍸 𝍸 𝍸	𝍸 𝍸 𝍸 𝍸

Results of $\overline{X}$'s, n = 2

1.0	1.5	2.0	2.5	3.0	3.5	4.0	4.5	5.0	5.5	6.0
||||	𝍸 |||	𝍸 𝍸 ||	𝍸 𝍸 𝍸 |	𝍸 𝍸 𝍸 𝍸	𝍸 𝍸 𝍸 𝍸 ||||	𝍸 𝍸 𝍸 𝍸	𝍸 𝍸 𝍸 |	𝍸 𝍸 ||	𝍸 |||	||||

FIGURE 5-20 **Dice illustration of central limit theorem.**

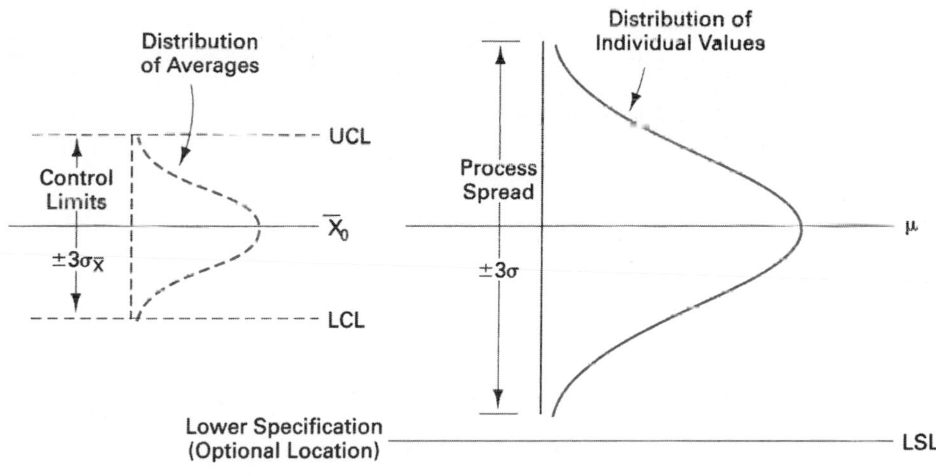

FIGURE 5-21 Relationship of limits, specifications, and distributions.

lated to any of the other features in the figure. The control limits, process spread, distribution of averages, and distribution of individual values are interdependent. They are determined by the process, whereas the specifications have an optional location. Control charts cannot determine if the process is meeting specifications.

Process Capability and Tolerance

Hereafter the process spread will be referred to as the process capability and is equal to 6σ. Also, the difference between specifications is called the tolerance. When the tolerance is established by the design engineer without regard to the spread of the process, undesirable situations can result. Three situations are possible: (1) when the process capability is less than the tolerance, (2) when the process capability is equal to the tolerance, and (3) when the process capability is greater than the tolerance.

Case I: $6\sigma < \text{USL} - \text{LSL}$. This situation, where the process capability (6σ) is less than the tolerance (USL - LSL), is the most desirable case. Figure 5-22 illustrates this ideal relationship by showing the distribution of individual values (X's), the $\overline{X}$ control chart limits, and distribution of averages ($\overline{X}$'s). The process is in control at (a). Since the tolerance is appreciably greater than the process capability, no difficulty is encountered even when there is a substantial shift in the process average, as shown at (b). This shift has resulted in an out-of-control condition as shown by the plotted points. However, no waste is produced because the

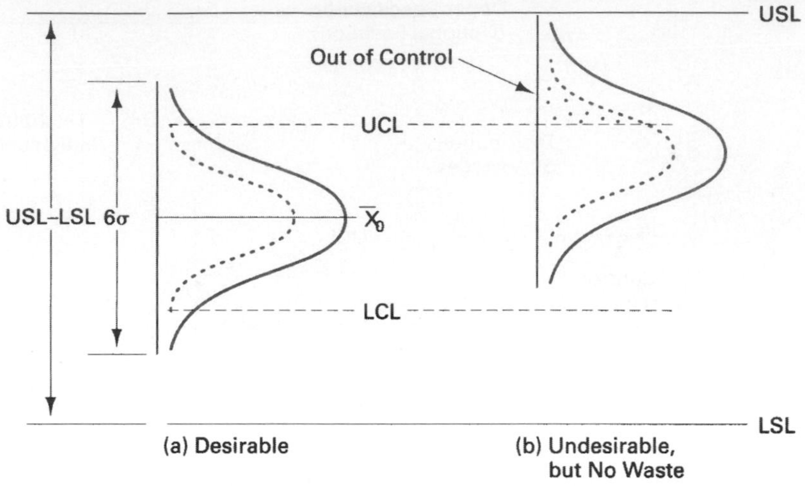

FIGURE 5-22 **Case I 6σ < USL − LSL.**

distribution of individual values (X's) has not exceeded the upper specification. Corrective action is required to bring the process into control.

Case II: 6σ = USL − LSL. Figure 5-23 illustrates this case where the process capability is equal to the tolerance. The frequency distribution of X's at (a) represents a natural pattern of variation. However, when there is a shift in the process average, as indicated at (b), the individual values (X's) exceed the specifications. As long as the process remains in control, no nonconforming product is produced; however, when the process is out of control as indicated at (b), nonconforming

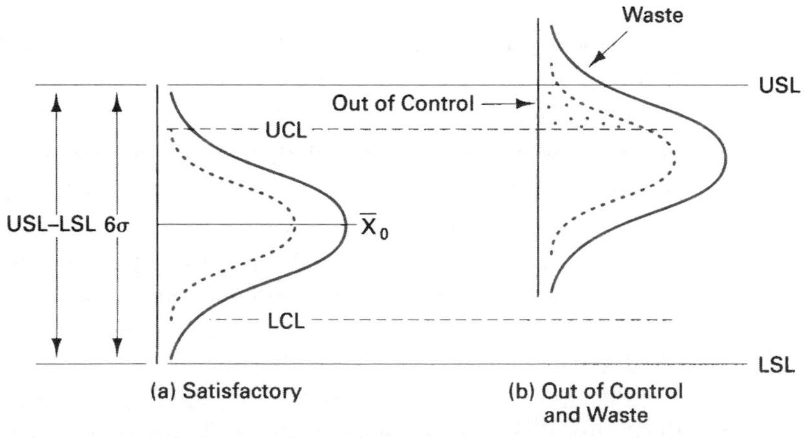

FIGURE 5-23 **Case II 6σ = USL − LSL.**

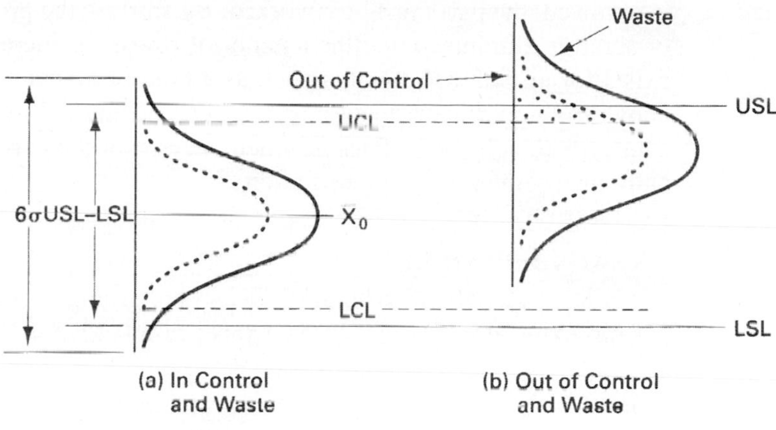

FIGURE 5-24 Case III $6\sigma >$ USL − LSL.

product is produced. Therefore, assignable causes of variation must be corrected as soon as they occur.

Case III: $6\sigma >$ USL − LSL. When the process capability is greater than the tolerance, an undesirable situation exists. Figure 5-24 illustrates this case. Even though a natural pattern of variation is occurring, as shown by the frequency distribution of X's at (a), some of the individual values are greater than the upper specification and are less than the lower specification. This case presents the unique situation where the process is in control as shown by the control limits and frequency distribution of $\overline{X}$'s, but nonconforming product is produced. In other words, the process is not capable of manufacturing a product that will meet the specifications. When the process changes as shown at (b), the problem is much worse.

When this situation occurs, 100% inspection is necessary to eliminate the nonconforming product.

One solution is to discuss with the design engineer the possibility of increasing the tolerance. This solution may require reliability studies with mating parts to determine if the product can function with an increased tolerance. Selective assembly might also be considered by the engineer.

A second possibility is to change the process dispersion so that a more peaked distribution occurs. To obtain a substantial reduction in the standard deviation might require new material, a more experienced operator, retraining, a new or overhauled machine, or possibly automatic in-process control.

Another solution is to shift the process average so that all of the nonconforming product occurs at one tail of the frequency distribution, as indicated in Figure 5-24(b). To illustrate this solution, assume that a shaft is being ground to tight specifications. If too much metal is removed, the part is scrapped; if too little is removed, the part must be reworked. By shifting the process average the amount of scrap is eliminated and the amount of rework is increased. A similar situation exists for an internal member such as a hole or keyway except that scrap occurs above the upper specification and rework occurs below the lower specification. This type of solution is feasible when the cost of the part is sufficient to economically justify the reworking operation.

EXAMPLE PROBLEM

Location pins for workholding devices are ground to a diameter of 12.50 mm (approximately 1/2 in.), with a tolerance of $\pm$0.05 mm. If the process is centered at 12.50 mm () and the dispersion is 0.02 mm (σ), what percent of the product must be scrapped and what percent can be reworked? How can the process center be changed to eliminate the scrap? What is the rework percentage?

The techniques for solving this problem were given in Chapter 3 and are shown below.

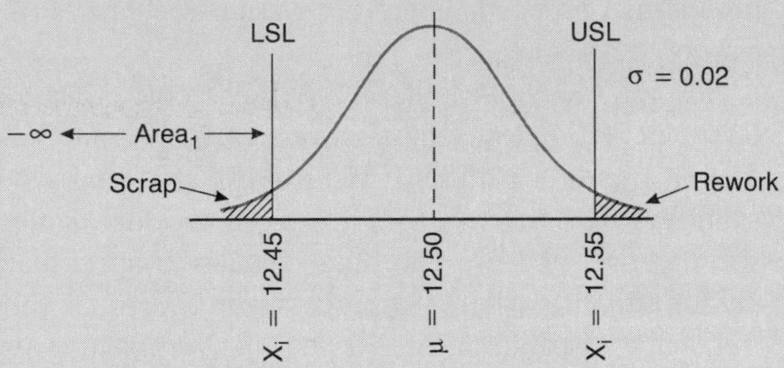

$$\text{USL} = \mu + 0.05 = 12.50 + 0.05 = 12.55 \text{ mm}$$

$$\text{LSL} = \mu - 0.05 = 12.50 - 0.05 = 12.45 \text{ mm}$$

$$Z = \frac{X_i - \mu}{\sigma}$$

$$= \frac{12.45 - 12.50}{0.02}$$

$$= -2.50$$

From Table A of the Appendix for a Z value of -2.50:

$$\text{Area}_1 = 0.0062 \text{ or } 0.62\% \text{ scrap}$$

Since the process is centered between the specifications and a symmetrical distribution is assumed, the rework percentage will be equal to the scrap percentage of 0.62%. The second part of the problem is solved using the following sketch:

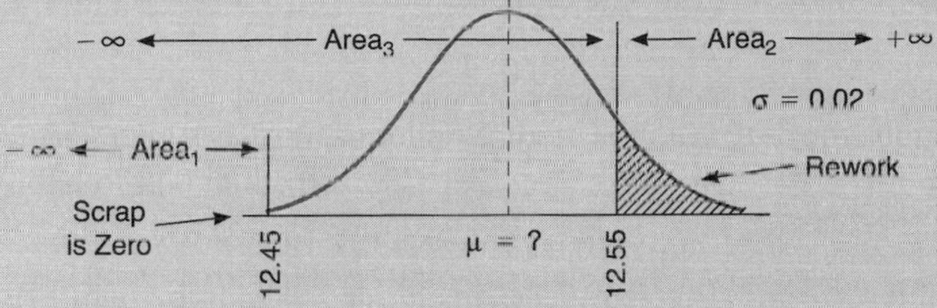

If the amount of scrap is to be zero, then $\text{Area}_1 = 0$. From Table A, the closest value to an Area_1 value of zero is 0.00017, which has a Z value of -3.59. Thus,

$$Z = \frac{X_i - \mu}{\sigma}$$

$$-3.59 = \frac{12.45 - \mu}{0.02}$$

$$\mu = 12.52\text{mm}$$

The percentage of rework is obtained by first determining Area_3.

$$Z = \frac{X_i - \mu}{\sigma}$$

$$= \frac{12.55 - 12.52}{0.02}$$

$$= +1.50$$

From Table A, $\text{Area}_3 = 0.9332$ and

$$\text{Area}_2 = \text{Area}_T - \text{Area}_3$$

$$= 1.0000 - 0.9332$$

$$= 0.0668, \text{ or } 6.68\%$$

The amount of rework is 6.68%, which, incidentally, is considerably more than the combined rework and scrap percentage (1.24%) when the process is centered.

The preceding analysis of the process capability and the specifications was made utilizing an upper and a lower specification. Many times there is only one specification and it may be either the upper or lower. A similar and much simpler analysis would be for a single specification limit.

PROCESS CAPABILITY

The true process capability cannot be determined until the $\overline{X}$ and R charts have achieved the optimal quality improvement without a substantial investment for new equipment or equipment modification. Process capability is equal to $6\sigma_0$ when the process is in statistical control.

In the example problem for the $\overline{X}$ and R charts, the quality improvement process began in January with $\sigma_0 = 0.038$. The process capability is $6\sigma = (6)(0.038) = 0.228$ mm. By July, $\sigma_0 = 0.030$, which gives a process capability of 0.180 mm. This is a 20% improvement in the process capability, which in most situations would be sufficient to solve a quality problem.

It is frequently necessary to obtain the process capability by a quick method rather than by using the $\overline{X}$ and R charts. This method assumes the process is stable or in statistical control, which may or may not be the case. The procedure is

1. Take 25 subgroups of size 4 for a total of 100 measurements.
2. Calculate the range, R, for each subgroup.
3. Calculate the average range, $\overline{R} = \Sigma R/g = \Sigma R/25$.
4. Calculate the estimate of the population standard deviation

$$\hat{\sigma}_0 = \frac{\overline{R}}{d_2}$$

   where d_2 is obtained from Table B and is 2.059 for $n = 4$.
5. Process capability will equal $6\sigma_0$.

Remember that this technique does not give the true process capability and should be used only if circumstances require its use. Also, more than 25 subgroups can be used to improve accuracy.

EXAMPLE PROBLEM

An existing process is not meeting the Rockwell-C specifications. Determine the process capability based on the range values for 25 subgroups of size 4. Data are 7, 5, 5, 3, 2, 4, 5, 9, 4, 5, 4, 7, 5, 7, 3, 4, 4, 5, 6, 4, 7, 7, 5, 5, and 7.

$$\overline{R} = \frac{\Sigma R}{g} = \frac{129}{25} = 5.16$$

$$\sigma_0 = \frac{\overline{R}}{d_2} = \frac{5.16}{2.059} = 2.51$$

$$6\sigma_0 = (6)(2.51) = 15.1$$

The process capability can also be obtained by using the standard deviation. Statistical control of the process is assumed. The procedure is

1. Take 25 subgroups of size 4 for a total of 100 measurements.
2. Calculate the sample standard deviation, s, for each subgroup.
3. Calculate the average sample standard deviation, $\bar{s} = \Sigma\, s/g = \Sigma\, s/25$.
4. Calculate the estimate of the population standard deviation

$$\hat{\sigma}_0 = \bar{s}/c_4$$

   where c_4 is obtained from Table B and is 0.9213 for $n = 4$.
5. Process capability will equal $6\sigma_0$.

More than 25 subgroups will improve the accuracy.

EXAMPLE PROBLEM

A new process is started, and the sum of the sample standard deviations for 25 subgropus of size 4 is 105. Determine the process capability,

$$\bar{s} = \frac{\Sigma s}{g} = \frac{105}{25} = 4.2$$

$$\sigma_0 = \frac{\bar{s}}{c_4} = \frac{4.2}{0.9213} = 4.56$$

$$6\sigma_0 = (6)(4.56) = 27.4$$

Either the range or the standard deviation method can be used. A histogram should be constructed to graphically present the process capability. Actually, a minimum of 50 measurements is required for a histogram. Therefore, the histograms made from the same data that were used to calculate the process capability should adequately represent the process at that time.

Process capability and the tolerance are combined to form a *capability index*, defined as

$$C_p = \frac{\text{USL} - \text{LSL}}{6\sigma_0}$$

where C_p = capability index
USL − LSL = upper specification − lower specification, or tolerance
$6\sigma_0$ = process capability

If the capability index is 1.00, we have the case II situation discussed in the preceding section; if the ratio is greater than 1.00, we have the case I situation, which is desirable; and if the ratio is less than 1.00, we have the case III situation, which is undesirable, Figure 5-25 shows these three cases.

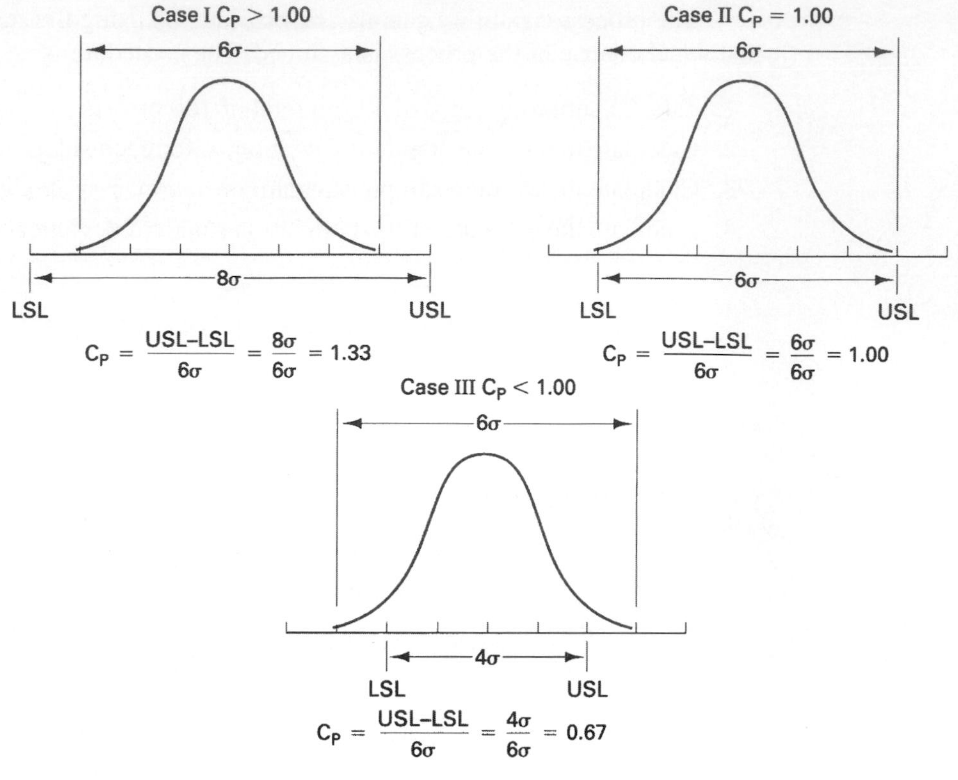

FIGURE 5-25 Capability index and three cases.

EXAMPLE PROBLEM

Assume that the specifications are 6.50 and 6.30 in the depth of keyway problem. Determine the capability index before ($\sigma_0 = 0.038$) and after ($\sigma_0 = 0.030$) improvement.

$$C_p = \frac{\text{USL} - \text{LSL}}{6\sigma_0} = \frac{6.50 - 6.30}{6(0.038)} = 0.88$$

$$C_p = \frac{\text{USL} - \text{LSL}}{6\sigma_0} = \frac{6.50 - 6.30}{6(0.030)} = 1.11$$

In the example problem the improvement in quality resulted in a desirable capability index (case I). The minimum capability index is frequently established at 1.33. Below this value, design engineers may be required to seek approval from manufacturing before the product can be released to production. A capability index of 1.33 is considered by most companies to be a de facto standard with even larger values desired.

Using the capability index[11] concept, we can measure quality provided the process is centered. The larger the capability index, the better the quality. We should strive to make the capability index as large as possible. This is accomplished by having realistic specifications and continual striving to improve the process capability.

The capability index does not measure process performance in terms of the nominal or target value. This measure is accomplished using C_{pk}, which is defined as

$$C_{pk} = \frac{\text{Min}\{(\text{USL} - \overline{X}) \text{ or } (\overline{X} - \text{LSL})\}}{3\sigma}$$

EXAMPLE PROBLEM

Determine C_{pk} for the previous example problem (USL = 6.50, LSL = 6.30, and $\sigma = 0.030$) when the average is 6.45.

$$C_{pk} = \frac{\text{Min}\{(\text{USL} - \overline{X}) \text{ or } (\overline{X} - \text{LSL})\}}{3\sigma}$$

$$= \frac{\text{Min}\{(6.50 - 6.45) \text{ or } (6.45 - 6.30)\}}{3(0.030)}$$

$$= \frac{0.05}{0.090} = 0.56$$

Find C_{pk} when the average is 6.38.

$$C_{pk} = \frac{\text{Min}\{(\text{USL} - \overline{X}) \text{ or } (\overline{X} - \text{LSL})\}}{3\sigma}$$

$$= \frac{\text{Min}\{(6.50 - 6.38) \text{ or } (6.38 - 6.30)\}}{3(0.030)}$$

$$= \frac{0.08}{0.090} = 0.89$$

---

[11]Another measure of the capability is called the *capability ratio*, which is defined as

$$C_r = \frac{6\sigma_0}{\text{USL} - \text{LSL}}$$

The only difference between the two measures is the change in the numerator and denominator. They are used for the same purpose; however, the interpretation is different. The de facto standard for a capability ratio is 0.75 with even smaller values desired. In both cases the de facto standard is established with the tolerance at $8\sigma_0$. To avoid misinterpretation between two parties, they should be sure which process capability measure is being used. In this book the capability index is used.

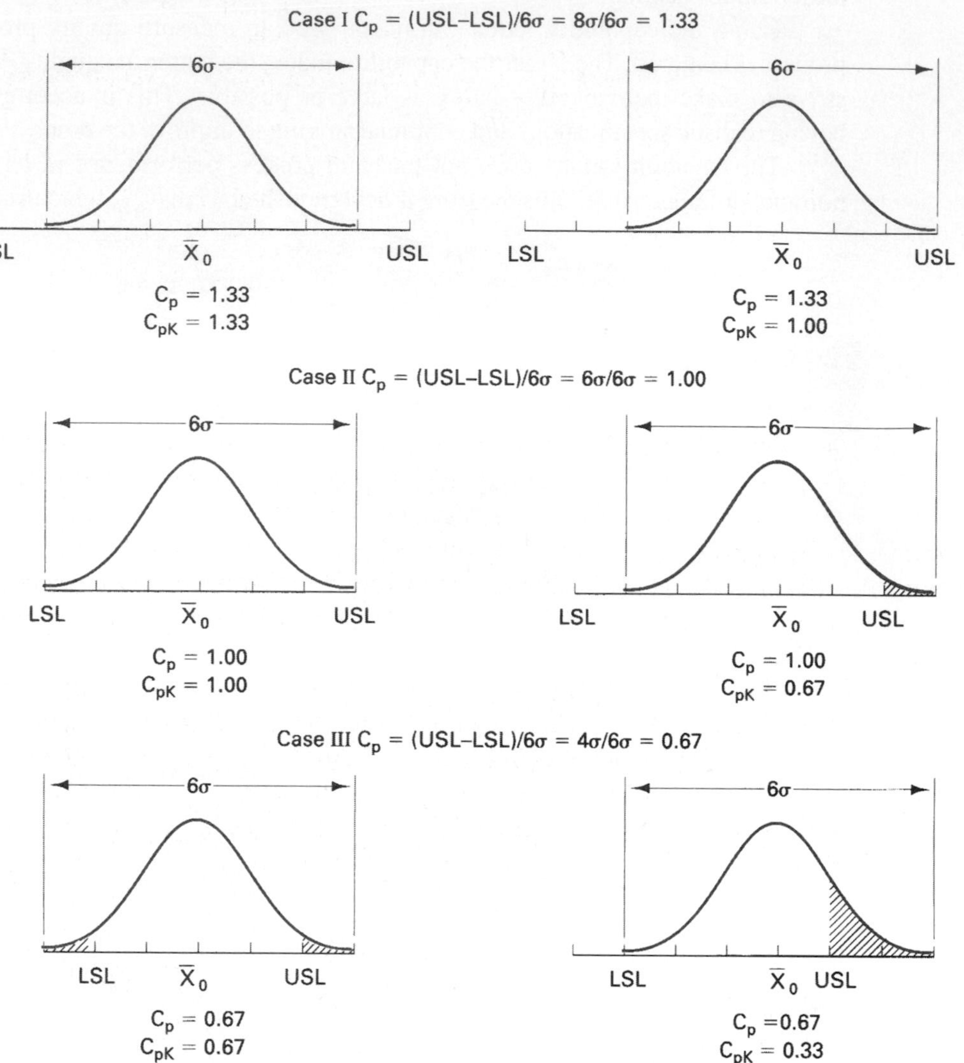

FIGURE 5-26 **C_p and C_{pk} values for the three cases.**

Figure 5-26 illustrates C_p and C_{pk} values for a process that is centered and one that is off center by 1σ for the three cases. Comments concerning C_p and C_{pk} are as follows:

1. The C_p value does not change as the process center changes.
2. $C_p = C_{pk}$ when the process is centered.

3. C_{pk} is always equal to or less than C_p.
4. A C_{pk} value of 1.00 is a de facto standard. It indicates that the process is producing product that conforms to specifications.
5. A C_{pk} value less than 1.00 indicates that the process is producing product that does not conform to specifications.
6. A C_p value less than 1.00 indicates that the process is not capable.
7. A C_{pk} value of zero indicates the average is equal to one of the specification limits.
8. A negative C_{pk} value indicates that the average is outside the specifications.

SIX SIGMA

As previously stated, standard deviation is the best measure of process variability because the smaller the standard deviation, the less variability in the process. If we can reduce sigma, σ, to the point that the specifications are at ± 6, then 99.9999998% of the product or service will be between specifications, and the nonconformance rate will be 0.002 parts/million with a C_p value of 2.0. Figure 5-27 illustrates this situation and Table 5-5 provides information about when it is possible to establish the specifications at other values.

According to the *Six-Sigma* philosophy, processes rarely stay centered—the center tends to "shift" above and below the target, . Figure 5-28 shows a process that is normally distributed, but has shifted within a range of 1.5σ above and 1.5 below the target. For the diagrammed situation, 99.9996600% of the product or service will be between specifications and the nonconformance rate will be 3.4 ppm. This off-center situation gives a process capability index (C_{pk}) of 1.5 Table 5-6

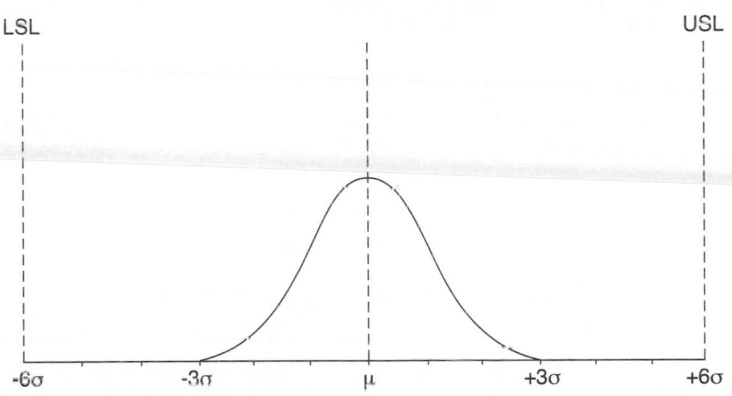

FIGURE 5-27 Nonconformance Rate When Process is Centered

TABLE 5-5 Nonconformance Rate and Process Capability When the Process is Centered

SPECIFICATION LIMIT	PERCENT CONFORMANCE	NONCONFORMANCE RATE (PPM)	PROCESS CAPABILITY (C_P)
$\pm 1\sigma$	68.7	317300	0.33
$\pm 2\sigma$	95.45	485500	0.67
$\pm 3\sigma$	99.73	2700	1.00
$\pm 4\sigma$	99.9937	63	1.33
$\pm 5\sigma$	99.999943	0.57	1.67
$\pm 6\sigma$	99.9999998	0.002	2.00

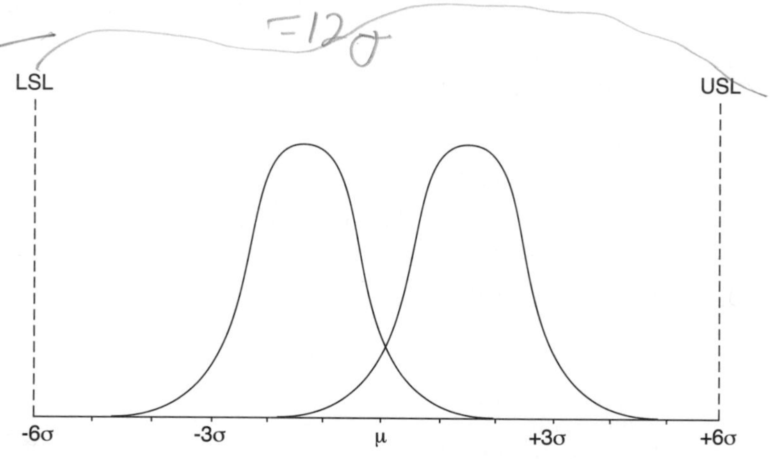

FIGURE 5-28 Nonconformance Rate When Process is Off-Center $\pm 1.5\sigma$.

TABLE 5-6 Nonconformance Rate and Process Capability When the Process is Off-Center $\pm 1.5\sigma$.

SPECIFICATION LIMIT	PERCENT CONFORMANCE	NONCONFORMANCE RATE (PPM)	PROCESS CAPABILITY (C_{PK})
$\pm 1\sigma$	30.23	697700	−0.167
$\pm 2\sigma$	69.13	308700	0.167
$\pm 3\sigma$	93.32	66810	0.500
$\pm 4\sigma$	99.3790	6210	0.834
$\pm 5\sigma$	99.97670	2330	1.167
$\pm 6\sigma$	99.9996600	3.4	1.500

shows the percent between specifications, the nonconformance rate, and process capability for different specification limit locations. The magnitude and type of shift is a matter of discovery and should not be assumed ahead of time.

Achieving Six-Sigma will not be easy and should only be attempted for critical quality characteristics. The economics of the situation must be evaluated.

DIFFERENT CONTROL CHARTS

The basic control charts for variables were discussed in previous sections. While most of the quality control activity for variables is concerned with the $\overline{X}$ and R chart or the $\overline{X}$ and s chart, there are other charts which find application in some situations. These charts are discussed briefly in this section.

Charts for Better Operator Understanding

Since production personnel have difficulty understanding the relationships between averages, individual values, control limits, and specifications, various charts have been developed to overcome this difficulty.

1. *Placing individual values on the chart.* This technique plots both the individual values and the subgroup average and is illustrated in Figure 5-29. A small dot represents an individual value and a larger circle represents the subgroup average. In some cases, an individual value and a subgroup average are identical, in which case the small dot is located inside the circle. When two individual values are identical, the two dots are placed side by side. A further refinement of the chart can be made by the addition of upper and lower specification lines; however, this practice is not recommended. In fact, the plotting of individual values is an unnecessary activity that can be overcome by proper operator training.

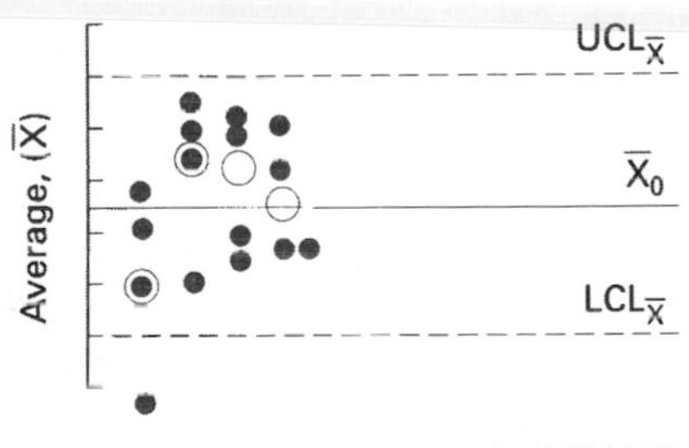

FIGURE 5-29 Chart showing a technique for plotting individual values and subgroup averages.

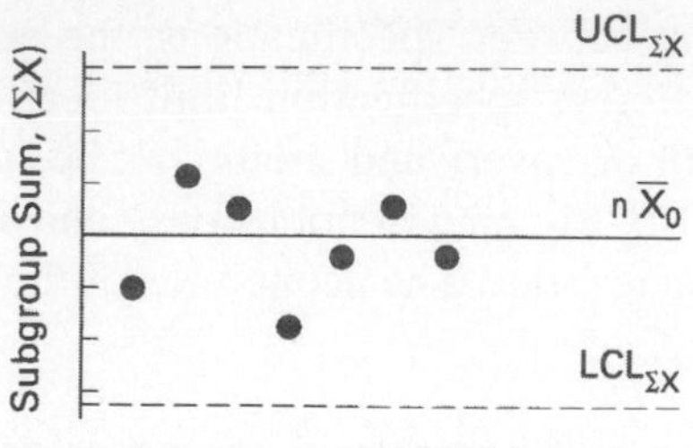

FIGURE 5-30 Subgroup sum chart.

2. *Chart for subgroup sums*. This technique plots the subgroup sum, $\boldsymbol{\sigma} X$, rather than the subgroup average, $\overline{X}$. Since the values on the chart are of a different magnitude than the specifications, there is no chance for confusion. Figure 5-30 shows a subgroup sum chart, which is an $\overline{X}$ chart with the scale magnified by the subgroup size, n. The central line is $n\overline{X}_0$ and the control limits are obtained by the formulas

$$\mathrm{UCL}_{\Sigma X} = n(\mathrm{UCL}_{\overline{X}})$$

$$\mathrm{LCL}_{\Sigma X} = n(\mathrm{LCL}_{\overline{X}})$$

This chart is mathematically equal to the $\overline{X}$ chart and has the added advantage of simpler calculations. Only addition and subtraction are required.

Chart for Variable Subgroup Size

Every effort should be made to keep the subgroup size constant. Occasionally, however, because of lost material, laboratory tests, production problems, or inspection mistakes, the subgroup size varies. When this situation occurs, the control limits will vary with the subgroup size. As the subgroup size, n, increases, the control limits become narrower; as the subgroup size decreases, the control limits become wider apart (Figure 5-31). This fact is confirmed by an analysis of the control limit factors A. D_1, and D_2, which are a function of the subgroup size and which are part of the control limit formulas. Control limits will also vary for the R chart.

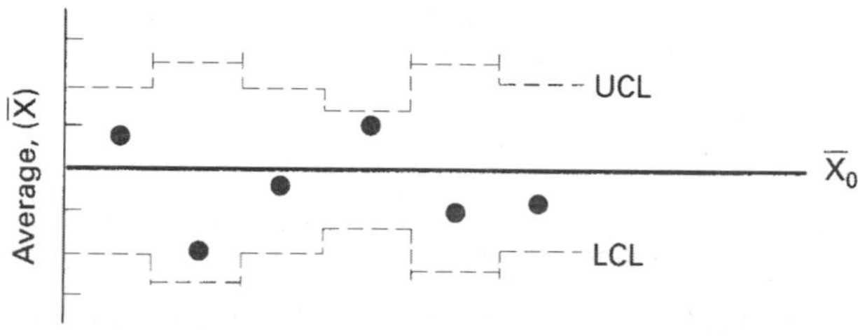

FIGURE 5-31 Chart for variable subgroup size.

One of the difficulties associated with a chart for variable subgroup size is the need to make a number of control limit calculations. A more serious difficulty involves the task of explaining to production people the reason for the different control limits. Therefore, this type of chart should be avoided.

Chart for Trends

When the plotted points of a chart have an upward or downward trend, it can be attributed to an unnatural pattern of variation or to a natural pattern of variation such as tool wear. In other words, as the tool wears, a gradual change in the average is expected and considered to be normal. Figure 5-32 illustrates a chart for a trend that reflects die wear. As the die wears, the measurement gradually increases until it reaches the upper reject limit. The die is then replaced or reworked.

Since the central line is on a slope, its equation must be determined. This is best accomplished using the least-squares method of fitting a line to a set of points. The equation for the trend line, using the slope-intercept form, is

$$\overline{X} = a + bG$$

where $\overline{X}$ = subgroup average and represents the vertical axis

G = subgroup number and represents the horizontal axis

a = point on the vertical axis where the line intercepts the vertical axis

$$a = \frac{(\Sigma\,\overline{X})(\Sigma\,G^2) - (\Sigma\,G)(\Sigma\,G\overline{X})}{g\Sigma\,G^2 - (\Sigma\,G)^2}$$

b = the slope of the line

$$b = \frac{g\Sigma\,G\overline{X} - (\Sigma\,G)(\Sigma\,\overline{X})}{g\Sigma\,G^2 - (\Sigma\,G)^2}$$

g = number of subgroups

The coefficients of a and b are obtained by establishing columns for G, $\overline{X}$, $G\overline{X}$, and G^2, as illustrated in Table 5-7; determining their sums; and inserting the sums in the equation.

Once the trend-line equation is known, it can be plotted on the chart by assuming values of G and calculating $\overline{X}$. When two points are plotted, the trend line is drawn between them. The control limits are drawn on each side of the trend line a distance (in the perpendicular direction) equal to $A_2\overline{R}$ or $A\sigma_0$.

The R chart will generally have the typical appearance shown in Figure 5-7. However, the dispersion may also be increasing.

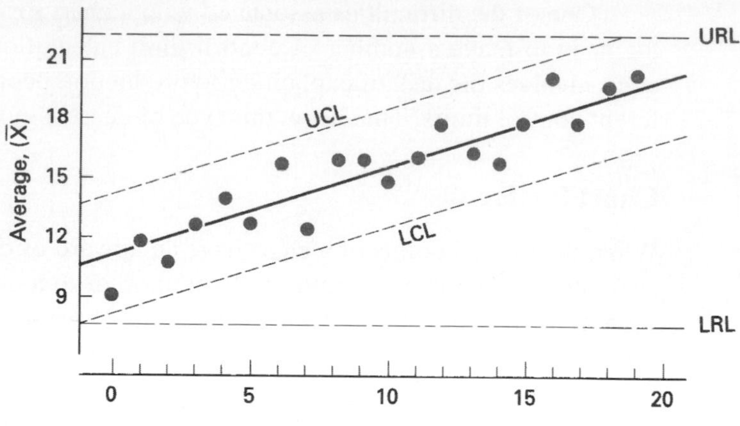

FIGURE 5-32 **Chart for trend.**

TABLE 5-7 **Least-Squares Calculations for Trend Line**

SUBGROUP NUMBER G	SUBGROUP AVERAGE $\overline{X}$	PRODUCT OF G AND $G\overline{X}$ $G\overline{X}$	G^2
1	9	9	1
2	11	22	4
3	10	30	9
.	.	.	.
.	.	.	.
.	.	.	.
.	.	.	.
g			
ΣG	$\Sigma\overline{X}$	$\Sigma G\overline{X}$	ΣG^2

Chart for Moving Average and Moving Range

In some situations a chart is used to combine a number of individual values and plot them on a control chart. This type is referred to as a moving-average and moving-range chart and is quite common in the chemical industry, where only one reading is possible at a time. Table 5-8 illustrates the technique. In the development of Table 5-8, no calculations are made until the third period when the sum of the three values is posted to the three-period moving-sum column ($35 + 26 + 28 = 89$). The average and range are calculated ($\overline{X} = \frac{89}{3} = 29.6$) ($R = 35 - 26 = 9$) and

TABLE 5-8 Calculations of Moving Average and Moving Range

VALUE	THREE-PERIOD MOVING SUM	X	R
35	—	—	—
26	—	—	—
28	89	29.6	9
32	86	28.6	6
36	96	32.0	8
.	.	.	.
.	.	.	.
.	.	.	.
.	.	.	.
.	.	.	.
		$\Sigma\overline{X} =$	$\Sigma R =$

posted to the $\overline{X}$ and R columns. Subsequent calculations are accomplished by adding a new value and dropping the earliest one; therefore, 32 is added and 35 is dropped, making the sum $26 + 28 + 32 = 86$. The average and range calculations are $\overline{X} = \frac{86}{3} = 28.6$ and $R = 32 - 26 = 6$. Once the columns for $\overline{X}$ and R are completed, the charts are developed and used in the same manner as regular $\overline{X}$ and R charts.

The discussion above used a time period of 3 h; the time period could have been 2 h, 5 days, 3 shifts, and so on.

In comparing the moving-average and moving-range charts with conventional charts, it is observed that an extreme reading has a greater effect on the former charts. This is true because an extreme value is used a number of times in the calculations.

Chart for Median and Range

A simplified variable control chart that minimizes calculations is the median and range. The data are collected in the conventional manner and the median, Md, and range, R, of each subgroup are found. When using manual methods, these values are arranged in ascending or descending order. The median of the subgroup medians or grand median, Md_{Md}, and the median of the subgroup range, R_{Md}, are found by counting to the midpoint value. The median control limits are determined from the formulas

$$\text{UCL}_{\text{Md}} = \text{Md}_{\text{Md}} + A_5 R_{\text{Md}}$$

$$\text{LCL}_{\text{Md}} = \text{Md}_{\text{Md}} - A_5 R_{\text{Md}}$$

TABLE 5-9 Factors for Computing 3σ Control Limits for Median and Range Charts from the Median Range

SUBGROUP SIZE	A_5	D_5	D_6	D_3
2	2.224	0	3.865	0.954
3	1.265	0	2.745	1.588
4	0.829	0	2.375	1.978
5	0.712	0	2.179	2.257
6	0.562	0	2.055	2.472
7	0.520	0.078	1.967	2.645
8	0.441	0.139	1.901	2.791
9	0.419	0.187	1.850	2.916
10	0.369	0.227	1.809	3.024

Source: Extracted by permission from P. C. CLifford, "Control Without Calculations," *Industrial Quality Control,* 15. No. 6 (May 1959), 44.

where Md_{Md} = grand median (median of the medians); Md_0 can be substituted in the formula

A_5 = factor for determining the 3σ control limits (see Table 5-9)

R_{Md} = median of subgroup ranges

The range control limits are determined from the formulas

$$\text{UCL}_R = \text{D}_6 R_{\text{Md}}$$

$$\text{LCL}_R = \text{D}_5 R_{\text{Md}}$$

where D_5 and D_6 are factors for determining the 3σ control limits based on R_{Md} and are found in Table 5-9. An estimate of the population standard deviation can be obtained from $\sigma = R_{\text{Md}}/d_3$.

The principal benefits of the median chart are (1) less arithmetic, (2) easier to understand, and (3) can be easily maintained by the operators. However, the median chart fails to grant any weight to extreme values in a subgroup.

When these charts are maintained by operating personnel a subgroup size of 3 is recommended. For example, consider the three values 36, 39, and 35. The Md is 36 and R is 4—all three values are used. Figure 5-33 is an example of a median chart. Subgroup sizes of 5 give a better chart; however, using the manual method the operator will have to order the data before determining the median. While these charts are not as sensitive to variation as the $\overline{X}$ and R, they can be quite effective, especially after quality improvement has been obtained and the process is in a monitoring phase. An unpublished Master's thesis study showed little difference in the effectiveness of the Md and R charts when compared to the $\overline{X}$ and R charts.

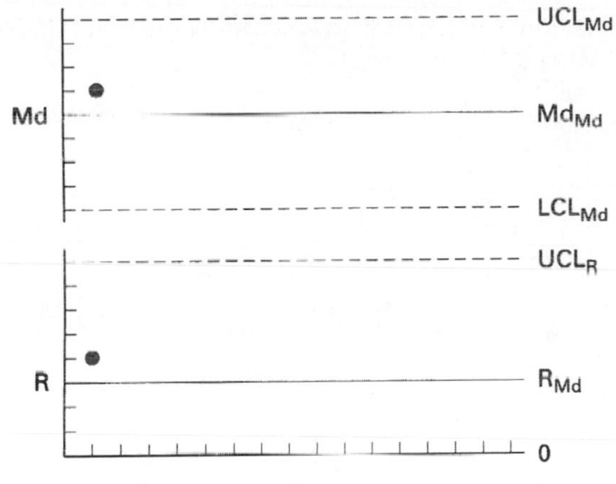

FIGURE 5-33 **Control charts for median and range.**

Chart for Individual Values

In many situations only one measurement is taken on a quality characteristic. This may be due to the fact that it is too expensive or too time consuming, there are too few items to inspect, or it may not be possible. In such cases an X chart will provide some information from limited data, whereas an $\overline{X}$ chart would provide no information or information only after considerable delay to obtain sufficient data. Figure 5-34 illustrates an X chart.

Formulas for the trial central line and control limits are

$$\overline{X} = \frac{\Sigma X}{g} \qquad \overline{R} = \frac{\Sigma R}{g}$$

$$\text{UCL}_x = \overline{X} + 2.660\overline{R} \qquad \text{UCL}_R = 3.267\overline{R}$$

$$\text{LCL}_x = \overline{X} - 2.660\overline{R} \qquad \text{LCL}_R = (0)\overline{R}$$

These formulas require the moving range technique with a subgroup size of 2.[12] To obtain the first range point, the value of X_1 is subtracted from X_2; to obtain the second point, X_2 is subtracted from X_3; and so forth. Each individual value is used for two different points except for the first and last: therefore, the name "moving" range. The range points should be placed between the subgroup number on the R chart since they are obtained from both values, or they can be placed at the second point.

These range points are averaged to obtain $\overline{R}$. Note that g for obtaining $\overline{R}$ will be one less than g for obtaining $\overline{X}$.

[12]This technique is the simplest approach to X and R charts. Other techniques are given on pages 23-15 to 23-17 of Juran, *Quality Control Handbook*, 3rd ed.

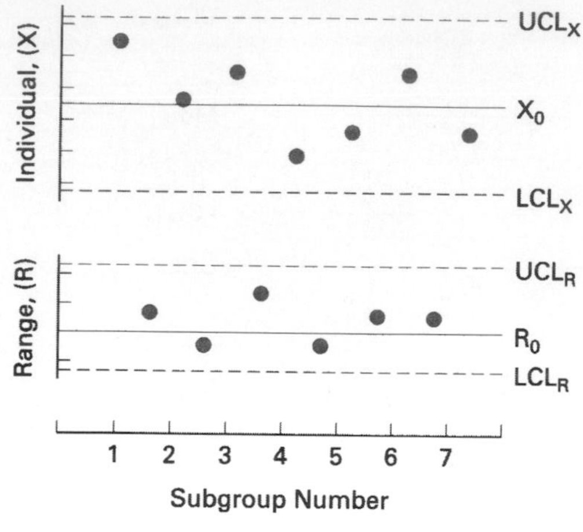

FIGURE 5-34 Control charts for individual values and moving range.

Formulas for the revised central line and control limits are

$$X_0 = \overline{X}_{\text{new}} \qquad R_0 = \overline{R}_{\text{new}}$$

$$\text{UCL}_x = X_0 + 3\sigma_0 \qquad \text{UCL}_R = (3.686)\sigma_0$$

$$\text{LCL}_x = X_0 - 3\sigma_0 \qquad \text{LCL}_R = (0)\sigma_0$$

where $\sigma_0 = 0.8865R_0$.

The X chart has the advantage of being easier for production personnel to understand and of providing a direct comparison with specifications. It does have the disadvantages of (1) requiring too many subgroups to indicate an out-of-control condition. (2) not summarizing the data as well as $\overline{X}$, and (3) distorting the control limits when the distribution is not normal. To correct for the last disadvantage, tests for normality should be used since the central limit theorem will not be applicable. Unless there is an insufficient amount of data, the $\overline{X}$ chart is recommended.

Charts with Non-Acceptance Limits

Non-Acceptance limits have the same relationship to averages as specifications have to individual values. Figure 5-35 shows the relationship of non-acceptance limits, control limits, and specifications for the three cases discussed in the section on specifications. The upper and lower specifications are shown in Figure 5-35 to illustrate the technique and are not included in actual practice.

In case I the non-acceptance limits are greater than the control limits, which is a desirable situation, since an out-of-control condition will not result in nonconforming product. Case II shows the situation where the non-acceptance limits are equal to the control limits; therefore, any out-of-control situations will result in

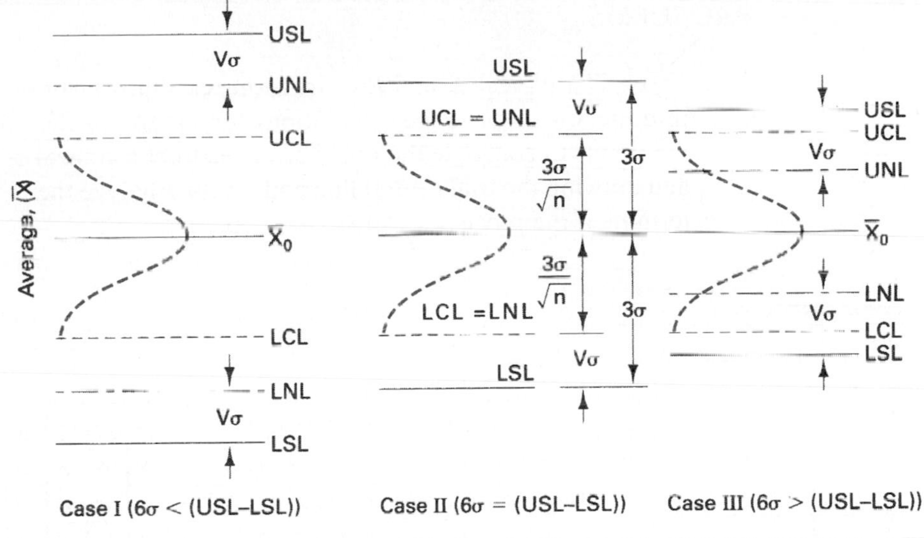

FIGURE 5-35 Relationship of non-acceptance limits, control limits, and specifications.

nonconforming product being manufactured. Case III illustrates the situation where the non-acceptance limits are inside the control limits, and therefore some nonconforming product will be manufactured even when the process is in control.

The figure shows that the non-acceptance limits are a prescribed distance from the specifications. This distance is equal to $V\sigma$, where V varies with the subgroup size and is equal to the value $3 - 3/\sqrt{n}$. The formula for V was derived from case II, because in that situation the control limits are equal to the non-acceptance limits.

Control limits tell what the process is capable of doing, and reject limits tell when the product is conforming to specifications. This can be a valuable tool for the quality professional and perhaps the first-line supervisor. Posting of non-acceptance limits for operating personnel should be avoided since they will be confusing and may lead to unnecessary adjustment. Also, the operator is only responsible to maintain the process between the control limits.

COMPUTER PROGRAM

Using EXCEL the software in the diskette inside the back cover will solve for $\overline{X}$ and R Charts, Md and R charts, X and MR Charts, and process capability. Their file names are *X-bar & R Charts, Md & R Charts, X & MR Charts* and *Process Capability*.

PROBLEMS

1. Given is a typical $\overline{X}$ and R chart form with information on acid content in milliliters. Complete the calculations for subgroups 22, 23, 24, and 25. Plot the points to complete the run chart. Construct a control chart by calculating and drawing the trial central line and limits. Analyze the plotted points to determine if the process is stable.

VARIABLES CONTROL CHART | DEPT/AREA: | CHART ID: *Problem 1*

PART ID: | OPERATION ID: | CHARACTERISTIC: *Acid Content*

CHECK METHOD: | NOMINAL VALUE: *0.70 ml* | TOLERANCE: *± 0.20*

SAMPLE READINGS	1	2	3	4	5	6	7	8	9	10	11	12	13	14	15	16	17	18	19	20	21	22	23	24	25
1	.85	.75	.80	.65	.75	.60	.80	.70	.75	.60	.80	.75	.70	.65	.85	.80	.70	.70	.65	.65	.55	.75	.80	.65	.65
2	.65	.85	.80	.75	.70	.75	.75	.60	.85	.70	.75	.85	.70	.70	.75	.75	.85	.60	.65	.60	.50	.65	.65	.60	.70
3	.65	.75	.75	.60	.65	.75	.65	.75	.85	.60	.90	.85	.75	.85	.80	.75	.75	.70	.85	.60	.65	.65	.75	.65	.70
4	.70	.85	.70	.70	.80	.70	.75	.75	.80	.80	.50	.65	.70	.75	.80	.80	.70	.70	.65	.65	.80	.80	.65	.60	.60
SUM, ΣX	2.85	3.20	3.05	2.70	2.90	2.80	2.95	2.80	3.25	2.70	2.95	3.10	2.85	2.95	3.20	3.10	3.00	2.70	2.80	2.50	2.50				
AVERAGE, $\overline{X}$	.71	.80	.76	.68	.73	.70	.74	.70	.81	.68	.74	.78	.71	.74	.80	.78	.75	.68	.70	.63	.63				
RANGE, R	.20	.10	.10	.15	.15	.15	.15	.15	.10	.20	.40	.20	.05	.20	.10	.05	.15	.10	.20	.05	.30				

2. Control charts for $\overline{X}$ and R are to be established on a certain dimension part, measured in millimeters. Data were collected in subgroup sizes of 6 and are given below. Determine the trial central line and control limits. Assume assignable causes and revise the central line and limits.

SUBGROUP NUMBER	$\overline{X}$	R	SUBGROUP NUMBER	$\overline{X}$	R
1	20.35	0.34	14	20.41	0.36
2	20.40	0.36	15	20.45	0.34
3	20.36	0.32	16	20.34	0.36
4	20.65	0.36	17	20.36	0.37
5	20.20	0.36	18	20.42	0.73
6	20.40	0.35	19	20.50	0.38
7	20.43	0.31	20	20.31	0.35
8	20.37	0.34	21	20.39	0.38
9	20.48	0.30	22	20.39	0.33
10	20.42	0.37	23	20.40	0.32
11	20.39	0.29	24	20.41	0.34
12	20.38	0.30	25	20.40	0.30
13	20.40	0.33			

3. The following table gives the average and range in kilograms for tensile tests on an improved plastic cord. The subgroup size is 4. Determine the trial central line and control limits. If any points are out of control, assume assignable causes and calculate revised limits and central line.

SUBGROUP NUMBER	$\overline{X}$	R	SUBGROUP NUMBER	$\overline{X}$	R
1	476	32	14	482	22
2	466	24	15	506	23
3	484	32	16	496	23
4	466	26	17	478	25
5	470	24	18	484	24
6	494	24	19	506	23
7	486	28	20	476	25
8	496	23	21	485	29
9	488	24	22	490	25
10	482	26	23	463	22
11	498	25	24	469	27
12	464	24	25	474	22
13	484	24			

4. Rework Problem 2 assuming subgroup sizes of 3, 4, and 5. How do the control limits compare?

5. Control charts for $\overline{X}$ and R are kept on the weight in kilograms of a color pigment for a batch process. After 25 subgroups with a subgroup size of 4, $\Sigma\overline{X} = 52.08$ kg (114.8 lb), $\Sigma R = 11.82$ kg (26.1 lb). Assuming the process is in a state of control, compute the $\overline{X}$ and R chart central line and control limits for the next production period.

do with formula
from pg 190-191

6. Control charts for $\overline{X}$ and s are to be established on the Brinell hardness of hardened tool steel in kilograms per square millimeter. Data for subgroup sizes of 8 are shown below. Determine the trial central line and control limits for the $\overline{X}$ and s charts. Assume that the out-of-control points have assignable causes. Calculate the revised limits and central line.

SUBGROUP NUMBER	$\overline{X}$	s	SUBGROUP NUMBER	$\overline{X}$	s
1	540	26	14	551	24
2	534	23	15	522	29
3	545	24	16	579	26
4	561	27	17	549	28
5	576	25	18	508	23
6	523	50	19	569	22
7	571	29	20	574	28
8	547	29	21	563	33
9	584	23	22	561	23
10	552	24	23	548	25
11	541	28	24	556	27
12	545	25	25	553	23
13	546	26			

7. Control charts for X and s are maintained on the resistance in ohms of an electrical part. The subgroup size is 6. After 25 subgroups, $\Sigma\overline{X} = 2046.5$ and $\Sigma s = 17.4$. If the process is in statistical control, what are the control limits and central line?

8. Rework Problem 6 assuming a subgroup size of 3.

9. Copy the s chart of Figure 5-8 on transparent paper. Place this copy on top of the R chart of Figure 5-5 and compare the pattern of variation.

10. In filling bags of nitrogen fertilizer, it is desired to hold the average overfill to as low a value as possible. The lower specification limit is 22.00 kg (48.50 lb), the population mean weight of the bags is 22.73 kg (50.11 lb), and the population standard deviation is 0.80 kg (1.76 lb). What percent of the bags contain less than 22 kg? If it is permissible for 5% of the bags to be below 22 kg, what would be the average weight? Assume a normal distribution.

11. Plastic strips that are used in a sensitive electronic device are manufactured to a maximum specification of 305.70 mm (approximately 12 in.) and a minimum specification of 304.55 mm. If the strips are less than the minimum specification, they are scrapped; if greater than the maximum specification, they are reworked. The part dimensions are normally distributed with a population mean of 305.20 mm and a standard deviation of 0.25 mm.

What percentage of the product is scrap? What percentage is rework? How can the process be centered to eliminate all but 0.1% of the scrap? What is the rework percentage now?

12. A company that manufactures oil seals found the population mean to be 49.15 mm (1.935 in.), the population standard deviation to be 0.51 mm (0.020 in.), and the data to be normally distributed. If the ID of the seal is below the lower specification limit of 47.80 mm, the part is reworked. However, if above the upper specification limit of 49.80 mm, the seal is scrapped. (a) What percentage of the seals are reworked? What percentage are scrapped? (b) For various reasons the process average is changed to 48.50 mm. With this new mean or process center, what percentage of the seals is reworked? What percentage is scrapped? If rework is economically feasible, is the change in the process center a wise decision?

13. The historical data of Problem 37 has a subgroup size of 3. Time is not available to collect data for a process capability study using a subgroup size of 4. Determine the process capability using first 25 subgroups. Use a d_2 value for $n = 3$.

14. Repeat Problem 13 using the last 25 subgroups and compare the results.

15. Determine the process capability of the case-hardening process of Problem 6.

16. Determine the process capability of the tensile tests of the improved plastic cord of Problem 3.

17. What is the process capability of:
    (a) Problem 2?
    (b) Problem 5?

18. Determine the capability index before ($\sigma_0 = 0.038$) and after ($\sigma_0 = 0.030$) improvement for the chapter example problem using specifications of 6.40 $\pm$ 0.15 mm.

19. A new process is started, and the sum of the sample standard deviations for 25 subgroups of size 4 is 750. If the specifications are 700 $\pm$ 80, what is the process capability index? What action would you recommend?

20. What is the C_{pk} value after improvement for Problem 18 when the process center is 6.40? When the process center is 6.30? Explain.

21. What is the C_{pk} value for the information in Problem 19 when the process average is 700, 740, 780, and 820? Explain.

22. Determine the revised central line and control limits for a subgroup sum chart using the data of:
    (a) Problem 2
    (b) Problem 3

23. Determine the trial central line and control limits for a moving-average and moving-range chart using a time period of 3. Data in liters are as follows: 4.56, 4.65, 4.66, 4.34, 4.65, 4.40, 4.50, 4.55, 4.69, 4.29, 4.58, 4.71, 4.61, 4.66, 4.46, 4.70, 4.65, 4.61, 4.54, 4.55, 4.54, 4.54, 4.47, 4.64, 4.72, 4.47, 4.66, 4.51, 4.43, 4.34. Are there any out-of-control points?

24. Repeat Problem 23 using a time period of 4. What is the difference in the central line and control limits? Are there any out-of-control points?

25. The Get-Well Hospital has completed a quality improvement project on the time to admit a patient using $\overline{X}$ and R charts. They now wish to monitor the activity using median and range charts. Determine the central line and control limits with the latest data in minutes as given below.

SUBGROUP NUMBER	OBSERVATION			SUBGROUP NUMBER	OBSERVATION		
	X_1	X_2	X_3		X_1	X_2	X_3
1	6.0	5.8	6.1	13	6.1	6.9	7.4
2	5.2	6.4	6.9	14	6.2	5.2	6.8
3	5.5	5.8	5.2	15	4.9	6.6	6.6
4	5.0	5.7	6.5	16	7.0	6.4	6.1
5	6.7	6.5	5.5	17	5.4	6.5	6.7
6	5.8	5.2	5.0	18	6.6	7.0	6.8
7	5.6	5.1	5.2	19	4.7	6.2	7.1
8	6.0	5.8	6.0	20	6.7	5.4	6.7
9	5.5	4.9	5.7	21	6.8	6.5	5.2
10	4.3	6.4	6.3	22	5.9	6.4	6.0
11	6.2	6.9	5.0	23	6.7	6.3	4.6
12	6.7	7.1	6.2	24	7.4	6.8	6.3

26. Determine the trial central line and control limits for median and range charts for the data of Table 5-2. Assume assignable causes for any out-of-control points and determine the revised central line and control limits. Compare the pattern of variation with the $\overline{X}$ and R charts in Figure 5-4.

27. An X and R chart is to be maintained on the pH value for the swimming pool water of a leading motel. One reading is taken each day for 30 days. Data are: 7.8, 7.9, 7.7, 7.6, 7.4, 7.2, 6.9, 7.5, 7.8, 7.7, 7.5, 7.8, 8.0, 8.1, 8.0, 7.9, 8.2, 7.3, 7.8, 7.4, 7.2, 7.5, 6.8, 7.3, 7.4, 8.1, 7.6, 8.0, 7.4, and 7.0. Plot the data on graph paper, determine the trial central line and limits, and evaluate the variation.

28. Determine upper and lower reject limits for the $\overline{X}$ chart of Problem 2. The specifications are 20.40 ± 0.25. Compare these limits to the revised control limits.

29. Repeat Problem 28 for specifications of 20.40 ± 0.30.

30. A new process is starting, and there is the possibility that the process temperature will give problems. Eight readings are taken each day at 8:00 A.M., 10:00 A.M., 12:00 noon, 2:00 P.M., 4:00 P.M., 6:00 P.M., 8:00 P.M., and 10:00 P.M. Prepare a run chart and evaluate the results.

DAY	TEMPERATURE (0° C)							
Monday	78.9	80.0	79.6	79.9	78.6	80.2	78.9	78.5
Tuesday	80.7	80.5	79.6	80.2	79.2	79.3	79.7	80.3
Wednesday	79.0	80.6	79.9	79.6	80.0	80.0	78.6	79.3
Thursday	79.7	79.9	80.2	79.2	79.5	80.3	79.0	79.4
Friday	79.3	80.2	79.1	79.5	78.8	78.9	80.0	78.8

31. The viscosity of a liquid is checked every half-hour during one 3-shift day. Prepare a histogram with 5 cells and the midpoint value of the first cell equal to 29 and evaluate the distribution. Prepare a run chart and evaluate the distribution again. What does the run chart indicate? Data are: 39, 42, 38, 37, 41, 40, 38, 36, 40, 36, 35, 38, 34, 35 37, 36, 39, 34, 38, 36, 32, 37, 35, 34, 33, 35, 32, 32, 38, 34, 37, 35, 35, 34, 31, 33, 35, 32, 36, 31, 29, 33, 32, 31, 30, 32, 32, and 29.

32. Using the software in the diskette solve
(a) Problem 1
(b) Problem 25
(c) Problem 27

33. Using EXCEL write a template for moving average and moving range charts for 3 periods and determine the charts using the data from
(a) Problem 23
(b) Problem 30
(c) Problem 31

34. Using EXCEL write a template for $\overline{X}$ and a charts and determine the charts for Problem 1.

35. Using the software in the diskette determine a X and MR chart for the data of
(a) Problem 30
(b) Problem 31

36. Using the software in the diskette determine the process capability of cypress bark bags in kilograms for the data below. Also determine the C_p and C_{pk} for an USL of 130 Kg and a LSL of 75 Kg.

SUBGROUP	X_1	X_2	X_3	X_4
1	95	90	93	120
2	76	81	81	83
3	107	80	87	95

SUBGROUP	X_1	X_2	X_3	X_4
4	83	77	87	90
5	105	93	95	103
6	88	76	95	97
7	100	87	100	103
8	97	91	92	94
9	90	91	95	101
10	93	79	91	94
11	106	97	100	90
12	89	91	80	82
13	92	83	95	75
14	87	90	100	98
15	97	95	95	90
16	82	106	99	101
17	100	95	95	90
18	81	94	97	90
19	98	101	87	89
20	78	96	100	72
21	91	91	87	89
22	76	91	106	80
23	95	97	100	93
24	92	99	97	94
25	92	85	90	90

37. Using the software in the diskette determine the $\overline{X}$ and R Charts for the data below on shampoo weights in Kilograms.

SUBGROUP NUMBER	X_1	X_2	X_3	SUBGROUP NUMBER	X_1	X_2	X_3
1	6.01	6.01	5.97	16	6.00	5.98	6.02
2	5.99	6.03	5.99	17	5.97	6.01	5.97
3	6.00	5.96	6.00	18	6.02	5.99	6.02
4	6.01	5.99	5.99	19	5.99	5.98	6.01
5	6.05	6.00	6.00	20	6.01	5.98	5.99
6	6.00	5.94	5.99	21	5.97	5.95	5.99
7	6.04	6.02	6.01	22	6.02	6.00	5.98
8	6.01	5.98	5.99	23	5.98	5.99	6.00
9	6.00	6.00	6.01	24	6.02	6.00	5.98
10	5.98	5.99	6.03	25	5.97	5.99	6.02
11	6.00	5.98	5.96	26	6.00	6.02	5.99
12	5.98	5.99	5.99	27	5.99	5.96	6.01
13	5.97	6.01	6.00	28	5.99	6.02	5.98
14	6.01	6.03	5.99	29	5.99	5.98	5.96
15	6.00	5.98	6.01	30	5.97	6.01	5.98

6 ADDITIONAL SPC TECHNIQUES FOR VARIABLES

INTRODUCTION

The previous chapter covered basic information on variable control charts, which are a fundamental aspect of Statistical Process Control (SPC). For the most part, that discussion concentrated on long production runs of discrete parts. This chapter augments that material by providing information on continuous and batch processes, short runs, and gage control.

CONTINUOUS AND BATCH PROCESSES

Continuous Processes

One of the best examples of a continuous process is depicted by the paper-making process. Paper-making machines are very long, with some exceeding the length of a football field, and wide, over 18 feet, and run at speeds of over 3600 feet/minute.

They operate 24 hours per day, 7 days a week, and stop only for scheduled maintenance or emergencies. Briefly, the paper-making process begins with wood chips being converted to wood pulp by chemical or mechanical means. The pulp is washed, treated, and refined until it consists of 99% water and 1% pulp. It then flows to the headbox of the paper-making machine, which is shown in Figure 6-1. Pulp flows onto a moving wire screen, and water drains to form a wet mat. The mat passes through pressing rollers and a drying section to remove more water. After calendering to produce a hard, smooth surface, the web is wound into large rolls.

Statistical process control on the web is shown in Figure 6-2. Observed values are taken in the machine direction (md) or cross-machine direction (cd) by either sensors or manually after a roll is complete. Average and range values for machine direction and cross-machine direction are different.

The flow of pulp at the headbox is controlled by numerous valves; therefore, from the viewpoint of SPC, we need a variables control chart for each valve. For example, if the headbox has 48 valves, 48 md control charts are needed to control each valve. This type of activity is referred to as multiple stream output.

In this particular process, a cd control chart would have little value for control of paper caliper. It might have some value for overall moisture control, since a control chart could indicate the need to increase or decrease the temperature of the drying rolls. The customer might be more interested in a cd control chart, since any out-of-control conditions could affect the performance of the paper on the customer's equipment.

It is extremely important for the practitioner to be knowledgeable about the process and have definite objectives for the control chart. In many continuous processes, it is extremely difficult to obtain samples from a location that can effectively control the process. In such cases, sensors may be helpful to collect data, compare to control limits, and automatically control the process.

Group Chart

This type of control chart eliminates the need for a chart for each stream. A single chart controls all the streams; however, it does not eliminate the need for measurements at each stream.

Data are collected in the same manner as outlined in Chapter 5, i.e., 25 subgroups for each stream. From this information, the central line and control limits are calculated. The plotted points for the $\overline{X}$ chart are the highest and the lowest averages, $\overline{X}_h$ and $\overline{X}_1$, and for the R chart, the highest range, R_h. Each stream or spindle is given a number, and it is recorded with a plotted point.

Of course, any out-of-control situation would call for corrective action. In addition, we have the out-of-control situation when the same stream gives the highest or lowest value r times in succession. Table 6–1 gives practical r values for the number of streams.

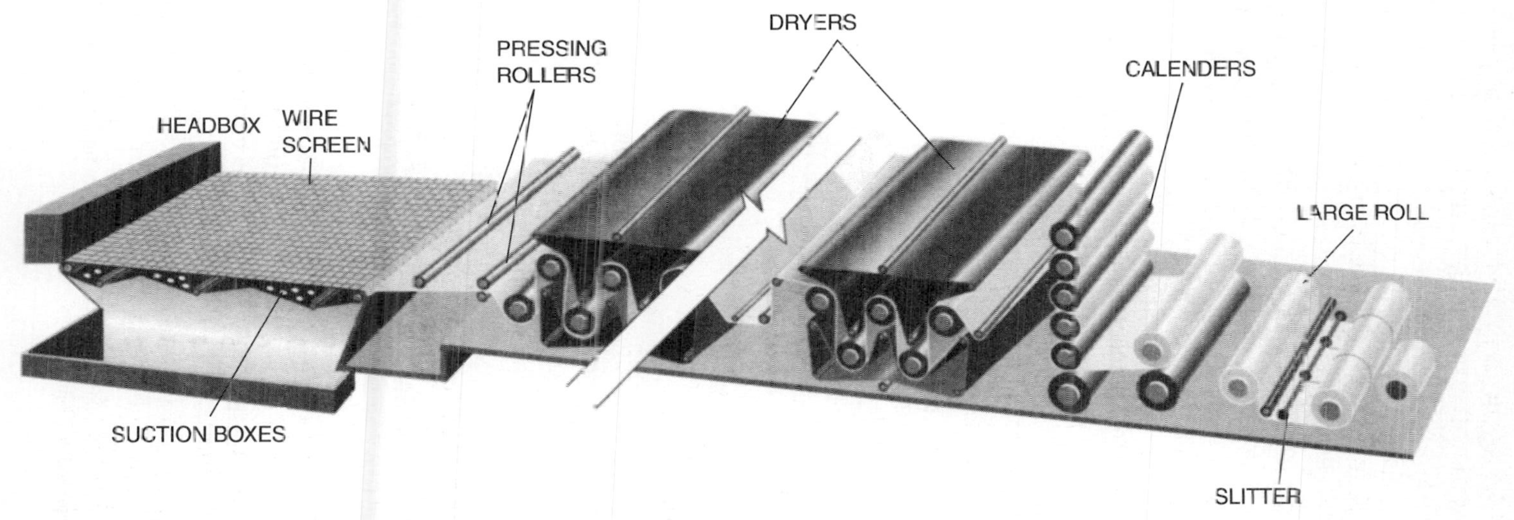

FIGURE 6-1 Paper-making machine (Adapted from *The New Book of Knowledge*, 1969 edition. Copyright 1969 by Grolier Incorporated. Reprinted by permission.)

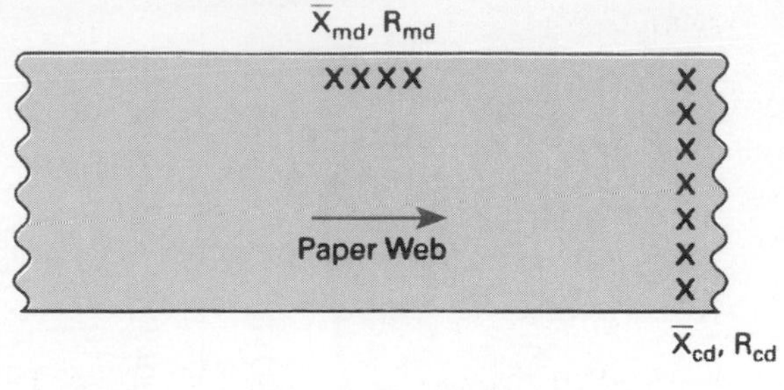

FIGURE 6-2 Paper web and observed valves for md and cd control charts.

TABLE 6-1 Suggested r Values for the Number of Streams

NUMBER OF STREAMS	*r*
2	9
3	7
4	6
5–6	5
7–10	4
11–27	3
Over 27	2

EXAMPLE PROBLEM

Assume a four-spindle filling machine as shown in Figure 6-3 and a subgroup size of 3. Determine the number of sub groups needed to establish the central lines and control limits. Also determine the number of times in succession one of the spindles can be plotted before an out-of-control situation occurs.

$$25 \text{ per spindle} \times 4 \text{ spindles} = 100 \text{ subgroups of 3 each}$$

From the table, $r = 6$.

This technique is applicable to machines, test equipment, operators, or suppliers as long as the following three criteria are met: each stream has the same target, same variation, and the variations are as close to normal as required by conventional $\overline{X}$ and R charts.[1]

[1]For more information, see L. S. Nelson, "Control Chart for Multiple Stream Processes," *Journal of Quality Technology,* 18, No. 4 (October 1986) pp. 255–256.

FIGURE 6-3 Example of multiple streams: a four-spindle filling machine.

Batch Processes

Many products are manufactured in batches such as paint, adhesives, soft drinks, bread, soup, iron, etc. Statistical process control of batches has two forms: within-batch variation and between-batch variation.

Within-batch variation can be very minimal for many liquids that are under agitation, heat, pressure, or any combination thereof. For example, the composition of a product such as perfume might be quite uniform throughout the batch. Thus, only one observed value of a particular quality characteristic can be obtained. In this case, an X and R chart for individuals would be an appropriate SPC technique. Each batch in a series of batches would be plotted on the control chart.

Some liquid products such as soup will exhibit within-batch variation. Observed values (samples) need to be obtained at different locations within the batch, which may be difficult or impossible to accomplish. If samples can be obtained, then $\overline{X}$ and R charts, or similar charts, are appropriate. Sometimes it is necessary to obtain samples from the next operation, which is usually packaging and requires an appropriate location for measuring a volume or weight-fill characteristic. Care must be exercised to ensure that you are measuring within-batch variation because the volume or weight-fill characteristic is a discrete process.

Batch-to-batch variation does not always occur. Because of the nature of some products, there is only one batch. In other words, a customer orders a product to a particular specification and never repeats the order. When there are repetitive batches of the same product, batch-to-batch variation can be charted in the same manner as discrete processes.

Many products are manufactured by combinations of continuous, batch, and discrete processes. For example, in the paper-making process that was previously

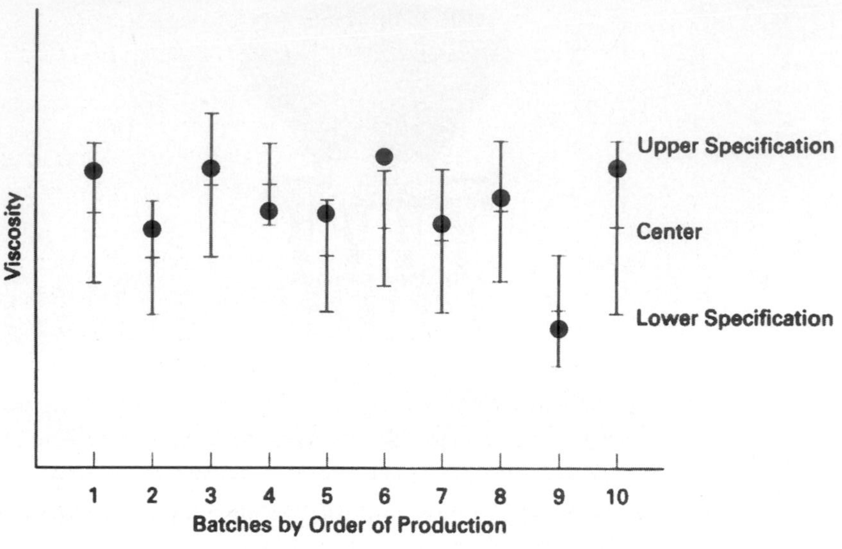

FIGURE 6-4 Batch chart for different batches with different specifications.

described, the pulping process is by batch in giant pressure cookers called digesters; the actual paper-making process is continuous; and the paper rolls are a discrete process.

Batch Chart

Many processing plants are designed to produce a few basic products to customer specifications. While the ingredients and process are essentially the same, the specifications will change with each customer's batch. Figure 6-4 shows a run chart for batch viscosity. The solid point represents the viscosity value, and the vertical line represents the specification range. A cursory analysis of the batches shows that eight of the 10 plotted points are on the high end of the specification. This information may lead to a minor adjustment so that the viscosity of future batches will be closer to the center of each batch specification. Batch charts, such as this one for the other quality characteristics, can provide information for effective quality improvement.

The batch chart is not a control chart. It might more appropriately be called a run chart.

MULTI-VARI CHART

This chart is a useful tool for detecting different types of variation that are found in products and services. Frequently, the chart will lead to a problem solution much faster than other techniques. Some of the processes that lend themselves to this chart are inside and outside diameters, molds with multiple cavities, and adhesion strength.

The multi-vari chart concept is shown in Figure 6-5. It uses a vertical line to show the range of variation of the observed values within a single piece or service. Types of variation are shown at (a) within a unit, (b) unit-to-unit, and (c) time-to-time.

Within-unit variation occurs within a single unit such as porosity in a casting, surface roughness, or cavities within a mold. Unit-to-unit variation occurs between consecutive units drawn from a process, batch-to-batch variations, and lot-to-lot variations. Time-to-time variation occurs from hour-to-hour, shift-to-shift, day-to-day, and week-to-week.

The procedure is to select three to five consecutive units, plot the highest and lowest observed value of each piece, and draw a line between them. After a period time, usually one hour or less, the process is repeated until about 80% of the variation of the process is captured.

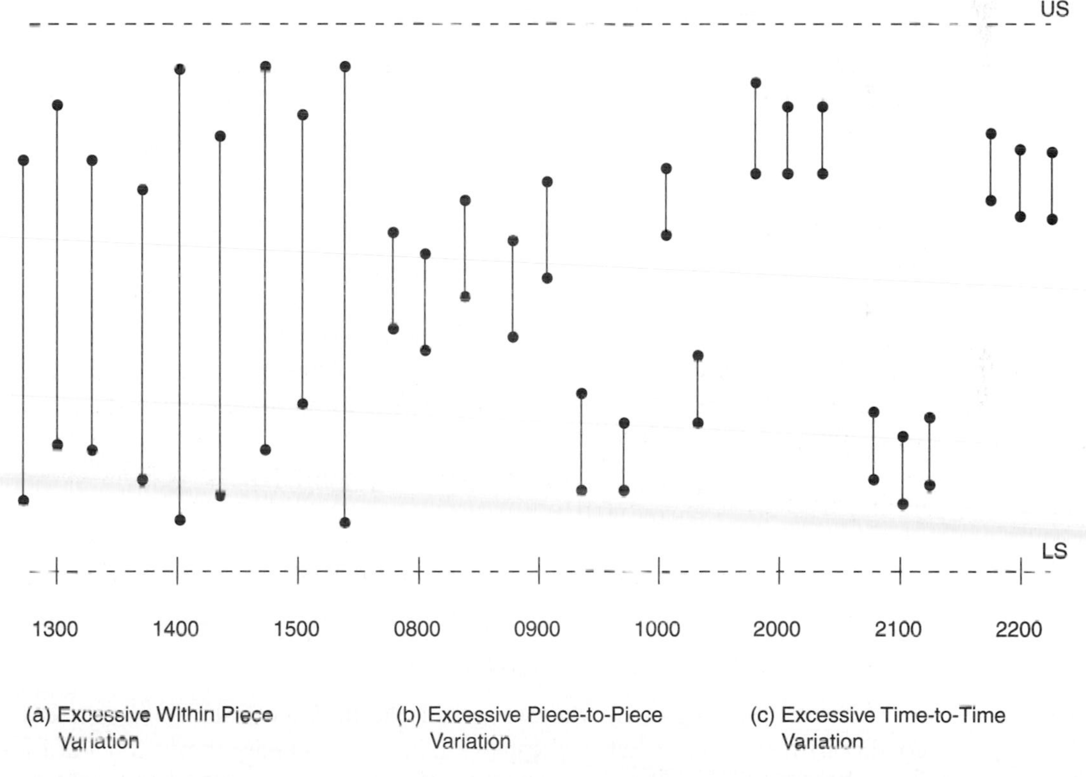

FIGURE 6-5 Multi-Vari Chart

SHORT-RUN SPC

Introduction

In many processes, the run is completed before the central line and control limits can be calculated. This fact is especially true for a job shop with small lot sizes. Furthermore, as companies practice just-in-time production, short runs are becoming more common.

Possible solutions to this problem are basing the chart on specifications, deviation chart, $\overline{Z}$ and W charts, Z and MW charts, precontrol, and percent tolerance precontrol. This section discusses these charting techniques.

Specification Chart

This type of chart gives some measure of control and a method of quality improvement. The central line and the control limits are established using the specifications.

Assume that the specifications call for 25.00 ± 0.12 mm. Then the central line, $\overline{X}_0 = 25.00$. The difference between the upper specification and the lower specification (USL − LSL) is 0.24 mm, which is the spread of the process under the case II situation ($C_p = 1.00$). Thus,

$$C_p = \frac{\text{USL} - \text{LSL}}{6\sigma}$$

$$\sigma = \frac{\text{USL} - \text{LSL}}{6C_p}$$

$$= \frac{25.12 - 24.88}{6(1.00)}$$

$$= 0.04$$

Figure 6-6 shows the relationship between the tolerance (USL − LSL) and the process capability for the case II situation, which was described in Chapter 5. Thus, for $n = 4$,

$$\text{URL}_{\overline{X}} = \overline{X}_0 + A\sigma = 25.00 + 1.500(0.04) = 25.06$$

$$\text{LRL}_{\overline{X}} = \overline{X}_0 - A\sigma = 25.00 - 1.500(0.04) = 24.94$$

$$R_0 = d_2\sigma = (2.059)(0.04) = 0.08$$

$$\text{URL}_R = D_2\sigma = (4.698)(0.04) = 0.19$$

$$\text{LRL}_R = D_1\sigma = (0)(0.04) = 0$$

These limits represent what we would like the process to do (as a maximum condition) rather than what it is capable of doing. Actually these limits are reject limits as discussed in the previous chapter; however, the method of calculation is slightly different.

We now have a chart that is ready to use for the first piece produced. Interpretation of the chart is the difficult part. Figure 6-7 shows the plotted point pattern for three situations. At (a) is the case II situation that was used to determine the reject limits. If the process has a $C_p = 1.00$, the plotted points will form a normal curve

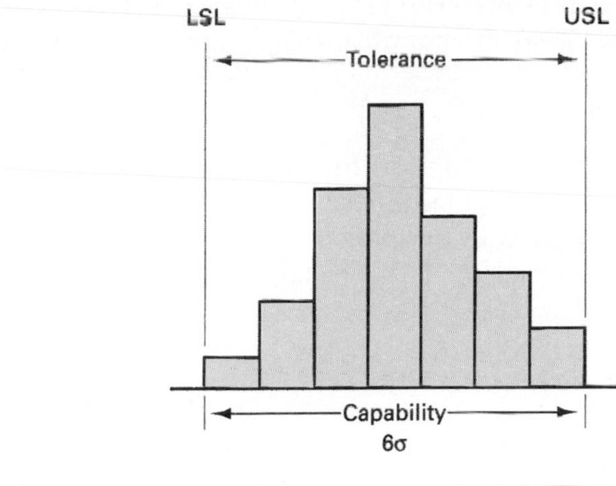

FIGURE 6-6 Relationship of tolerance and capability for the case II situation.

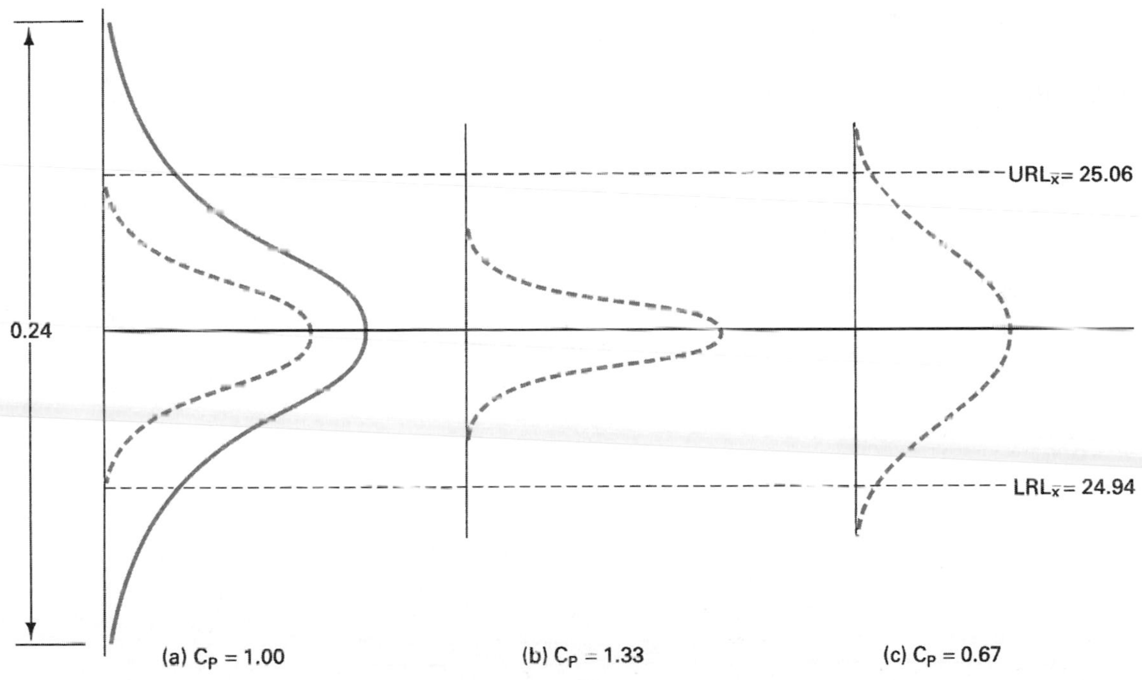

FIGURE 6-7 Comparison for different capabilities.

within the limits. If the process is quite capable, as illustrated at (b), with a $C_p = 1.33$, then the plotted points will be compact about the central line. The most difficult interpretation occurs at (c) where the process is not capable. For example, if a plotted point falls outside the limits, it could be due to an assignable cause or due to the process not being capable. Since the actual C_p value is unknown until there are sufficient plotted points, personnel need to be well trained in process variation. They must closely observe the pattern to know when to adjust and when not to adjust the machine.

Deviation Chart

Figure 6-8 shows a deviation chart for individuals (X's). It is identical to an X chart (see Chapter 5), except the plotted point is the deviation from the target. For example, at time 0130, a test was taken for the carbon equivalent (CE) of iron melt with a resulting value of 4.38. The target is 4.35; therefore, the deviation is

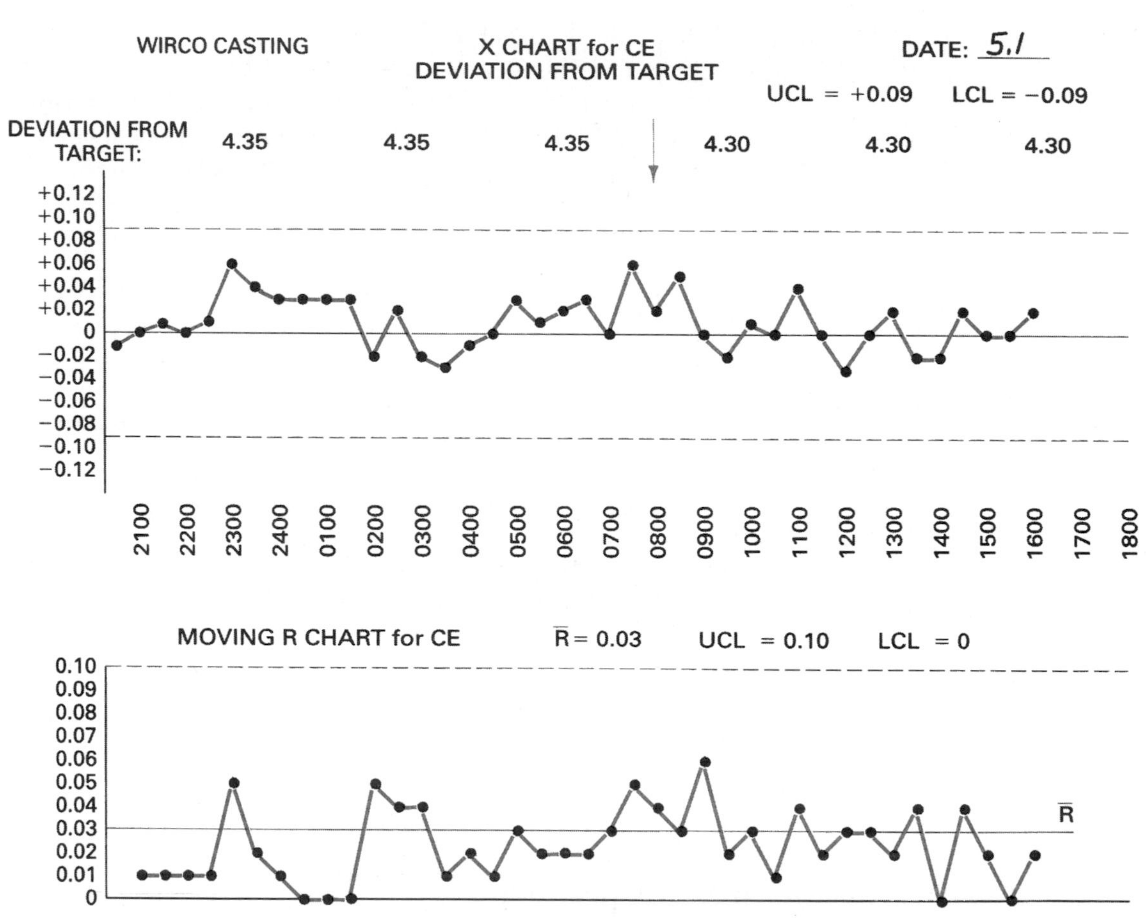

FIGURE 6-8 Deviation chart for individuals (X's) and moving range (R's).

4.38 − 4.35 = 0.03. This value is posted to the chart. There is no change to the R chart; it still uses the moving-range technique.

Even though the target changes, the central line for the X chart is always zero (0). Therefore, the chart can accommodate short runs with different targets. The figure shows that the CE target changes from 4.35 to 4.30. Use of this technique requires the variances (s^2) of the different targets or nominals to be identical. This requirement is verified by an Analysis of Variance (ANOVA) or the following rule of thumb:

$$\frac{\overline{R}_{\text{Process}}}{\overline{R}_{\text{Total}}} \leq 1.3$$

where $\overline{R}_{\text{Process}}$ = average range of the process

$\overline{R}_{\text{Total}}$ = average range for all of the processes

EXAMPLE PROBLEM

The average range for all of the iron melt processes with different CE targets is 0.03. For the process with a targeted CE value of 4.30, the average range is 0.026. Can this process use the deviation technique? What about the process for a targeted CE value of 4.40 with an average range of 0.038?

$$\frac{\overline{R}_{4.30}}{\overline{R}_{\text{Total}}} = \frac{0.026}{0.03} = 0.87 \text{ (ok)}$$

$$\frac{\overline{R}_{4.40}}{\overline{R}_{\text{Total}}} = \frac{0.038}{0.03} = 1.27 \text{ (ok)}$$

The deviation technique is also applicable to $\overline{X}$ and R charts. Data are collected as deviations from the target; otherwise the technique is the same as discussed in Chapter 5.

EXAMPLE PROBLEM

A lathe turns rough diameters between 5 mm and 50 mm and runs last less than two hours. Material and depth of cut do not change. Determine the central line and control limits. Data are

SUB-GROUP	TARGET	X_1	X_2	X_3	X_4	$\overline{X}$	R
1	28.500	0	+.005	−.005	0	0	.010
⋮	⋮	⋮	⋮	⋮	⋮	⋮	⋮
15	45.000	0	−.005	0	−.005	−.0025	.005
⋮	⋮	⋮	⋮	⋮	⋮	⋮	⋮
25	17.000	+.005	0	0	+.005	+.0025	.005
					Σ	+.020	.175

$$\overline{\overline{X}} = \Sigma\overline{X}/g = 0.020/25 = 0.0008$$

(Note: $\overline{X}_0 = 0$ since the central line must be zero.)

$$\overline{R} = \Sigma R/g = 0.175/25 = 0.007$$

$$UCL_{\overline{X}} = \overline{X}_0 + A_2\overline{R} = 0 + .729(.007) = +.005$$

$$LCL_{\overline{X}} = \overline{X}_0 - A_2\overline{R} = 0 - .729(.007) = -.005$$

$$UCL_R = D_4\overline{R} = 2.282(.007) = .016$$

$$LCL_R = D_3\overline{R} = 0(.007) = 0$$

Deviation charts are also called difference, nominal, or target charts. The disadvantage of this type of chart is the requirement that the variation from process to process be relatively constant. If the variation is too great, as judged by the rule of thumb discussed earlier in this section, then a $\overline{Z}$ or Z chart can be used.

$\overline{Z}$ and *W* Charts

These charts are very good for short runs. The central line and control limits are derived from the traditional formulas. Looking at the *R* chart first, we have

R Chart Inequality	$LCL_R < R < UCL_R$
Substituting the formulas	$D_3\overline{R} < R < D_4\overline{R}$
Dividing by $\overline{R}$	$D_3 < \dfrac{R}{\overline{R}} < D_4$

Figure 6-9 shows the UCL $= D_4$ and the LCL $= D_3$. The central line is equal to 1.00 because it occurs when $R = \overline{R}$. This chart is called a *W* chart, and the plotted point is

$$W = R/\text{Target } \overline{R}$$

The control limits D_3 and D_4 are independent of $\overline{R}$; however, they are a function of *n*, which must be constant.

Looking at the $\overline{X}$ chart, we have

$\overline{X}$ Chart Inequality is	$LCL_{\overline{X}} < \overline{X} < UCL_{\overline{X}}$
Substituting the formulas	$\overline{\overline{X}} - A_2\overline{R} < \overline{X} < \overline{\overline{X}} + A_2\overline{R}$
Subtract $\overline{\overline{X}}$	$-A_2\overline{R} < \overline{X} - \overline{\overline{X}} < +A_2\overline{R}$
Divide by $\overline{R}$	$-A_2 < \dfrac{\overline{X} - \overline{\overline{X}}}{\overline{R}} < +A_2$

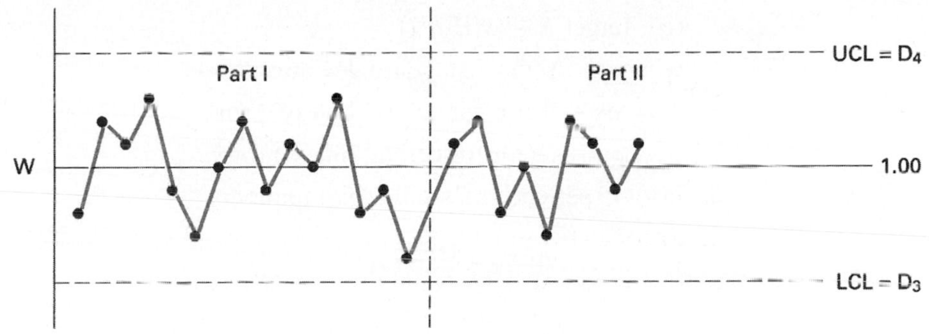

FIGURE 6-9 ***W* chart.**

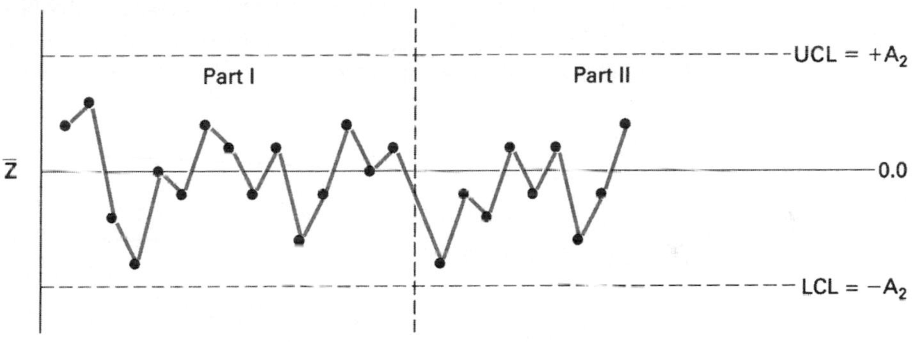

FIGURE 6-10 $\overline{Z}$ **chart.**

Figure 6-10 shows UCL $= +A_2$ and LCL $= -A_2$. The central line is equal to 0.0 because it occurs when $\overline{X} - \overline{\overline{X}} = 0$, which is the perfect situation. This chart is called a $\overline{Z}$ chart, and the plotted point is

$$\overline{Z} = (\overline{X} - \text{Target } \overline{\overline{X}})/\text{Target } \overline{R}$$

The control limits $+A_2$ and $-A_2$ are independent of $\overline{R}$; however, they are a function of *n*, which must be constant.

Target $\overline{\overline{X}}$ and $\overline{R}$ values for a given part are determined by

1. Prior control charts
2. Historical data:

   (a) Target $\overline{\overline{X}} = \Sigma\overline{X}/m$

   where m = number of measurements

(b) Target $\overline{R} = s(d_2/c_4)$

where s = sample standard deviation for m

d_2 = factor for central line ($\overline{R}$) for n

c_4 = factor for central line (s) for m

3. Prior experience on similar part numbers

4. Specifications[2]

(a) Target $\overline{\overline{X}}$ = Nominal print specification

(b) Target $\overline{R} = \dfrac{d_2(\text{USL} - \text{LSL})}{6C_p}$

EXAMPLE PROBLEM

Determine the central lines and control limits for a $\overline{Z}$ and W chart with a subgroup size of 3. If the target $\overline{\overline{X}}$ is 4.25 and the target $\overline{R}$ is .10, determine the plotted points for three subgroups.

SUBGROUP	X_1	X_2	X_3	$\overline{X}$	R
1	4.33	4.35	4.32	4.33	.03
2	4.28	4.38	4.22	4.29	.16
3	4.26	4.23	4.20	4.23	.06

From Table B for $n = 3$, $D_3 = 0$, $D_4 = 2.574$, and $A_2 = 1.023$.

$$\overline{Z}_1 = \frac{\overline{X} - \text{Target } \overline{\overline{X}}}{\text{Target } \overline{R}} = \frac{4.33 - 4.25}{0.10} = +0.80$$

$$\overline{Z}_2 = \quad = \frac{4.29 - 4.25}{0.10} = +0.40$$

$$\overline{Z}_3 = \quad = \frac{4.23 - 4.25}{0.10} = -0.20$$

$$W_1 = \frac{R}{\text{Target } \overline{R}} = \frac{0.03}{0.10} = 0.3$$

$$W_2 = \quad = \frac{0.16}{0.10} = 1.6$$

$$W_3 = \quad = \frac{0.06}{0.10} = 0.6$$

[2]*SPC for Short Production Runs,* prepared for US Army Annament Munitions and Chemical Command by Davis R. Bothe, International Quality Institute, Inc., 1988.

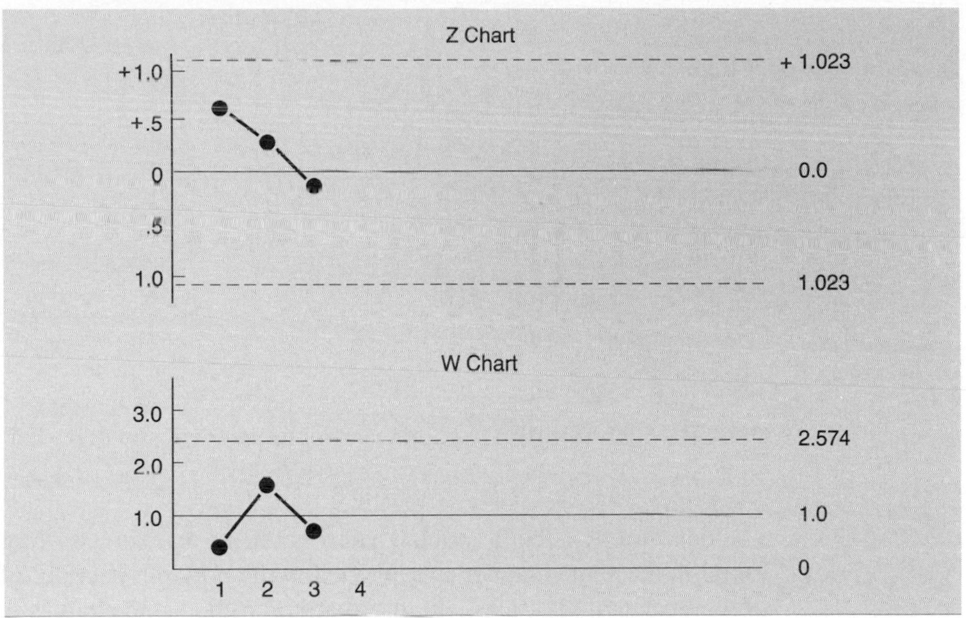

In addition to the advantage of short runs, the $\overline{Z}$ and W charts provide the opportunity to convey enhanced information. On these same charts we can plot

1. Different quality characteristics such as length and width
2. An operator's daily performance
3. The entire part history, thereby providing statistical evidence of the quality for the customer

It must be remembered, however, that the subgroup size must remain constant. The basic disadvantage is the fact that the plotted points are more difficult to calculate.

Z and *MW* Charts

The traditional $\overline{X}$ and R charts have their counterparts in $\overline{Z}$ and W charts. The X (individual) and MR charts, where MR is the moving range of the X values, have their counterparts in Z and MW charts, where MW is the moving range of the Z values. The concept is the same.

Figure 6-11 gives the control limits for the Z and MW charts. These limits and central lines are always the values shown in the figure. The limits are based on a moving range of Z as explained in Chapter 5, under the topic Chart for Individual Values. The plot points for the Z and MW charts are

$$Z = (X - \text{Target } \overline{X})/\text{Target } \overline{R}$$

$$MW_{i+1} = Z_i - Z_{i+1}$$

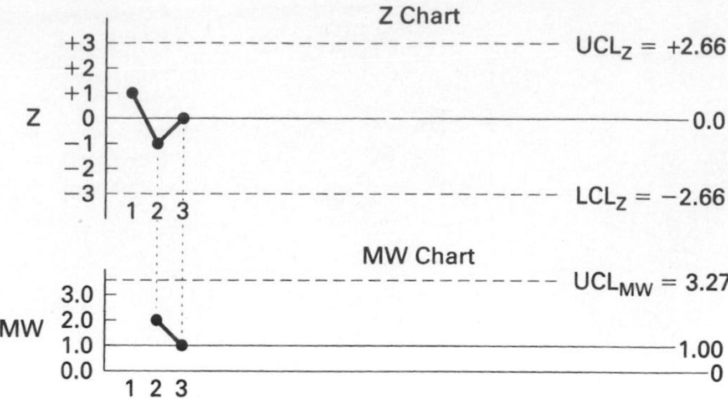

FIGURE 6-11 Central lines and control limits for Z and MW charts.

The derivation of the Z and MW chart is left as an exercise. Targets $\overline{X}$ and $\overline{R}$ are found in the same manner as explained in the previous section. Also, the information about including multiple information on the same chart is the same as for $\overline{Z}$ and W charts. The MW chart uses the absolute value.

EXAMPLE PROBLEM

Graph the Z and MW charts and plot the points for a target $\overline{X}$ of 39.0 and a target $\overline{R}$ of 0.6. Individual values for four subgroups are 39.6, 40.5, 38.2, 39.0.

$$Z_1 = \frac{X - \text{Target } \overline{X}}{\text{Target } \overline{R}} = \frac{39.6 - 39.0}{0.6} = +1.00$$

$$Z_2 = \frac{X - \text{Target } \overline{X}}{\text{Target } \overline{R}} = \frac{40.5 - 39.0}{0.6} = +2.50$$

$$Z_3 = \frac{X - \text{Target } \overline{X}}{\text{Target } \overline{R}} = \frac{38.2 - 39.0}{0.6} = -1.33$$

$$Z_4 = \frac{X - \text{Target } \overline{X}}{\text{Target } \overline{R}} = \frac{39.0 - 39.0}{0.6} = 0$$

$$MW_2 = |Z_1 - Z_2| = |1.00 - 2.50| = 1.50$$

$$MW_3 = |Z_2 - Z_3| = |2.50 - (-1.33)| = 3.83$$

$$MW_4 = |Z_3 - Z_4| = |-1.33 - 0| = 1.33$$

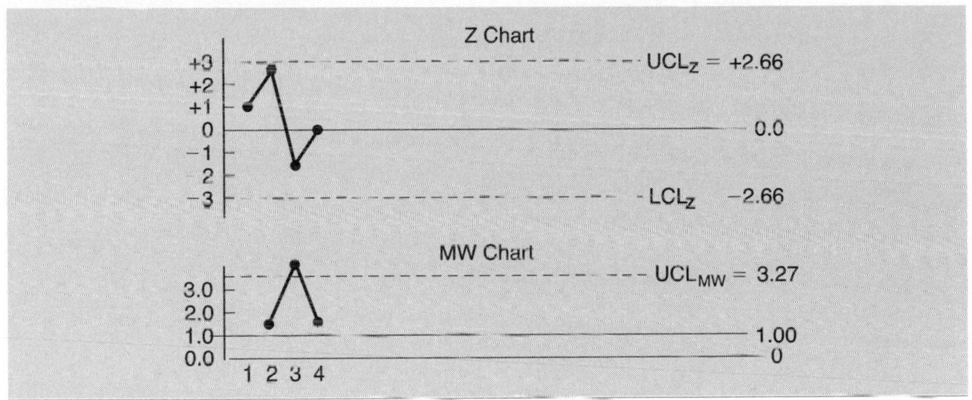

Precontrol

Control charts for variables, notably the $\overline{X}$ and R charts, are excellent for problem solving. They do, however, have certain disadvantages when used by operating personnel to monitor a process after a project team has improved the process:

On short runs, the process is often completed before the operators have time to calculate the limits.

Operators may not have the time or ability to make the necessary calculations.

Frequently, operators are confused about specifications and control limits. This fact is especially true when a process is out of control but waste is not being produced.

Precontrol corrects these disadvantages as well as offers some advantages of its own.

The first step in the process is to be sure that the process capability is less than the specifications. Therefore, a capability index, C_p, of 1.00 or more, preferably more, is required. It is management's responsibility to ensure that the process is capable of meeting the specifications. Next, precontrol (PC) lines are established to divide the tolerance into five zones as shown in Figure 6-12(a). These PC lines are located halfway between the nominal value and the outside limits of the tolerance as given by USL for upper specifications and LSL for lower specifications. The center zone is one half the print tolerance and is called the green area. On each side are the yellow zones, and each amounts to one-fourth of the total tolerance. Outside the specifications are the red zones. The colors make the procedure simple to understand and apply.

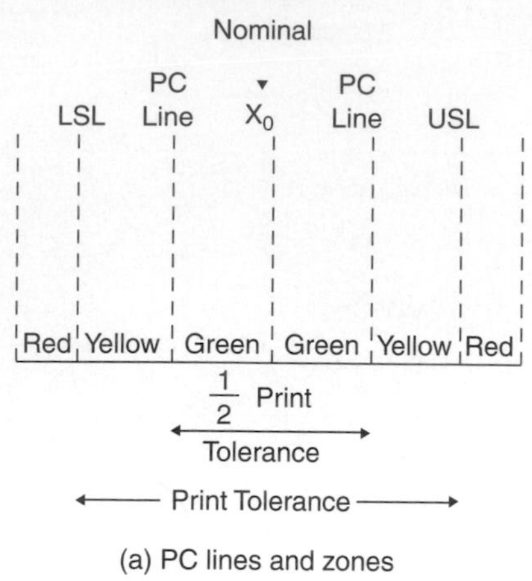

(a) PC lines and zones

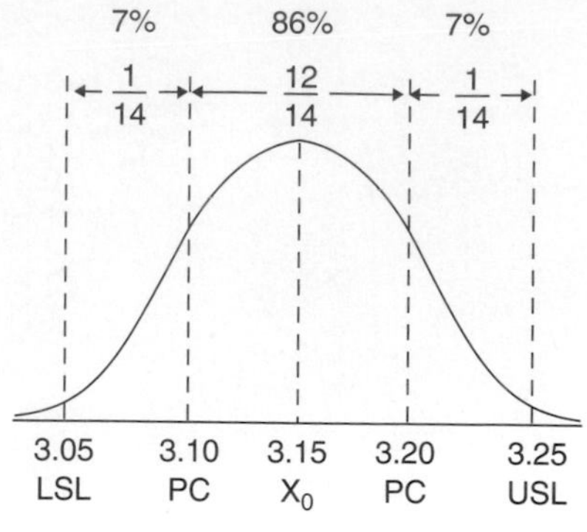

(b) Probability when $C_p = 1.00$ and $C_{pk} = 1.00$

FIGURE 6-12 Precontrol lines.

For a specification of 3.15 ± 0.10 mm, the calculations are

1. Divide tolerance by 4: 0.20/4 = 0.05
2. Add value to lower specification, 3.05:

$$PC = 3.05 + 0.05 = 3.10$$

3. Subtract value from upper specification, 3.25:

$$PC = 3.25 - 0.05 = 3.20$$

Thus, the two PC lines are located at 3.10 and 3.20 mm. These values are shown in (b).

The statistical foundation of precontrol is shown in Figure 6-12(b). First, the process capability is equal to the specifications and is centered as indicated by $C_p = 1.00$ and $C_{pk} = 1.00$. For a normal distribution, 86% of the parts (12 out of 14) will fall between the PC lines, which is the green zone, and 7 percent of the parts (1 out of 14) will fall between the PC line and the specifications, which are the two yellow zones. As the capability index increases, the chance of a part falling in the yellow zone decreases. Also, with a large capability index ($C_p = 1.33$ is considered a de facto standard), distributions that depart from normal are easily accommodated.

The precontrol procedure has two stages: start-up and run. These stages are shown in Figure 6-13. One part is checked, and the results can fall in one of the three color zones. If the part is outside specifications (red zone), the process is stopped and reset. If the part is between the PC lines and the specifications (yellow

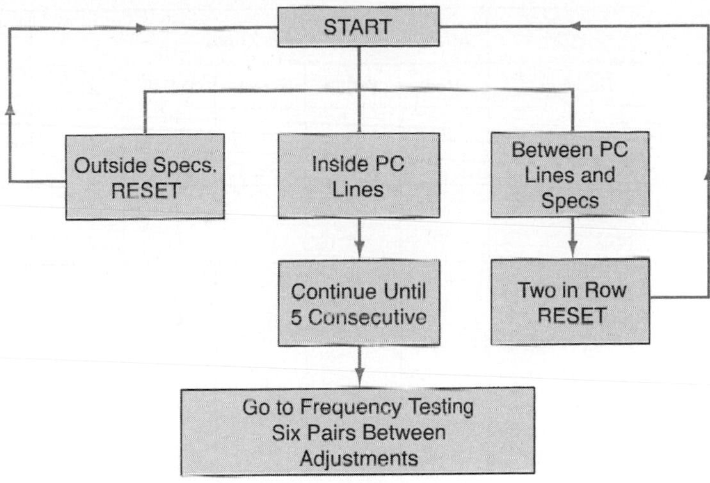

FIGURE 6-13 Precontrol procedure.

zone), a second part is tested; if the second part is in the yellow zone, the process is stopped and reset. If the part falls between the PC lines (green zone), additional parts are tested until five consecutive parts are in the green zone. Operators become quite adept at "nudging" the setting when a reset is required.

Once there are five consecutive parts in the green zone, the run, or frequency testing, stage commences. Frequency testing is the evaluation of pairs of parts. The frequency rule is to sample six pairs between adjustments, and Table 6-2 gives the time between measurements for various adjustment frequencies. As can be seen by the table, there is a linear relationship between the two variables. Thus, if, on the average, an adjustment is made every 6 h, the time between measurement of pairs is 60 mins. The time between adjustments is determined by the operator and supervisor based on historical information.

TABLE 6-2 Frequency of Measuring

TIME BETWEEN ADJUSTMENTS, HOURS	TIME BETWEEN MEASUREMENT, MINUTES
1	10
2	20
3	30
4	40
.	.
.	.
.	.

Decision	Color Zones					Probability
	Red	Yellow	Green	Yellow	Red	
Stop, Go to Start-up	A					nil
					A	nil
Stop, Get help		A		B		1/14 * 1/14 = 1/196
		B		A		1/14 * 1/14 = 1/196
Adjust, Go to Start-up		A,B				1/14 * 1/14 = 1/196
				A,B		1/14 * 1/14 = 1/196
Continue		A	A,B			12/14 * 12/14 = 144/196
		B	B			1/14 * 12/14 = 12/196
			A			1/14 * 12/14 = 12/196
			A	B		12/14 * 1/14 = 12/196
			B	A		12/14 * 1/14 = 12/196
	LSL	PC	X_0	PC	USL	Total = 196/196

↑ Nominal (Target)

FIGURE 6-14 Run decision and probability.

Figure 6-14 shows the decision rules for the measured pairs (designated A, B) for the different color zone possibilities:

1. Where a part falls in the red zone, the process is shut down, reset, and the procedure returned to the start-up stage.

2. Where an A, B pair falls in opposite yellow zones, the process is shut down and help is requested, since this may require a more sophisticated adjustment.

3. Where an A, B pair falls in the same yellow zone, the process is adjusted and the procedure returned to the start-up stage.

4. Where one or both A and B fall in the green zone, the process continues to run.

On the right side of the figure is the probability that a particular A, B pair will occur.

Precontrol is made even easier to use by painting the measuring instrument in green, yellow, and red at the appropriate places. In this way, the operator knows when to go, apply caution, or stop.

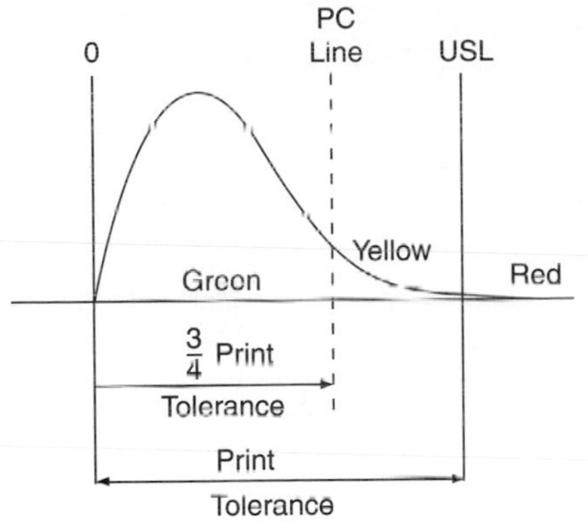

FIGURE 6-15 Precontrol for single specifications.

Precontrol can be used for single specifications, as shown by Figure 6-15. In these cases the green area is established at 3/4 of the print tolerance. The figure shows a skewed distribution, which would most likely occur with an out-of-round characteristic where the target is zero.

Precontrol can also be used for attributes. Appropriately colored "go/no-go" gages that designate the precontrol lines are issued to the operator along with the usual gages for the upper and lower specifications. Precontrol is also used for visual characteristics by assigning visual standards for the PC lines.

The advantages of precontrol are as follows:

1. It is applicable to short production runs as well as long production runs.
2. No recording, calculating, or plotting of data is involved. A precontrol chart can be used if the consumer desires statistical evidence of process control (see Figure 6-16).
3. It is applicable to start up so the process is centered on the target.
4. It works directly with the tolerance rather than easily misunderstood control limits.
5. It is applicable to attributes.
6. It is simple to understand, so training is very easy.

While the precontrol technique has a lot of advantages, we must remember that it is only a monitoring technique. Control charts are used for problem solving, since they have the ability to improve the process by correcting assignable causes and testing improvement ideas. Also, the control chart is more appropriate for process capability and detecting process shifts.

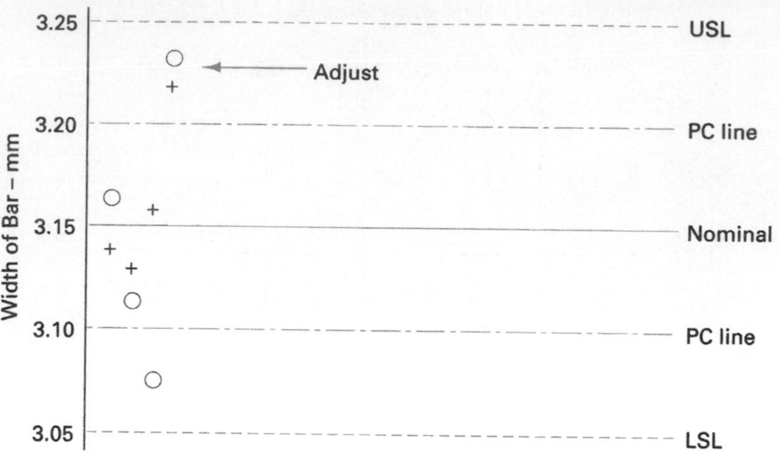

FIGURE 6-16 Precontrol chart.

In summary, precontrol means better management and operator understanding, operator responsibility for quality, reduced rejects, reduced adjustments, reduced operator frustration, and a subsequent increase in morale. These benefits have been realized for many different types of processes.

Percent Tolerance Precontrol Chart[3]

The concepts of Z charts, with its ability to accommodate more than one quality characteristic, and precontrol, with its simplicity, can be combined into one technique by use of percent tolerance precontrol chart (PTPCC). Recall from page 255 that the plotted point of the Z chart is

$$Z = (X - \text{target } \overline{X})/\text{Target } \overline{R}$$

Using a similar logic, an individual measurement, X, can be transformed to percent deviation from the target or nominal by

$$X^* = (X - \text{nominal})/[(\text{USL} - \text{LSL})/2]$$

Where X^* = deviation from nominal as a percent (decimal) of tolerance

$(\text{USL} - \text{LSL})/2 = \frac{1}{2}$ the print tolerance and is the target $\overline{R}$ for the precontrol concept

[3]S. K. Vermani, "SPC Modified With Percent Tolerance Precontrol Charts", *Quality Progress* (October 2000): 43–48.

Some examples will show the use of the formula:

1. The specification of part number 1234 is 2.350 ± 0.005 and an inspected measured value is 2.3485.

$$X^* = (X - \text{nominal})/[(\text{USL} - \text{LSL})/2]$$
$$= (2.3485 - 2.350)/[(2.345 - 2.355)/2]$$
$$= -0.3 \text{ or } -30\%$$

2. The specification of part number 5678 is 0.5000 ± 0.0010 and an inspected measured value is 0.4997.

$$X^* = (X - \text{nominal})/[(\text{USL} - \text{LSL})/2]$$
$$= (0.4997 - 0.5000)/[(0.5010 - 0.4990)/2]$$
$$= -0.3 \text{ or } -30\%$$

3. The specification of part number 1234 is 2.350 ± 0.005 and an inspected measured value is 2.351.

$$X^* = (X - \text{nominal})/[(\text{USL} - \text{LSL})/2]$$
$$= (2.351 - 2.350)/[(2.345 - 2.355)/2]$$
$$= 0.2 \text{ or} 20\%$$

Note that both Example 1 and Example 2 have the same percent deviation (−30%) from the nominal even though the tolerance is much different. The negative value indicates that the observed value is below the nominal. Comparing Example 1 and 3, it is seen that while both have the same nominal and tolerance, Example 1 is 30% below the nominal and Example 2 is 20% above the nominal.

A spreadsheet can be set up to calculate the values and generate the plotted points on the PTPCC. Figure 6-17 shows the calculations and Figure 6-18 shows

Percent Tolerance Precontrol Chart Data

Machine Number: *Mill* 21 Date *24 JAN 04*

Part No.	Time	USL	LSL	Nominal	Part 1	Part 2	Plot 1 (%)	Plot 2 (%)	Status
1234	0800 h	2.355	2.345	2.350	2.3485	2.3510	−30.0	20.0	ok
	0830 h	2.355	2.345	2.350	2.3480	2.3490	−40.0	−20.0	ok
	0900 h	2.355	2.345	2.350	2.3500	2.3530	0.0	60.0	ok (1Y)
5678	1000 h	0.5010	0.4990	0.5000	0.4997	0.5002	−30.0	20.0	ok
	1030 h	0.5010	0.4990	0.5000	0.5006	0.4997	60.0	−30.0	ok (1Y)
	1100 h	0.5010	0.4990	0.5000	0.5000	0.5003	0.0	30.0	ok
	1130 h	0.5010	0.4990	0.5000	0.4994	0.4992	−60.0	−80.0	2Y

FIGURE 6-17 Calculations for Percent Tolerance Precontrol Chart

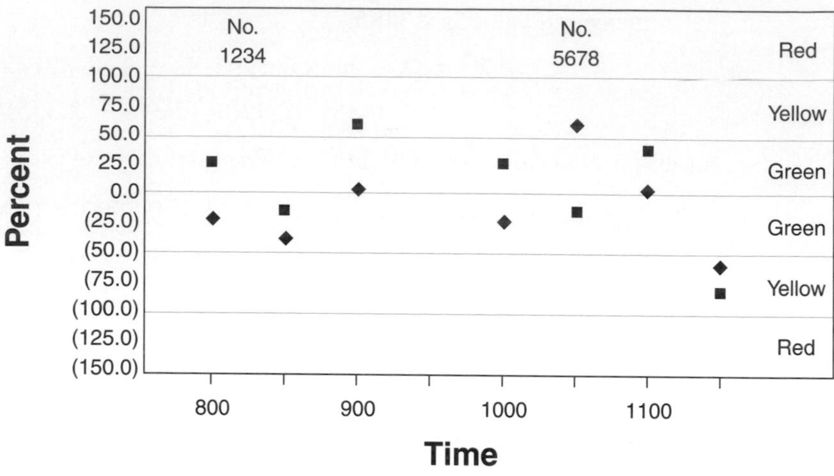

FIGURE 6-18 **Percent Tolerance Precontrol Chart**

the plotted points. Two parts are shown on the PTPCC; however, there could be as many parts or part features as space would permit. Each part or part feature could have different nominal values and tolerances. Different parts can be plotted on the same chart. In fact, the entire part history of its different operations can be plotted on one chart.

The rules of control, which were defined in the previous section, are applicable. A review of this particular PTPCC shows that the process went out of control at 1130 h with two readings in the yellow zone.

GAGE CONTROL[4]

SPC requires accurate and precise data (see the section on data collection—Chapter 3); however, all data have measurement errors. Thus, an observed value has two components:

$$\text{Observed Value} = \text{True Value} + \text{Measurement Error}$$

[4]This section is extracted by permission from Bruce W. Price, Task Force Coordinator, *Fundamental Statistical Process Control* (Troy, Michigan: Automotive Industry Action Group), pp. 119–129.

The previous chapter discussed variation that occurs due to the process and the measurement, thus

$$\text{Total Variation} = \text{Product Variation} + \text{Measurement Variation}$$

Measurement variation is further divided into repeatability, which is due to equipment variation, and reproducibility, which is due to appraiser (inspector) variation. It is called GR&R, or Gage Repeatability and Reproducibility.

Before we perform calculations to evaluate GR&R, it is necessary to calibrate the gage. Calibration must be performed either in-house or by an independent laboratory. It must be accomplished so that it is traceable to reference standards of known accuracy and stability such as those of the National Institute for Standards and Technology (NIST). For industries or products where such standards do not exist, the calibration must be traceable to developed criteria.

There are different GR&R techniques. We will discuss the average and range technique recommended by the Automobile Industry Action Group.

Data Collection

The number of parts, appraisers, or trials can vary, but 10 parts, two or three appraisers, and two or three trials are considered optimum. Readings are taken by randomizing the part sequence on each trial. For example, on the first trial, each appraiser would measure the part characteristic in the following order: 4, 7, 5, 9, 1, 6, 2, 10, 8, and 3. On the second trial, the order might be 2, 8, 6, 4, 3, 7, 9, 10, 1, and 5. A table of random numbers, as given in Table D in the Appendix, should be used to determine the order on each trial.

Calculations

While the order of taking measurements is random, the calculations are performed by part and appraiser. Calculations are as follows:

1. The average and range are calculated for each part by an appraiser.
2. The values in Step 1 are averaged to obtain:

$$\overline{R}_a, \overline{R}_b, \overline{R}_c, \overline{\overline{X}}_a, \overline{\overline{X}}_b, \overline{\overline{X}}_c$$

3. The values in Step 2 are used to obtain:

$$\overline{\overline{R}} \text{ and } \overline{\overline{X}}_{\text{Diff}} \quad \text{Where } \overline{\overline{X}}_{\text{Diff}} = \overline{\overline{X}}_{\text{Max}} - \overline{\overline{X}}_{\text{Min}}$$

4. The UCL and LCL for the range are determined in the same manner as given in Chapter 5

$$\text{UCL}_R = D_4\overline{\overline{R}} \quad \text{LCL}_R = D_3\overline{R}$$

where D_3 and D_4 are obtained from Table B in the Appendix for subgroup sizes of 2 or 3.

Any range value (R_a, R_b, or R_c) that is out of control should be discarded and the above calculations repeated where appropriate, or the readings should be retaken for that appraiser and part and the above calculations repeated where appropriate.

5. Determine $\overline{\overline{X}}$ for each part, and from this information, calculate the range.

$$R_p = \overline{\overline{X}}_{\text{Max}} - \overline{\overline{X}}_{\text{Min}}$$

Analysis of Results

The analysis will estimate the variation and percent of process variation for the total measurement system and its component's repeatability, reproducibility, and part-to-part variation. The equations and sequence of analysis are as follows:

1. Repeatability

$$\text{EV} = r\overline{\overline{R}}$$

where EV = Equipment Variation (repeatability)

r = 4.56 for 2 trials and 3.05 for 3 trials

2. Reproducibility

$$\text{AV} = \sqrt{(k\overline{\overline{X}}_{\text{Diff}})^2 - (\text{EV}^2/n\,r)}$$

where AV = Appraiser Variation (reproducibility)

k = 3.65 for 2 appraisers and 2.70 for 3 appraisers

n = number of parts

r = number of trials

If a negative value occurs under the square root sign, the AV value defaults to zero.

3. Repeatability and Reproducibility

$$\text{R \& R} = \sqrt{\text{EV}^2 + \text{AV}^2}$$

where R&R = Repeatability and Reproducibility

4. Part Variation

$$\text{PV} = jR_{\text{p}}$$

where PV = Part Variation

R_{p} = range of the part averages

j = dependent on number of parts

Part	2	3	4	5	6	7	8	9	10
j	3.65	2.70	2.30	2.08	1.93	1.82	1.74	1.67	1.62

5. Total Variation

$$TV = \sqrt{R\,\&\,R^2 + PV^2}$$

where TV = Total Variation

6. The percent of the total variation is calculated using the equations below. Note that the percent consumed by each factor will not total 100%.

$$\%EV = 100(EV/TV)$$

$$\%AV = 100(AV/TV)$$

$$\%R\&R = 100(R\&R/TV)$$

$$\%PV = 100(PV/TV)$$

Evaluation

If repeatability is large compared to reproducibility, the reasons may be

1. The gage needs maintenance.
2. The gage should be redesigned to be more rigid.
3. The clamping or location for gaging needs to be improved.
4. There is excessive within-part variation.

If reproducibility is large compared to repeatability, the reasons may be

1. The operator needs to be better trained in how to use and read the gage.
2. Calibrations on the gage are not legible.
3. A fixture may be needed to help the operator use the gage consistently.

Guidelines for acceptance of GR&R (%R&R) are

Under 10% error	Gage system is satisfactory.
10% to 30% error	May be acceptable based upon importance of application, cost of gage, cost of repairs, etc.
Over 30% error	Gage system is not satisfactory. Identify the causes and take corrective action.

EXAMPLE PROBLEM

Given the data on the next page for readings by two appraisers on five parts with three trials, determine if the measurement system is acceptable. The readings were randomized. Calculations are in boldface.

	PART NUMBER				
	1	**2**	**3**	**4**	**5**
Appraiser A					
Trial 1	.34	.50	.42	.44	.26
Trial 2	.42	.56	.46	.48	.30
Trial 3	.38	.48	.40	.38	.28
$\bar{X}$	**.38**	**.51**	**.43**	**.43**	**.28**
R	**.08**	**.08**	**.06**	**.10**	**.04**
Appraiser B					
Trial 1	.28	.54	.38	.46	.30
Trial 2	.32	.48	.42	.44	.28
Trial 3	.24	.44	.34	.40	.36
$\bar{X}$	**.28**	**.49**	**.38**	**.43**	**.31**
R	**.08**	**.10**	**.08**	**.06**	**.08**

$$\bar{R}_a = (.08 + .08 + .06 + .10 + .04)/5 = .07$$

$$\bar{R}_b = (.08 + .10 + .08 + .06 + .08)/5 = .08$$

$$\bar{\bar{X}}_a = (.38 + .51 + .43 + .43 + .28)/5 = .41$$

$$\bar{\bar{X}}_b = (.28 + .49 + .38 + .43 + .31)/5 = .38$$

$$\bar{\bar{R}} = (.07 + .08)/2 = .08$$

$$\bar{\bar{X}}_{\text{Diff}} = .41 - .38 = .03$$

$$\text{UCL}_R = 2.574 * .08 = .21$$

$$\text{LCL}_R = 0$$

None of the range values are out of control.

$$\bar{\bar{X}}_1 = (.38 + .28)/2 = .33$$

$$\bar{\bar{X}}_2 = (.51 + .49)/2 = .50$$

$$\bar{\bar{X}}_3 = (.43 + .38)/2 = .41$$

$$\bar{\bar{X}}_4 = (.43 + .43)/2 = .43$$

$$\bar{\bar{X}}_5 = (.28 + .31)/2 = .30$$

$$R_p = .50 - .30 = .20$$

$$\text{EV} = 3.05 * .08 = .24$$

$$AV = \sqrt{(3.65 * .03)^2 - (.24^2/5 * 3)} = .09$$

$$R \& R = \sqrt{.24^2 + .09^2} = .26$$

$$PV = 2.08^*.20 = .42$$

$$TV = \sqrt{.26^2 + .42^2} = .49$$

%EV = 49% %AV = 18%

%R & R = 53% %PV = 86%

The gage system is not satisfactory. The equipment variation (repeatability) is quite large in relation to the appraiser variation (reproducibility).

Comments

If the process variation is known and its value is based on 6*s*, then it can be used to calculate TV and PV using the following equations:

$$TV = 5.15(\text{process variation}/6)$$

$$PV = \sqrt{TV^2 - R \& R^2}$$

If an analysis based on percent of tolerance is preferred, the tolerance value is substituted for TV in the denominator of the %EV, %AV, %R&R, and %PV equations.

Environmental conditions such as temperature, humidity, air cleanliness, and electric discharge can influence the GR&R results.

COMPUTER PROGRAM

Using EXCEL the software in the diskette inside the back cover will solve for the $\overline{Z}$ and *W* charts, for PTPCC, and for *GR & R*. Their file names are *Z-bar & W charts*, *PTPCC* and *GR & R* respectively.

PROBLEMS

1. Determine the number of subgroups needed to establish the central lines and control limits for $\overline{X}$ and *R* charts with a subgroup size of 2 for an 8-spindle filling machine. How many times in succession can one of the spindles be plotted?

2. Determine the number of subgroups needed to establish the central lines and control limits for *X* and moving *R* charts for a paper towel process that is controlled by 24 valves. How many times in succession can one of the valves be plotted?

3. Given the data below in μ inches for the surface roughness of a grinding operation, construct a multi-vari chart and analyze the results.

TIME	0700 h			1400 h			2100 h		
Part No.	20	21	22	82	83	84	145	146	147
	38	26	31	32	19	29	10	28	14
Surface	28	08	30	25	29	09	05	11	15
Roughness	30	31	38	16	20	18	26	38	04
Measure	37	20	22	22	21	16	32	30	38
	39	44	35	30	28	24	29	38	10

4. Given below are data for a lathe turning operation. The part is 15 cm long and has a target diameter of 60.000 mm ± 0.012. One measurement is taken at each end and three consecutive parts are measured every 30 minutes. Construct a multi-vari chart and analyze the process. The data are in deviations from the target of 60.000, thus 60.003 is coded 3 and 59.986 is coded −14.

TIME	PART 1	PART 2	PART 3
0800 h	−7/10	−2/13	9/18
0830 h	2/15	5/14	2/14
0900 h	0/14	3/15	−7/15
0930 h	−23/−5	−20/−6	−14/1
1000 h	−20/−8	−22/−7	−11/10
1030 h	−15/9	−18/6	−14/5
1100 h	−9/8	−13/4	−12/8
1130 h	−19/1	−14/9	−13/12

5. Determine the central line and limits for a short production run that will be completed in 3 h. The specifications are 25.0 ± 0.3 Ω. Use $n = 4$.

6. Determine the central line and limits for a short production run that will be completed in 1 h. The specifications are 3.40 ± .05 mm. Use $n = 3$.

7. A 5-stage progressive die has four critical dimensions. $\overline{X}$ is 25.30, 14.82, 105.65, and 58.26 mm and $\overline{R}$ is .06, .05, .07, and .06 for the dimensions. Can a deviation chart be used? Let $\overline{R}_{\text{Total}}$ equal the average of the four range values.

8. Determine the central lines and control limits for $\overline{Z}$ and W charts with a subgroup size of 2 and draw the graphs. If the target $\overline{\overline{X}}$ is 1.50 and the target $\overline{R}$ is 0.08, determine the plotted points for three subgroups.

SUBGROUP	X_1	X_2
1	1.55	1.59
2	1.43	1.53
3	1.30	1.38

9. Determine the central lines and control limits for $\overline{Z}$ and W charts with a subgroup size of 3 and draw the graphs. If the target $\overline{X}$ is 25.00 and target $\overline{R}$ is 0.05, determine the plotted points for the three subgroups below. Are any points out of control?

SUBGROUP	X_1	X_2	X_3
1	24.97	25.01	25.00
2	25.08	25.06	25.09
3	25.03	25.04	24.98

10. Using the information in the Chart for Individual Values (Chapter 5), derive the control limits and plotted points for the Z and MW charts.

11. Graph the central lines and control limits for Z and MW charts. The target $\overline{X}$ is 1.15, and the target $\overline{R}$ is 0.03. Plot the points for $X_1 = 1.20$, $X_2 = 1.06$, and $X_3 = 1.14$. Are there any out-of-control points?

12. Graph the central lines and control limits for Z and MW charts. The target $\overline{X}$ is 3.00, and the target $\overline{R}$ is 0.05. Plot the points for $X_1 = 3.06$, $X_2 = 2.91$, and $X_3 = 3.10$. Are any points out of control?

13. What are the PC lines for a process that has a nominal of 32.0° and a tolerance of ±1.0°C?

14. Determine the PC line for the concentricity of a shaft when the total indicator reading (TIR) tolerance is 0.06 mm and the target is 0. *Hint:* This problem is a one-sided tolerance; however, the green zone is still half the tolerance. Graph the results.

15. What is the probability of an A, B pair being green? Of an A, B pair having one yellow and one green?

16. If a process is adjusted every 3 h, how often should pairs of parts be measured?

17. Determine the PTPCC plot point for the following

	USL	LSL	NOMINAL	PART 1	PART 2
(a)	0.460	0.440	0.450	0.449	0.458
(b)	1.505	1.495	1.500	1.496	1.500
(c)	1.2750	1.2650	1.2700	1.2695	1.2732
(d)	0.7720	0.7520	0.7620	0.7600	0.7590

18. Construct a PTPCC chart for the data in Problem 4. Use the average value for Part 1 and Part 2. For example the data for 0800 h would be Part 1 (59.993 + 60.010)/2 = 60.0015 and Part 2 (59.998 + 60.013)/2 = 60.0055. Do not use Part 3.

19. Given the data below for readings by 3 appraisers on 6 parts with 2 trials, determine if the measurement system is acceptable. The readings were randomized.

	PART NUMBER					
	1	**2**	**3**	**4**	**5**	**6**
Appraiser A						
Trial 1	.65	1.00	.85	.85	.55	1.00
Trial 2	.60	1.00	.80	.95	.45	1.00
Appraiser B						
Trial 1	.55	1.05	.80	.80	.40	1.00
Trial 2	.55	.95	.75	.75	.40	1.05
Appraiser C						
Trial 1	.50	1.05	.80	.80	.45	1.00
Trial 2	.55	1.00	.80	.80	.50	1.05

20. Given the data below for readings by 2 appraisers on 4 parts with 3 trials, determine if the measurement system is acceptable. The readings were randomized.

	PART NUMBER			
	1	**2**	**3**	**4**
Appraiser A				
Trial 1	.55	.45	.60	.20
Trial 2	.55	.40	.60	.30
Trial 3	.50	.35	.55	.25
Appraiser B				
Trial 1	.55	.35	.60	.15
Trial 2	.50	.30	.55	.20
Trial 3	.45	.25	.50	.15

21. Using the software for $\overline{Z}$ and W charts solve
    (a) Problem 8
    (b) Problem 9

22. Develop a template for Z and MW charts. (Hint: Similar to X and MR charts)

23. Using the software for *GR & R* verify your answers to
    (a) Problem 19
    (b) Problem 20

24. In the laboratory perform a *GR & R* study with one of the measuring instruments and some parts.

7 FUNDAMENTALS OF PROBABILITY

BASIC CONCEPTS

Definition of Probability

The term *probability* has a number of synonyms, such as likelihood, chance, tendency, and trend. To the layperson, probability is a well known term that refers to the chance that something will happen. "I will probably play golf tomorrow" or "I will probably receive an A in this course" are typical examples. When the commentator on the evening news states, "The probability of rain tomorrow is 25%," the definition has been quantified. It is possible to define probability with extreme mathematical rigor; however, in this text we will define probability from a practical viewpoint as it applies to quality control.

If a nickel is tossed, the probability of a head is $\frac{1}{2}$ and the probability of a tail is $\frac{1}{2}$. A die, which is used in games of chance, is a cube with six sides and spots on each side from one to six. When the die is tossed on the table, the likelihood or probability of one spot is $\frac{1}{6}$, the probability of two spots is $\frac{1}{6}, \ldots$, the probability of six spots is $\frac{1}{6}$. Another example of probability is illustrated by the drawing of a card from a deck of cards. The probability of a spade is $\frac{13}{52}$, since there are 13 spades in a deck of cards that contains 52 total cards. For hearts, diamonds, and clubs, the other three suits in the deck, the probability is also $\frac{13}{52}$.

Figure 7-1 shows the probability distributions for the examples above. It is noted that the area of each distribution is equal to 1.000 ($\frac{1}{2} + \frac{1}{2} = 1.000$; $\frac{1}{6} + \frac{1}{6} + \frac{1}{6} + \frac{1}{6} + \frac{1}{6} + \frac{1}{6} = 1000$; and $\frac{13}{52} + \frac{13}{52} + \frac{13}{52} + \frac{13}{52} = 1.000$). It is recalled that the area under the normal distribution curve, which is a probability distribution, is also equal to 1.000. Therefore, the total probability of any situation will be equal to 1.000. The probability is expressed as a decimal, such as (1) the probability of heads is 0.500, which is expressed in symbols as [$P(h) = 0.500$], (2) the probability of a 3 on a die is 0.167 [$P(3) = 0.167$], and (3) the probability of a spade is 0.250 [$P(s) = 0.250$].

The probabilities given in the preceding examples will occur provided sufficient trials are made and provided there is an equal likelihood of the events occurring. In other words, the probability of a head (the event) will be 0.500 provided that the chance of a head or a tail is equal (equally likely). In most coins the equally likely condition is met; however, the addition of a little extra metal on one side would produce a biased coin and the equally likely condition could not be met. Similarly, an unscrupulous person might fix a die so that a three appears more often than one out of six times, or he might stack a deck of cards so that all the aces were at the top.

Returning to the example of a six-sided die, there are six possible outcomes(1, 2, 3, 4, 5, and 6). An *event* is a collection of outcomes. Thus, the event of a 2 or 4 occurring on a throw of the die has two outcomes, and the total number of outcomes is 6. The probability is obviously $\frac{2}{6}$, or 0.333.

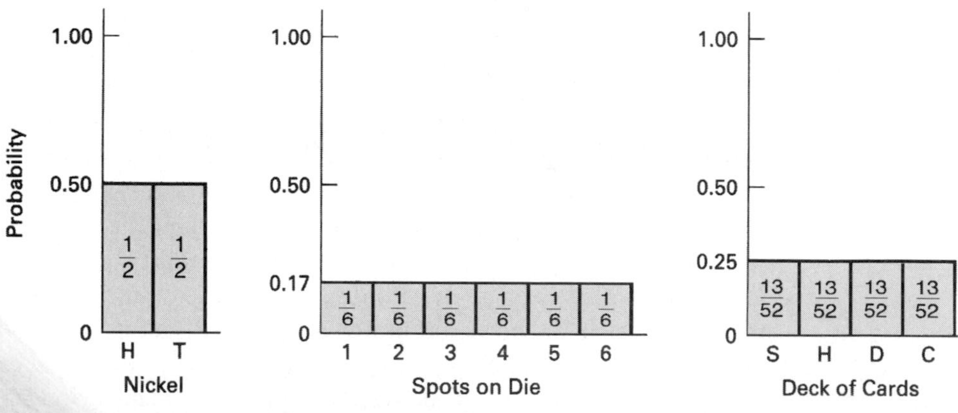

FIGURE 7-1 Probability distributions.

From the discussion above, a definition based on a frequency interpretation can be given. If an event A can occur in N_A outcomes out of a total of N possible and equally likely outcomes, then the probability that the event will occur is

$$P(A) = \frac{N_A}{N}$$

where $P(A)$ = probability of an event A occurring to 3 decimal places
N_A = number of successful outcomes of event A
N = total number of possible outcomes

This definition can be used when the number of outcomes is known or when the number of outcomes is found by experimentation.

EXAMPLE PROBLEM

A part is selected at random from a container of 50 parts that are known to have 10 nonconforming units. The part is returned to the container and a record of the number of trials and the number nonconforming is maintained. After 90 trials, 16 nonconforming units were recorded. What is the probability based on known outcomes and on experimental outcomes?

Known outcomes:

$$P(A) = \frac{N_A}{N} = \frac{10}{50} = 0.200$$

Experimental outcomes:

$$P(A) = \frac{N_A}{N} = \frac{16}{90} = 0.178$$

The probability calculated using known outcomes is the true probability, and the one calculated using experimental outcomes is different due to the chance factor. If, say, 900 trials were taken, the probability using experimental outcomes would be much closer since the chance factor would be minimized.

In most cases, the number nonconforming in the container would not be known; therefore, the probability with known outcomes cannot be determined. If we consider the probability using experimental outcomes to represent the sample and known outcomes to represent the population, there is the same relationship between sample and population that was discussed in Chapter 4.

The definition above is useful for finite situations where N_A, the number of successful outcomes, and N, total number of outcomes, are known or must be found experimentally. For an infinite situation, where $N = \infty$, the definition would always lead to a probability of zero. Therefore, in the infinite situation the probability of an event occurring is proportional to the population distribution. A discussion of this situation is given in the material on continuous and discrete probability distributions.

Theorems of Probability

Theorem 1. Probability is expressed as a number between 1,000 and 0, where a value of 1.000 is a certainty that an event will occur and a value of 0 is a certainty that an event will not occur.

Theorem 2. If $P(A)$ is the probability that event A will occur, then the probability that A will not occur, $P(\not{A})$, is $1.000 - P(A)$.

EXAMPLE PROBLEM

If the probability of finding an error on an income tax return is 0.04, what is the probability of finding an error-free or conforming return?

$$\begin{aligned} P(\not{A}) &= 1.000 - P(A) \\ &= 1.000 - 0.040 \\ &= 0.960 \end{aligned}$$

Therefore, the probability of finding a conforming income tax return is 0.960.

Before proceeding to the other theorems, it is appropriate to learn where they are applicable. In Figure 7-2, we see that if the probability of only one event is desired, then Theorem 3 or 4 is used, depending on whether the event is mutually exclusive or not. If the probability of two or more events is desired, then Theorem 6 or 7 is used, depending on whether the events are independent or not. Theorem 5 is

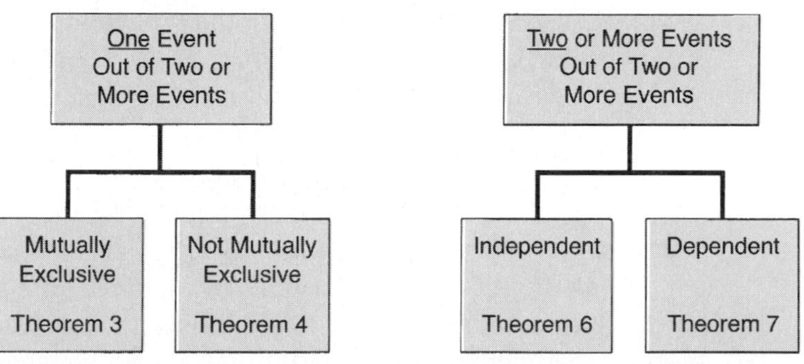

FIGURE 7-2 **When to use Theorems 3, 4, 6, and 7.**

TABLE 7-1 Inspection Results by Supplier

SUPPLIER	NUMBER CONFORMING	NUMBER NONCONFORMING	TOTAL
X	50	3	53
Y	125	6	131
Z	75	2	77
Total	250	11	261

not included in the figure, since it pertains to a different concept. Table 7-1 provides the data for the example problems for Theorems 3, 4, 6, and 7.

Theorem 3. If A and B are two mutually exclusive events, then the probability that either event A or event B will occur is the sum of their respective probabilities.

$$P(A \text{ or } B) = P(A) + P(B)$$

"Mutually exclusive" means that the occurrence of one event makes the other event impossible. Thus, if on one throw of a die a 3 occurred (event A), then event B, say, a 5, could not possibly occur.

Whenever an "or" is verbalized, the mathematical operation is usually addition, or as we shall see in Theorem 4, it can be subtraction. Theorem 3 was illustrated with two events—it is equally applicable for more than two [$P(A \text{ or } B \text{ or } \ldots \text{ or } F) = P(A) + P(B) + \cdots + P(F)$].

EXAMPLE PROBLEM

If the 261 parts described in Table 7–1 are contained in a box, what is the probability of selecting a random part produced by supplier X or by supplier Z?

$$P(X \text{ or } Z) = P(X) + P(Z)$$
$$= \frac{53}{261} + \frac{77}{261}$$
$$= 0.498$$

What is the probability of selecting a nonconforming part from supplier X or a conforming part from supplier Z?

$$P(\text{nc. } X \text{ or co. } Z) = P(\text{nc. } X) + P(\text{co. } Z)$$
$$= \frac{3}{261} + \frac{75}{261}$$
$$= 0.299$$

EXAMPLE PROBLEM

If the 261 parts described in Table 7–1 are contained in a box, what is the probability that a randomly selected part will be from supplier *Z*, a nonconforming unit from unit from supplier *X*, or a conforming part from supplier *Y*?

$$P(Z \text{ or nc. } X \text{ or co. } Y) = P(Z) + P(\text{nc. } X) + P(\text{co. } Y)$$

$$= \frac{77}{261} + \frac{3}{261} + \frac{125}{261}$$

$$= 0.785$$

Theorem 3 is frequently referred to as the *additive law of probability.*

Theorem 4. If event *A* and event *B* are not mutually exclusive events, then the probability of either event *A* or event *B* or both is given by

$$P(A \text{ or } B \text{ or both}) = P(A) + P(B) - P(\text{both})$$

Events that are not mutually exclusive have some outcomes in common.

EXAMPLE PROBLEM

If the 261 parts described in Table 7–1 are contained in a box, what is the probability that a randomly selected part will be from supplier *X* or a nonconforming unit?

$$P(X \text{ or nc. or both}) = P(X) + P(\text{nc.}) - P(X \text{ and nc.})$$

$$= \frac{53}{261} + \frac{11}{261} - \frac{3}{261}$$

$$= 0.234$$

In the example problem, there are three outcomes common to both events. The 3 nonconforming units of supplier *X* are counted twice as outcomes of $P(X)$ and of $P(\text{nc.})$; therefore, one set of three is subtracted out. This theorem is also applicable to more than two events. A Venn diagram is sometimes used to describe the not mutually exclusive concept, as shown in Figure 7-3. The circle on the left contains 53 units from supplier *X* and the circle on the right contains 11 nonconforming units. The 3 nonconforming units from supplier *X* are found where the two circles intersect.

Theorem 5. The sum of the probabilities of the events of a situation is equal to 1.000.

$$P(A) + P(B) + \cdots + P(N) = 1.000$$

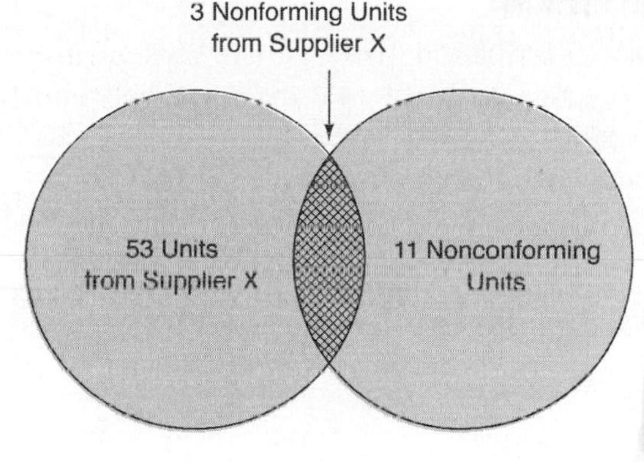

FIGURE 7-3 Venn diagram for Theorem 4 example problem.

This theorem was illustrated in Figure 7-1 for the coin-tossing, die-rolling, and card-drawing situations wherein the sum of the events equaled 1.000.

EXAMPLE PROBLEM

A health inspector examines 3 products in a subgroup to determine if they are acceptable. From past experience it is known that the probability of finding no nonconforming units in the sample of 3 is 0.990, the probability of 1 nonconforming unit in the sample of 3 is 0.006, and the probability of finding 2 nonconforming units in the sample of 3 is 0.003. What is the probability of finding 3 nonconforming units in the sample of 3?

There are tour, and only four, events to this situation: O nonconforming units, 1 nonconforming unit, 2 nonconforming units, and 3 nonconforming units.

$$P(0) + P(1) + P(2) + P(3) = 1.000$$

$$0.990 + 0.006 + 0.003 + P(3) = 1.000$$

$$P(3) = 0.001$$

Thus, the probability of 3 nonconforming units in the sample of 3 is 0.001.

Theorem 6. If A and B are independent events, then the probability of both A and B occurring is the product of their respective probabilities.

$$P(A \text{ and } B) = P(A) \times P(B)$$

An independent event is one where its occurrence has no influence on the probability of the other event or events. This theorem is referred to as the *multiplicative law of probabilities*. Whenever an "and" is verbalized, the mathematical operation is multiplication.

EXAMPLE PROBLEM

If the 261 parts described in Table 7-1 are contained in a box, what is the probability that 2 randomly selected parts will be from supplier *X* and supplier *Y*? Assume that the first part is returned to the box before the second part is selected (called *with replacement*)

$$
\begin{aligned}
P(X \text{ and } Y) &= P(X) \times P(Y) \\
&= \left(\frac{53}{261}\right)\left(\frac{131}{261}\right) \\
&= 0.102
\end{aligned}
$$

At first thought, the result of the example problem seems too low, but there are other possibilities such as *XX, YY, ZZ, YX, XZ, ZX, YZ*, and *ZY*. This theorem is applicable to more than two events.

Theorem 7. If *A* and *B* are *dependent* events, the probability of both *A* and *B* occurring is the product of the probability of *A* and the probability that if *A* occurred, then *B* will occur also.

$$P(A \text{ and } B) = P(A) \times P(B|A)$$

The symbol $P(B|A)$ is defined as the probability of event *B* provided that event *A* has occurred. A dependent event is one whose occurrence influences the probability of the other event or events. This theorem is sometimes referred to as the *conditional theorem*, since the probability of the second event depends on the result of the first event. It is applicable to more than two events.

EXAMPLE PROBLEM

Assume that in the preceding example problem the first part was not returned to the box before the second part was selected. What is the probability?

$$
\begin{aligned}
P(X \text{ and } Y) &= P(X) \times P(Y|X) \\
&= \left(\frac{53}{261}\right)\left(\frac{131}{260}\right) \\
&= 0.102
\end{aligned}
$$

Since the first part was not returned to the box, there was a total of only 260 parts in the box

What is the probability of choosing both parts from supplier *Z*?

$$P(Z \text{ and } Z) = P(Z) \times P(Z|Z)$$
$$= \left(\frac{77}{261}\right)\left(\frac{76}{260}\right)$$
$$= 0.086$$

Since the first part was from supplier Z, there are only 76 from supplier Z of the new total of 260 in the box.

To solve many probability problems it is necessary to use several theorems, as shown by the example below, which uses Theorems 3 and 6.

EXAMPLE PROBLEM

If the 261 parts described in Table 7–1 are contained in a box, what is the probability that two randomly selected parts (with replacement) will have one conforming part from supplier *X* and one conforming part from supplier *Y* or supplier *Z*?

$$P[\text{co. } X \text{ and } (\text{co. } Y \text{ or co. } Z)] = P(\text{co. } X)[P(\text{co. } Y) + P(\text{co. } Z)]$$
$$= \left(\frac{50}{261}\right)\left(\frac{125}{261} + \frac{75}{261}\right)$$
$$= 0.147$$

Counting of Events

Many probability problems, such as those where the events are uniform probability distributions, can be solved using counting techniques. There are three counting techniques that are quite often used in the computation of probabilities.

1. *Simple multiplication.* If an event *A* can happen in any of *a* ways or outcomes and, after it has occurred, another event *B* can happen in *b* ways or outcomes, the number of ways that both events can happen is *ab*.

EXAMPLE PROBLEM

A witness to a hit-and-run accident remembered the first 3 digits of the license plate out of 5 and noted the fact that the last 2 were numerals. How many owners of automobiles would the police have to investigate?

$$ab = (10)(10)$$
$$= 100$$

If the last 2 were letters, how many would need to be investigated?

$$ab = (26)(26)$$
$$= 676$$

2. *Permutations*. A *permutation* is an ordered arrangement of a set of objects. The permutations of the word *cup* are cup, cpu, upc, ucp, puc, and pcu. In this case there are 3 objects in the set and we arranged them in groups of 3 to obtain six permutations. This is referred to as a permutation of n objects taking r at a time where $n = 3$ and $r = 3$. How many permutations would there be for 4 objects taken 2 at a time? Using the word *fork* to represent the four objects, the permutations are fo, of, fr, rf, fk, kf, or, ro, ok, ko, rk, and kr. As the number of objects, n, and the number that are taken at one time, r, become larger, it becomes a tedious task to list all the permutations. The formula to find the number of permutations more easily is

$$P_r^n = \frac{n!}{(n-r)!}$$

where P_r^n = number of permutations of n objects taken r of them at a time (the symbol is sometimes written as ${}_nP_r$)

n = total number of objects

r = number of objects selected out of the total number

The expression $n!$ is read "n factorial" and means $n(n-1)(n-2)\cdots(1)$. Thus, $6! = 6 \cdot 5 \cdot 4 \cdot 3 \cdot 2 \cdot 1 = 720$. By definition, $0! = 1$.

EXAMPLE PROBLEM

How many permutations are there of 5 objects taken 3 at a time?

$$P_r^n = \frac{n!}{(n-r)!}$$

$$P_3^5 = \frac{5!}{(5-3)!} = \frac{5 \cdot 4 \cdot 3 \cdot 2 \cdot 1}{2 \cdot 1}$$
$$= 60$$

EXAMPLE PROBLEM

In the license plate example, suppose the witness further remembers that the numerals were not the same.

$$P_r^n = \frac{n!}{(n-r)!}$$

$$P_2^{10} = \frac{10!}{(10-2)!} = \frac{10 \cdot 9 \cdot 8 \cdot 7 \cdots 1}{8 \cdot 7 \cdots 1}$$

$$= 90$$

This problem could also have been solved by simple multiplication where $a = 10$ and $b = 9$. In other words, there are 10 ways for the first digit but only 9 ways for the second since duplicates are not permitted.

The symbol P is used for both permutation and probability. No confusion should result from this dual usage, since for permutations the superscript n and subscript r are used.

3. *Combinations*. If the way the objects are ordered is unimportant, then we have a *combination*. The word *cup* has *six* permutations when the 3 objects are taken 3 at a time. However, there is only *one* combination, since the same three letters are in a different order. The word *fork* has 12 permutations when the 4 letters are taken 2 at a time; but the number of combinations is fo, fr, fk, or, ok, and rk, which gives a total of six. The formula for the number of combinations is

$$C_r^n = \frac{n!}{r!(n-r)!}$$

where C_r^n = number of combinations of n objects taken r at a time (the symbol is sometimes written ${}_nC_r$ or $\binom{n}{r}$)

n = total number of objects

r = number of objects selected out of the total number

EXAMPLE PROBLEM

An interior designer has 5 different colored chairs and will use 3 in a living room arrangement. How many different combinations are possible?

$$C_r^n = \frac{n!}{r!(n-r)!}$$

$$C_3^5 = \frac{5!}{3!(5-3)!} = \frac{5 \cdot 4 \cdot 3 \cdot 2 \cdot 1}{3 \cdot 2 \cdot 1 \cdot 2 \cdot 1}$$

$$= 10$$

There is a symmetry associated with combinations such that $C_3^5 = C_2^5$, $C_1^4 = C_3^4$, $C_2^{10} = C_8^{10}$, and so on. Proof of this symmetry is left as an exercise.

The probability definition, the seven theorems, and the three counting techniques are all used to solve probability problems. Many hand calculators have permutation and combination functional keys that eliminate calculation errors provided that the correct keys are punched.

DISCRETE PROBABILITY DISTRIBUTIONS

When specific values such as the integers 0, 1, 2, 3 are used, then the probability distribution is discrete. Typical discrete probability distributions are hypergeometric, binomial, and Poisson.

Hypergeometric Probability Distribution

The *hypergeometric* probability distribution occurs when the population is finite and the random sample is taken without replacement. The formula for the hypergeometric is constructed of three combinations (total combinations, nonconforming combinations, and conforming combinations) and is given by

$$P(d) = \frac{C_d^D C_{n-d}^{N-D}}{C_n^N}$$

where $P(d)$ = probability of d nonconforming units in a sample of size n

C_n^N = combinations of all units

C_d^D = combinations of nonconforming units

C_{n-d}^{N-D} = combinations of conforming units

N = number of units in the lot (population)

n = number of units in the sample

D = number of nonconforming units in the lot

d = number of nonconforming units in the sample

$N - D$ = number of conforming units in the lot

$n - d$ = number of conforming units in the sample

The formula is obtained from the application of the probability definition, simple multiplication, and combinations. In other words, the numerator is the ways or outcomes of obtaining nonconforming units times the ways or outcomes of obtaining conforming units, and the denominator is the total possible ways or outcomes. Note that symbols in the combination formula have been changed to make them more appropriate for quality.

An example will make the application of this distribution more meaningful.

EXAMPLE PROBLEM

A lot of 9 thermostats located in a container has 3 nonconforming units. What is the probability of drawing one nonconforming unit in a random sample of 4?

For instructional purposes the following is a graphical illustration of the problem.

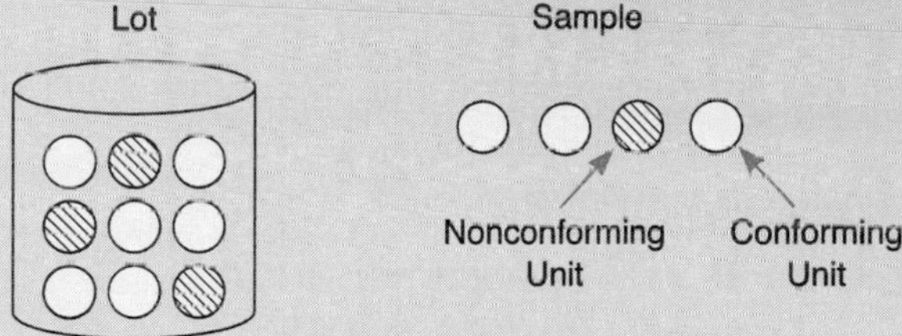

From the picture or from the statement of the problem, $N = 9$, $D = 3$, $n = 4$, and $d = 1$.

$$P(d) = \frac{C_d^D C_{n-d}^{N-D}}{C_n^N}$$

$$P(1) = \frac{C_1^3 C_{4-1}^{9-3}}{C_4^9}$$

$$= \frac{\dfrac{3!}{1!(3-1)!} \cdot \dfrac{6!}{3!(6-3)!}}{\dfrac{9!}{4!(9-4)!}}$$

$$= 0.476$$

Similarly, $P(0) = 0.119$, $P(2) = 0.357$, and $P(3) = 0.048$. Since there are only 3 nonconforming units in the lot, $P(4)$ is impossible. The sum of the probabilities must equal 1,000, and this is verified as follows:

$$P(T) = P(0) + P(1) + P(2) + P(3)$$

$$= 0.119 + 0.476 + 0.357 + 0.048$$

$$= 1.000$$

The complete probability distribution is given in Figure 7-4. As the parameters of the hypergeometric distribution change, the shape of the distribution changes, as illustrated by Figure 7-5. Therefore, each hypergeometric distribution has a unique shape based on *N*, *n*, and *D*. With hand calculators and computers, the efficient calculation of the distribution is not too difficult.

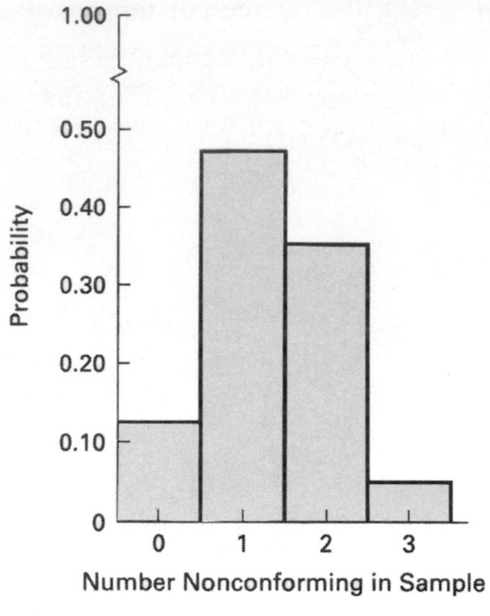

FIGURE 7-4 **Hypergeometric distribution for $N = 9$, $n = 4$, and $D = 3$.**

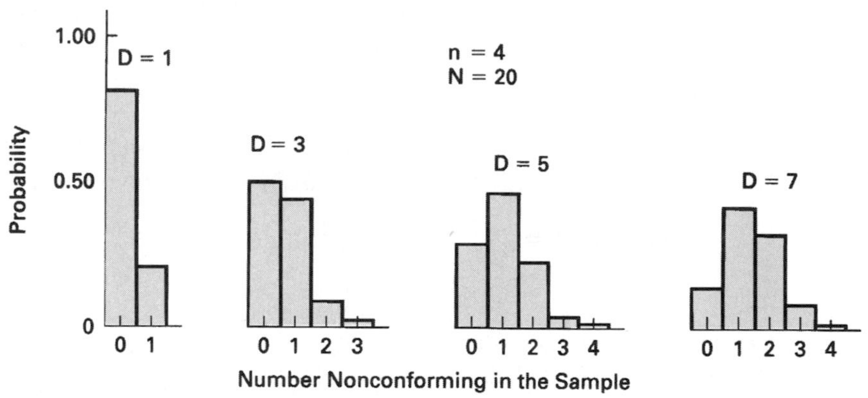

FIGURE 7-5 **Comparison of hypergeometric distributions with different fraction nonconforming in lot.**

Some solutions require an "or less" probability. In such cases the method is to add up the respective probabilities. Thus,

$$P(2 \text{ or less}) = P(2) + P(1) + P(0)$$

Similarly, some solutions require an "or more" probability and use the formulas

$$\begin{aligned} \text{P}(2 \text{ or more}) &= \text{P(T)} - \text{P}(1 \text{ or less}) \\ &= \text{P}(2) + \text{P}(3) + \cdots \end{aligned}$$

In the latter series, the number of terms to calculate is determined by the sample size, the number nonconforming in the lot, or when the value is less than 0.001. Thus, if the sample size is, say, 4, you would add $P(2)$, $P(3)$, and $P(4)$; if the number nonconforming in the lot was, say, 6, you would add $P(2)$, $P(3)$, $P(4)$, $P(5)$, and $P(6)$; and if $P(3) = 0.0009$, you would add $P(2)$ and $P(3)$.

Binomial Probability Distribution

The *binomial* probability distribution is applicable to discrete probability problems that have an infinite number of items or that have a steady stream of items coming from a work center. The binomial is applied to problems that have attributes, such as conforming or nonconforming, success or failure, pass or fail, and heads or tails. It corresponds to successive terms in the binomial expansion, which is

$$(p + q)^n = p^n + np^{n-1}q + \frac{n(n-1)}{2}p^{n-2}q^2 + \cdots + q^n$$

where p = probability of an event such as a nonconforming unit (proportion nonconforming)

$q = 1 - p$ = probability of a nonevent such as a conforming unit (proportion conforming)

n = number of trials or the sample size

Applying the expansions to the distribution of tails ($p = \frac{1}{2}, q = \frac{1}{2}$) resulting from an infinite number of throws of 11 coins at once, the expansion is

$$\left(\tfrac{1}{2} + \tfrac{1}{2}\right)^{11} = \left(\tfrac{1}{2}\right)^{11} + 11\left(\tfrac{1}{2}\right)^{10}\left(\tfrac{1}{2}\right) + 55\left(\tfrac{1}{2}\right)^{9}\left(\tfrac{1}{2}\right)^{2} + \cdots + \left(\tfrac{1}{2}\right)^{11}$$

$$= 0.001 + 0.005 + 0.027 + 0.080 + 0.161 + \cdots + 0.001$$

The probability distribution of the number of tails is shown in Figure 7-6. Since $p = q$, the distribution is symmetrical regardless of the value of n; however, when

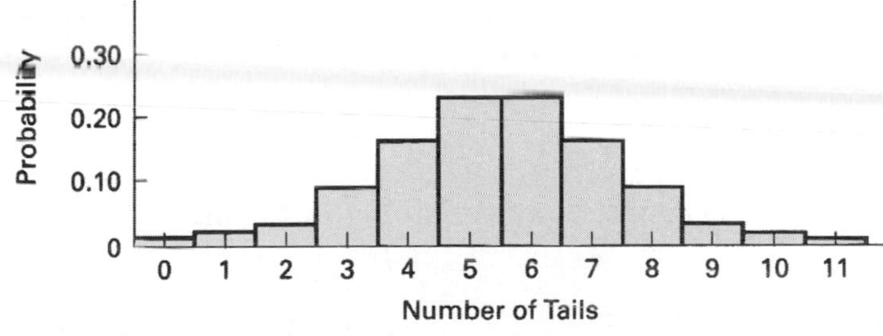

FIGURE 7-6 Distribution of the number of tails for an infinite number of tosses of 11 coins.

$p \neq q$, the distribution is asymmetrical. In quality work p is the proportion or fraction nonconforming and is usually less than 0.15.

In most cases in quality work, we are not interested in the entire distribution, only in one or two terms of the binomial expansion. The binomial formula for a single term is

$$P(d) = \frac{n!}{d!(n-d)!} p_0^d q_0^{n-d}$$

where $P(d)$ = probability of d nonconforming units

n = number in the sample

d = number nonconforming in the sample

p_0 = proportion (fraction) nonconforming in the population[1]

q_0 = proportion (fraction) conforming $(1 - p_0)$ in the population

Since the binomial is for the infinite situation, there is no lot size, N, in the formula.

EXAMPLE PROBLEM

A random sample of 5 hinges is selected from a steady stream of product from a punch press, and the proportion nonconforming is 0.10. What is the probability of 1 nonconforming unit in the sample? What is the probability of 1 or less? What is the probability of 2 or more?

$$q_0 = 1 - p_0 = 1.00 - 0.10 = 0.90$$

$$P(d) = \frac{n!}{d!(n-d)!} p_0^d q_0^{n-d}$$

$$P(1) = \frac{5!}{1!(5-1)!}(0.10^1)(0.90^{5-1})$$

$$= 0.328$$

What is the probability of 1 or less nonconforming units? To solve, we need to use the addition theorem and add $P(1)$ and $P(0)$.

$$P(d) = \frac{n!}{d!(n-d)!} p_0^d q_0^{n-d}$$

$$P(0) = \frac{5!}{0!(5-0)!}(0.10^0)(0.90^{5-0})$$

$$= 0.590$$

[1]Also standard or reference value; see Chapter 5.

Thus,

$$P(1 \text{ or less}) = P(0) + P(1)$$
$$= 0.590 + 0.328$$
$$= 0.918$$

What is the probability of 2 or more nonconforming units? Solution can be accomplished using the addition theorem and adding the probabilities of 2, 3, 4 and 5 nonconforming units.

$$P(2 \text{ or more}) = P(2) + P(3) + P(4) + P(5)$$

Or it can be accomplished by using the theorem that the sum of the probabilities is 1.

$$P(2 \text{ or more}) = P(T) - P(1 \text{ or less})$$
$$= 1.000 - 0.918$$
$$= 0.082$$

Calculations for two nonconforming units and three nonconforming units for the data in the example problem give $P(2) = 0.073$ and $P(3) = 0.008$. The complete distribution is shown as the graph on the left of Figure 7-7. Calculations for $P(4)$ and $P(5)$ give values less than 0.001 so they are not included in the graph.

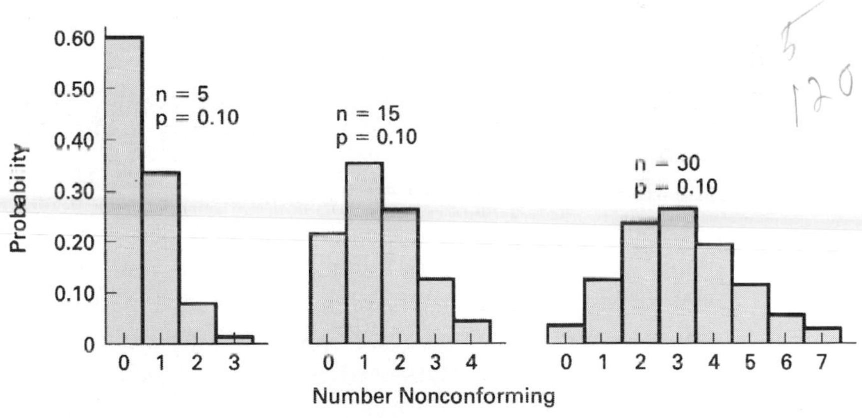

FIGURE 7-7 **Binomial distribution for various sample sizes when $p = 0.10$.**

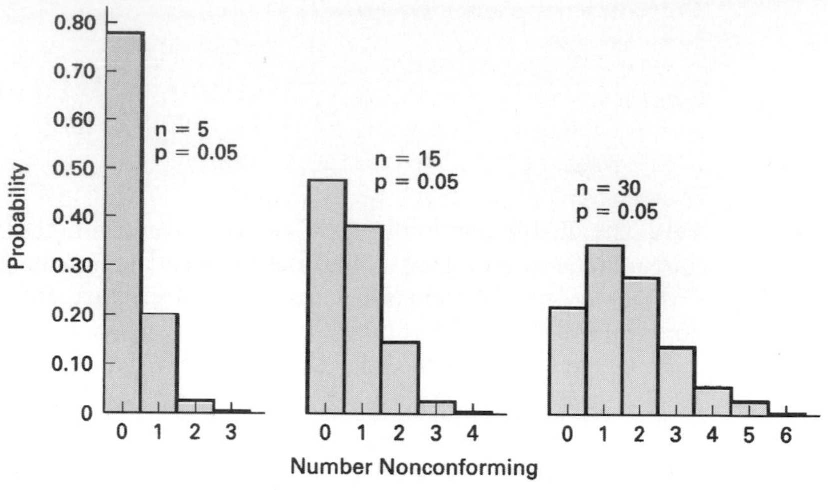

FIGURE 7-8 Binomial distributions for various sample sizes when $p = 0.05$.

Figure 7-7 illustrates the change in the distribution as the sample size increases for the proportion nonconforming of $p = 0.10$, and Figure 7-8 illustrates the change for $p = 0.05$. As the sample size gets larger, the shape of the curve will become symmetrical even though $p \neq q$. Comparing the distribution for $p = 0.10$ and $n = 15$ in Figure 7-7 with the distribution of $p = 0.05$, $n = 15$ in Figure 7-8, it is noted that for the same value of n, the larger the value of the proportion nonconforming p, the greater the symmetry of the distribution.

The shape of the distribution is always a function of the sample size, n, and the proportion nonconforming, p. Change either of these values and a different distribution results.

Tables are available for the binomial distribution. However, since three variables (n, p, and d) are needed, they require a considerable amount of space.

The calculator and computer can make the required calculations quite efficiently; therefore, there is no longer any need for the tables.

The binomial is used for the infinite situation but will approximate the hypergeometric under certain conditions that are discussed later in the chapter. It requires that there be two and only two possible outcomes (a nonconforming or a conforming unit) and that the probability of each outcome does not change. In addition, the use of the binomial requires that the trials are independent; that is, if a nonconforming unit occurs, then the chance of the next one being nonconforming neither increases nor decreases.

In addition, the binomial distribution is the basis for one of the control chart groups discussed in Chapter 8.

Poisson Probability Distribution

A third discrete probability distribution is referred to as the Poisson, named after Simeon Poisson, who described it in 1837. The distribution is applicable to many situations that involve observations per unit of time: for example, the count of cars arriving at a highway toll booth in 1-min intervals, the count of machine breakdowns in 1 day, and the count of shoppers entering a grocery store in 5-min intervals. The distribution is also applicable to situations involving observations per unit of amount: for example, the count of weaving nonconformities in 1000 m^2 of cloth, the count of nonconformities per lot of product, and the count of rivet nonconformities in a mobile home.

In each of the preceding situations, there are many equal opportunities for the occurrence of an event. Each rivet in a recreational vehicle has an equal opportunity to be a nonconformity; however, there will only be a few nonconformities out of the hundreds of rivets. The Poisson is applicable when n is quite large and p_0 is small.

The formula for the Poisson distribution is

$$P(c) = \frac{(np_0)^c}{c!} e^{-np_0}$$

where c = count, or number, of events of a given classification occurring in a sample, such as count of nonconformities, cars, customers, or machine breakdowns

np_0 = average count, or average number, of events of a given classification occurring in a sample

e = 2.718281

When the Poisson is used as an approximation to the binomial (to be discussed later in the chapter), the symbol c has the same meaning as d has in the binomial and hypergeometric formulas. Since c and np_0 have similar definitions, there is some confusion, which can be corrected by thinking of c as an individual value and np_0 as an average or population value.

Using the formula, a probability distribution can be determined. Suppose that the average count of cars that arrive at a highway toll booth in a 1-min interval is 2, then the calculations are

$$P(c) = \frac{(np_0)^c}{c!} e^{-np_0}$$

$$P(0) = \frac{(2)^0}{0!} e^{-2} = 0.135$$

$$P(1) = \frac{(2)^1}{1!} e^{-2} = 0.271$$

$$P(2) = \frac{(2)^2}{2!}e^{-2} = 0.271$$

$$P(3) = \frac{(2)^3}{3!}e^{-2} = 0.180$$

$$P(4) = \frac{(2)^4}{4!}e^{-2} = 0.090$$

$$P(5) = \frac{(2)^5}{5!}e^{-2} = 0.036$$

$$P(6) = \frac{(2)^6}{6!}e^{-2} = 0.012$$

$$P(7) = \frac{(2)^7}{7!}e^{-2} = 0.003$$

The resulting probability distribution is the one on the right in Figure 7-9. This distribution indicates the probability that a certain count of cars will arrive in any 1-min time interval. Thus, the probability of zero cars in any 1-min interval is 0.135, the probability of one car in any 1-min interval is 0.271, . . ., and the probability of seven cars in any 1-min interval is 0.003.

Figure 7-9 also illustrates the property that as np_0 gets larger, the distribution approaches symmetry. Other properties of the Poisson distribution are that the mean equals np_0 and the standard deviation equals $\sqrt{np_0}$.

Probabilities for the Poisson distribution for np_0 values of from 0.1 to 5.0 in intervals of 0.1 and from 6.0 to 15.0 in intervals of 1.0 are given in Table C in the

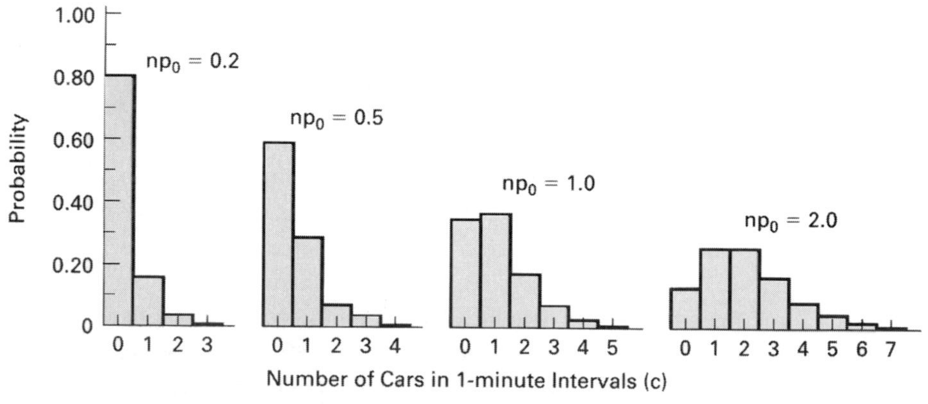

FIGURE 7-9 **Poisson probability distribution for various np_0 values.**

Appendix. Values in parentheses in the table are cumulative probabilities for obtaining "or less" answers. The use of this table simplifies the calculations, as illustrated in the following problem.

EXAMPLE PROBLEM

The average count of billing errors at a local bank per 8-h shift is 1.0. What is the probability of 2 billing errors? The probability of 1 or less? The probability of 2 or more?

From Table C for an np_0 value of 1.0:

$$P(2) = 0.184$$

$$P(1 \text{ or less}) = 0.736$$

$$\begin{aligned} P(2 \text{ or more}) &= 1.000 - P(1 \text{ or less}) \\ &= 1.000 - 0.736 \\ &= 0.264 \end{aligned}$$

The Poisson distribution can be used as an approximation for the binomial in some situations. Given below is an example problem that illustrates this concept.

EXAMPLE PROBLEM

If the probability that a heat-treating batch will be nonconforming is 0.01, what is the probability of 2 bad batches out of 250? What is the probability of 2 or less?

$$np_0 = (250)(0.01) = 2.5$$

From Table C, the intersection of the column with an np_0 value of 2.5 and the row with a c value of 2 gives

$$P(2) = 0.256 \qquad P(2 \text{ or less}) = 0.543$$

The answers using the binomial are

$$P(2) = 0.257 \qquad P(2 \text{ or less}) = 0.543$$

The Poisson probability distribution is the basis for attribute control charts and for acceptance sampling, which are discussed in subsequent chapters. In addition to quality applications, the Poisson distribution is used in other industrial situations, such as accident frequencies, computer simulation, operations research, and work sampling.

From a theoretical viewpoint a discrete probability distribution should use a bar graph. However, it is a common practice (and the one followed for the figures in this book) to use the histogram.

Other discrete probability distributions are the uniform, geometric, and negative binomial. The uniform distribution was illustrated in Figure 7-1. From an application viewpoint it is the one used to generate a random number table. The geometric and negative binomial are used in reliability studies for discrete data.

CONTINUOUS PROBABILITY DISTRIBUTIONS

When measurable data such as meters, kilograms, and ohms are used, the probability distribution is continuous. While there are many continuous probability distributions, only the normal is of sufficient importance to warrant a detailed discussion in an introductory text.

Normal Probability Distribution

The *normal curve* is a continuous probability distribution. Solutions to probability problems that involve continuous data can be solved using the normal probability distribution. In Chapter 4, techniques were learned to determine the percentage of the data that were above a certain value, below a certain value, or between two values. These same techniques are applicable to probability problems, as illustrated in the following example problem.

EXAMPLE PROBLEM

If the operating life of an electric mixer, which is normally distributed, has a mean of 2200 h and standard deviation of 120 h, what is the probability that a single electric mixer will fail to operate at 1900 h or less?

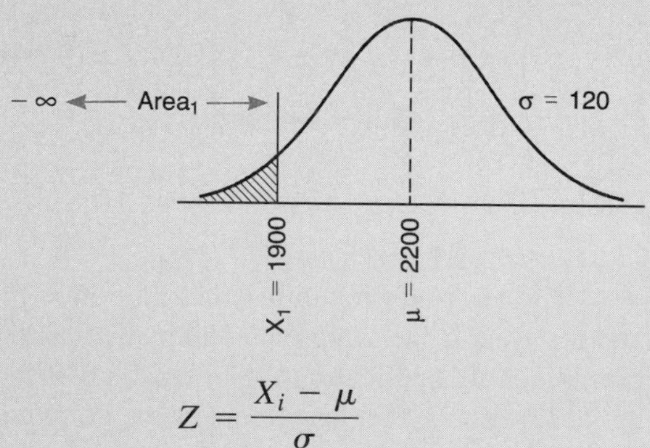

$$Z = \frac{X_i - \mu}{\sigma}$$

$$= \frac{1900 - 2200}{120}$$

$$= -2.5$$

From Table A of the Appendix, for a Z value of -2.5, $\text{area}_1 = 0.0062$. Therefore, the probability of an electric mixer failing is

$$P(\text{failure at 1900 h or less}) = 0.0062$$

The answer in the problem could have been stated as, "The percent of items less than 1900 h is 0.62%." Therefore, the areas under the normal curve can be treated as either a probability value or a relative frequency value.

Under certain conditions the normal probability distribution will approximate the binomial probability distribution. These conditions will be discussed later in the chapter. For the present, we are concerned with the problem-solving technique which is illustrated in the next example problem.

EXAMPLE PROBLEM

Find the probability of getting 2, 3, or 4 tails in 12 tosses of a coin by the normal approximation to the binomial. The problem is shown graphically below. The required probability area is crosshatched and can be approximated by the normal curve, which is shown as a dashed line. Since the data must be continuous for the normal curve, the probability of obtaining 2 to 4 tails is considered to be from 1.5 to 4.5.

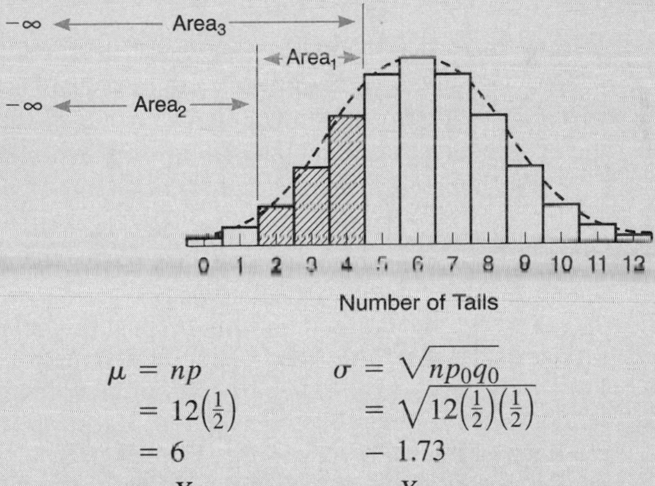

$$\mu = np = 12\left(\tfrac{1}{2}\right) = 6 \qquad \sigma = \sqrt{np_0q_0} = \sqrt{12\left(\tfrac{1}{2}\right)\left(\tfrac{1}{2}\right)} = 1.73$$

$$Z_2 = \frac{X_i - \mu}{\sigma} \qquad Z_3 = \frac{X_i - \mu}{\sigma}$$

$$= \frac{1.5 - 6}{1.73} \qquad = \frac{4.5 - 6}{1.73}$$

$$= -2.60 \qquad = -0.87$$

From Table A, for $Z_2 = -2.60$, $\text{area}_2 = 0.0047$, and for $Z_3 = -0.87$, $\text{area}_3 = 0.1922$.

$$\begin{aligned} \text{Area}_1 &= \text{Area}_3 - \text{Area}_2 \\ &= 0.1922 - 0.0047 \\ &= 0.1875 \end{aligned}$$

Thus, the required probability is

$$P(2, 3 \text{ or } 4 \text{ tails}) = 0.1875$$

Other Continuous Probability Distributions

Of the many other continuous probability distributions only two are of significant importance to mention their practical applications. The exponential probability distribution is used in reliability studies when there is a constant failure rate, and the Weibull is used when the time to failure is not constant. These two distributions are discussed in Chapter 11.

DISTRIBUTION INTERRELATIONSHIP

With so many distributions, it is sometimes difficult to know when they are applicable. Certainly, since the Poisson can be easily calculated using Table C, it should be used whenever appropriate. Figures 7-5, 7-8, and 7-9 show a similarity among the hypergeometric, binomial, and Poisson distributions.

The hypergeometric is used for finite lots of stated size N. It can be approximated by the binomial when $n/N \leqq 0.10$; or by the Poisson when $n/N \leqq 0.10$, $p_0 \leqq 0.10$, and $np_0 \leqq 5$; or by the normal when $n/N \leqq 0.10$ and the normal approximates the binomial.

The binomial is used for infinite situations or when there is a steady stream of product so that an infinite situation is assumed. It can be approximated by the Poisson when $p_0 \leqq 0.10$ and $np_0 \leqq 5$. The normal curve is an excellent approximation when p_0 is close to 0.5 and $n \geqq 10$. As np_0 deviates from 0.5, the approximation is still good as long as $np_0 \geqq 5$ and n increases to 50 or more for values of p_0 as low as 0.10 and as high as 0.90. Since the binomial calculation time is not too much different than the normal calculation time, there is little advantage to using the normal as an approximation.

The information given above can be considered to provide approximation guidelines rather than absolute laws. Approximations are better the farther the data are from the limiting values. *For the most part the efficiency of the calculator and computer have made the use of approximations obsolete.*

COMPUTER PROGRAM

The EXCEL software has the capability of performing calculations under the function wizard as follows:

PERMUT (n, r)
COMBIN (n, r)
HYPERGEOMDIST (d, n, D, N)
BINOMDIST (d, n, p, TRUE (cumulative) or FALSE (individual))
POISSON (c, np, TRUE (cumulative) or FALSE (individual))

All are located under the heading, *Statistical*, except COMBIN which is under the heading, *Math. & Trig*.

PROBLEMS

1. If an event is certain to occur, what is its probability? If an event will not occur, what is its probability?
2. What is the probability that you will live forever? What is the probability that an octopus will fly?
3. If the probability of obtaining a 3 on a 6-sided die is 0.167, what is the probability of obtaining any number but a 3?
4. Determine an event that has a probability of 1.000.
5. The probability of drawing a pink chip from a bowl of different-colored chips is 0.35, the probability of a blue chip is 0.46, the probability of a green chip is 0.15, and the probability of a purple chip is 0.04. What is the probability of a blue or a purple chip? What is the probability of a pink or a blue chip?
6. At any hour in a hospital intensive care unit the probability of an emergency is 0.247. What is the probability that there will be tranquility for the staff?
7. If a hotel has 20 king-size beds, 50 queen-size beds, 100 double beds, and 30 twin beds available, what is the probability that you will be given a queen-size or a twin bed when you register?

8. A ball is drawn at random from a container that holds 8 yellow balls numbered 1 to 8, 6 orange balls numbered 1 to 6, and 10 gray balls numbered 1 to 10. What is the probability of obtaining an orange ball or a ball numbered 5 or an orange ball numbered 5 in a draw of one ball? What is the probability of a gray ball or a ball numbered 8 or a gray ball numbered 8 in a draw of one ball?

9. If the probability of obtaining 1 nonconforming unit in a sample of 2 from a large lot of neoprene gaskets is 0.18 and the probability of 2 nonconforming units 0.25, what is the probability of 0 nonconforming units?

10. Using the information in Problem 9, find the probability of obtaining 2 nonconforming units on the first sample of 2 and 1 nonconforming unit on the second sample of 2. What is the probability of 0 nonconforming units on the first sample and 2 nonconforming units on the second? The first gasket selected is returned to the lot before the second one is selected.

11. A basket contains 34 heads of lettuce, 5 of which are spoiled. If a sample of 2 is drawn and not replaced, what is the probability that both will be spoiled?

12. If a sample of 1 can be drawn from an automatic storage and retrieval rack with three different storage racks and 6 different trays in each rack, what is the number of different ways of obtaining the sample of one?

13. A small model-airplane motor has 4 starting components: key, battery, wire, and glow plug. What is the probability that the system will work if the probability that each component will work is as follows: key (0.998), battery (0.997), wire (0.999), and plug (0.995)?

14. An inspector has to inspect products from 3 machines in one department, 5 machines in another, and 2 machines in a third. The quality manager wants to vary the inspector's route. How many different ways are possible?

15. If in the example problem with the hit-and-run driver there were one numeral and one letter, how many automobiles would need to be investigated?

16. A sample of 3 is selected from 10 people on a Caribbean cruise. How many permutations are possible?

17. From a lot of 90 airline tickets a sample of 8 is selected. How many permutations are possible?

18. A sample of 4 is selected from a lot of 20 piston rings. How many different sample combinations are possible?

19. From a lot of 100 hotel rooms, a sample of 3 is selected for audit. How many different sample combinations are possible?

20. A sample of 2 is selected from a tray of 20 bolts. How many different sample combinations are possible?

21. In the Illinois lottery, the numbers available are 1 to 54. On Saturday night 6 numbers are selected. How many different combinations are possible?

22. In the Illinois lottery each participant selects two sets of 6 numbers. What is the probability of having all 6 numbers?

23. The game of KENO has 80 numbers and you select 15. What is the probability of having all 15 numbers and winning the jackpot?

24. An automatic garage-door opener has 12 switches that can be set on or off. Both the transmitter and receiver are set the same and the owner has the option of setting 1 to 12 switches. What is the probability that another person with the same model transmitter could open the door?

25. Compare the answers of C_3^5 with C_2^5, C_1^4 with C_3^4, and C_2^{10} with C_8^{10}. What conclusion can you draw?

26. Calculate C_0^6, C_0^{10}, and C_0^{25}. What conclusion can you draw?

27. Calculate C_3^3, C_9^9, and C_{35}^{35}. What conclusion can you draw?

28. Calculate C_1^7, C_1^{12}, and C_1^{18}. What conclusion can you draw?

29. A random sample of 4 insurance claims is selected from a lot of 12 that has 3 nonconforming units. Using the hypergeometric distribution, what is the probability that the sample will contain exactly 0 nonconforming units? 1 nonconforming unit? 2 nonconforming units? 3 nonconforming units? 4 nonconforming units?

30. A finite lot of 20 digital watches is 20% nonconforming. Using the hypergeometric distribution, what is the probability that a sample of 3 will contain 2 nonconforming watches?

31. In Problem 30 what is the probability of obtaining 2 or more nonconforming units? What is the probability of 2 or less nonconforming units?

32. A steady stream of income tax returns has a proportion nonconforming of 0.03. What is the probability of obtaining 2 nonconforming units from a sample of 20? Use the binomial distribution.

33. Find the probability, using the binomial distribution, of obtaining 2 or more nonconforming units when sampling 5 computers from a batch known to be 6% nonconforming.

34. Using the binomial distribution, find the probability of obtaining 2 or less nonconforming restaurants in a sample of 9 when the lot is 15% nonconforming.

35. What is the probability of guessing correctly exactly 4 answers on a true-false examination that has 9 questions? Use the binomial distribution.

36. An injection molder produces golf tees that are 15.0% nonconforming. Using the normal distribution as an approximation to the binomial, find the probability that, in a random sample of 300 golf tees, 34 or less are nonconforming.

37. A random sample of 10 automotive bumpers is taken from a stream of product that is 5% nonconforming. Using the Poisson as an approximation to the binomial distribution, determine the probability of 2 nonconforming automotive bumpers. Compare the result with the binomial distribution.

38. If the probability is 0.08 that a single article is nonconforming, what is the probability that a sample of 20 will contain 2 or less nonconforming units? Use the Poisson as an approximation to the binomial distribution.

39. Using the data from Problem 38, determine the probability of 2 or more nonconforming units.

40. A sample of 10 washing machines is selected from a finite lot of 100. If $p_0 = 0.08$, what is the probability of 1 nonconforming washing machine in the sample? Is the Poisson a good approximation?

41. A lot of 15 has 3 nonconforming units. What is the probability that a sample of 3 will have 1 nonconforming unit? Would the Poisson be a good approximation?

42. A sample of 3 medicine bottles is taken from a tray of 30 bottles. If the tray is 10% nonconforming, what is the probability of 1 nonconforming medicine bottle in the sample? Is the binomial a good approximation?

43. A steady stream of light bulbs has a fraction nonconforming of 0.09. If 67 are sampled, what is the probability of 3 nonconforming units? Is the Poisson a good approximation?

44. Using the EXCEL function wizard solve some of the problems and compare your answers with those obtained using the calculator or by hand.

45. Using EXCEL construct the graph showing the entire distribution for:
    (a) Hypergeometric where n = 4, N = 20, and D = 5
    (b) Binomial where n = 15 and p = 0.05
    (c) Poisson where np = 1.0

8 CONTROL CHARTS FOR ATTRIBUTES[1]

INTRODUCTION

Attribute

An attribute was defined in Chapter 4 and is repeated to refresh the reader's memory. The term *attribute,* as used in quality, refers to those quality characteristics that conform to specifications or do not conform to specifications.

Attributes are used:

1. Where measurements are not possible—for example, visually inspected items such as color, missing parts, scratches, and damage.
2. Where measurements can be made but are not made because of time, cost, or need. In other words, although the diameter of a hole can be measured with an inside micrometer, it may be more convenient to use a "go–no go" gage and determine if it conforms or does not conform to specifications.

[1]The information in this chapter is based on ANSI/ASQc B1–B3—1996.

Where an attribute does not conform to specifications, various descriptive terms are used. A *nonconformity* is a departure of a quality characteristic from its intended level or state that occurs with a severity sufficient to cause an associated product or service not to meet a specification requirement. The definition of a *defect* is similar, except it is concerned with satisfying intended normal, or reasonably foreseeable, usage requirements. *Defect* is appropriate for use when evaluation is in terms of usage, and *nonconformity* is appropriate for conformance to specifications.

The term *nonconforming unit* is used to describe a unit of product or service containing at least one nonconformity. *Defective* is analogous to defect and is appropriate for use when a unit of product or service is evaluated in terms of usage rather than conformance to specifications.

In this book we are using the terms *nonconformity* and *nonconforming unit*. This practice avoids the confusion and misunderstanding that occurs with *defect* and *defective* in product-liability lawsuits.

Limitations of Variable Control Charts

Variable control charts are excellent means for controlling quality and subsequently improving it; however, they do have limitations. One obvious limitation is that these charts cannot be used for quality characteristics which are attributes. The converse is not true, since a variable can be changed to an attribute by stating that it conforms or does not conform to specifications. In other words, nonconformities such as missing parts, incorrect color, and so on, are not measureable and a variable control chart is not applicable.

Another limitation concerns the fact that there are many variables in a manufacturing entity. Even a small manufacturing plant could have as many as 1000 variable quality characteristics. Since an $\overline{X}$ and R chart is needed for each characteristic, 1000 charts would be required. Clearly, this would be too expensive and impractical. A control chart for attributes can minimize this limitation by providing overall quality information at a fraction of the cost.

Types of Attribute Charts

There are two different groups of control charts for attributes. One group of charts is for nonconforming units. It is based on the binomial distribution. A proportion, p, chart shows the proportion nonconforming in a sample or subgroup. The proportion is expressed as a fraction or a percent. Similarly we could have charts for proportion conforming, and they too could be expressed as a fraction or a percent. Another chart in the group is for the number nonconforming, a np chart,[2] and it too could also be expressed as number conforming.

---

[2]The ANSI/ASQc B1–3—1996 standard uses the symbol *pn;* however, current practice uses *np*.

The other group of charts is for nonconformities. It is based on the Poisson distribution. A c chart shows the count of nonconformities in an inspected unit such as an automobile, bolt of cloth, or roll of paper. Another closely related chart is the u chart, which is for the count of nonconformities per unit.

Much of the information on control charts for attributes is similar to that given in Chapter 5. The reader is referred to the sections on "State of Control" and "Analysis of Out-of-Control Condition."

CONTROL CHARTS FOR NONCONFORMING UNITS

Introduction

The p chart is used for data that consist of the proportion of the number of occurrences of an event to the total number of occurrences. It is used in quality to report the fraction or percent nonconforming in a product, quality characteristic, or group of quality characteristics. As such, the *fraction* nonconforming is the proportion of the number nonconforming in a sample or subgroup to the total number in the sample or subgroup. In symbolic terms the formula is

$$p = \frac{np}{n}$$

where p = proportion or fraction nonconforming in the sample or subgroup

n = number in the sample or subgroup

np = number nonconforming in the sample or subgroup

np = # of BAD IN SAMPLE

EXAMPLE PROBLEM

During the first shift, 450 inspections are made of book-of-the month shipments and 5 nonconforming units are found. Production during the shift was 15,000 units. What is the fraction nonconforming?

$$p = \frac{np}{n} = \frac{5}{450} = 0.011$$

The fraction nonconforming, p, is usually small, say, 0.10 or less. Except in unusual circumstances, values greater than 0.10 would indicate that the organization is in serious difficulty and that measures more drastic than a control chart are required. Because the fraction nonconforming is very small, the subgroup sizes must be quite large to produce a meaningful chart.

The p chart is an extremely versatile control chart. It can be used to control one quality characteristic, as is done with the $\overline{X}$ and R chart; to control a group of quality characteristics of the same type or of the same part; or to control the entire product. The p chart can be established to measure the quality produced by a work center, by a department, by a shift, or by an entire plant. It is frequently used to report the performance of an operator, group of operators, or management as a means of evaluating their quality performance.

The subgroup size of the p chart can be either variable or constant. A constant subgroup size is preferred; however, there may be many situations such as changes in mix and 100% automated inspection, where the subgroup size changes.

Objectives

The objectives of nonconforming charts are to:

1. *Determine the average quality level.* Knowledge of the quality average is essential as a benchmark. This information provides the process capability in terms of attributes.

2. *Bring to the attention of management any changes in the average.* Once the average quality (proportion nonconforming) is known, changes, either increasing or decreasing, become significant.

3. *Improve the product quality.* In this regard a p chart can motivate operating and management personnel to initiate ideas for quality improvement. The chart will tell whether the idea is an appropriate or inappropriate one. A continual and relentless effort must be made to improve the quality.

4. *Evaluate the quality performance of operating and management personnel.* Supervisors of activities and especially the Chief Executive Officer (CEO) should be evaluated by a chart for nonconforming units. Other functional areas, such as engineering, sales, finance, and so on, may find a chart for nonconformities more applicable for evaluation purposes.

5. *Suggest places to use $\overline{X}$ and R charts.* Even though the cost of computing and charting $\overline{X}$ and R charts is more than the chart for nonconforming units, the $\overline{X}$ and R charts are much more sensitive to variations and are more helpful in diagnosing causes. In other words, the chart for nonconforming units suggests the source of difficulty, and the $\overline{X}$ and R chart finds the cause.

6. *Determine acceptance criteria of a product before shipment to the customer.* Knowledge of the proportion nonconforming provides management with information on whether or not to release an order.

These objectives indicate the scope and value of a nonconforming chart.

p-Chart Construction for Constant Subgroup Size

The general procedures that apply to variable control charts also apply to the p chart.

1. *Select the quality characteristic(s).* The first step in the procedure is to determine the use of the control chart. A p chart can be established to control the proportion nonconforming of (a) a single quality characteristic, (b) a group of quality characteristics, (c) a part, (d) an entire product, or (e) a number of products. This establishes a hierarchy of utilization so that any inspections applicable for a single quality characteristic also provide data for other p charts, which represent larger groups of characteristics, parts, or products.

A p chart can also be established for performance control of an (a) operator, (b) work center, (c) department, (d) shift, (e) plant, or (f) corporation. Using the chart in this manner, comparisons may be made between like units. It is also possible to evaluate the quality performance of a unit. A hierarchy of utilization exists so that data collected for one chart can also be used on a more all-inclusive chart.

The use for the chart or charts will be based on securing the greatest benefit for a minimum of cost. One chart should measure the CEO's quality performance.

2. *Determine the subgroup size and method.* The size of the subgroup is a function of the proportion nonconforming. If a part has a proportion nonconforming, p, of 0.001 and a subgroup size, n, of 1000, then the average number nonconforming, np, would be one per subgroup. This would *not* make a good chart, since a large number of values, posted to the chart, would be zero. If a part has a proportion nonconforming of 0.15 and a subgroup size of 50, the average number of nonconforming would be 7.5, which would make a good chart.

Therefore, the selection of the subgroup size requires some preliminary observations to obtain a rough idea of the proportion nonconforming and some judgment as to the average number of nonconforming units that will make an adequate graphical chart. A minimum size of 50 is suggested as a starting point. Inspection can either be by audit or on-line. Audits are usually done in a laboratory under optimal conditions. On-line provides immediate feedback for corrective action.

3. *Collect the data.* The quality technician will need to collect sufficient data for at least 25 subgroups, or the data may be obtained from historical records. Perhaps the best source is from a check sheet designed by a project team. Table 8–1 gives the inspection results for the blower motor in an electric hair dryer for the motor department. For each subgroup the proportion nonconforming is calculated by the formula $p = np/n$. The quality technician reported that subgroup 19 had an abnormally large number of nonconforming units, owing to faulty contacts.

TABLE 8-1 Inspection Results of Hair Dryer Blower Motor, Motor Department, May

SUBGROUP NUMBER	NUMBER INSPECTED n	NUMBER NONCONFORMING np	PROPORTION NONCONFORMING p
1	300	12	0.040
2	300	3	0.010
3	300	9	0.030
4	300	4	0.013
5	300	0	0.0
6	300	6	0.020
7	300	6	0.020
8	300	1	0.003
9	300	8	0.027
10	300	11	0.037
11	300	2	0.007
12	300	10	0.033
13	300	9	0.030
14	300	3	0.010
15	300	0	0.0
16	300	5	0.017
17	300	7	0.023
18	300	8	0.027
19	300	16	0.053
20	300	2	0.007
21	300	5	0.017
22	300	6	0.020
23	300	0	0.0
24	300	3	0.010
25	300	2	0.007
Total	7500	138	

The data can be plotted as a run chart, as shown in Figure 8-1. A run chart shows the variation in the data; however, we need statistical limits to determine if the process is stable.

This type of chart is very effective during the start-up phase of a new item or process when the process is very erratic. Also, many organizations prefer to use this type of chart to measure quality performance rather than a control chart.

Since the run chart does not have limits, it is not a control chart. This fact does not reduce its effectiveness in many situations.

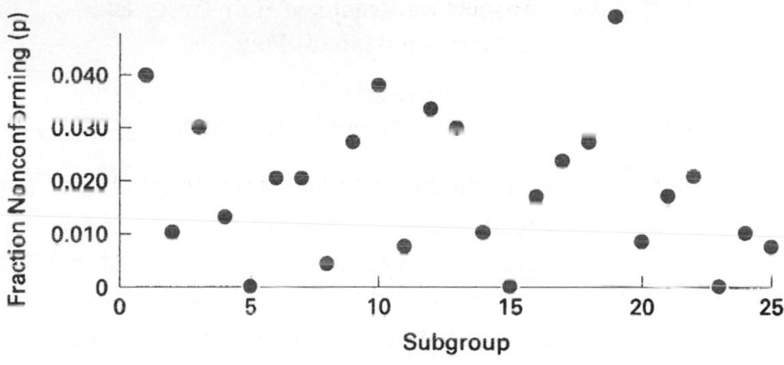

FIGURE 8-1 Run chart for the data of Table 8-1.

4. *Calculate the trial central line and control limits.* The formula for the trial control limits is given by

$$UCL = \overline{p} + 3\sqrt{\frac{\overline{p}(1-\overline{p})}{n}}$$

$$LCL = \overline{p} - 3\sqrt{\frac{\overline{p}(1-\overline{p})}{n}}$$

where $\overline{p}$ = average proportion nonconforming for many subgroups
n = number inspected in a subgroup

The average proportion nonconforming, $\overline{p}$, is the central line and is obtained by the formula $\overline{p} = \Sigma np/\Sigma n$. Calculations for the 3σ trial control limits using the data on the electric hair dryer are as follows:

P average →

$$\overline{p} = \frac{\Sigma np}{\Sigma n} = \frac{138}{7500} = 0.018$$

$$UCL = \overline{p} + 3\sqrt{\frac{\overline{p}(1-\overline{p})}{n}}$$

$$= 0.018 + 3\sqrt{\frac{0.018(1-0.018)}{300}}$$

Know

$$= 0.041$$

$$LCL = \overline{p} - 3\sqrt{\frac{\overline{p}(1-\overline{p})}{n}}$$

$$= 0.018 - 3\sqrt{\frac{0.018(1 - 0.018)}{300}}$$

$$= -0.005 \text{ or } 0.0$$

Calculations for the lower control limit resulted in a *negative* value, which is a theoretical result. In practice, a negative proportion nonconforming would be impossible. Therefore, the lower control limit value of -0.005 is changed to zero.

When the lower control limit is positive, it may in some cases be changed to zero. If the *p* chart is to be viewed by operating personnel, it would be difficult to explain why a proportion nonconforming that is below the lower control limit is out of control. In other words, performance of exceptionally good quality would be classified as out of control. To avoid the need to explain this situation to operating personnel, the lower control limit is changed from a positive value to zero. When the *p* chart is to be used by quality personnel and by management, a positive lower control limit is left unchanged. In this manner exceptionally good performance (below the lower control limit) will be treated as an out-of-control situation and investigated for an assignable cause. It is hoped that the assignable cause will indicate how the situation can be repeated.

The central line, $\overline{p}$, and the control limits are shown in Figure 8-2; the proportion nonconforming, *p*, from Table 8-1 is also posted to that chart. This chart is used to determine if the process is stable and is not posted. It is important to recognize that the central line and control limits were determined from the data.

5. *Establish the revised central line and control limits.* In order to determine the revised 3σ control limits, the standard or reference value for the proportion noncon-

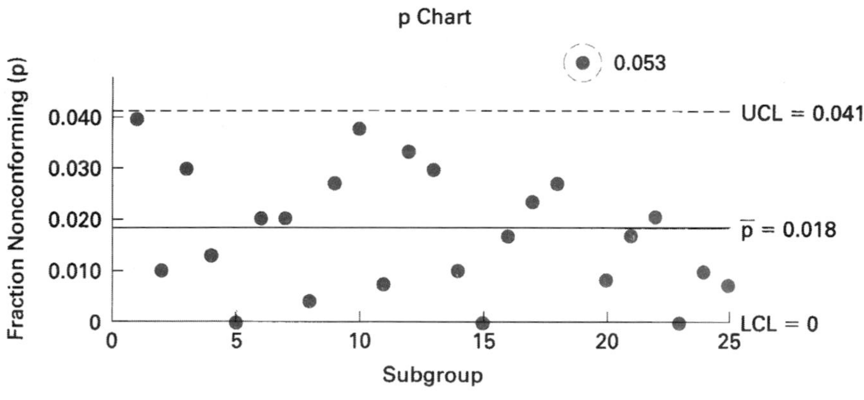

FIGURE 8-2 ***p* chart to illustrate the trial central line and control limits using the data from Table 8-1.**

forming, p_0, needs to be determined. If an analysis of the chart of step 4 above shows good control (a stable process), then $\bar{p}$ can be considered to be representative of that process. Therefore, the best estimate of p_0 at this time is $\bar{p}$, and $p_0 = \bar{p}$.

Most industrial processes, however, are not in control when first analyzed, and this fact is illustrated in Figure 8-2 by subgroup 19, which is above the upper control limit and therefore out of control. Since subgroup 19 has an assignable cause, it can be discarded from the data and a new $\bar{p}$ computed with all of the subgroups except 19. The calculations can be simplified by using the formula

$$\bar{p}_{new} = \frac{\Sigma np - np_d}{\Sigma n - n_d}$$

where np_d = number nonconforming in the discarded subgroups

n_d = number inspected in the discarded subgroups

In discarding data it must be remembered that only those subgroups with assignable causes are discarded. Those subgroups without assignable causes are left in the data. Also, out-of-control points below the lower control limit are not discarded, since they represent exceptionally good quality. If the out-of-control point on the low side is due to an inspection error, it should be discarded.

With an adopted standard or reference value for the proportion nonconforming, p_0, the revised control limits are given by

$$p_0 = \bar{p}_{new}$$

$$\text{UCL} = p_0 + 3\sqrt{\frac{p_0(1 - p_0)}{n}}$$

$$\text{LCL} = p_0 - 3\sqrt{\frac{p_0(1 - p_0)}{n}}$$

where p_0, the central line, represents the reference or standard value for the fraction nonconforming. These formulas are for the control limits for three standard deviations from the central line p_0.

Thus, for the preliminary data in Table 8-1, a new $\bar{p}$ is obtained by discarding subgroup 19.

$$\bar{p}_{new} = \frac{\Sigma np - np_d}{\Sigma n - n_d}$$

$$= \frac{138 - 16}{7500 - 300}$$

$$= 0.017$$

Since $\overline{p}_{new}$ is the best estimate of the standard or reference value, $p_0 = 0.017$. The revised control limits for the p chart are obtained as follows:

$$\begin{aligned}
\text{UCL} &= p_0 + 3\sqrt{\frac{p_0(1 - p_0)}{n}} \\
&= 0.017 + 3\sqrt{\frac{0.017(1 - 0.017)}{300}} \\
&= 0.039 \\
\text{LCL} &= p_0 - 3\sqrt{\frac{p_0(1 - p_0)}{n}} \\
&= 0.017 - 3\sqrt{\frac{0.017(1 - 0.017)}{300}} \\
&= -0.005 \text{ or } 0.0
\end{aligned}$$

The revised control limits and the central line, p_0, are shown in Figure 8-3. This chart, without the plotted points, is posted in an appropriate place, and the proportion nonconforming, p, for each subgroup is plotted as it occurs.

6. *Achieve the objective*. The first five steps are planning. The last step involves action and leads to the achievement of the objective. The revised control limits were based on data collected in May. Some representative values of inspection results for the month of June are shown in Figure 8-3. Analysis of the June results shows that the quality improved. This improvement is expected, since the posting of a quality control chart usually results in improved quality. Using the

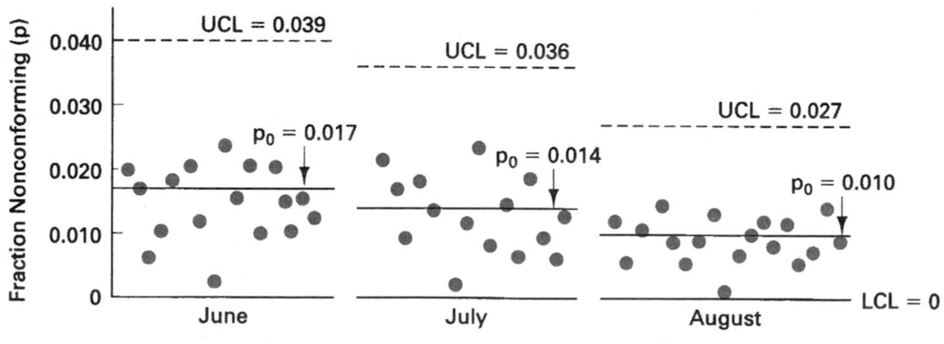

FIGURE 8-3 Continuing use of the it p chart for representative values of the proportion nonconforming, *p*.

June data, a better estimate of the proportion nonconforming is obtained. The new value ($p_0 = 0.014$) is used to obtain the UCL of 0.036.

During the latter part of June and the entire month of July, various quality improvement ideas generated by a project team are tested. These ideas are new shellac, change in wire size, stronger spring, $\overline{X}$ and R charts on the armature, and so on. In testing ideas there are three criteria: a minimum of 25 subgroups are required, the 25 subgroups can be compressed in time as long as no sampling bias occurs, and only one idea can be tested at one time. The control chart will tell whether the idea improves the quality, reduces the quality, or has no effect on the quality. The control chart should be located in a conspicuous place so operating personnel can view it.

Data from July are used to determine the central line and control limits for August. The pattern of variation for August indicates that no further improvement resulted. However, a 41% improvement occurred from June (0.017) to August (0.010). At this point, we have obtained considerable improvement testing the ideas of the project team. While this improvement is very good, we must continue our relentless pursuit of quality improvement—1 out of every 100 is still nonconforming. Perhaps a detailed failure analysis or technical assistance from product engineering will lead to additional ideas that can be evaluated. A new project team may help.

Quality improvement is never terminated. Efforts may be redirected to other areas based on need and/or resources available.

Some Comments on *p* Charts

Like the $\overline{X}$ and R chart, the p chart is most effective if it is posted where operating and quality personnel can view it. Also, like the $\overline{X}$ and R chart, the control limits are usually three standard deviations from the central value. Therefore, approximately 99% of the plotted points, p, will fall between the upper and lower control limits.

A state of control for a p chart is treated in a manner similar to that described in Chapter 5. The reader may wish to briefly review that section. A control chart for subgroup values of p will aid in disclosing the occasional presence of assignable causes of variation in the manufacturing process. The elimination of these assignable causes will lower p_0 and, therefore, have a positive effect on spoilage, production efficiency, and cost per unit. A p chart will also indicate long-range trends in quality, which will help to evaluate changes in personnel, methods, equipment, tooling, materials, and inspection techniques.

If the population proportion nonconforming, ϕ, is known, it is not necessary to calculate the trial control limits. This is a considerable timesaver, since $p_0 = \phi$, which allows the p chart to be introduced immediately. Also, p_0 may be assigned a desired value—in which case the trial control limits are not necessary.

Since the p chart is based on the binomial distribution, there must be a constant chance of selecting a nonconforming product. In some production operations, if one nonconforming unit occurs, all product that follows will be nonconforming until the condition is corrected. This type of condition also occurs in batch processes where the entire batch is nonconforming or when an error is made in dimensions, color, and so on. In such cases a constant chance of obtaining a nonconforming unit does not occur, and therefore, the p chart is not suitable.

Presentation Techniques

The information in the preceding example is presented as a fraction nonconforming. It could also be presented in percent nonconforming, fraction conforming, or percent conforming. All four techniques convey the same information, as shown by Figure 8-4. The two lower figures show opposite information from the respective upper figures.

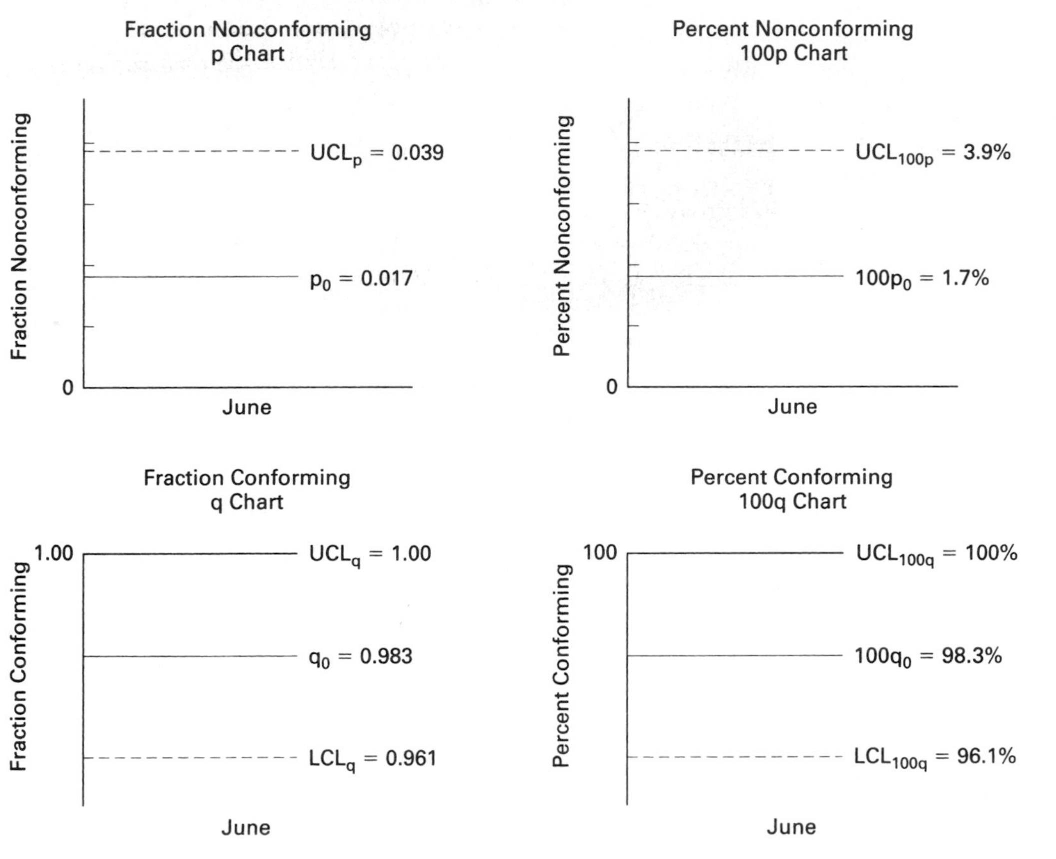

FIGURE 8-4 Different techniques for presenting *p* chart information.

TABLE 8-2 Calculating Central Line and Limits for the Different Presentation Techniques

	FRACTION NONCONFORMING	PERCENT NONCONFORMING	FRACTION CONFORMING	PERCENT CONFORMING
Central line	p_o	$100p_o$	$q_o = 1 - p_o$	$100q_o = 100(1 - p_o)$
Upper control limit	UCL_p	$100(UCL_p)$	$UCLq = 1 - LCL_p$	$100(UCLq)$
Lower control limit	LCL_p	$100(LCL_p)$	$LCLq = 1 - UCL_p$	$100(LCLq)$

Table 8-2 shows the equations for calculating the central line and control limits for the three techniques as a function of p_0.

Many organizations are taking the positive approach and using either of the two conforming presentation techniques. The use of the chart and the results will be the same no matter which chart is used.

p-Chart Construction for Variable Subgroup Size

Whenever possible, *p* charts should be developed and used with a constant subgroup size. This situation is not possible when the *p* chart is used for 100% inspection of output that varies from day to day. Also, data for *p*-chart use from sampling inspection might vary for a variety of reasons. Since the control limits are a function of the subgroup size, *n*, the control limits will vary with the subgroup size. Therefore, they need to be calculated for each subgroup.

While a variable subgroup size is undesirable, it does exist and must be handled. The procedures of data collection, trial central line and control limits, and revised central line and control limits are the same as those for a *p* chart with constant subgroup size. An example without steps 1 and 2 will be used to illustrate the procedure.

Step 3. Collect the data. A computer modem manufacturer has collected data from the final test of the product for the end of March and all of April. Subgroup size was one day's inspection results. The inspection results for 25 subgroups are shown in the first three columns of Table 8-3: subgroup designation, number inspected, and number nonconforming. A fourth column for the fraction nonconforming is calculated using the formula $p = np/n$. The last two columns are for the upper and lower control limit calculations, which are discussed in the next section.

The variation in the number inspected per day can be due to a number of reasons. Machines may have breakdowns or not be scheduled. Product models may have different production requirements, which will cause day-to-day variations.

TABLE 8-3 Preliminary Data of Computer Modem Final Test and Control Limits for Each Subgroup

SUBGROUP	NUMBER INSPECTED n	NUMBER NONCONFORMING np	FRACTION NONCONFORMING p	LIMIT UCL	LCL
March 29	2,385	55	0.023	0.029	0.011
30	1,451	18	0.012	0.031	0.009
31	1,935	50	0.026	0.030	0.010
April 1	2,450	42	0.017	0.028	0.012
2	1,997	39	0.020	0.029	0.011
5	2,168	52	0.024	0.029	0.011
6	1,941	47	0.024	0.030	0.010
7	1,962	34	0.017	0.030	0.010
8	2,244	29	0.013	0.029	0.011
9	1,238	53	0.043	0.032	0.008
12	2,289	45	0.020	0.029	0.011
13	1,464	26	0.018	0.031	0.009
14	2,061	47	0.023	0.029	0.011
15	1,667	34	0.020	0.030	0.010
16	2,350	31	0.013	0.029	0.011
19	2,354	38	0.016	0.029	0.011
20	1,509	28	0.018	0.031	0.009
21	2,190	30	0.014	0.029	0.011
22	2,678	113	0.042	0.028	0.012
23	2,252	58	0.026	0.029	0.011
26	1,641	34	0.021	0.030	0.010
27	1,782	19	0.011	0.030	0.010
28	1,993	30	0.015	0.030	0.010
29	2,382	17	0.007	0.029	0.011
30	2,132	46	0.022	0.029	0.011
	50,515	1,015			

For the data in Table 8-3, there was a low on April 9 of 1238 inspections because the second shift did not work and a high on April 22 of 2678 inspections because of overtime in one work center.

Step 4. Determine the trial central line and control limits. Control limits are calculated using the same procedures and formulas as for a constant subgroup. However, since the subgroup size changes each day, limits must be calculated for each day. First, the average fraction nonconforming, which is the central line, must be determined, and it is

$$\overline{p} = \frac{\Sigma np}{\Sigma n} = \frac{1015}{50,515} = 0.020$$

Using $\overline{p}$, the control limits for each day can be obtained. For March 29, the limits are

$$\text{UCL}_{29} = \overline{p} + 3\sqrt{\frac{\overline{p}(1 - \overline{p})}{n_{29}}}$$

$$= 0.020 + 3\sqrt{\frac{0.020(1 - 0.020)}{2385}}$$

$$= 0.029$$

$$\text{LCL}_{29} = \overline{p} - 3\sqrt{\frac{\overline{p}(1 - \overline{p})}{n_{29}}}$$

$$= 0.020 - 3\sqrt{\frac{0.020(1 - 0.020)}{2385}}$$

$$= 0.011$$

For March 30, the control limits are

$$\text{UCL}_{30} = \overline{p} + 3\sqrt{\frac{\overline{p}(1 - \overline{p})}{n_{30}}}$$

$$= 0.020 + 3\sqrt{\frac{0.020(1 - 0.020)}{1451}}$$

$$= 0.031$$

$$\text{LCL}_{30} = \overline{p} - 3\sqrt{\frac{\overline{p}(1 - \overline{p})}{n_{30}}}$$

$$= 0.020 - 3\sqrt{\frac{0.020(1 - 0.020)}{1451}}$$

$$= 0.009$$

The control limit calculations above are repeated for the remaining 23 subgroups. Since n is the only variable that is changing, it is possible to simplify the calculations as follows:

$$\text{CL's} = \overline{p} \pm \frac{3\sqrt{\overline{p}(1 - \overline{p})}}{\sqrt{n}}$$

$$= 0.020 \pm \frac{3\sqrt{0.020(1 - 0.020)}}{\sqrt{n}}$$

$$= 0.020 \pm \frac{0.42}{\sqrt{n}}$$

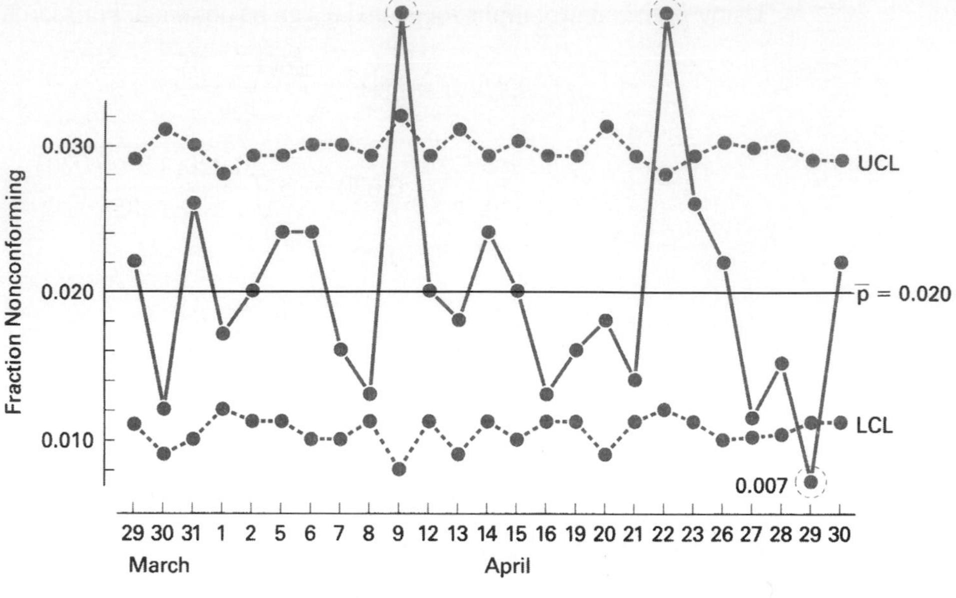

FIGURE 8-5 Preliminary data, central line, and trial control limits.

Using this technique the calculations are much quicker. The control limits for all 25 subgroups are shown in columns 5 and 6 of Table 8-3. A graphical illustration of the trial control limits, central line, and subgroup values is shown in Figure 8-5.

Note that as the subgroup size gets larger, the control limits are closer together; as the subgroup size gets smaller, the control limits become wider apart. This fact is apparent from the formula and by comparing the subgroup size, n, with its UCL and LCL.

Step 5. Establish revised central line and control limits. A review of Figure 8-5 shows that an out-of-control situation is present on April 9, April 22, and April 29. There was a problem with the wave solder on April 9 and April 22. Also, it was found that on April 29 the testing instrument was out of calibration. Since all these out-of-control points have assignable causes, they are discarded. A new $\overline{p}$ is obtained as follows:

$$\overline{p}_{\text{new}} = \frac{\Sigma np - np_d}{\Sigma n - n_d}$$

$$= \frac{1015 - 53 - 113 - 17}{50{,}515 - 1238 - 2678 - 2382}$$

$$= 0.019$$

TABLE 8-4 Inspection Results for May 3, 4, and 5

SUBGROUP	NUMBER INSPECTED	NUMBER NONCONFORMING
May 3	1535	31
4	2262	28
5	1872	45

Since this value represents the best estimate of the standard or reference value of the fraction nonconforming, $p_0 = 0.019$.

The fraction nonconforming, p_0, is used to calculate upper and lower control limits for the next period, which is the month of May. However, the limits cannot be calculated until the end of each day, when the subgroup size, n, is known. This means that the control limits are never known ahead of time. Table 8-4 shows the inspection results for the first three working days in May. Control limits and the fraction nonconforming for May 3 are as follows:

$$p_{\text{May 3}} = \frac{np}{n} = \frac{31}{1535} = 0.020$$

$$\begin{aligned} \text{UCL}_{\text{May 3}} &= p_0 + 3\sqrt{\frac{p_0(1 - p_0)}{n_{\text{May 3}}}} \\ &= 0.019 + 3\sqrt{\frac{0.019(1 - 0.019)}{1535}} \\ &= 0.029 \end{aligned}$$

$$\begin{aligned} \text{LCL}_{\text{May 3}} &= p_0 - 3\sqrt{\frac{p_0(1 - p_0)}{n_{\text{May 3}}}} \\ &= 0.019 - 3\sqrt{\frac{0.019(1 - 0.019)}{1535}} \\ &= 0.009 \end{aligned}$$

The upper and lower control limits and the fraction nonconforming for May 3 are posted to the p chart as illustrated in Figure 8-6. In a similar manner, calculations are made for May 4 and 5 and the results posted to the chart.

The chart is continued until the end of May, using $p_0 = 0.019$. Since an improvement usually occurs after introduction of a chart, a better estimate of p_0 will probably be obtained at the end of May using that month's data. In the future the value of p_0 should be evaluated periodically.

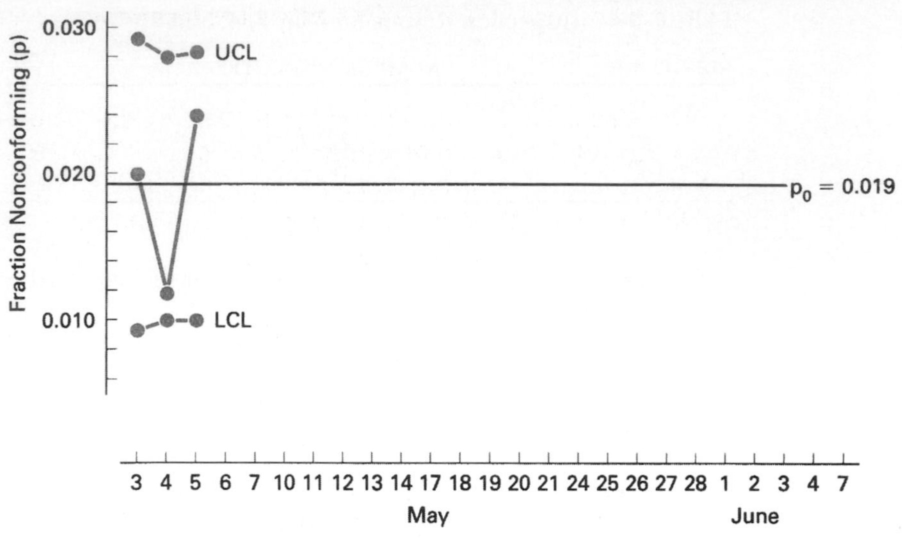

FIGURE 8-6 Control limits and fraction nonconforming for first three working days in May.

If p_0 is known, the process of data collection and trial control limits is not necessary. This saves considerable time and effort.

Since some confusion occurs among p_0, $\overline{p}$, and p, their definitions will be repeated:

1. p is the proportion (fraction) nonconforming in a single subgroup. It is posted to the chart but is *not* used to calculate the control limits.
2. $\overline{p}$ is the average proportion (fraction) nonconforming of many subgroups. It is the sum of the number nonconforming divided by the sum of the number inspected and is used to calculate the trial control limits.
3. p_0 is the standard or reference value of the proportion (fraction) nonconforming based on the best estimate of $\overline{p}$. It is used to calculate the revised control limits. It can be specified as a desired value.
4. ϕ is the population proportion (fraction) nonconforming. When this value is known, it can be used to calculate the limits, since $p_0 = \phi$.

Minimizing the Effect of Variable Subgroup Size

When the control limits vary from subgroup to subgroup, it presents an unattractive chart that is difficult to explain to operating personnel. It is also difficult to explain that control limits are calculated at the end of each day or time period rather than ahead of time. There are two techniques that minimize the effect of the variable subgroup size.

1. *Control limits for an average subgroup size.* By using an average subgroup size, one limit can be calculated and placed on the control chart. The average group size can be based on the anticipated production for the month or the previous month's inspections. As an example, the average number inspected for the preliminary data in Table 8-3 would be

$$n_{av} = \frac{\Sigma n}{g} = \frac{50{,}515}{25} = 2020.6, \quad \text{say, } 2000$$

Using a value of 2000 for the subgroup size, n, and $p_0 = 0.019$, the upper and lower control limits become

$$\begin{aligned} \text{UCL} &= p_0 + 3\sqrt{\frac{p_0(1 - p_0)}{n_{av}}} \\ &= 0.019 + 3\sqrt{\frac{0.019(1 - 0.019)}{2000}} \\ &= 0.028 \end{aligned}$$

$$\begin{aligned} \text{LCL} &= p_0 - 3\sqrt{\frac{p_0(1 - p_0)}{n_{av}}} \\ &= 0.019 - 3\sqrt{\frac{0.019(1 - 0.019)}{2000}} \\ &= 0.010 \end{aligned}$$

These control limits are shown in the p chart of Figure 8-7 along with the fraction nonconforming, p, for each day in May.

When an average subgroup size is used, there are four situations that occur between the control limits and the individual fraction nonconforming values.

Case I. This case occurs when a point (subgroup fraction nonconforming) falls inside the limits and its subgroup size is smaller than the average subgroup size. The data for May 6, $p = 0.011$ and $n = 1828$, represent this case. Since the May 6 subgroup size (1828) is less than the average of 2000, the control limits for May 6 will be wider apart than the control limits for the average subgroup size. Therefore, in this case individual control limits are not needed. If p is in control when $n = 2000$, it must also be in control when $n = 1828$.

Case II. This case occurs when a point (subgroup fraction nonconforming) falls inside the average limits and its subgroup size is larger than the average subgroup

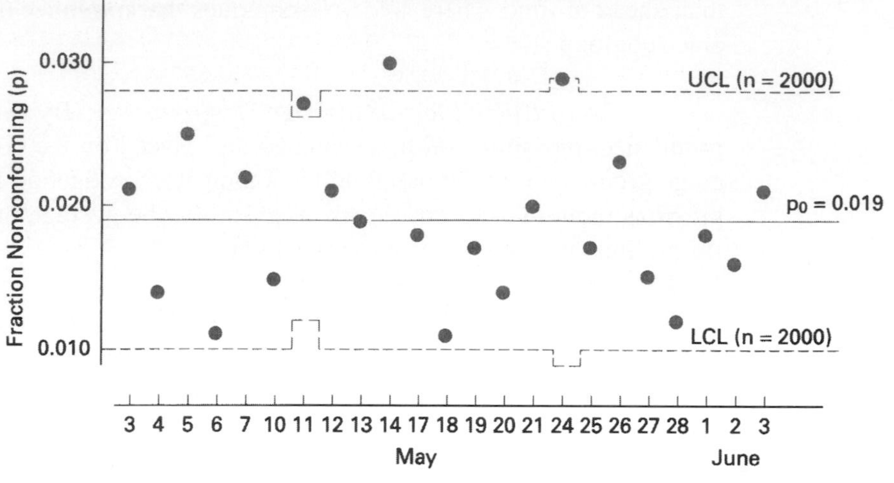

FIGURE 8-7 Chart for May data illustrating use of an average subgroup size.

size. The data for May 11, $p = 0.027$ and $n = 2900$, illustrate this case. Since the May 11 subgroup size is greater than the average subgroup size, the control limits for May 11 will be closer together than the control limits for the average subgroup size. Therefore, when there is a substantial difference in the subgroup size, individual control limits are calculated. For May 11 the values for the upper and lower control limits are 0.026 and 0.012, respectively. These individual control limits are shown in Figure 8-7. It is seen that the point is beyond the individual control limit and so we have an out-of-control situation.

Case III. This case occurs when a point (subgroup fraction nonconforming) falls outside the limits and its subgroup size is larger than the average subgroup size: The data for May 14, $p = 0.030$ and $n = 2365$, illustrate this case. Since the May 14 subgroup size (2365) is greater than the average of 2000, the control limits for May 14 will be narrower than the control limits for the average subgroup size. Therefore, in this case individual control limits are not needed. If p is out of control when $n = 2000$, it must also be out of control when $n = 2365$.

Case IV. This case occurs when a point (subgroup fraction nonconforming) falls outside the limits and its subgroup size is less than the average subgroup size. The data for May 24, $p = 0.029$ and $n = 1590$, illustrate this case. Since the May 24 subgroup size (1590) is less than the average of 2000, the control limits for May 24 will be wider apart than the control limits for the average subgroup size. Therefore, when there is a substantial difference in the subgroup size, individual control limits are calculated. For May 24 the values for the upper and lower control limits are 0.029 and 0.009, respectively. These individual control limits are shown in Figure 8-7. It is seen that the point is on the individual control limit and is assumed to be in control.

It is not always necessary to calculate the individual control limits in cases II and IV. Only when the value of p is close to the control limits is it necessary to determine the individual limits. For this example problem, p values within, say, $\pm$ 0.002 of the original limits should be checked. Since approximately 5% of the p values will be close to the control limits, few p values will need to be evaluated.

In addition, it is not necessary to calculate individual control limits as long as the subgroup size does not deviate substantially from the average, say, 15%. For this example, subgroup sizes of from 1700 to 2300 would be satisfactory and not need to have individual limit calculations.

Actually, when the average subgroup size is used, individual control limits are determined infrequently—about once every 3 months.

2. *Control limits for different subgroup sizes.* Another technique, which has been found to be effective, is to establish control limits for different subgroup sizes. Figure 8-8 illustrates such a chart. Using the different control limits and the four cases described previously, the need to calculate individual control limits would be rare. For example, the subgroup for July 16 with 1150 inspections is in control, and the subgroup for July 22 with 3500 inspections is out of control.

An analysis of Figure 8-8 shows that the relationship of the control limits to the subgroup size, n, is exponential rather than linear. In other words, the control limit lines are not equally spaced for equal subdivisions of the subgroup size, n. This type of chart can be effective when there are extreme variations in subgroup size.

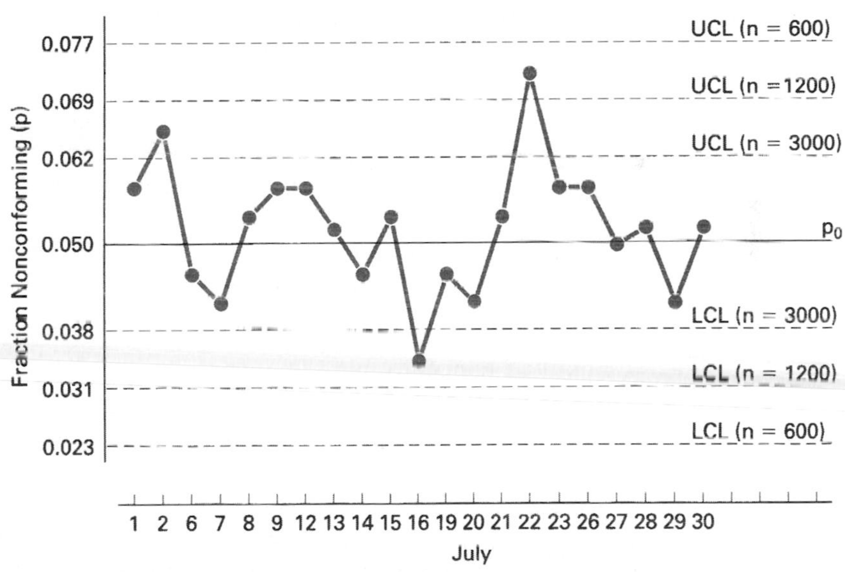

FIGURE 8-8 ***p* chart illustrating central line and control limits for different subgroup sizes.**

Number Nonconforming Chart

The number nonconforming chart (np chart) is almost the same as the p chart; however, you would not use both for the same objective.

The np chart is easier for operating personnel to understand than the p chart. Also, inspection results are posted directly to the chart without any calculations.

If the subgroup size is allowed to vary, the central line and the control limits will vary, which presents a chart that is almost meaningless. Therefore, one limitation of an np chart is the requirement that the subgroup size be constant. The sample size should be shown on the chart so viewers have a reference point.

Since the number nonconforming chart is mathematically equivalent to the proportion nonconforming chart, the central line and control limits are changed by a factor of n. Formulas are

$$\text{Central line} = np_0$$
$$\text{Control limits} = np_0 \pm 3\sqrt{np_0(1 - p_0)}$$

If the fraction nonconforming p_0 is unknown, then it must be determined by collecting data, calculating trial control limits, and obtaining the best estimate of p_0. The trial control limits formulas are obtained by substituting $\overline{p}$ for p_0 in the formulas above. An example problem illustrates the technique.

EXAMPLE PROBLEM

A government agency samples 200 documents per day from a daily lot of 6000. From past records the standard or reference value for the fraction nonconforming p_0, is 0.075.

Central line and control limit calculations are

$$np_0 = 200(0.075) = 15.0$$

$$\begin{aligned} \text{UCL} &= np_0 + 3\sqrt{np_0(1 - p_0)} \\ &= 15 + 3\sqrt{15(1 - 0.075)} \\ &= 26.2 \end{aligned} \qquad \begin{aligned} \text{LCL} &= np_0 - 3\sqrt{np_0(1 - p_0)} \\ &= 15 - 3\sqrt{15(1 - 0.075)} \\ &= 3.8 \end{aligned}$$

Since the number nonconforming is a whole number, the limit values should be whole numbers; however, they can be left as fractions. This practice prevents a plotted point from falling on a control limit. Of course the central line is a fraction. The control chart is shown in Figure 8-9 for 4 weeks in October.

Process Capability

The process capability of a variable was described in Chapter 5. For an attribute this process is much simpler. In fact, the process capability is the central line of the control chart.

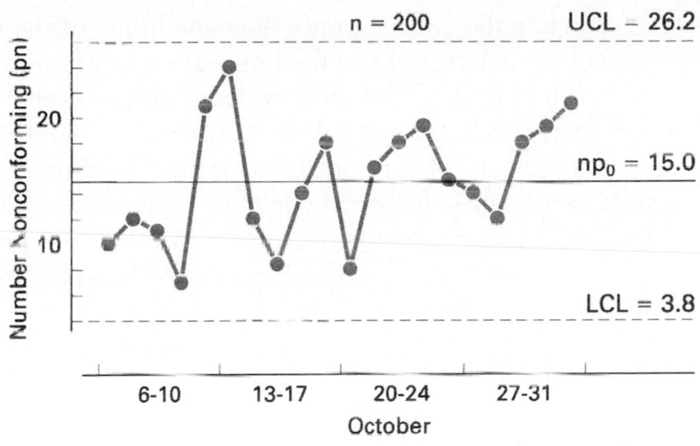

FIGURE 8-9 Number nonconforming chart (*np* chart).

Figure 8-10 shows a percent nonconforming chart for first-run automobile water leaks with a central line of 5.0%. The 5.0% value is the process capability, and the plotted points vary from the capability within the control limits. This variation occurs in a random manner but follows the binomial distribution.

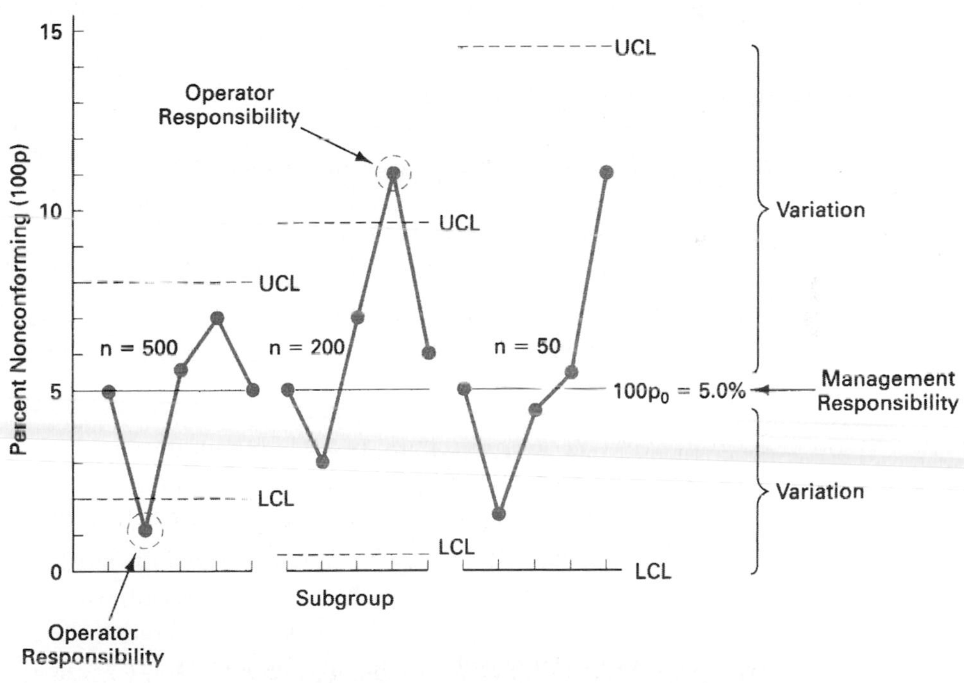

FIGURE 8-10 Process capability explanation and responsibility.

While the control limits show the limits of the variation of the capability, it should be understood that the limits are a function of the subgroup size. This fact is shown in Figure 8-10 for subgroup sizes of 500, 200, and 50. As the subgroup size increases, the control limits become closer to the central line.

Management is responsible for the capability. If the 5% value is not satisfactory, then management must initiate the procedures and provide the resources to take the necessary corrective action. As long as operating personnel (operators, firstline supervisors, and maintenance workers) are maintaining the plotted points within the control limits, they are doing what the process is capable of doing. When the plotted point is outside the control limit, operating personnel are usually responsible. A plotted point below the lower control limit is due to exceptionally good quality. It should be investigated to determine the assignable cause so that, if it is not due to an inspection error, it can be repeated.

CONTROL CHARTS FOR COUNT OF NONCONFORMITIES

Introduction

The other group of attribute charts is the nonconformity charts. While a p chart controls the proportion nonconforming of the product or service, the nonconformities chart controls the count of nonconformities within the product or service. Remember, an item is classified as a nonconforming unit whether it has one or many nonconformities. There are two types of charts: count of nonconformities (c) chart and count of nonconformities per unit (u) chart.

Since these charts are based on the Poisson distribution, two conditions must be met. First, the average count of nonconformities must be much less than the total possible count of nonconformities. In other words, the opportunity for nonconformities is large, whereas the chance of a nonconformity at any one location is very small. This situation is typified by the rivets on a commercial airplane, where there are a large number of rivets but a small chance of any one rivet being a nonconformity. The second condition specifies that the occurrences are independent. In other words, the occurrence of one nonconformity does not increase or decrease the chance of the next occurrence being a nonconformity. For example, if a typist types an incorrect letter there is an equal likelihood of the next letter being incorrect. Any beginning typist knows that this is not always the case because if the hands are not on the home keys, the chance of the second letter being incorrect is almost a certainty.

Other places where a chart of nonconformities meets the two conditions are: imperfections in a large roll of paper, typographical errors on a printed page, rust spots on steel sheets, seeds or air pockets in glassware, adhesion defects per 1000 square feet of corrugated board, mold marks on fiberglass canoes, billing errors, and errors on forms.

Like nonconforming unit charts, the control limits for charts for nonconformities are usually based on 3σ from the central line. Therefore, approximately 99% of the subgroup values will fall within the limits. It is suggested that the reader review the section "State of Control" in Chapter 5, since much of that information is applicable to the nonconformity charts.

Objectives

While the charts for count of nonconformities are not as inclusive as the $\overline{X}$ and R charts or the p charts, they still have a number of applications, some of which have been mentioned.

The objectives of charts for count of nonconformities are to:

1. Determine the average quality level as a benchmark or starting point. This information gives the initial process capability.
2. Bring to the attention of management any changes in the average. Once the average quality is known, any change becomes significant.
3. Improve the product quality. In this regard a chart for count of nonconformities can motivate operating and management personnel to initiate ideas for quality improvement. The chart will tell whether the idea is an appropriate or inappropriate one. A continual and relentless effort must be made to improve the quality.
4. Evaluate the quality performance of operating and management personnel. As long as the chart is in control, operating personnel are performing satisfactorily. Since the charts for count of nonconformities are usually applicable to errors, they are very effective in evaluating the quality of the functional areas of finance, sales, customer service, and so on.
5. Suggest places to use the $\overline{X}$ and R charts. Some applications of the charts for count of nonconformities lend themselves to more detailed analysis by $\overline{X}$ and R charts.
6. Provide information concerning the acceptability of the product prior to shipment.

These objectives are almost identical to those for nonconforming charts. Therefore, the reader is cautioned to be sure that the appropriate group of charts is being used.

Because of the limitations of the charts for count of nonconformities, many organizations do not have occasion for their use.

c-Chart Construction

The procedures for the construction of a c chart are the same as those for the p chart. If the count of nonconformities, c_0, is unknown, it must be found by collecting data, calculating trial control limits, and obtaining the best estimate.

1. *Select the quality characteristic(s).* The first step in the procedure is to determine the use of the control chart. Like the p chart, it can be established to control (a) a single quality characteristic, (b) a group of quality characteristics, (c) a part, (d) an entire product, or (e) a number of products. It can also be established for performance control of (a) an operator, (b) a work center, (c) a department, (d) a shift, (e) a plant, or (f) a corporation. The use for the chart or charts will be based on securing the greatest benefit for a minimum of cost.

2. *Determine the subgroup size and method.* The size of a c chart is one inspected unit. An inspected unit could be one airplane, one case of soda cans, one gross of pencils, one bundle of Medicare applications, one stack of labels, and so forth. The method of obtaining the sample can either be by audit or on-line.

3. *Collect the data.* Data were collected on the count of nonconformities of a blemish nature for fiberglass canoes. These data were collected during the first and second weeks of May by inspecting random production samples. Data are shown in Table 8-5 for 25 canoes, which is the minimum number of subgroups needed for trial control limit calculations. Note that canoes MY132 and MY278 both had production difficulties.

4. *Calculate the trial central line and control limits.* The formulas for the trial control limits are

$$UCL = \bar{c} + 3\sqrt{\bar{c}}$$

$$LCL = \bar{c} - 3\sqrt{\bar{c}}$$

Know

where $\bar{c}$ is the average count of nonconformities for a number of subgroups. The value of $\bar{c}$ is obtained from the formula $\bar{c} = \Sigma c/g$ where g is the number of subgroups and c is the count of nonconformities. For the data in Table 8-5, the calculations are:

$$\bar{c} = \frac{\Sigma c}{g} = \frac{141}{25} = 5.64$$

$$\begin{aligned} UCL &= \bar{c} + 3\sqrt{\bar{c}} & LCL &= \bar{c} - 3\sqrt{\bar{c}} \\ &= 5.64 + 3\sqrt{5.64} & &= 5.64 - 3\sqrt{5.64} \\ &= 12.76 & &= -1.48. \text{ or } 0 \end{aligned}$$

Since a lower control limit of -1.48 is impossible, it is changed to zero. The upper control limit of 12.76 is left as a fraction so that the plotted point, which is a whole

TABLE 8-5 Count of Blemish Nonconformities (c) by Canoe Serial Number

SERIAL NUMBER	COUNT OF NONCONFORMITIES	COMMENT	SERIAL NUMBER	COUNT OF NONCONFORMITIES	COMMENT
MY102	7		MY198	3	
MY113	6		MY208	2	
MY121	6		MY222	7	
MY125	3		MY235	5	
MY132	20	Mold sticking	MY241	7	
MY143	8		MY258	2	
MY150	6		MY259	8	
MY152	1		MY264	0	
MY164	0		MY267	4	
MY166	5		MY278	14	Fell off skid
MY172	14		MY281	4	
MY184	3		MY288	5	
MY185	1				
			Total	$\Sigma c = 141$	

number, cannot lie on the control limit. Figure 8-11 illustrates the central line, $\bar{c}$, the control limits, and the count of nonconformities, c, for each canoe of the preliminary data.

5. *Establish the revised central line and control limits.* In order to determine the revised 3σ control limits, the standard or reference value for the count of defects, c_0, is needed. If an analysis of the preliminary data shows good control, then $\bar{c}$ can be considered to be representative of that process $c_0 = \bar{c}$. Usually, however, an analysis of the preliminary data does not show good control, as illustrated in Figure 8-11. A better estimate of $\bar{c}$ (one that can be adopted for c_0) can be obtained by discarding out-of-control values with assignable causes. Low values that do not have an assignable cause represent exceptionally good quality. The calculations can be simplified by using the formula

$$\bar{c}_{new} = \frac{\Sigma c - c_d}{g - g_d}$$

where c_d = count of nonconformities in the discarded subgroups

g_d = number of discarded subgroups

Once an adopted standard or reference value is obtained, the revised 3σ control limits are found using the formulas

$$UCL = c_0 + 3\sqrt{c_0}$$

$$LCL = c_0 - 3\sqrt{c_0}$$

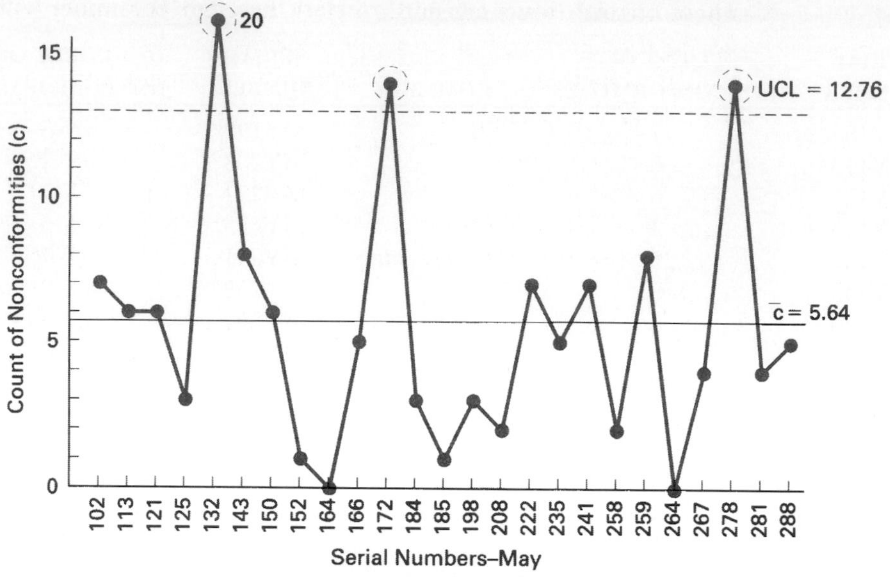

FIGURE 8-11 Control chart for count of nonconformities (*c* chart), using preliminary data.

where c_0 is the reference or standard value for the count of nonconformities. The count of nonconformities, c_0, is the central line of the chart and is the best estimate using the available data. It equals $\bar{c}_{new}$.

Using the information from Figure 8-11 and Table 8-5, revised limits can be obtained. An analysis of Figure 8-11 shows that canoe numbers 132, 172, and 278 are out of control. Since canoes 132 and 278 have an assignable cause (see Table 8-5), they are discarded; however, canoe 172 may be due to a chance cause, it is not discarded. Therefore, $\bar{c}_{new}$ is obtained as follows:

$$\bar{c}_{new} = \frac{\Sigma c - c_d}{g - g_d} = \frac{141 - 20 - 14}{25 - 2} = 4.65$$

Since $\bar{c}_{new}$ is the best estimate of the central line, $c_0 = 4.65$. The revised control limits for the c chart are:

$$\text{UCL} = c_0 + 3\sqrt{c_0} = 4.65 + 3\sqrt{4.65} = 11.1 \qquad \text{LCL} = c_0 - 3\sqrt{c_0} = 4.65 - 3\sqrt{4.65} = -1.82, \text{ or } 0$$

These control limits are used to start the chart beginning with canoes produced during the third week of May and are shown in Figure 8-12.

If c_0 had been known, the data collection and trial control limit phase would have been unnecessary.

6. *Achieve the objective.* The reason for the control chart is to achieve one or more of the previously stated objectives. Once the objective is reached, the chart is discontinued or inspection activity is reduced and resources are allocated to another quality problem. Some of the objectives, however, such as the first one, can be ongoing.

As with the other types of control charts, an improvement in the quality is expected after the introduction of a chart. At the end of the initial period, a better estimate of the number of nonconformities can be obtained. Figure 8-12 illustrates the change in c_0 and in the control limits for August as the chart is continued in use. Quality improvement resulted from the evaluation of ideas generated by the project

Chart for Canoe Blemish Nonconformities
Model–17S

Type of Nonconformity																														
Scratches	1		2		2		3			1						2					1	2	1		1			1		
Paint Imperfections						1	2	1						1			3				1				1					3
Indentations	1		2						2					1					1		1	2							1	
Scuff Marks	1	1	5		3	4	3	5	2	2				1	1	2		1	2		1	3	1	5	2	1	2	2		3
Total	3	1	9	0	5	5	8	6	4	3			0	6	1	4	3	1	3	0	4	7	2	5	4	1	2	3	1	6
Serial Number	305	310	321	354	373	409	441	469	485	487			129	150	178	185	209	230	260	283	303	321	347	359	407	471	485	493	564	589

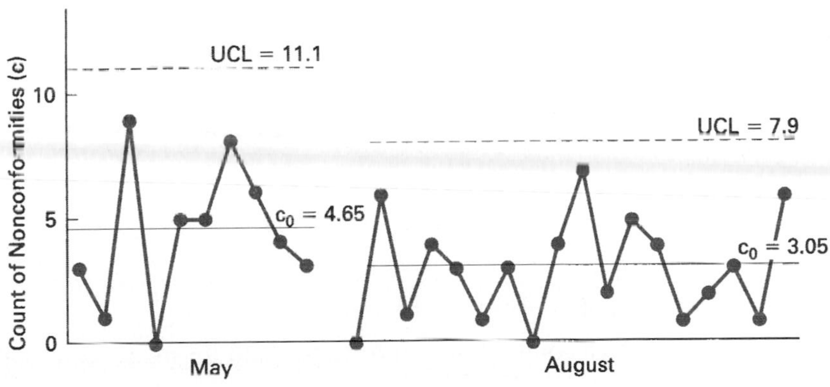

FIGURE 8-12 ***c* chart for canoe blemish nonconformities.**

team such as attaching small pieces of carpet to the skids, faster-drying ink, worker training programs, and so on. The control chart shows whether the idea improves the quality, reduces the quality, or does not change the quality. A minimum of 25 subgroups is needed to evaluate each idea. The subgroups can be taken as often as practical, as long as they are representative of the process. Only one idea should be evaluated at a time.

Figure 8-12 also illustrates a technique for reporting the number of nonconformities of individual quality characteristics, and the graph reports the total. This is an excellent technique for presenting the total picture and one that is accomplished with little additional time or cost. It is interesting to note that the serial numbers of the canoes that were selected for inspection were obtained from a random-number table.

The control chart should be placed in a conspicuous place where it can be viewed by operating personnel.

Chart for Count of Nonconformities/Unit (*u* Chart)[3]

The *c* chart is applicable where the subgroup size is an inspected unit of one such as a canoe, an airplane, 1000 square feet of cloth, a ream of paper, 100 income tax forms, and a keg of nails. The inspected unit can be any size that meets the objective; however, it must be constant. Recall that the subgroup size, *n*, is not in the calculations because its value is 1. When situations arise where the subgroup size varies, then the *u* chart (count of nonconformities/unit) is the appropriate chart. The *u* chart can also be used when the subgroup size is constant.

The *u* chart is mathematically equivalent to the *c* chart. It is developed in the same manner as the *c* chart, with the collection of 25 subgroups, calculation of trial central line and control limits, acquisition of an estimate of the standard or reference count of nonconformities per unit, and calculation of the revised limits. Formulas used for the procedure are

$$u = \frac{c}{n} \qquad \bar{u} = \frac{\Sigma c}{\Sigma n}$$

$$\text{UCL} = \bar{u} + 3\sqrt{\frac{\bar{u}}{n}} \qquad \text{LCL} = \bar{u} - 3\sqrt{\frac{\bar{u}}{n}}$$

where c = count of nonconformities in a subgroup

n = number inspected in a subgroup

u = count of nonconformities/unit in a subgroup

$\bar{u}$ = average count of nonconformities/unit for many subgroups

[3]The chart is not included in ANSI/ASQC B1–B3—1996.

Revised control limits are obtained by substituting u_0 in the trial control limit formula. The u chart will be illustrated by an example.

Each day a clerk inspects the waybills of a small overnight air freight company for errors. Because the number of waybills varies from day to day, a u chart is the appropriate technique. If the number of waybills was constant, either the c or u chart would be appropriate. Data are collected as shown in Table 8-6. The date,

TABLE 8-6 Count of Nonconformities per Unit for Waybills

DATE	NUMBER INSPECTED n	COUNT OF NONCONFORMITIES c	NONCONFORMITIES PER UNIT u	UCL	LCL
Jan. 30	110	120	1.09	1.51	0.89
31	82	94	1.15	1.56	0.84
Feb. 1	96	89	.93	1.53	0.87
2	115	162	1.41	1.50	0.90
3	108	150	1.39	1.51	0.89
4	56	82	1.46	1.64	0.76
6	120	143	1.19	1.50	0.90
7	98	134	1.37	1.53	0.87
8	102	97	.95	1.53	0.87
9	115	145	1.26	1.50	0.90
10	88	128	1.45	1.55	0.85
11	71	83	1.16	1.59	0.81
13	95	120	1.26	1.54	0.86
14	103	116	1.13	1.52	0.88
15	113	127	1.12	1.51	0.89
16	85	92	1.08	1.56	0.84
17	101	140	1.39	1.53	0.87
18	42	60	1.19	1.70	0.70
20	97	121	1.25	1.53	0.87
21	92	108	1.17	1.54	0.86
22	100	131	1.31	1.53	0.87
23	115	119	1.03	1.50	0.90
24	99	93	.94	1.53	0.87
25	57	88	1.54	1.64	0.76
27	89	107	1.20	1.55	0.85
28	101	105	1.04	1.53	0.87
Mar. 1	122	143	1.17	1.49	0.91
2	105	132	1.26	1.52	0.88
3	98	100	1.02	1.53	0.87
4	48	60	1.25	1.67	0.73
Total	2823	3389			

number inspected, and count of nonconformities are obtained and posted to the table. The count of nonconformities per unit, u, is calculated and posted. Also, because the subgroup size varies, the control limits are calculated for each subgroup.

Data for 5 weeks at 6 days per week are collected for a total of 30 subgroups. Although only 25 subgroups are required, this approach eliminates any bias that could occur from the low activity that occurs on Saturday. The calculation for the trial central line is

$$\bar{u} = \frac{\Sigma c}{\Sigma n} = \frac{3389}{2823} = 1.20$$

Calculations for the trial control limits and the plotted point, u, must be made for each subgroup. For January 30 they are

$$\begin{aligned} \text{UCL}_{\text{Jan }30} &= \bar{u} + 3\sqrt{\frac{\bar{u}}{n}} & \text{LCL}_{\text{Jan }30} &= \bar{u} - 3\sqrt{\frac{\bar{u}}{n}} \\ &= 1.20 + 3\sqrt{\frac{1.20}{110}} & &= 1.20 - 3\sqrt{\frac{1.20}{110}} \\ &= 1.51 & &= 0.89 \end{aligned}$$

$$u_{\text{Jan }30} = \frac{c}{n} = \frac{120}{110} = 1.09$$

These calculations must be repeated for 29 subgroups and the values posted to the table.

A comparison of the plotted points with the upper and lower control limits in Figure 8-13 shows that there are no out-of-control values. Therefore, $\bar{u}$ can be considered the best estimate of u_0 and $u_0 = 1.20$. A visual inspection of the plotted points indicates a stable process. This situation is somewhat unusual at the beginning of control-charting activities.

To determine control limits for the next 5-week period, we can use an average subgroup size in the same manner as the variable subgroup size of the p chart. A review of the chart shows that the control limits for Saturday are much wider apart than for the rest of the week. This condition is due to the smaller subgroup size. Therefore, it appears appropriate to establish separate control limits for Saturday. Calculations are as follows:

$$n_{\text{Sat. avg.}} = \frac{\Sigma n}{g} = \frac{(56 + 71 + 42 + 57 + 48)}{5} = 55$$

$$\begin{aligned} \text{UCL} &= u_0 + 3\sqrt{\frac{u_0}{n}} & \text{LCL} &= u_0 - 3\sqrt{\frac{u_0}{n}} \\ \text{UCL} &= 1.20 + 3\sqrt{\frac{1.20}{55}} & \text{LCL} &= 1.20 - 3\sqrt{\frac{1.20}{55}} \\ &= 1.64 & &= 0.76 \end{aligned}$$

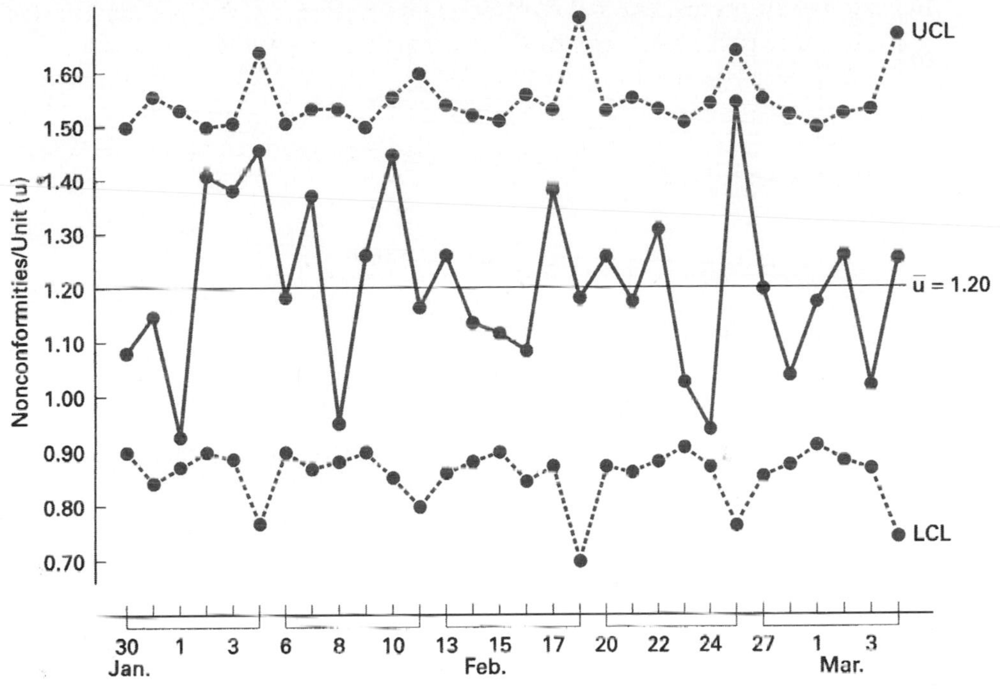

FIGURE 8-13 ***u* chart for errors on waybills.**

$$n_{\text{daily avg.}} = \frac{\Sigma n}{g} = \frac{2823 - 274}{25} = 102, \quad \text{say, } 100$$

$$\text{UCL} = u_0 + 3\sqrt{\frac{u_0}{n}} \qquad \text{LCL} = u_0 - 3\sqrt{\frac{u_0}{n}}$$

$$= 1.20 + 3\sqrt{\frac{1.20}{100}} \qquad = 1.20 - 3\sqrt{\frac{1.20}{100}}$$

$$= 1.53 \qquad = 0.87$$

The control chart for the next period is shown in Figure 8-14. When the subgroup is a day's inspections, the true control limits will need to be calculated about once every 3 months.

The control chart can now be used to achieve the objective. If a project team is involved, it can test ideas for quality improvement.

The *u* chart is identical to the *c* chart in all aspects except two. One difference is the scale, which is continuous for a *u* chart but discrete for the *c* chart. This difference provides more flexibility for the *u* chart since the subgroup size can vary. The other difference is the subgroup size, which is 1 for the *c* chart.

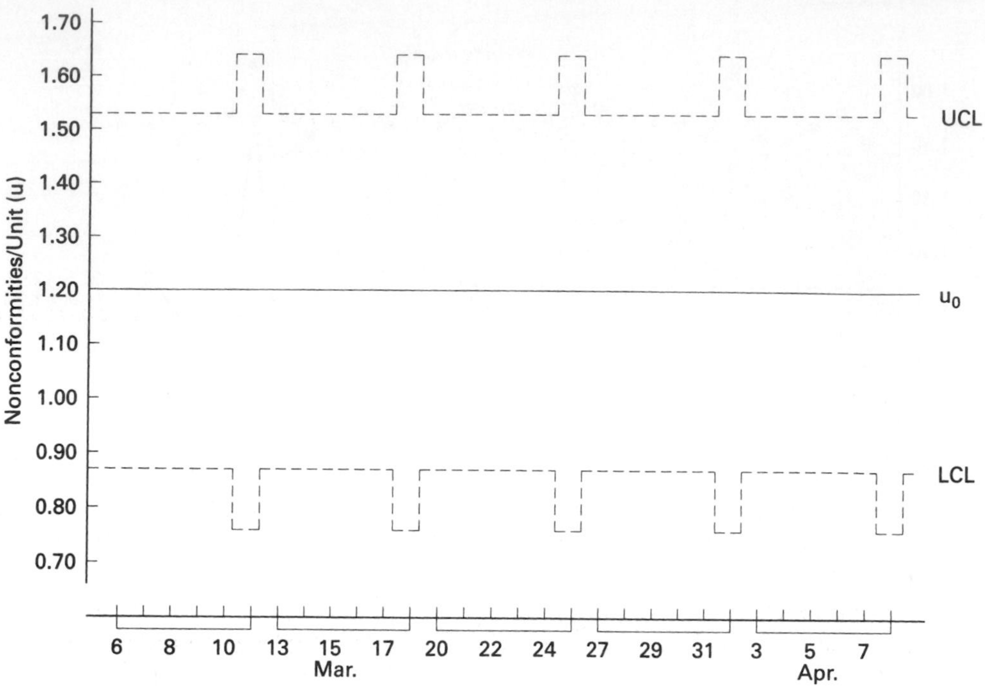

FIGURE 8-14 ***u* chart for next period.**

The *u* chart is limited in that we do not know the location of the nonconformities. For example, in Table 8-6, February 4 has 82 nonconformities out of 56 inspected for a value of 1.46. All 82 nonconformities could have been counted on one unit.

Final Comments

Process capability for nonconformities is treated in a manner similar to nonconforming units. The reader is referred to Figure 8-10.

Figure 8-15 shows when to use the various attribute charts. First you need to decide whether to chart nonconformities or nonconforming units. Next you need to determine whether the subgroup size will be constant or will vary. These two decisions give the appropriate chart.

A QUALITY RATING SYSTEM

Introduction

In the attribute charts of the preceding section, all nonconformities and nonconforming units had the same weight, regardless of their seriousness. For example, in

		Attribute Chart	
		Nonconforming Units	Nonconformities
Sample Size	Constant	np	c (n = 1)
	Constant or Varies	p	u

FIGURE 8-15 When to use the various attribute charts.

the evaluation of desk chairs, one chair might have 5 nonconformities, all related to the surface finish, while another chair might have 1 nonconformity, a broken leg. The usable chair with 5 trivial nonconformities has 5 times the influence on the attribute chart as the unusable chair with 1 serious nonconformity. This situation presents an incorrect evaluation of the product quality. A quality rating system will correct this deficiency.

There are many situations where it is desirable to compare the performance of operators, shifts, plants, or vendors. In order to compare quality performance, a quality rating system is needed to classify, weigh, and evaluate nonconformities.

Nonconformity Classification

Nonconformities and, for that matter, nonconforming units, are classified according to their severity. One system groups nonconformities into three classes:

1. *Critical nonconformities.* A critical nonconformity is a nonconformity that judgment and experience indicate is likely to result in hazardous or unsafe conditions for individuals using, maintaining, or depending upon the product, or a nonconformity that judgment and experience indicate is likely to prevent performance of the function of the product.

2. *Major nonconformities.* A major nonconformity is a nonconformity, other than critical, that is likely to result in failure or to reduce materially the usability of the product for its intended purpose.

3. *Minor nonconformities.* A minor nonconformity is a nonconformity that is not likely to reduce materially the usability of the product for its intended purpose. Minor nonconformities are usually associated with appearance.

To summarize, a critical nonconformity *will* affect usability; a major nonconformity *might* affect usability; and a minor nonconformity *will not* affect usability of the unit.

Other classification systems use four classes or two classes, depending on the complexity of the product. A catastrophic class is sometimes used.

Once the classifications are determined, the weights to assign to each class can be established. While any weights can be assigned to the classifications, 9 points for a critical, 3 points for a major, and 1 point for a minor are usually considered to be satisfactory since a major is three times as important as a minor and a critical is three times as important as a major.

Control Chart[4]

Control charts are established and plotted for count of demerits per unit. A demerit per unit is given by the formula

$$D = w_c u_c + w_{ma} u_{ma} + w_{mi} u_{mi}$$

Where D = demerits per unit

w_c, w_{ma}, w_{mi} = weights for the three classes—critical, major, and minor

u_c, u_{ma}, u_{mi} = count of nonconformities per unit in each of the three classes—critical, major, and minor

When w_c, w_{ma}, and w_{mi} are 9, 3, and 1, respectively, the formula is

$$D = 9u_c + 3u_{ma} + 1u_{mi}$$

The D values calculated from the formula are posted to the chart for each subgroup.

The central line and the 3σ control limits are obtained from the formulas

$$D_0 = 9u_{0c} + 3u_{0ma} + 1u_{0mi}$$

$$\sigma_{0u} = \sqrt{\frac{9^2 u_{0c} + 3^2 u_{0ma} + 1^2 u_{0mi}}{n}}$$

$$\text{UCL} = D_0 + 3\sigma_{0u} \qquad \text{LCL} = D_0 - 3\sigma_{0u}$$

where u_{0c}, u_{0ma}, and u_{0mi} represent the standard nonconformities per unit for the critical, major, and minor classifications, respectively. The nonconformities per unit for the critical, major, and minor classifications are obtained by separating the nonconformities into the three classifications and treating each as a separate u chart.

EXAMPLE PROBLEM

Assuming that a 9:3:1 three-class weighting system is used, determine the central line and control limits when $u_{0c} = 0.08$, $u_{0ma} = 0.5$, $u_{0mi} = 3.0$ and $n = 40$. Also calculate the demerits per unit for May 25 when critical nonconformities are 2. major nonconformities are 26, and minor nonconformities are

[4]The demerit chart is not included in ANSI/ASQ/B1–B3—1996.

160 for the 40 units inspected on that day. Is the May 25 subgroup in control or out of control?

$$D_0 = 9u_{0c} + 3u_{0ma} + 1u_{0mi}$$
$$= 9(0.08) + 3(0.5) + 1(3.0)$$
$$= 5.2$$

$$\sigma_{0u} = \sqrt{\frac{9^2u_{0c} + 3^2u_{0ma} + 1^2u_{0mi}}{n}}$$
$$= \sqrt{\frac{81(0.08) + 9(0.5) + 1(3.0)}{40}}$$
$$= 0.59$$

$$\text{UCL} = D_0 + 3\sigma_{0u} \qquad \text{LCL} = D_0 - 3\sigma_{0u}$$
$$= 5.2 + 3(0.59) \qquad = 5.2 - 3(0.59)$$
$$= 7.0 \qquad = 3.4$$

The central line and control limits are illustrated in Figure 8-16. Calculations for the May 25 subgroup are

$$D_{\text{May 25}} = 9u_c + 3u_{ma} + 1u_{mi}$$
$$= 9\left(\frac{2}{40}\right) + 3\left(\frac{26}{40}\right) + 1\left(\frac{160}{40}\right)$$
$$= 6.4 \text{ (in control)}$$

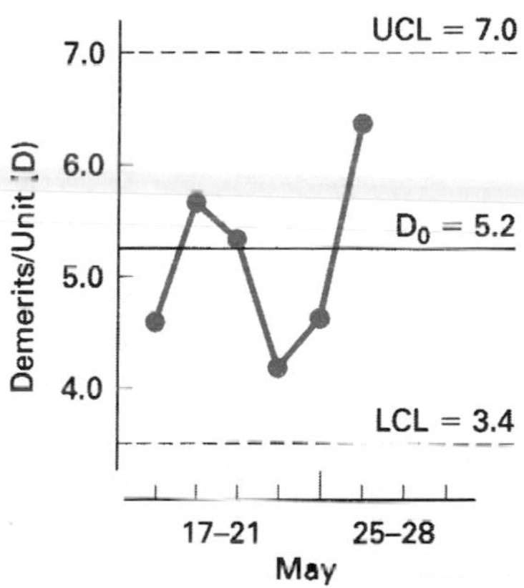

FIGURE 8-16 Demerit-per-unit chart (*D* chart).

Quality rating systems based on demerits per unit are useful for performance control and can be an important feature of a total quality system.

COMPUTER PROGRAM

Using EXCEL the software in the diskette inside the back cover will solve for the four charts in this chapter. Their file names are *p-chart, np-chart, c-chart*, and *u-chart*.

PROBLEMS

1. On page 340 is a typical attribute chart form with information concerning 2-L soda bottles.
   (a) Calculate the proportion nonconforming for subgroups 21, 22, 23, 24, and 25. Construct a run chart.
   (b) Calculate the trial central line and control limits. Draw these values on the chart.
   (c) If it is assumed that any out-of-control points have assignable causes, what central line and control limits should be used for the next period?

2. Determine the trial central line and control limits for a p chart using the following data, which are for the payment of dental insurance claims. Plot the values on graph paper and determine if the process is stable. If there are any out-of-control points, assume an assignable cause and determine the revised central line and control limits.

SUBGROUP NUMBER	NUMBER INSPECTED	NUMBER NONCONFORMING	SUBGROUP NUMBER	NUMBER INSPECTED	NUMBER NONCONFORMING
1	300	3	14	300	6
2	300	6	15	300	7
3	300	4	16	300	4
4	300	6	17	300	5
5	300	20	18	300	7
6	300	2	19	300	5
7	300	6	20	300	0
8	300	7	21	300	2
9	300	3	22	300	3
10	300	0	23	300	6
11	300	6	24	300	1
12	300	9	25	300	8
13	300	5			

3. The supervisor is not sure about the best way to display the quality performance determined in Problem 2. Calculate the central line and limits for the other methods of presentation.

4. After achieving the objective in the example problem concerning the hair dryer motor, it is decided to reduce the sample size to 80. What are the central line and control limits?

5. Fifty motor generators are inspected per day from a stable process. The best estimate of the fraction nonconforming is 0.076. Determine the central line and control limits. On a particular day, 5 nonconforming generators were discovered. Is this in control or out of control?

6. Inspection results of video-of-the-month shipments to customers for 25 consecutive days are given in the table. What central line and control limits should be established and posted if it is assumed that any out-of-control points have assignable causes? The number of inspections each day is constant and equals 1750.

DATE	NUMBER NONCONFORMING	DATE	NUMBER NONCONFORMING
July 6	47	July 23	37
7	42	26	39
8	48	27	51
9	58	28	44
12	32	29	61
13	38	30	48
14	53	Aug. 2	56
15	68	3	48
16	45	4	40
19	37	5	47
20	57	6	25
21	38	9	35
22	53		

7. The performance of the first shift is reflected in the inspection results of electric carving knives. Determine the trial central line and control limits for each subgroup. Assume that any out-of-control points have assignable causes and determine the standard value for the fraction nonconforming for the next production period.

DATE	NUMBER INSPECTED	NUMBER NONCONFORMING	DATE	NUMBER INSPECTED	NUMBER NONCONFORMING
Sept. 6	500	5	Sept. 23	525	10
7	550	6	24	650	3
8	700	8	27	675	8

ATTRIBUTES CONTROL CHART p ☒ np ☐ u ☐ c ☐ CHART ID:

PART ID: *2 LITER-BOTTLE* OPERATION ID: *NEW PACKAGING LINE* DEPT/AREA: *PACKAGING*

CHECK METHOD: *VISUAL* CHARACTERISTIC: *CASE PACKING DEFECTS*

DAY:	1	2	3	4	5	6	7	8	9	10	11	12	13	14	15	16	17	18	19	20	21	22	23	24	25					
SAMPLE (n)	400	400	400	400	400	400	400	400	400	400	400	400	400	400	400	400	400	400	400	400	400	400	400	400	400					
NUMBER (np, c)	43	21	14	20	15	16	8	12	18	4	6	12	5	4	3	8	7	31	8	6	4	7	9	6	10					
PROPORTION (p, u)	.108	.053	.035	.050	.038	.040	.020	.030	.045	.010	.015	.030	.013	.010	.008	.020	.018	.078	.020	.015										

AVG = UCL = LCL =

DATE	NUMBER INSPECTED	NUMBER NONCONFORMING	DATE	NUMBER INSPECTED	NUMBER NONCONFORMING
9	625	9	28	450	23
10	700	7	29	500	2
13	550	8	30	375	3
14	450	16	Oct. 1	550	8
15	600	6	4	600	7
16	475	9	5	700	4
17	650	6	6	660	9
20	650	7	7	450	8
21	550	8	8	500	6
22	525	7	11	525	1

8. Daily inspection results for the model 305 electric range assembly line are given in the table. Determine trial control limits for each subgroup. Assume that any out-of-control points have assignable causes, and determine the standard value for the fraction nonconforming for December.

DATE AND SHIFT	NUMBER INSPECTED	NUMBER NONCONFORMING	DATE AND SHIFT	NUMBER INSPECTED	NUMBER NONCONFORMING
Nov. 8 I	171	31	Nov. 17 I	165	16
II	167	6	II	170	35
9 I	170	8	18 I	175	12
II	135	13	II	167	6
10 I	137	26	19 I	141	50
II	170	30	II	159	26
11 I	45	3	22 I	181	16
II	155	11	II	195	38
12 I	195	30	23 I	165	33
II	180	36	II	140	21
15 I	181	38	24 I	162	18
II	115	33	II	191	22
16 I	165	26	25 I	139	16
II	189	15	II	181	27

9. Control limits are to be established based on the average number inspected from the information in Problem 8. What are these control limits and the central line? Describe the cases where individual control limits will need to be calculated.

10. Control charts are to be established on the manufacture of backpack frames. The revised fraction nonconforming is 0.08. Determine control limit lines for inspection rates of 1000 per day, 1500 per day, and 2000 per day. Draw the control chart. Why are the control limits unequally spaced?

11. Determine the revised central line and control limits for a percent nonconforming chart for the information in:
(a) Problem 2
(b) Problem 6

12. From the information of Problem 2, determine the revised central line and control limits for an *np* chart.

13. From the information of Problem 6, determine the revised central line and control limits for an *np* chart. Which chart is more meaningful to operating personnel?

14. An *np* chart is to be established on a painting process that is in statistical control. If 35 pieces are to be inspected every 4 hours and the fraction nonconforming is 0.06, determine the central line and control limits.

15. Determine the revised central line and control limits for *fraction conforming*, *percent conforming*, and *number conforming* charts for the information in:
(a) Problem 2
(b) Problem 6

16. Find the process capability for:
(a) Problem 6
(b) Problem 7
(c) Problem 10

17. The count of surface nonconformities in 1000 square meters of 20-kg kraft paper is given in the table. Determine the trial central line and control limits and the revised central line and control limits, assuming that out-of-control points have assignable causes.

LOT NUMBER	COUNT OF NONCONFORMITIES	LOT NUMBER	COUNT OF NONCONFORMITIES
20	10	36	2
21	8	37	12
22	6	38	0
23	6	39	6
24	2	40	14
25	10	41	10
26	8	42	8
27	10	43	6
28	0	44	2
29	2	45	14
30	8	46	16
31	2	47	10

LOT NUMBER	COUNT OF NONCONFORMITIES	LOT NUMBER	COUNT OF NONCONFORMITIES
32	20	48	2
33	10	49	6
34	6	50	3
35	30		

18. A leading bank has compiled the data in the table showing the count of nonconformities for 1000 accounting transactions per day during December and January. What control limits and central line are recommended for the control chart for February? Assume any out-of-control points have assignable causes.

COUNT OF NONCONFORMITIES	COUNT OF NONCONFORMITIES
8	17
19	14
14	9
18	7
11	15
16	22
8	19
15	38
21	12
8	13
23	5
10	2
9	16

19. A quality technician has collected data on the count of rivet nonconformities in 4-m travel trailers. After 30 trailers, the total count of nonconformities is 316. Trial control limits have been determined, and a comparison with the data shows no out-of-control points. What is the recommendation for the central line and the revised control limits for a count of nonconformities chart?

20. One hundred product labels are inspected every day for surface nonconformities. Results for the past 25 days are 22, 29, 25, 17, 20, 16, 34, 11, 31, 29, 15, 10, 33, 23, 27, 15, 17, 17, 19, 22, 23, 27, 29, 33, and 21. Plot the points on graph paper (run chart) and determine if the process is stable. Determine the trial central line and control limits.

21. Determine the trial control limits and revised control limits for a u chart using the data in the table for the surface finish of rolls of white paper. Assume any out-of-control points have assignable causes.

LOT NUMBER	SAMPLE SIZE	TOTAL NONCONFORMITIES	LOT NUMBER	SAMPLE SIZE	TOTAL NONCONFORMITIES
1	10	45	15	10	48
2	10	51	16	11	35
3	10	36	17	10	39
4	9	48	18	10	29
5	10	42	19	10	37
6	10	5	20	10	33
7	10	33	21	10	15
8	8	27	22	10	33
9	8	31	23	11	27
10	8	22	24	10	23
11	12	25	25	10	25
12	12	35	26	10	41
13	12	32	27	9	37
14	10	43	28	10	28

22. A warehouse distribution activity has been in statistical control, and control limits are needed for the next period. If the subgroup size is 100, the total count of nonconformities is 835, and the number of subgroups is 22, what are the new control limits and central line?

23. Construct a control chart for the data in the table for empty bottle inspections of a soft-drink manufacturer. Assume assignable causes for any points that are out of control.

NUMBER OF BOTTLES	CHIPS, SCRATCHES, OTHER	FOREIGN MATERIAL ON SIDES	FOREIGN MATERIAL ON BOTTOM	TOTAL NONCONFORMITIES
40	9	9	27	45
40	10	1	29	40
40	8	0	25	33
40	8	2	33	43
40	10	6	46	62
52	12	16	51	79
52	15	2	43	60
52	13	2	35	50
52	12	2	59	73
52	11	1	42	54
52	15	15	25	55
52	12	5	57	74
52	14	2	27	43
52	12	7	42	61
40	11	2	30	43

NUMBER OF BOTTLES	CHIPS, SCRATCHES, OTHER	FOREIGN MATERIAL ON SIDES	FOREIGN MATERIAL ON BOTTOM	TOTAL NONCONFORMITIES
40	9	4	19	32
40	5	6	34	45
40	8	11	14	33
40	3	9	38	50
40	9	9	10	28
52	13	8	37	58
52	11	5	30	46
52	14	10	47	71
52	12	3	41	56
52	12	2	28	42

24. Assuming that a 10:5:1 demerit weighting system is used, determine the central line and control limits when $u_c = 0.11$, $u_{ma} = 0.70$, $u_{mi} = 4.00$, and $n = 50$. If the subgroup inspection results for a particular day are 1 critical, 35 major, and 110 minor nonconformities, determine if the results are in control or out of control.

25. Solve the following problems using the software on the diskette inside the back cover:
(a) Problem 2
(b) Problem 13
(c) Problem 17
(d) Problem 21

26. Prepare an EXCEL template for the four charts to ensure that the LCL is always 0.

27. Write an EXCEL template for a D chart.

28. Determine the equations for a number conforming chart, nq.

9 LOT-BY-LOT ACCEPTANCE SAMPLING BY ATTRIBUTES[1]

FUNDAMENTAL CONCEPTS

Importance

In recent years, acceptance sampling has declined in importance as statistical process control has assumed a more prominent role in the quality function. However, acceptance sampling still has a place in the entire body of knowledge that constitutes quality science. In addition to statistical acceptance sampling discussed in this chapter and the next, there are several other acceptance sampling practices such as fixed percentage, occasional random checks, and 100% inspection.

[1]This chapter is based on ANSI/ASQ S2—1995.

Description

Lot-by-lot acceptance sampling by attributes is the most common type of sampling. With this type of sampling, a predetermined number of units (sample) from each lot is inspected by attributes. If the number of nonconforming units is less than the prescribed minimum, the lot is accepted; if not, the lot is not accepted. Acceptance sampling can be used either for the number of nonconforming units or for nonconformities per unit. To simplify the presentation in this chapter, the number of nonconforming units is used; however, it is understood that the information is also applicable to nonconformities per unit. Sampling plans are established by severity (critical, major, minor) or on a demerit-per-unit basis.

A single sampling plan is defined by the lot size, N, the sample size, n, and the acceptance number, c. Thus, the plan

$$N = 9000$$

$$n = 300$$

$$c = 2$$

means that a lot of 9000 units has 300 units inspected. And if two or fewer nonconforming units are found in the 300-unit sample, the lot is accepted. If three or more nonconforming units are found in the 300-unit sample, the lot is not accepted.

Acceptance sampling can be performed in a number of different situations where there is a consumer-producer relationship. The consumer and producer can be from two different organizations, two plants within the same organization, or two departments within the same organization's facility. In any case, there is always the problem of deciding whether to accept or not accept the product.

Acceptance sampling of the product is most likely to be used in one of five situations:

1. When the test is destructive (such as a test on an electrical fuse or a tensile test), sampling is necessary; otherwise, all of the product will be destroyed by testing.
2. When the cost of 100% inspection is high in relation to the cost of passing a nonconforming unit.
3. When there are many similar units to be inspected, sampling will frequently produce as good, if not better, results than 100% inspection. This is true because with manual inspection, fatigue and boredom cause a higher percentage of nonconforming material to be passed than would occur on the average using a sampling plan.
4. When information concerning producer's quality, such as $\overline{X}$ and R, p or c charts, and C_{pk}, is not available.
5. When automated inspection is not available.

Advantages and Disadvantages of Sampling

When sampling is compared with 100% inspection, it has the following advantages:

1. Places responsibility for quality in the appropriate place rather than on inspection, thereby encouraging rapid improvement in the product.
2. Is more economical owing to fewer inspections (fewer inspectors) and less handling damage during inspection.
3. Upgrades the inspection job from monotonous piece-by-piece decisions to lot-by-lot decisions.
4. Applies to destructive testing.
5. Entire lots are not accepted rather than the return of a few nonconforming units, thereby giving stronger motivation for improvement.

Inherent disadvantages of acceptance sampling are:

1. There are certain risks of not accepting conforming lots and accepting nonconforming lots.
2. More time and effort is devoted to planning and documentation.
3. Less information is provided about the product, although there is usually enough.
4. There is no assurance given that the entire lot conforms to specifications.

Types of Sampling Plans

There are four types of sampling plans: single, double, multiple, and sequential. In the single sampling plan, one sample is taken from the lot and a decision to accept or not accept the lot is made based on the inspection results of that sample. This type of sampling plan was described earlier in the chapter.

Double sampling plans are somewhat more complicated. On the initial sample, a decision, based on the inspection results, is made whether (1) to accept the lot, (2) to not accept the lot, or (3) to take another sample. If the quality is very good the lot is accepted on the first sample and a second sample is not taken; if the quality is very poor the lot is not accepted on the first sample and a second sample is not taken. Only when the quality level is neither very good nor very bad is a second sample taken.

If a second sample is required, the results of that inspection and the first inspection are used to make a decision. A double sampling plan is defined by

N = lot size

n_1 = sample size on the first sample

c_1 = acceptance number on the first sample (sometimes the symbol Ac is used)

r_1 = non-acceptance number on the first sample (sometimes the symbol *Re* is used)

n_2 = sample size on the second sample

c_2 = acceptance number for *both* samples

r_2 = non-acceptance number for *both* samples

If values are not given for r_1 and r_2, they are equal to $c_2 + 1$.

An illustrative example will help to clarify the double sampling plan: $N = 9000$, $n_1 = 60$, $c_1 = 1$, $r_1 = 5$, $n_2 = 150$, $c_2 = 6$, and $r_2 = 7$. An initial sample (n_1) of 60 is selected from the lot (N) of 9000 and inspected. One of the following judgments is made:

1. If there are 1 or fewer nonconforming units (c_1), the lot is accepted.

2. If there are 5 or more nonconforming units (r_1), the lot is not accepted.

3. If there are 2, 3, or 4 nonconforming units, no decision is made and a second sample is taken.

A second sample of 150 (n_2) from the lot (N) is inspected, and one of the following judgments is made:

1. If there are 6 or fewer nonconforming units (c_2) in both samples, the lot is accepted. This number (6 or fewer) is obtained by 2 in the first sample and 4 or fewer in the second sample, by 3 in the first sample and 3 or fewer in the second sample, or by 4 in the first sample and 2 or fewer in the second sample.

2. If there are 7 or more nonconforming units (r_2) in both samples, the lot is not accepted. This number (7 or more) is obtained by 2 in the first sample and 5 or more in the second sample, by 3 in the first sample and 4 or more in the second sample, or by 4 in the first sample and 3 or more in the second sample.

A multiple sampling plan is a continuation of double sampling in that three, four, five, or as many samples as desired can be established. Sample sizes are much smaller. The technique is the same as that described for double sampling; therefore, a detailed description is not given. Multiple sampling plans of ANSI/ASQ Z1.4 use seven samples. An example of a multiple sampling plan with four samples is illustrated later in this chapter.

In sequential sampling, items are sampled and inspected one after another. A cumulative record is maintained, and a decision is made to accept or not accept the lot as soon as there is sufficient cumulative evidence. Additional information on sequential sampling is given in the next chapter.

All four types of sampling plans can give the same results; therefore, the chance of a lot being accepted under a single sampling plan is the same under the appropriate double, multiple, or sequential sampling plan. Thus, the type of plan for a particular unit is based on factors other than effectiveness. These factors are simplicity, administrative costs, quality information, number of units inspected, and psychological impact.

Perhaps the most important factor is simplicity. In this regard, single sampling is the best and sequential sampling the poorest.

Administrative costs for training, inspection, record keeping, and so on, are least for single sampling and greatest for sequential sampling.

Single sampling provides more information concerning the quality level in each lot than double sampling and much more than multiple or sequential sampling.

In general, the number of units inspected is greatest under single sampling and least under sequential. An ASN curve, shown later in the chapter, illustrates this concept.

A fifth factor concerns the psychological impact of the four types of sampling plans. Under single sampling there is no second chance; however, in double sampling, if the first sample is borderline, a second chance is possible by taking another sample. Many producers like the second-chance psychology provided by the double sample. In multiple and sequential sampling there are a number of "second chances"; therefore, the psychological impact is less than with double sampling.

Careful consideration of the five factors is necessary to select a type of sampling plan that will be best for the particular situation.

Formation of Lots

Lot formation can influence the effectiveness of the sampling plan. Guidelines are as follows:

1. Lots should be homogeneous, which means that all product in the lot is produced by the same machine, same operator, same input material, and so on. When product from different sources is mixed, the sampling plan does not function properly. Also, it is difficult to take corrective action to eliminate the source of nonconforming product.

2. Lots should be as large as possible. Since sample sizes do not increase as rapidly as lot sizes, a lower inspection cost results with larger lot sizes. For example, a lot of 2000 would have a sample size of 125 (6.25%), but an equally effective sampling plan for a lot of 4000 would have a sample size of 200 (5.00%). When an organization starts a just-in-time procurement philosophy, the lot sizes are usually reduced to a 2- or 3-day supply. Thus, the relative amount inspected and the inspection costs will increase. The benefits to just-in-time are far greater than the increase in inspection costs; therefore, smaller lot sizes are to be expected.

The reader is cautioned not to confuse the packaging requirements for shipment and materials handling with the concept of a homogeneous lot. In other words, a lot may consist of a number of packages and may also consist of a number of shipments. If two different machines and/or two different operators are included in a shipment, they are separate lots and should be so identified. The reader should also be aware that partial shipments of a homogeneous lot can be treated as if they are homogeneous lots.

Sample Selection

The sample units selected for inspection should be representative of the entire lot. All sampling plans are based on the premise that each unit in the lot has an equal likelihood of being selected. This is referred to as *random sampling*.

The basic technique of random sampling is to assign a number to each unit in the lot. Then a series of random numbers is generated that tells which of the numbered units are to be sampled and inspected. Random numbers can be generated from a computer, electronic hand calculator, 20-sided random-number die, numbered chips in a bowl, and so on. They may be used to select the sample or to develop a table of random numbers.

A random-number table is shown in Table D of the Appendix. A portion of Table D is reproduced here as Table 9–1. To use the table it is entered at any location and numbers are selected sequentially from one direction, such as up, down, left, or right. Any number that is not appropriate is discarded. For locating convenience, this table is established with 5 digits per column. It could have been established with 2, 3, 6, or any number per column. In fact, the digits could have run across the page with no spaces, but that format would make the table difficult to read. Any number of digits can be used for a random number.

An example will help to illustrate the technique. Assume that a lot of 90 units has been assigned numbers from 1 to 90 and it is desired to select a sample of 9. A two-digit number is selected at random, as indicated by the number 53. Numbers are selected downward and the first three numbers are 53, 15, and 73. Starting at the top of the next column the numbers 45, 30, 06, 27, and 96 are obtained. The number 96 is too high and is discarded. The next numbers are 52 and 82. Units with the numbers 53, 15, 73, 45, 30, 06, 27, 52, and 82 comprise the sample.

TABLE 9-1 Random Numbers.

74972	38712	36401	45525	40640	16281	13554	79945
75906	91807	56827	30825	40113	08243	08459	28364
29002	46453	25653	06543	27340	10493	60147	15702
80033	69828	88215	27191	23756	54935	13385	22782
25348	04332	18873	96927	64953	99337	68689	03263

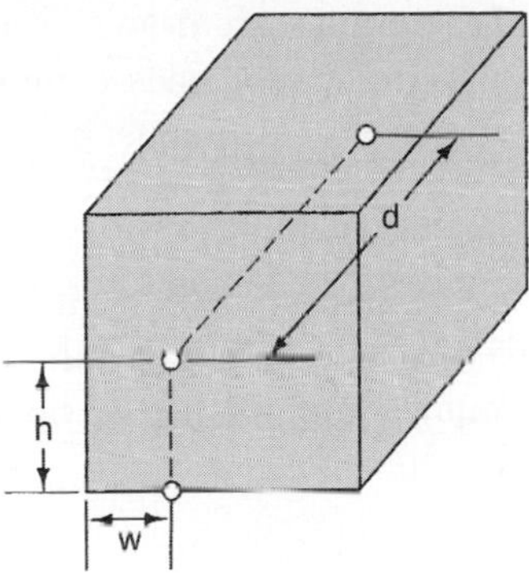

FIGURE 9-1 Location and random numbers.

Many products have serial numbers that can be used as the assigned number. This practice avoids the difficult process of assigning numbers to each unit. In many situations, units are systematically packed in a container and the assigned number can be designated by the location. A three-digit number would represent the width, height, and depth in a container as shown in Figure 9-1. Thus, the random number 328 could specify the unit located at the third row, second level, and eighth unit from the front. For fluid or other well-mixed products, the sample can be taken from any location, since the product is presumed to be homogeneous.

It is not always practical to assign a number to each unit, utilize a serial number, or utilize a locational number. Stratification of the lot or package with samples drawn from each stratum can be an effective substitute for random sampling. The technique is to divide the lot or package into strata or layers as shown in Figure 9-2. Each stratum is further subdivided into cubes, as illustrated by stratum

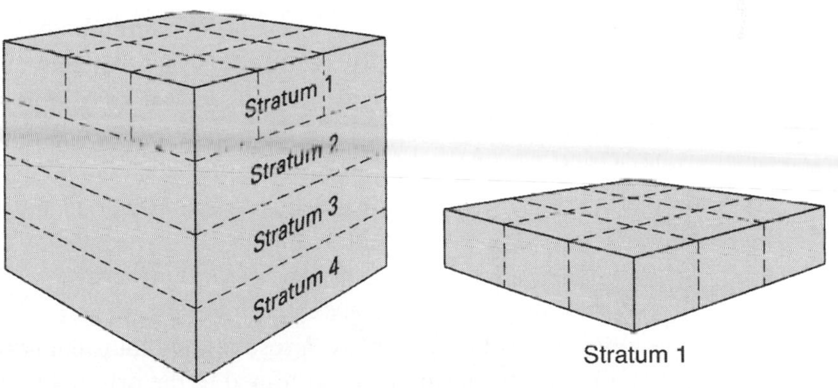

FIGURE 9-2 Dividing a lot for stratified sampling.

1. Within each cube, samples are drawn from the entire volume. The dividing of the lot or package into strata and cubes within each stratum is an imaginary process done by the inspector. By this technique, pieces are selected from all locations in the lot or package.

Unless an adequate sampling method is used, a variety of biases can occur. An example of a biased sample occurs when the operator makes sure that units on the top of a lot are the best quality, and the inspector selects the sample from the same location. Adequate supervision of operators and inspectors is necessary to ensure that no bias occurs.

Non-Accepted Lots

Once a lot has not been accepted, there are a number of courses of action that can be taken.

1. The non-accepted lot can be passed to the production facilities and the nonconforming units sorted by production personnel. This action is not a satisfactory alternative since it defeats the purpose of sampling inspection and slows production. However, if the units are badly needed, there may be no other choice.

2. The non-accepted lot can be rectified at the consumer's plant by personnel from either the producer's or the consumer's plant. Although shipping costs are saved, there is a psychological disadvantage, since all the consumer's personnel are aware that producer *X* had product that was not accepted. This fact may be used as a crutch to explain poor performance when using producer *X*'s material at a future time. In addition, space at the consumer's plant must be provided for personnel to perform the sorting operation.

3. The non-accepted lot can be returned to the producer for rectification. This is the only appropriate course of action, since it results in long-run improvement in the quality. Since shipping costs are paid in both directions, cost becomes a motivating factor to improve the quality. Also, when the lot is sorted in the producer's plant, all the employees are aware that consumer *Y* expects to receive a quality product. This, too, is a motivating factor for quality improvement the next time an order is produced for consumer *Y*. This course of action may require the production line to be shut down, which would be a loud and clear signal to the supplier and operating personnel that quality is important.

It is assumed that non-accepted lots will receive 100% inspection and the non-conforming units discarded. A resubmitted lot is not normally reinspected, but if it is, the inspection should be confined to the original nonconformity. Since the nonconforming units are discarded, a resubmitted lot will have fewer units than the original.

STATISTICAL ASPECTS

OC Curve for Single Sampling Plans

An excellent evaluation technique is an *operating characteristic* (OC) *curve*. In judging a particular sampling plan, it is desirable to know the probability that a lot submitted with a certain percent nonconforming, $100p_0$, will be accepted. The OC curve will provide this information, and a typical OC curve is shown in Figure 9-3. When the percent nonconforming is low, the probability of the lot being accepted is large and decreases as the percent nonconforming increases.

The construction of an OC curve can be illustrated by a concrete example. A single sampling plan has a lot size $N = 3000$, a sample size $n = 89$, and an acceptance number $c = 2$. It is assumed that the lots are from a steady stream of product that can be considered infinite, and therefore the binomial probability distribution can be used for the calculations. Fortunately, the Poisson is an excellent approximation to the binomial for almost all sampling plans; therefore, the Poisson is used for determining the probability of the acceptance of a lot.

In graphing the curve with the variables $100\ P_a$ (percent of lots accepted) and $100p_0$ (percent nonconforming), one value, $100p_0$, will be assumed and the other calculated. For illustrative purposes we will assume a $100p_0$ value of, say, 2%, which gives an np_0 value of

$$p_0 = 0.02$$

$$np_0 = (89)(0.02) = 1.8$$

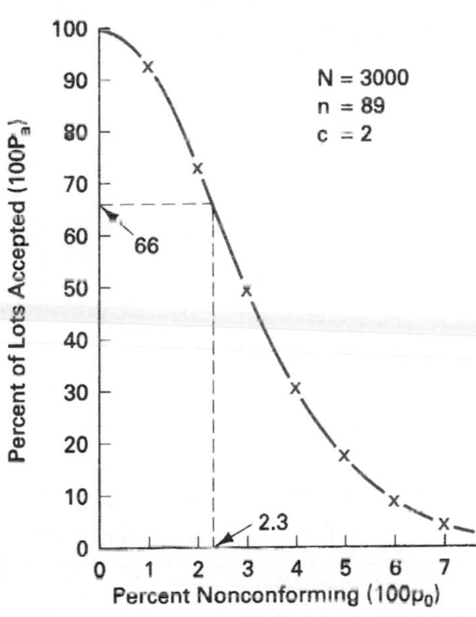

FIGURE 9-3 OC curve for the single sampling plan $N = 3000$, $n = 89$, and $c = 2$.

Acceptance of the lot is based on the acceptance number $c = 2$ and is possible when there are 0 nonconforming units in the sample, 1 nonconforming unit in the sample, or 2 nonconforming units in the sample. Thus

$$\begin{aligned} P_a &= P_0 + P_1 + P_2 \\ &= P_2 \text{ or less} \\ &= 0.731 \text{ or } 100P_a = 73.1\% \end{aligned}$$

The P_a value is obtained from Table C for $c = 2$ and $np_0 = 1.8$.

A table can be used to assist with the calculations, as shown in Table 9-2. The curve is terminated when the P_a value is close to 0.05. Since $P_a = 0.055$ for $100p_0 = 7\%$, it is not necessary to make any calculations for values greater than 7%. Approximately 7 points are needed to describe the curve with a greater concentration of points where the curve changes direction.

Information from the table is plotted to obtain the OC curve shown in Figure 9-3. The steps are: (1) assume p_0 value, (2) calculate np_0 value, (3) attain P_a values from the Poisson table using the applicable c and np_0 values, (4) plot point ($100p_0$, $100P_a$), and (5) repeat 1, 2, 3, and 4 until a smooth curve is obtained.

To make the curve more readable, the label Percent of Lots (expected to be) Accepted is used rather than Probability of Acceptance.

Once the curve is constructed, it shows the chance of a lot being accepted for a particular incoming quality. Thus, if the incoming process quality is 2.3% nonconforming, the percent of the lots that are expected to be accepted is 66%. Similarly, if 55 lots from a process that is 2.3% nonconforming are inspected using this sampling plan, 36 [(55)(0.66) = 36] will be accepted and 19 [55 − 36 = 19] will be unacceptable.

TABLE 9-2 Probabilities of Acceptance for the Single Sampling Plan: $n = 89$, $c = 2$.

ASSUMED PROCESS QUALITY		SAMPLE SIZE, n		PROBABILITY OF ACCEPTANCE	PERCENT OF LOTS ACCEPTED
p_0	$100p_0$		np_0	P_a	$100P_a$
0.01	1.0	89	0.9	0.938	93.8
0.02	2.0	89	1.8	0.731	73.1
0.03	3.0	89	2.7	0.494	49.4
0.04	4.0	89	3.6	0.302	30.2
0.05	5.0	89	4.5	0.174	17.4
0.06	6.0	89	5.3	0.106*	10.6
0.07	7.0	89	6.2	0.055*	5.5

*By interpolation.

This OC curve is unique to the sampling plan defined by $N = 3000$, $n = 89$, and $c = 2$. If this sampling plan does not give the desired effectiveness, then the sampling plan should be changed and a new OC curve constructed and evaluated.

OC Curve for Double Sampling Plans

The construction of an OC curve for double sampling plans is somewhat more involved since two curves must be determined. One curve is for the probability of acceptance on the first sample; the second curve is the probability of acceptance on the combined samples.

A typical OC curve is shown in Figure 9-4 for the double sampling plan $N = 2400$, $n_1 = 150$, $c_1 = 1$, $r_1 = 4$, $n_2 = 200$, $c_2 = 5$, and $r_2 = 6$. The first step in the construction of the OC curve is to determine the equations. If there is one or fewer nonconforming unit on the first sample, the lot is accepted. Symbolically, the equation is

$$(P_a)_I = (P_{1 \text{ or less}})_I$$

To obtain the equation for the second sample, the number of different ways in which the lot can be accepted is determined. A second sample is taken only if there are 2 or 3 nonconforming units on the first sample. If there is 1 or less, the lot is

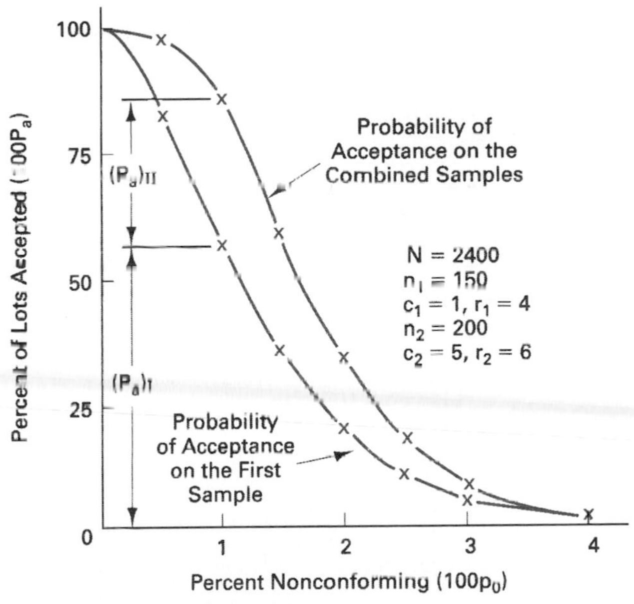

FIGURE 9-4 OC curve for double sampling plan.

accepted; if there are 4 or more, the lot is not accepted. Therefore, the lot can be accepted by obtaining

1. Two nonconforming units on the first sample *and* 3 or less nonconforming units on the second sample, *or*
2. Three nonconforming units on the first sample *and* 2 or less nonconforming units on the second sample.

The *and's* and *or's* are emphasized above to illustrate the use of the additive and multiplicative theorems, which were discussed in Chapter 7. Where an *and* occurs, multiply, and where an *or* occurs, add, and the equation becomes

$$(P_a)_{\text{II}} = (P_2)_{\text{I}}\,(P_{3\text{ or less}})_{\text{II}} + (P_3)_{\text{I}}\,(P_{2\text{ or less}})_{\text{II}}$$

Roman numerals are used as a subscript for the sample number. The equations derived above are applicable only to this double sampling plan; another plan will require a different set of equations. Figure 9-5 graphically illustrates the technique. Note that the number of nonconforming units in each term in the second equation

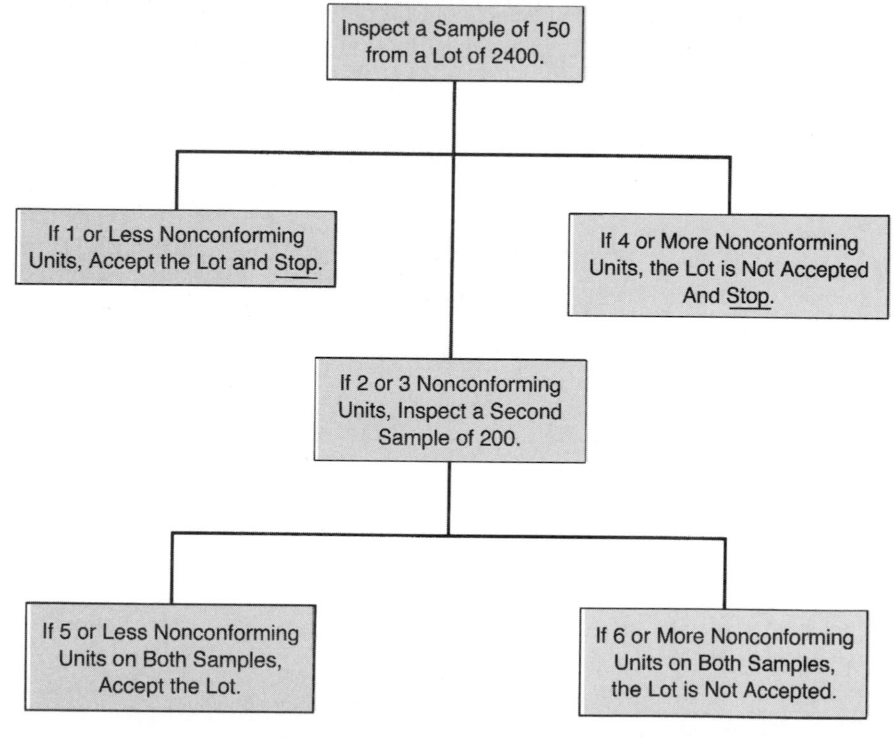

FIGURE 9-5 Graphical description of the double sampling plan: $N = 2400$, $n_1 = 150$, $c_1 = 1$. $r_1 = 4$, $n_2 = 200$, $c_2 = 5$, and $r_2 = 6$.

is equal to or less than the acceptance number, c_2. By combining the equations, the probability of acceptance for the combined samples is obtained:

$$(P_a)_{\text{combined}} = (P_a)_{\text{I}} + (P_a)_{\text{II}}$$

Once the equations are obtained, the OC curves are found by assuming various P_0 values and calculating the respective first and second sample P_a values. For example, using Table C of the Appendix and assuming a p_0 value of 0.01 ($100p_0 = 1.0$),

$$(np_0)_{\text{I}} = (150)(0.01) = 1.5 \qquad (np_0)_{\text{II}} = (200)(0.01) = 2.0$$

$$(P_a)_{\text{I}} = (P_{1\text{ or less}})_{\text{I}} = 0.558$$

$$(P_a)_{\text{II}} = (P_2)_{\text{I}}(P_{3\text{ or less}})_{\text{II}} + (P_3)_{\text{I}}(P_{2\text{ or less}})_{\text{II}}$$

$$(P_a)_{\text{II}} = (0.251)(0.857) + (0.126)(0.677)$$

$$(P_a)_{\text{II}} = 0.300$$

$$(P_a)_{\text{combined}} = (P_a)_{\text{I}} + (P_a)_{\text{II}}$$

$$(P_a)_{\text{combined}} = 0.558 + 0.300$$

$$(P_a)_{\text{combined}} = 0.858$$

These results are illustrated in Figure 9-4. When the two sample sizes are different, the np_0 values are different, which can cause a calculating error. Another source of error is neglecting to use the "or less" probabilities. Calculations are usually to three decimal places. The remaining calculations for other points on the curve are:

For $P_0 = 0.005$ ($100P_0 = 0.5$),

$$(nP_0)_{\text{I}} = (150)(0.005) = 0.75 \qquad (np_0)_{\text{II}} = (200)(0.005) = 1.00$$

$$(P_a)_{\text{I}} = 0.826$$

$$(P_a)_{\text{II}} = (0.133)(0.981) + (0.034)(0.920) = 0.162$$

$$(P_a)_{\text{combined}} = 0.988$$

For $p_0 = 0.015$ ($100p_0 = 1.5$),

$$(np_0)_{\text{I}} = (150)(0.015) = 2.25 \qquad (np_0)_{\text{II}} = (200)(0.015) = 3.00$$

$$(P_a)_{\text{I}} = 0.343$$

$$(P_a)_{\text{II}} = (0.266)(0.647) + (0.200)(0.423) = 0.257$$

$$(P_a)_{\text{combined}} = 0.600$$

For $p_0 = 0.020$ ($100p_0 = 2.0$),

$$(np_0)_{\text{I}} = (150)(0.020) = 3.00 \qquad (np_0)_{\text{II}} = (200)(0.020) = 4.00$$

$$(P_a)_{\text{I}} = 0.199$$

$$(P_a)_{\text{II}} = (0.224)(0.433) + (0.224)(0.238) = 0.150$$

$$(P_a)_{\text{combined}} = 0.349$$

For $p_0 = 0.025$ $(100p_0 = 2.5)$,

$$(np_0)_{\text{I}} = (150)(0.025) = 3.75 \qquad (np_0)_{\text{II}} = (200)(0.025) = 5.00$$

$$(P_a)_{\text{I}} = 0.112$$

$$(P_a)_{\text{II}} = (0.165)(0.265) + (0.207)(0.125) = 0.070$$

$$(P_a)_{\text{combined}} = 0.182$$

For $p_0 = 0.030$ $(100p_0 = 3.0)$,

$$(np_0)_{\text{I}} = (150)(0.030) = 4.5 \qquad (np_0)_{\text{II}} = (200)(0.030) = 6.0$$

$$(P_a)_{\text{I}} = 0.061$$

$$(P_a)_{\text{II}} = (0.113)(0.151) + (0.169)(0.062) = 0.028$$

$$(P_a)_{\text{combined}} = 0.089$$

For $p_0 = 0.040$ $(100p_0 = 4.0)$,

$$(np_0)_{\text{I}} = (150)(0.040) = 6.0 \qquad (np_0)_{\text{II}} = (200)(0.040) = 8.0$$

$$(P_a)_{\text{I}} = 0.017$$

$$(P_a)_{\text{II}} = (0.045)(0.043) + (0.089)(0.014) = 0.003$$

$$(P_a)_{\text{combined}} = 0.020$$

Similar to the construction of the OC curve for single sampling, points are plotted as they are calculated, with the last few calculations used for locations where the curve changes direction. Whenever possible, both sample sizes should be the same value to simplify the calculations and the inspector's job. Also, if r_1 and r_2 are not given, they are equal to $c_2 + 1$.

The steps are: (1) assume p_0 value, (2) calculate $(np_0)_{\text{I}}$ and $(np_0)_{\text{II}}$ values, (3) determine P_a value using the three equations and Table C, (4) plot points, and (5) repeat steps 1, 2, 3, and 4 until a smooth curve is obtained.

OC Curve for Multiple Sampling Plans

The construction of an OC curve for multiple sampling plans is more involved than for double or single sampling plans; however, the technique is the same. A multiple sampling plan with four levels is illustrated in Figure 9-6 and is specified as:

$$N = 3000$$

$$n_1 = 30 \qquad c_1 = 0 \qquad r_1 = 4$$

$$n_2 = 30 \qquad c_2 = 2 \qquad r_2 = 5$$

$$n_3 = 30 \qquad c_3 = 3 \qquad r_3 = 5$$

$$n_4 = 30 \qquad c_4 = 4 \qquad r_4 = 5$$

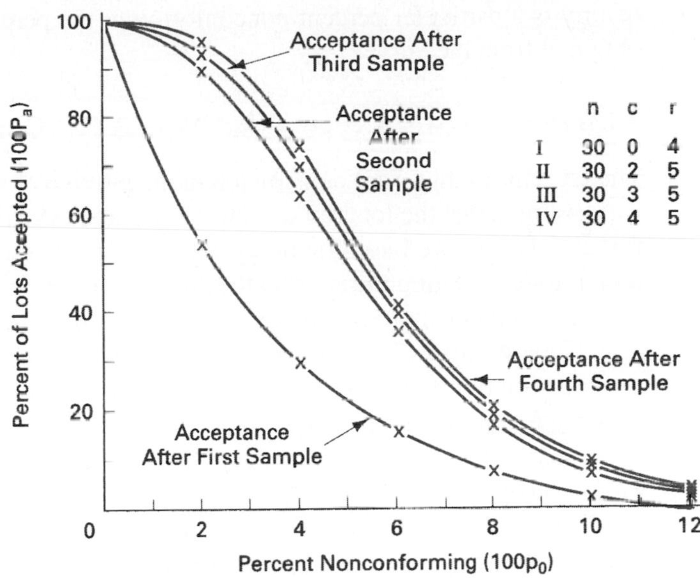

FIGURE 9-6 OC curve for a multiple sampling plan.

Equations for this multiple sampling plan are:

$$(P_a)_{\rm I} = (P_0)_{\rm I}$$

$$(P_a)_{\rm II} = (P_1)_{\rm I}(P_{1\text{ or less}})_{\rm II} + (P_2)_{\rm I}(P_0)_{\rm II}$$

$$(P_a)_{\rm III} = (P_1)_{\rm I}\,(P_2)_{\rm II}(P_0)_{\rm III} + (P_2)_{\rm I}(P_1)_{\rm II}(P_0)_{\rm III} + (P_3)_{\rm I}(P_0)_{\rm II}(P_0)_{\rm III}$$

$$\begin{aligned}(P_a)_{\rm IV} = {} & (P_1)_{\rm I}(P_2)_{\rm II}(P_1)_{\rm III}(P_0)_{\rm IV} + (P_1)_{\rm I}(P_3)_{\rm II}(P_0)_{\rm III}(P_0)_{\rm IV} \\ & + (P_2)_{\rm I}(P_1)_{\rm II}(P_1)_{\rm III}(P_0)_{\rm IV} + (P_2)_{\rm I}(P_2)_{\rm II}(P_0)_{\rm III}(P_0)_{\rm IV} \\ & + (P_3)_{\rm I}(P_0)_{\rm II}(P_1)_{\rm III}(P_0)_{\rm IV} + (P_3)_{\rm I}(P_1)_{\rm II}(P_0)_{\rm III}(P_0)_{\rm IV}\end{aligned}$$

Using the equations above and varying the fraction nonconforming, p_0, the OC curve of Figure 9–6 is constructed. This is a tedious task and one that is ideally suited for the computer.

Comment

An operating characteristic curve evaluates the effectiveness of a particular sampling plan. If that sampling plan is not satisfactory, as shown by the OC curve, another one should be selected and its OC curve constructed.

Since the process quality or lot quality is usually not known, the OC curve (as well as other curves in this chapter) are "what if" curves. In other words, if the quality is a particular percent nonconforming, the percent of lots accepted can be obtained from the curve.

Difference Between Type A and Type B OC Curves

The OC curves that were constructed in the previous sections are type B curves. It was assumed that the lots came from a continuous stream of product, and therefore the calculations are based on an infinite lot size. The binomial is the exact distribution for calculating the acceptance probabilities; however, the Poisson was used, since it is a good approximation. Type B curves are continuous.

Type A curves give the probability of accepting an isolated finite lot. With a finite situation the hypergeometric is used to calculate the acceptance probabilities. As the lot size of a type A curve increases, it approaches the type B curve and will become almost identical when the lot size is at least 10 times the sample size ($n/N \leqq 0.10$). A type A curve is shown in Figure 9-7, with the small open circles representing the discrete data and a discontinuous curve; however, the curve is drawn as a continuous one. Thus, a 4% value is impossible, since it represents 2.6 nonconforming units in the lot of 65 [(0.04)(65) = 2.6], but 4.6% nonconforming units are possible, as it represents 3 nonconforming units in the lot of 65 [(0.046)(65) = 3.0]. Therefore, the "curve" exists only where the small open circles are located.

In comparing the type A and type B curves of Figure 9-7, the type A curve is always lower than the type B curve. When the lot size is small in relation to the

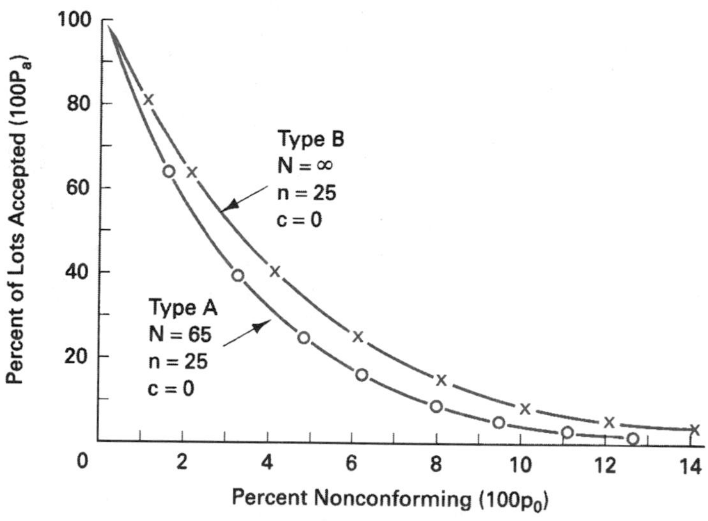

FIGURE 9-7 **Types A and B OC curves.**

sample size, the difference between the curves is significant enough to construct the type A curve.

Unless otherwise stated, all discussion of OC curves will be in terms of type B curves.

OC Curve Properties

Acceptance sampling plans with similar properties can give different OC curves. Four of these properties and the OC curve information are given in the information that follows:

1. *Sample size as a fixed percentage of lot size*. Prior to the use of statistical concepts for acceptance sampling, inspectors were frequently instructed to sample a fixed percentage of the lot. If this value is, say, 10% of the lot size, plans for lot sizes of 900, 300, and 90 are:

$$N = 900 \qquad n = 90 \qquad c = 0$$

$$N = 300 \qquad n = 30 \qquad c = 0$$

$$N = 90 \qquad n = 9 \qquad c = 0$$

Figure 9-8 shows the OC curves for the three plans, and it is evident that they offer different levels of protection. For example, for a process that is 5% nonconforming,

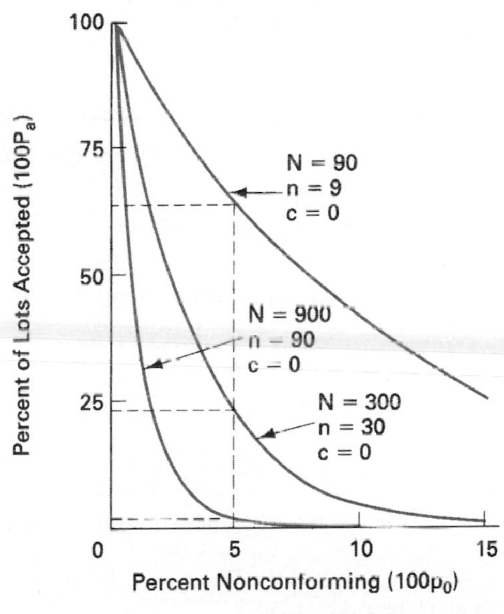

FIGURE 9-8 OC curves for sample sizes that are 10% of the lot size.

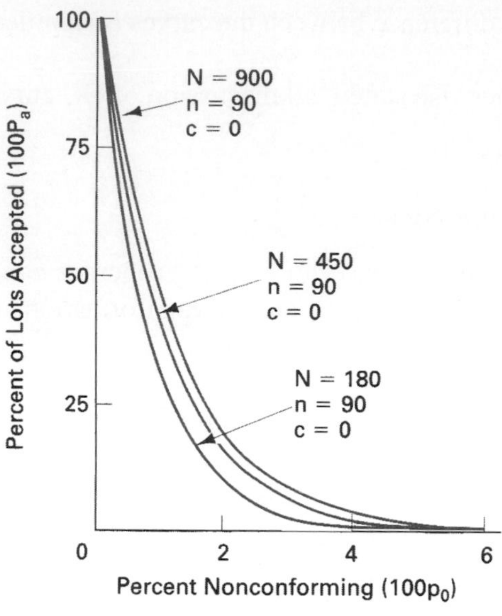

FIGURE 9-9 OC curves for fixed sample size (type A).

$100P_a = 2\%$ for lot sizes of 900, $100P_a = 22\%$ for lot sizes of 300, and $100P_a = 63\%$ for lot sizes of 90.

2. *Fixed sample size.* When a fixed or constant sample size is used, the OC curves are very similar. Figure 9-9 illustrates this property for the type A situation where $n = 10\%$ of N. Naturally, for type B curves or when $n < 10\%$ of N, the curves are identical. The sample size has more to do with the shape of the OC curve and the resulting quality protection than does the lot size.

3. *As sample size increases, the curve becomes steeper.* Figure 9-10 illustrates the change in the shape of the OC curve. As the sample size increases, the slope of the curve becomes steeper and approaches a straight vertical line. Sampling plans with large sample sizes are better able to discriminate between acceptable and unacceptable quality. Therefore, the consumer has fewer lots of unacceptable quality accepted and the producer fewer lots of acceptable quality rejected.

4. *As the acceptance number decreases, the curve becomes steeper.* The change in the shape of the OC curve as the acceptance number changes is shown in Figure 9-11. As the acceptance number decreases, the curve becomes steeper. This fact has frequently been used to justify the use of sampling plans with acceptance numbers of zero. However, the OC curve for $N = 2000$, $n = 300$, and $c = 2$, which is shown by the dashed line, is steeper than the plan with $c = 0$.

A disadvantage of sampling plans with $c = 0$ is that their curves drop sharply down rather than have a horizontal plateau before descending. Since this is the area of the producer's risk (discussed in the next section), sampling plans with $c = 0$

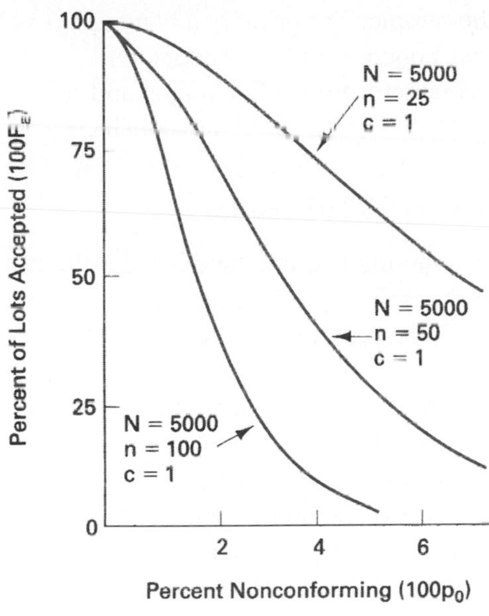

FIGURE 9-10 **OC curves illustrating change in sample size.**

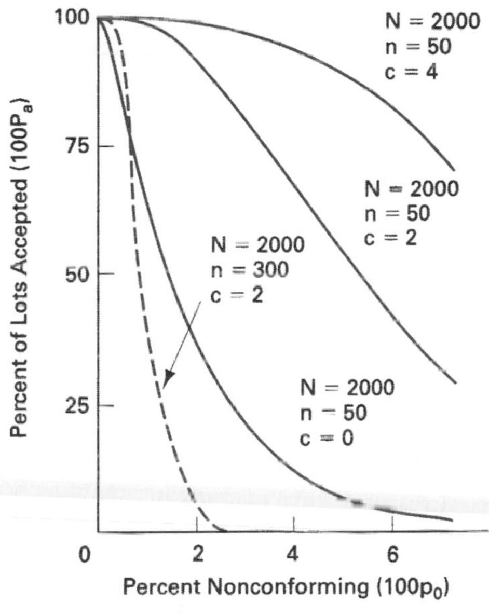

FIGURE 9-11 **OC curves illustrating change in acceptance number.**

are more demanding of the producer. Sampling plans with acceptance numbers greater than zero can actually be superior to those with zero; however, these require a larger sample size, which is more costly. In addition, many producers have a psychological aversion to plans that reject lots when only one nonconforming unit is found in the sample. The primary advantage of sampling plans with $c = 0$ is the perception that nonconforming product will not be tolerated and should be used for critical nonconformities. For major and minor nonconformities, acceptance numbers greater than zero should be considered.

Consumer-Producer Relationship

When acceptance sampling is used, there is a conflicting interest between the consumer and the producer. The producer wants all acceptable lots accepted, and the consumer wants no unacceptable lots accepted. Only an ideal sampling plan that has an OC curve that is a vertical line can satisfy both the producer and consumer. An "ideal" OC curve, as shown in Figure 9-12, can be achieved only with 100% inspection, and the pitfalls of this type of inspection were mentioned earlier in the chapter. Therefore, sampling carries risks of not accepting lots that are acceptable and of accepting lots that are unacceptable. Because of the seriousness of these risks, various terms and concepts have been standardized.

The *producer's risk,* which is represented by the symbol α, is the probability of non-acceptance of a conforming lot. This risk is frequently given as 0.05, but it can range from 0.001 to 0.10 or more, Since α is expressed in terms of the probability of non-acceptance, it cannot be located on an OC curve unless specified in terms of the probability of acceptance. This conversion is accomplished by subtracting from 1. Thus, $P_a = 1 - \alpha$, and for $\alpha = 0.05$, $P_a = 1 - 0.05 = 0.95$.

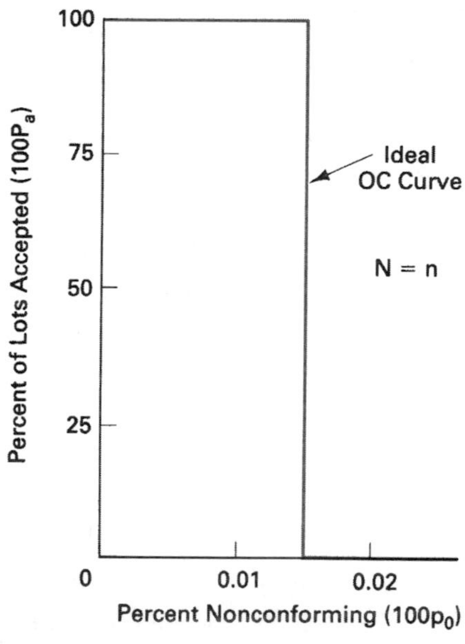

FIGURE 9-12 Ideal OC curve.

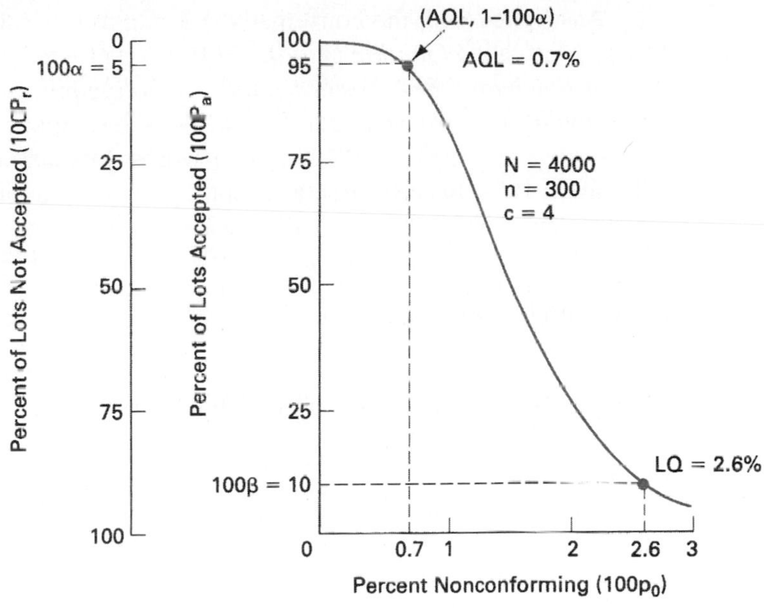

FIGURE 9-13 Consumer-producer relationship.

Figure 9-13 shows the producer's risk, α, or 0.05 on an imaginary axis labeled "Percent of Lots Not Accepted."

Associated with the producer's risk is a numerical definition of an acceptable lot, which is called *Acceptable Quality Level* (AQL). *The AQL is the maximum percent nonconforming that can be considered satisfactory for the purposes of acceptance sampling. It is a reference point on the OC curve and is not meant to convey to the producer that any percent nonconforming is acceptable. It is a statistical term and is not meant to be used by the general public.* The only way the producer can be guaranteed that a lot will be accepted is to have 0% nonconforming or to have the number nonconforming in the lot less than or equal to the acceptance number. In other words, the producer's quality goal is to meet or exceed the specifications so that no nonconforming units are present in the lot.

For the sampling plan $N = 4000$, $n = 300$, and $c = 4$, the AQL = 0.7% for 100 = 5%, as shown in Figure 9-13. In other words, product that is 0.7% nonconforming will have a non-acceptance probability of 0.05, or 5%. Or, stated another way, 1 out of 20 lots that are 0.7% nonconforming will not be accepted by the sampling plan.

The *consumer's risk,* represented by the symbol β, is the probability of acceptance of a nonconforming lot. This risk is frequently given as 0.10. Since β is expressed in terms of probability of acceptance, no conversion is necessary.

Associated with the consumer's risk is a numerical definition of a nonconforming lot, called *Limiting Quality* (LQ). *The LQ is the percent nonconforming in a lot or batch for which, for acceptance sampling purposes, the consumer wishes the probability of acceptance to be low.* For the sampling plan in Figure 9-13, the LQ = 2.6% for 100β = 10%. In other words, lots that are 2.6% nonconforming will have a 10% chance of being accepted. Or, stated another way, 1 out of 10 lots that are 2.6% nonconforming will be accepted by this sampling plan.

Average Outgoing Quality

The *Average Outgoing Quality* (AOQ) is another technique for the evaluation of a sampling plan. Figure 9-14 shows an AOQ curve for the sampling plan $N = 3000$, $n = 89$, and $c = 2$. This is the same plan as the one for the OC curve shown in Figure 9-3.

The information for the construction of an average outgoing quality curve is obtained by adding one column (an AOQ column) to the table used to construct an OC curve. Table 9-3 shows the information for the OC curve and the additional column for the AOQ curve. The average outgoing quality in percent nonconforming is determined by the formula AOQ = $(100p_0)(P_a)$. This formula does not account for the discarded nonconforming units; however, it is close enough for practical purposes and is simpler to use.

Note that to present a more readable graph, the AOQ scale is much larger than the incoming process quality scale. The curve is constructed by plotting the percent nonconforming ($100p_0$) with its corresponding AOQ value.

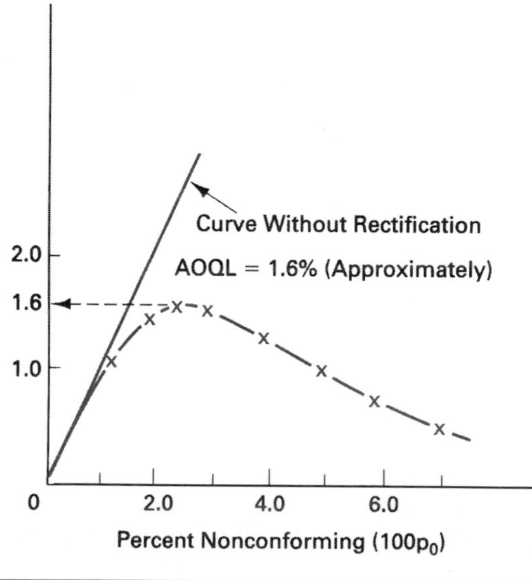

FIGURE 9-14 Average outgoing quality curve for the sampling plan $N = 3000$, $n = 89$, and $c = 2$.

TABLE 9-3 Average Outgoing Quality (AOQ) for the Sampling Plan $N = 3000$, $n = 89$, and $c = 2$.

PROCESS QUALITY $100p_0$	SAMPLE SIZE n	np_0	PROBABILITY OF ACCEPTANCE P_a	AOQ $(100p_0)(P_a)$
1.0	89	0.9	0.938	0.938
2.0	89	1.8	0.731	1.462
3.0	89	2.7	0.494	1.482
4.0	89	3.6	0.302	1.208
5.0	89	4.5	0.174	0.870
6.0	89	5.3	0.106	0.636
7.0	89	6.2	0.055	0.385
2.5*	89	2.2	0.623	1.558

*Additional point where curve changes direction.

The AOQ is the quality that leaves the inspection operation. It is assumed that any non-accepted lots have been rectified or sorted and returned with 100% good product. When rectification does not occur, the AOQ is the same as the incoming quality, and this condition is represented by the straight line in Figure 9–14.

Analysis of the curve shows that when the incoming quality is 2.0% nonconforming, the average outgoing quality is 1.46% nonconforming, and when the incoming quality is 6.0% nonconforming, the average outgoing quality is 0.64% nonconforming. Therefore, because non-accepted lots are rectified, the average outgoing quality is always better than the incoming quality. In fact, there is a limit that is given the name Average Outgoing Quality Limit (AOQL). Thus, for this sampling plan, as the percent nonconforming of the incoming quality changes, the average outgoing quality never exceeds the limit of approximately 1.6% nonconforming.

A better understanding of the concept of acceptance sampling can be obtained from an example. Suppose that over a period of time 15 lots of 3000 each are shipped by the producer to the consumer. The lots are 2% nonconforming and a sampling plan of $n = 89$ and $c = 2$ is used to determine acceptance. Figure 9-15 shows this information by a solid line. The OC curve for this sampling plan (Figure 9-3) shows that the percent of lots accepted for a 2% nonconforming lot is 73.1%. Thus, 11 lots ($15 \times 0.731 = 10.97$) are accepted by the consumer, as shown by the wavy line. Four lots are not accepted by the sampling plan and returned to the producer for rectification, as shown by the dashed line. These four lots receive 100% inspection and are returned to the consumer with 0% nonconforming, as shown by a dashed line.

A summary of what the consumer actually receives is shown at the bottom of the figure. Two percent, or 240, of the four rectified lots are discarded by the producer, which gives 11,760 rather than 12,000. The calculations show that the

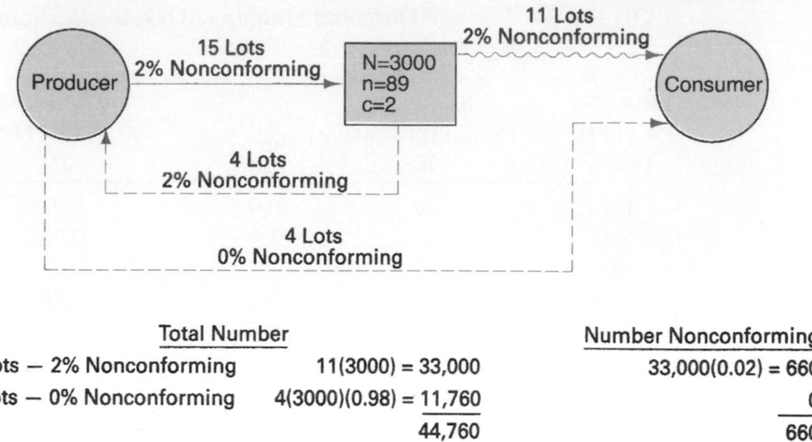

	Total Number	Number Nonconforming
11 Lots — 2% Nonconforming	11(3000) = 33,000	33,000(0.02) = 660
4 Lots — 0% Nonconforming	4(3000)(0.98) = 11,760	0
	44,760	660

$$\text{Percent Nonconforming (AOQ)} = \frac{660}{44{,}760} \times 100 = 1.47\%$$

FIGURE 9-15 How acceptance sampling works.

consumer actually receives 1.47% nonconforming, whereas the producer's quality is 2% nonconforming.

It should be emphasized that the acceptance sampling system works only when non-accepted lots are returned to the producer and rectified. The AQL for this particular sampling plan at $\alpha = 0.05$ is 0.9%; therefore, the producer at 2% nonconforming is not achieving the desired quality level.

The AOQ curve, in conjunction with the OC curve, provides two powerful tools for describing and analyzing acceptance sampling plans.

Average Sample Number

The Average Sample Number (ASN) is a comparison of the average amount inspected per lot by the consumer for single, double, multiple, and sequential sampling. Figure 9–16 shows the comparison for the four different but equally effective sampling plan types. In single sampling the ASN is constant and equal to the sample size, n. For double sampling the process is somewhat more complicated because a second sample may or may not be taken.

The formula for double sampling is

$$\text{ASN} = n_1 + n_2(1 - P_{\text{I}})$$

where P_{I} is the probability of a decision on the first sample. An example problem will illustrate the concept.

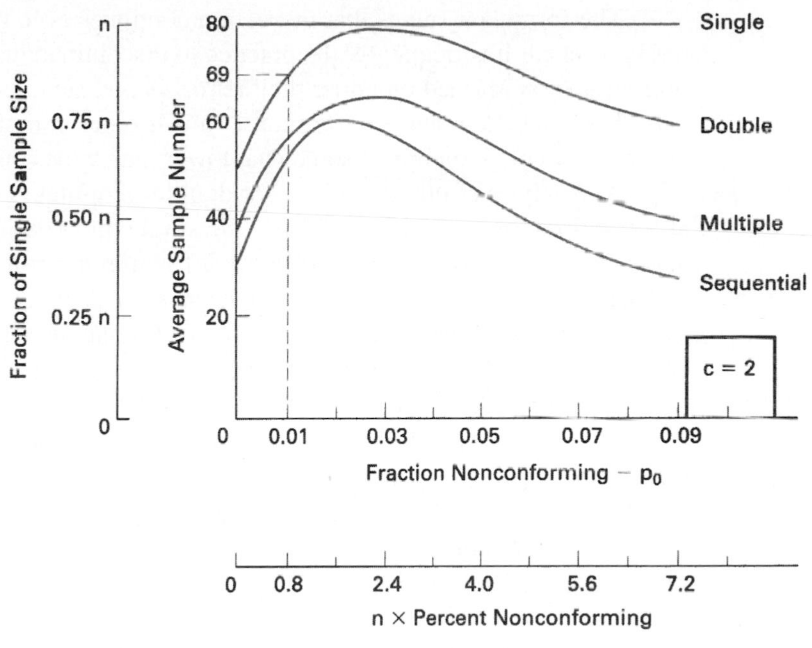

FIGURE 9-16 ASN curves for single, double, multiple, and sequential sampling.

EXAMPLE PROBLEM

Given the single sampling plan $n = 80$ and $c = 2$ and the equally effective double sampling plan $n_1 = 50$, $c_1 = 0$, $r_1 = 3$, $n_2 = 50$, $c_2 = 3$, and $r_2 = 4$, compare the ASN of the two by constructing their curves.

For single sampling, the ASN is the straight line at $n = 80$. For double sampling, the solution is

$$P_{\mathrm{I}} = P_0 + P_{3 \text{ or more}}$$

Assume that $p_0 = 0.01$; then $np_0 = 50(0.01) = 0.5$. From Appendix C:

$$P_0 = 0.607$$

$$P_{3 \text{ or more}} = 1 - P_{2 \text{ or less}} = 1 - 0.986 = 0.014$$

$$\begin{aligned} \text{ASN} &= n_1 + n_2\,(1 - (P_0 + P_{3 \text{ or more}})) \\ &= 50 + 50(1 - (0.607 + 0.014)) \\ &= 69 \end{aligned}$$

Repeating for different values of p_0, the double sampling plan is plotted as shown in Figure 9-16.

The formula assumes that inspection continues even after the rejection number is reached. It is frequently the practice to discontinue inspection after the rejection number is reached on either the first or second sample. This practice is called curtailed inspection, and the formula is much more complicated. Thus, the ASN curve for double sampling is somewhat lower than what actually occurs.

An analysis of the ASN curve for double sampling in Figure 9-16 shows that at a fraction nonconforming of 0.03, the single and double sampling plans have about the same amount of inspection. For fraction nonconforming less than 0.03, double sampling has less inspection because a decision to accept on the first sample is more likely. Similarly, for fraction nonconforming greater than 0.03, double sampling has less inspection because a decision not to accept on the first sample is more likely and a second sample is not required. It should be noted that in most ASN curves, the double sample curve does not get close to the single sample one.

Calculation of the ASN curve for multiple sampling is much more difficult than for double sampling. The formula is

$$\text{ASN} = n_1P_1 + (n_1 + n_2)P_{\text{II}} + \ldots + (n_1+n_2 + \ldots + n_k)P_k$$

where n_k is the sample size of the last level and P_k the probability of a decision at the last level.

Determining the probabilities of a decision at each level is quite involved—more so than for the OC curve since the conditional probabilities must also be determined.

Figure 9-16 shows the ASN curve for an equivalent multiple sampling plan with seven levels. As expected, the average amount inspected is much less than single or double sampling.

The reader may have been curious concerning the two extra scales in Figure 9-16. Since we are comparing equivalent sampling plans, the double and multiple plans can be related to the single sampling plans where $c = 2$ and n is the equivalent single sample size by the additional scales. To use the horizontal scale, multiply the single sample size n by the fraction nonconforming. The ASN value is found from the vertical scale by multiplying the scale fraction with the single sample size.

Figure 9–17, which is taken from ANSI/ASQ Z1.4—1993 (to be discussed), shows a number of ASN curve comparisons indexed by the acceptance number, c. These curves assume no curtailment of inspection and are approximate to the extent that they are based on the Poisson distribution, and that the sample sizes for double and multiple sampling are assumed to be $0.631n$ and $0.25n$, respectively. Therefore, these curves can be used to find the amount inspected per lot for different percent nonconforming without having to make the calculations. The arrow indicates the location of the AQL.

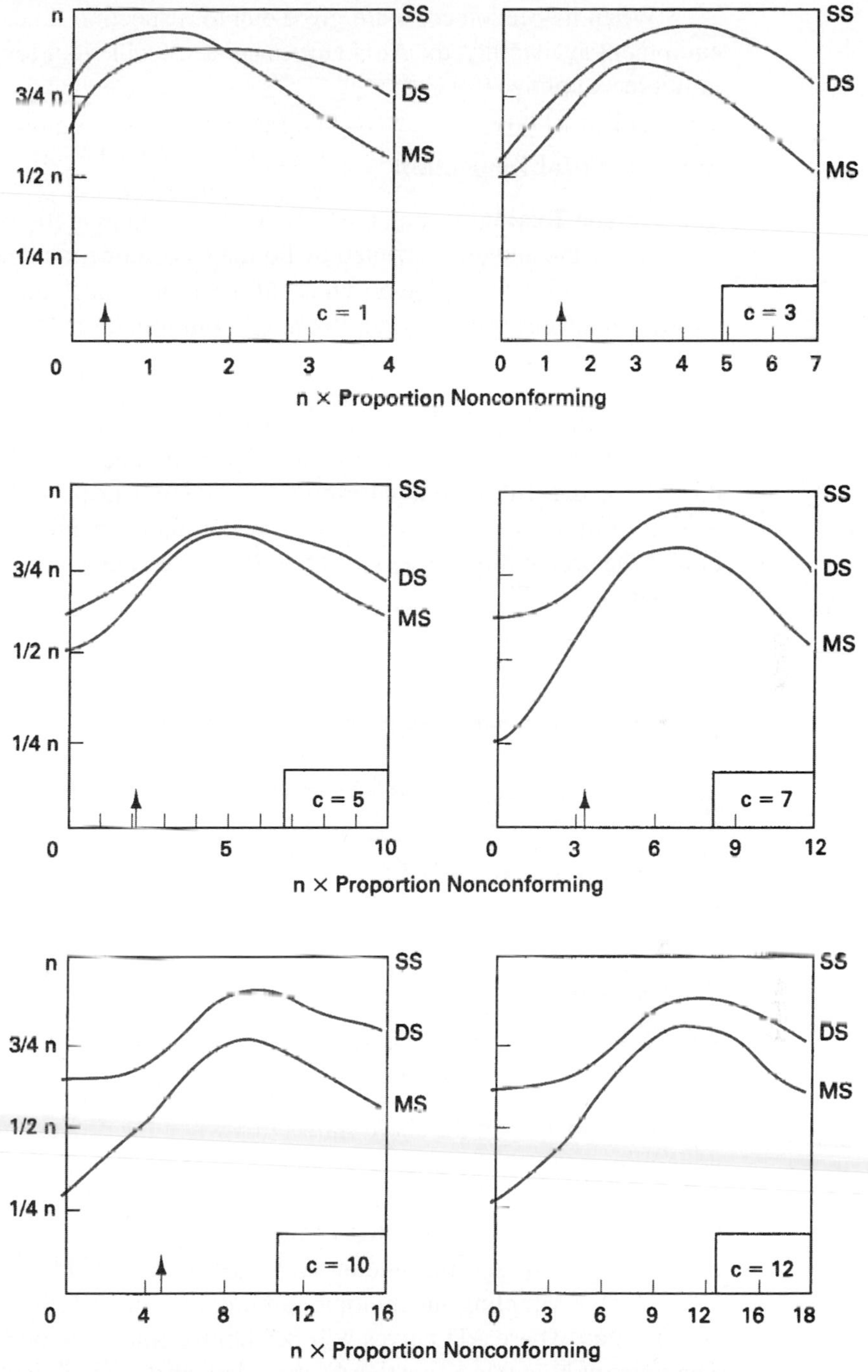

FIGURE 9-17 Typical ASN curves from ANSI/ASQ Z1.4—1993.

When inspection costs are great due to inspection time, equipment costs, or equipment availability, the ASN curves are a valuable tool for justifying double or multiple sampling.

Average Total Inspection

The Average Total Inspection (ATI) is another technique for evaluating a sampling plan. ATI is the amount inspected by both the consumer and the producer. Like the ASN curve, it is a curve that provides information on the amount inspected and not on the effectiveness of the plan. For single sampling, the formula is

$$\text{ATI} = n + (1 - P_a)(N - n)$$

It assumes that rectified lots will receive 100% inspection. If lots are submitted with 0% nonconforming, the amount inspected is equal to n, and if lots are submitted that are 100% nonconforming, the amount inspected is equal to N. Since neither of these possibilities is likely to occur, then the amount inspected is a function of the probability of rejection $(1 - P_a)$. An example problem will illustrate the calculation.

EXAMPLE PROBLEM

Determine the ATI curve for the single sampling plan $N = 3000$, $n = 89$, and $c = 2$.

Assume that $p_0 = 0.02$. From the OC curve (Figure 9-3), $P_a = 0.731$.

$$\begin{aligned} \text{ATI} &= n + (1 - P_a)(N - n) \\ &= 89 + (1 - 0.731)(3000 - 89) \\ &= 872 \end{aligned}$$

Repeat for other p_0 values until a smooth curve is obtained, as shown in Figure 9-18.

Examination of the curve shows that when the process quality is close to 0% nonconforming, the average total amount inspected is close to the sample size n. When process quality is very poor, at, say, 9% nonconforming, most of the lots are not accepted, and the ATI curve becomes asymptotic to 3000. As the percent nonconforming increases, the amount inspected by the producer dominates the curve.

Double sampling and multiple sampling formulas for the ATI curves are more complicated. These ATI curves will be slightly below the one for single sampling. The amount below is a function of the ASN curve, which is the amount inspected by the consumer, and this amount is usually very small in relation to the ATI, which is dominated by the amount inspected by the producer. From a practical viewpoint,

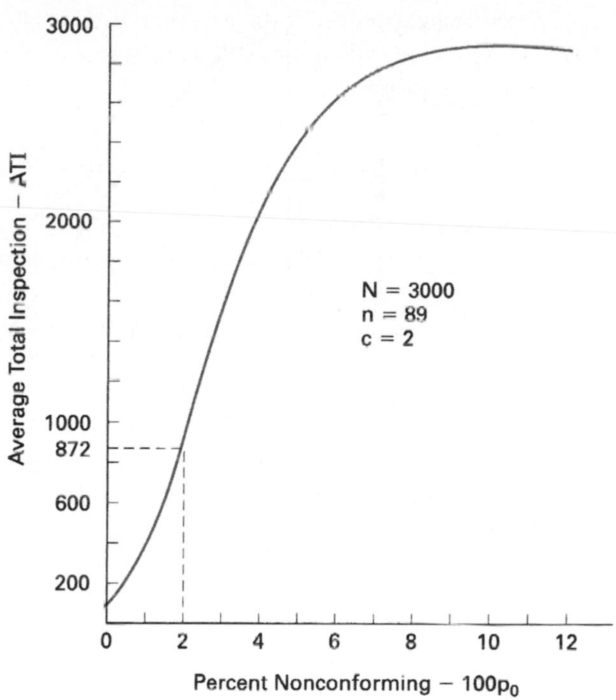

FIGURE 9-18 ATI curve for $N = 3000$, $n = 89$, and $c = 2$.

the ATI curves for double and multiple sampling are not necessary since the equivalent single sampling curve will convey a good estimate.

SAMPLING PLAN DESIGN

Sampling Plans for Stipulated Producer's Risk

When the producer's risk α and its corresponding Acceptable Quality Level (AQL) are specified, a sampling plan or, more precisely, a family of sampling plans can be determined. For a producer's risk, α, of, say, 0.05 and an AQL of 1.2%, the OC curves for a family of sampling plans as shown in Figure 9-19 are obtained. Each of the plans passes through the point defined by $100P_a = 95\%$ ($100\alpha = 5\%$) and $p_{0.95} = 0.012$. Therefore, each of the plans will ensure that product 1.2% nonconforming will not be accepted 5% of the time or, conversely, accepted 95% of the time.

The sampling plans are obtained by assuming a value for c and finding its corresponding np_0 value from Table C. When np_0 and p_0 are known, the sample size n is obtained. In order to find the np_0 values using Table C, interpolation is required.

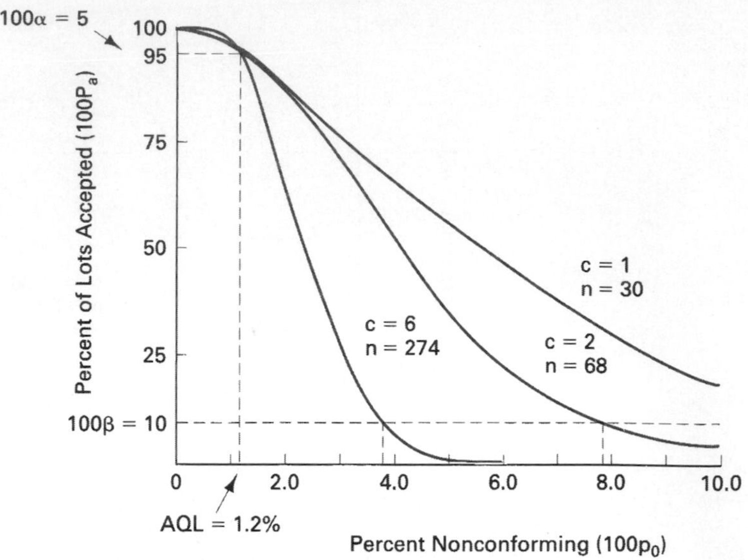

FIGURE 9-19 **Single sampling plans for stipulated producer's risk and AQL.**

To eliminate the interpolation operation, np_0 values for various α and β values are reproduced in Table 9-4. In this table, c is cumulative, which means that a c value of 2 represents 2 or less.

Calculations to obtain the three sampling plans of Figure 9-19 are as follows:

$$P_a = 0.95 \qquad p_{0.95} = 0.012$$

For $c = 1$, $np_{0.95} = 0.355$ (from Table 9-4) and

$$n = \frac{np_{0.95}}{p_{0.95}} = \frac{0.355}{0.012} = 29.6. \text{ or } 30$$

For $c = 2$, $np_{0.95} = 0.818$ (from Table 9-4) and

$$n = \frac{np_{0.95}}{p_{0.95}} = \frac{0.818}{0.012} = 68.2. \text{ or } 68$$

For $c = 6$, $np_{0.95} = 3.286$ (from Table 9-4) and

$$n = \frac{np_{0.95}}{p_{0.95}} = \frac{3.286}{0.012} = 273.9 \text{ or } 274$$

The sampling plans for $c = 1$, $c = 2$, and $c = 6$ were arbitrarily selected to illustrate the technique. Construction of the OC curves is accomplished by the methods given at the beginning of the chapter.

TABLE 9-4 ***np* Values for Corresponding *c* Values and Typical Producers' and Consumers' Risks**

c	$P_a = 0.99$ ($\alpha = 0.01$)	$P_a = 0.95$ ($\alpha = 0.05$)	$P_a = 0.90$ ($\alpha = 0.10$)	$P_a = 0.10$ ($\beta = 0.10$)	$P_a = 0.05$ ($\beta = 0.05$)	$P_a = 0.01$ ($\beta = 0.01$)	RATIO OF $p_{0.10}/p_{0.95}$
0	0.010	0.051	0.105	2.303	2.996	4.605	44.890
1	0.149	0.355	0.532	3.890	4.744	6.638	10.946
2	0.436	0.818	1.102	5.322	6.296	8.406	6.509
3	0.823	1.366	1.745	6.681	7.754	10.045	4.890
4	1.279	1.970	2.433	7.994	9.154	11.605	4.057
5	1.785	2.613	3.152	9.275	10.513	13.108	3.549
6	2.330	3.286	3.895	10.532	11.842	14.571	3.206
7	2.906	3.981	4.656	11.771	13.148	16.000	2.957
8	3.507	4.695	5.432	12.995	14.434	17.403	2.768
9	4.130	5.426	6.221	14.206	15.705	18.783	2.618
10	4.771	6.169	7.021	15.407	16.962	20.145	2.497
11	5.428	6.924	7.829	16.598	18.208	21.490	2.397
12	6.099	7.690	8.646	17.782	19.442	22.821	2.312
13	6.782	8.464	9.470	18.958	20.668	24.139	2.240
14	7.477	9.246	10.300	20.128	21.886	25.446	2.177
15	8.181	10.035	11.135	21.292	23.098	26.743	2.122

Source: Extracted by permission from J. M. Cameron, "Tables for Constructing and for Computing the Operating Characteristics of Single-Sampling Plans," *Industry Quality Control,* 9, No. 1 (July 1952), 39.

While all the plans provide the same protection for the producer, the consumer's risk, at, say, $\beta = 0.10$, is quite different. From Figure 9-19 for the plan $c = 1, n = 30$, product that is 13% nonconforming will be accepted 10% ($\beta = 0.10$) of the time; for the plan $c = 2, n = 68$, the product that is 7.8% nonconforming will be accepted 10% ($\beta = 0.10$) of the time; and, for the plan $c = 6, n = 274$, product that is 3.8% nonconforming will be accepted 10% ($\beta = 0.10$) of the time. From the consumer's viewpoint the latter plan provides better protection; however, the sample size is greater, which increases the inspection cost. The selection of the appropriate plan to use is a matter of judgment, which usually involves the lot size. This selection would also include plans for $c = 0, 3, 4, 5, 7$, and so forth.

Sampling Plans for Stipulated Consumer's Risk

When the consumer's risk β and its corresponding Limiting Quality (LQ) are specified, a family of sampling plans can be determined. For a consumer's risk, β, of, say, 0.10 and a LQ of 6.0%, the OC curves for a family of sampling plans as shown in Figure 9-20 are obtained. Each of the plans pass through the point defined by $P_a = 0.10$ ($\beta = 0.10$) and $p_{0.10} = 0.060$. Therefore, each of the plans will ensure that product 6.0% nonconforming will be accepted 10% of the time.

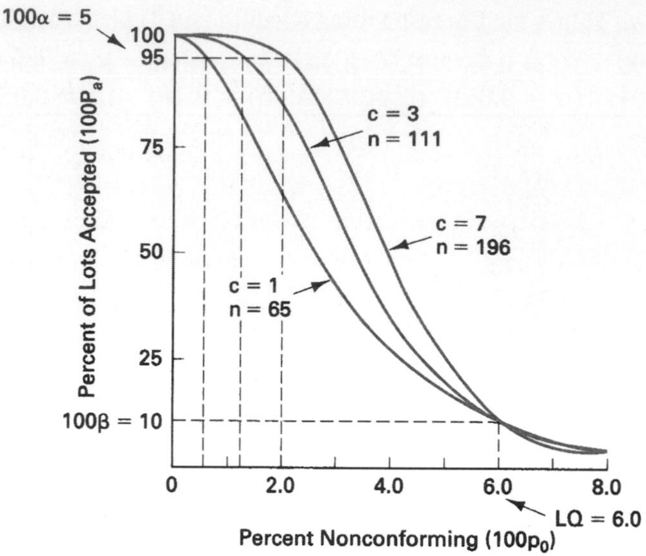

FIGURE 9-20 Single sampling plans for stipulated consumer's risk and LQ.

The sampling plans are determined in the same manner as used for a stipulated producer's risk. Calculations are as follows:

$$P_a = 0.10 \; p_{0.10} = 0.060$$

For $c = 1$, $np_{0.10} = 3.890$ (from Table 9-4) and

$$n = \frac{np_{0.10}}{p_{0.10}} = \frac{3.890}{0.060} = 64.8, \text{ or } 65$$

For $c = 3$, $np_{0.10} = 6.681$ (from Table 9-4) and

$$n = \frac{np_{0.10}}{p_{0.10}} = \frac{6.681}{0.060} = 111.4, \text{ or } 111$$

For $c = 7$, $np_{0.10} = 11.771$ (from Table 9-4) and

$$n = \frac{np_{0.10}}{p_{0.10}} = \frac{11.771}{0.060} = 196.2, \text{ or } 196$$

The sampling plans for $c = 1$, $c = 3$, and $c = 7$ were arbitrarily selected to illustrate the technique. Construction of the OC curves is accomplished by the method given at the beginning of the chapter.

While all the plans provide the same protection for the consumer, the producer's risk, at, say, $\alpha = 0.05$, is quite different. From Figure 9–20 for the plan

$c = 1, n = 65$, product that is 0.5% nonconforming will not be accepted 5% ($100\alpha = 5\%$) of the time; for the plan $c = 3, n = 111$, product that is 1.2% nonconforming will not be accepted 5% ($100\alpha = 5\%$) of the time; and for the plan $c = 7, n = 196$, product that is 2.0% nonconforming will not be accepted 5% ($\alpha = 0.05$) of the time. From the producer's viewpoint the latter plan provides better protection; however, the sample size is greater, which increases the inspection costs. The selection of the appropriate plan is a matter of judgment, which usually involves the lot size. This selection would also include plans for $c = 0, 2, 4, 5, 6, 8$, and so forth.

Sampling Plans for Stipulated Producer's and Consumer's Risk

Sampling plans are also stipulated for both the consumer's risk and the producer's risk. It is difficult to obtain an OC curve that will satisfy both conditions. More than likely there will be four sampling plans that are close to meeting the consumer's and producer's stipulations. Figure 9–21 shows four plans that are close to meeting the stipulations of $\alpha = 0.05$, AQL $= 0.9$ and $\beta = 0.10$, LQ $= 7.8$. The OC curves of two plans meet the consumer's stipulation that product which is 7.8% nonconforming (LQ) will be accepted 10% ($\beta = 0.10$) of the time and comes close to the producer's stipulation. These two plans are shown by the dashed lines in Figure 9-21 and are $c = 1, n = 50$ and $c = 2, n = 68$. The two other plans exactly meet the producer's stipulation that product which is 0.9% nonconforming (AQL) will not be accepted 5% ($\alpha = 0.05$) of the time. These two plans are shown by the solid lines and are $c = 1, n = 39$ and $c = 2, n = 91$.

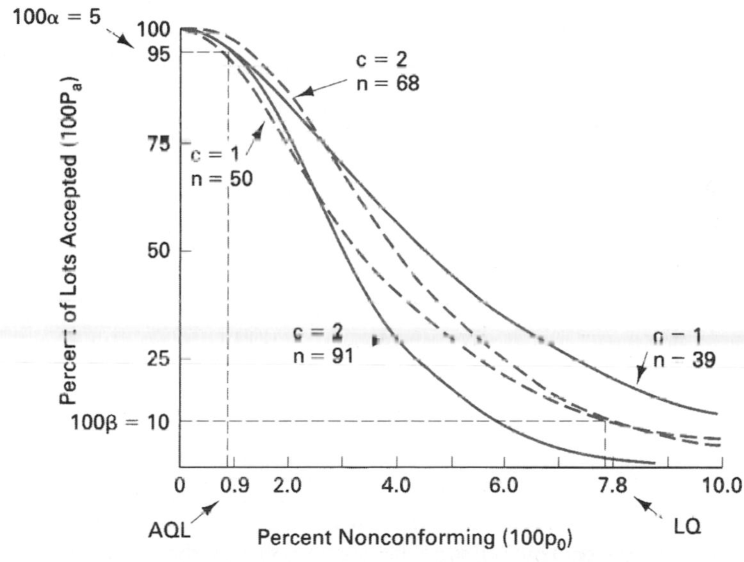

FIGURE 9-21 Sampling plans for stipulated producer's and consumer's risk.

In order to determine the plans, the first step is to find the ratio of $p_{0.10}/p_{0.95}$, which is

$$\frac{p_{0.10}}{p_{0.95}} = \frac{0.078}{0.009} = 8.667$$

From the ratio column of Table 9-4, the ratio of 8.667 falls between the row for $c = 1$ and the row for $c = 2$. Thus, plans that exactly meet the consumer's stipulation of LQ = 7.8% for $\beta = 0.10$ are

For $c = 1$,

$$p_{0.10} = 0.078$$

$$np_{0.10} = 3.890 \text{ (from Table 9-4)}$$

$$n = \frac{np_{0.10}}{p_{0.10}} = \frac{3.890}{0.078} = 49.9, \text{ or } 50$$

For $c = 2$,

$$p_{0.10} = 0.078$$

$$np_{0.10} = 5.322 \text{ (from Table 9-4)}$$

$$n = \frac{np_{0.10}}{p_{0.10}} = \frac{5.322}{0.078} = 68.2, \text{ or } 68$$

Plans that exactly meet the producer's stipulation of AQL = 0.9% for $\alpha = 0.05$ are

For $c = 1$,

$$p_{0.95} = 0.009$$

$$np_{0.95} = 0.355 \text{ (from Table 9-4)}$$

$$n = \frac{np_{0.95}}{p_{0.95}} = \frac{0.355}{0.009} = 39.4, \text{ or } 39$$

For $c = 2$,

$$P_{0.95} = 0.009$$

$$np_{0.95} = 0.818 \text{ (from Table 9-4)}$$

$$n = \frac{np_{0.95}}{p_{0.95}} = \frac{0.818}{0.009} = 90.8, \text{ or } 91$$

Construction of the OC curve follows the method given at the beginning of the chapter.

Which of the four plans to select is based on one of four additional criteria. The first additional criterion is the stipulation that the plan with the lowest sample size be selected. The plan with the lowest sample size is one of the two with the lowest acceptance number. Thus, for the example problem, only the two plans for $c = 1$ are calculated, and $c = 1, n = 39$ is the sampling plan selected. A second additional criterion is the stipulation that the plan with the greatest sample size be selected. The plan with the greatest sample size is one of two with the largest acceptance number. Thus, for the example problem, only the two plans for $c = 2$ are calculated, and $c = 2, n = 91$ is the sampling plan selected.

A third additional criterion is the stipulation that the plan exactly meets the consumer's stipulation and comes as close as possible to the producer's stipulation. The two plans that exactly meet the consumer's stipulation are $c = 1, n = 50$ and $c = 2, n = 68$. Calculations to determine which plan is closest to the producer's stipulation of AQL = 0.9%, $\alpha = 0.05$ are

For $c = 1, n = 50$,

$$p_{0.95} = \frac{np_{0.95}}{n} = \frac{0.355}{50} = 0.007$$

For $c = 2, n = 68$,

$$p_{0.95} = \frac{np_{0.95}}{n} = \frac{0.818}{68} = 0.012$$

Since $p_{0.95} = 0.007$ is closest to the stipulated value of 0.009, the plan of $c = 1, n = 50$ is selected.

The fourth additional criterion for the selection of one of the four sampling plans is the stipulation that the plan exactly meet the producer's stipulation and comes as close as possible to the consumer's stipulation. The two plans that are applicable are $c = 1, n = 39$ and $c = 2, n = 91$. Calculations to determine which is the closest to the consumer's stipulation of LQ = 7.8%, $\beta = 0.10$ are

For $c = 1, n = 39$,

$$p_{0.10} = \frac{np_{0.10}}{n} = \frac{0.3890}{39} = 0.100$$

For $c = 2, n = 91$,

$$p_{0.10} = \frac{np_{0.10}}{n} = \frac{5.322}{91} = 0.058$$

Since $p_{0.10} = 0.058$ is closest to the stipulated value of 0.078, the plan of $c = 2, n = 91$ is selected.

Some Comments

The previous discussions have concerned single sampling plans. Double and multiple sampling plan design, although more difficult, would follow similar techniques.

In the previous discussion, a producer's risk of 0.05 and a consumer's risk of 0.10 were used to illustrate the technique. The producer's risk is usually set at 0.05 but can be as small as 0.01 or as high as 0.15. The consumer's risk is usually set at 0.10 but can be as low as 0.01 or as high as 0.20.

Sampling plans can also be specified by the Average Outgoing Quality Limit (AOQL). If an AOQL of 1.5% for an incoming quality of, say, 2.0% is stipulated, the probability of acceptance is

$$\text{AOQL} = (100p_0)(P_a)$$

$$1.5 = 2.0P_a$$

$$P_a = 0.75 \text{ or } 100P_a = 75\%$$

Figure 9-22 shows a family of OC curves for various sampling plans which satisfy the AOQL criteria.

To design a sampling plan, some initial stipulations are necessary by the producer, consumer, or both. These stipulations are decisions based on historical data, experimentation, or engineering judgment. In some cases the stipulations are negotiated as part of the purchasing contract.

The task of designing a sampling plan system is a tedious one. Fortunately, sampling plan systems are available. One such system that is almost universally used for the acceptance of product is ANSI/ASQ Z1.4—1993. This system is an

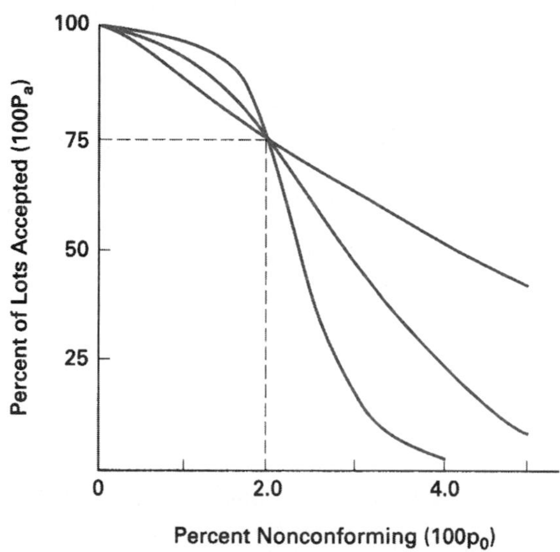

FIGURE 9-22 AOQL sampling plans.

AQL, or producer's risk system. Another system, DodgeRomig, uses the LQ or consumer's risk and AOQL methods for determining the sampling plan. These systems and others are discussed in the next chapter.

COMPUTER PROGRAM

The software in the diskette inside the back cover will solve OC and AOQ curves for single sampling plans using EXCEL. Its file name is *OC Curve*.

PROBLEMS

1. A real estate firm evaluates incoming selling agreement forms using the single sampling plan $N = 1500$, $n = 110$, and $c = 3$. Construct the OC curve using about 7 points.
2. A doctor's clinic evaluates incoming disposable cotton-tipped applicators using the single sampling plan $N = 8000$, $n = 62$, and $c = 1$. Construct the OC curve using about 7 points.
3. Determine the equation for the OC curve for the sampling plan $N = 10{,}000$, $n_1 = 200$, $c_1 = 2$, $r_1 = 6$, $n_2 = 350$, $c_2 = 6$, and $r_2 = 7$. Construct the curve using about 5 points.
4. Determine the equation for the OC curve for the following sampling plans:
   (a) $N = 500$, $n_1 = 50$, $c_1 = 0$, $r_1 = 3$, $n_2 = 70$, $c_2 = 2$, and $r_2 = 3$
   (b) $N = 6000$, $n_1 = 80$, $c_1 = 2$, $r_1 = 4$, $n_2 = 160$, $c_2 = 5$, and $r_2 = 6$
   (c) $N = 22{,}000$, $n_1 = 260$, $c_1 = 5$, $r_1 = 9$, $n_2 = 310$, $c_2 = 8$, and $r_2 = 9$
   (d) $N = 10{,}000$, $n_1 = 300$, $c_1 = 4$, $r_1 = 9$, $n_2 = 300$, and $c_2 = 8$
   (e) $N = 800$, $n_1 = 100$, $c_1 = 0$, $r_1 = 5$, $n_2 = 100$, and $c_2 = 4$
5. For the sampling plan of Problem 1, determine the AOQ curve and the AOQL.
6. For the sampling plan of Problem 2, determine the AOQ curve and the AOQL.
7. A major U.S. automotive manufacturer is using a sampling plan of $n = 200$ and $c = 0$ for all lot sizes. Construct the OC and AOQ curves. Graphically determine the AQL value for $\alpha = 0.05$ and the AOQL value.
8. A leading computer firm uses a sampling plan of $n = 50$ and $c = 0$ regardless of lot sizes. Construct the OC and AOQ curves. Graphically determine the AQL value for $\alpha = 0.05$ and the AOQL value.
9. Construct the ASN curves for the single sampling plan $n = 200$, $c = 5$ and the equally effective double sampling plan $n_1 = 125$, $c_1 = 2$, $r_1 = 5$, $n_2 = 125$, $c_2 = 6$, and $r_2 = 7$. Compare with Figure 9-17.

10. Construct the ASN curves for the single sampling plan $n = 80$, $c = 3$ and the equally effective double sampling plan $n_1 = 50$, $c_1 = 1$, $r_1 = 4$, $n_2 = 50$, $c_2 = 4$, and $r_2 = 5$. Compare with Figure 9-17.

11. Construct the ATI curve for $N = 500$, $n = 80$, and $c = 0$.

12. Construct the ATI curve for $N = 10{,}000$, $n = 315$, and $c = 5$.

13. Determine the AOQ curve and the AOQL for the single sampling plan $N = 16{,}000$, $n = 280$, and $c = 4$.

14. Using $c = 1$, $c = 5$, and $c = 8$, determine 3 sampling plans which ensure that product 0.8% nonconforming will be rejected 5.0% of the time.

15. For $c = 3$, $c = 6$, and $c = 12$, determine the sampling plans for AQL = 1.5% and $\alpha = 0.01$.

16. A bed-sheet supplier and a large motel system have decided to evaluate product in lots of 1000 using an AQL of 1.0% with a probability of nonacceptance of 0.10. Determine sampling plans for $c = 0$, 1, 2, and 4. How would you select the most appropriate plan?

17. For a consumer's risk of 0.10 and a LQ of 6.5%, determine the sampling plans for $c = 2$, 6, and 14.

18. If product that is 8.3% nonconforming is accepted 5% of the time, determine three sampling plans which meet this criteria. Use $c = 0$, 3, and 7.

19. A manufacturer of loudspeakers has decided that product 2% nonconforming will be accepted with a probability of 0.01. Determine single sampling plans for $c = 1$, 3, and 5.

20. Construct the OC and AOQ curves for the $c = 3$ plan of Problem 19.

21. A single sampling plan is desired with a consumer's risk of 0.10 of accepting 3.0% nonconforming product and a producer's risk of 0.05 of not accepting 0.7% nonconforming product. Select the plan with the lowest sample size.

22. The producer's risk is defined by $\alpha = 0.05$ for 1.5% nonconforming product, and the consumer's risk is defined by $\beta = 0.10$ for 4.6% nonconforming product. Select a sampling plan that exactly meets the producer's stipulation and comes as close as possible to the consumer's stipulation.

23. For the information of Problem 21, select the plan that exactly meets the consumer's stipulation and comes as close as possible to the producer's stipulation.

24. For the information of Problem 22, select the plan with the smallest sample size.

25. Given $p_{0.10} = 0.053$ and $p_{0.95} = 0.014$, determine the single sampling plan that exactly meets the consumer's stipulation and comes as close as possible to the producer's stipulation.

26. For the information of Problem 25, select the plan that meets the producer's stipulation and comes as close as possible to the consumer's stipulation.

27. If a single sampling plan is desired with an AOQL of 1.8% at an incoming quality of 2.6%, what is the common point on the OC curves for a family of sampling plans that meet the AOQL and $100p_0$ stipulation?

28. Using the software in the diskette solve:
    (a) Problem 1 and 5
    (b) Problem 2 and 6

29. Using the software in the diskette, copy the template to a new sheet and change the increment for the data points from 0.0025 to 0.002. Resolve Problems 28 a. and b. and compare results.

30. Using EXCEL write a program for:
    (a) OC curve for double sampling
    (b) AOQ curve for double sampling
    (c) ASN curve for single and double sampling
    (d) ATI curve for single and double sampling

10 ACCEPTANCE SAMPLING SYSTEMS

This chapter covers three different types of acceptance sampling plans: (1) lot-by-lot acceptance sampling for attributes, (2) continuous production acceptance sampling for attributes, and (3) acceptance sampling for variables. In this chapter it is helpful to distinguish among:

1. An individual sampling plan that states the lot size, sample size or sizes, and the acceptance criteria.
2. A sampling scheme that is a combination of sampling plans with switching rules and possibly a provision for discontinuance.
3. A sampling system that is a collection of sampling schemes.

LOT-BY-LOT ACCEPTANCE SAMPLING PLANS FOR ATTRIBUTES

ANSI/ASQ Z1.4—1993[1]

Introduction

An acceptance sampling plan for lot-by-lot inspection by attributes for use by the government was first devised in 1942 by a group of engineers at Bell Telephone Laboratories. It was designated JAN-STD-105. Since that time, there have been five revisions, with the last one being designated MIL-STD-105E. In 1973, it was adopted by the International Organization for Standardization and designated International Standard ISO/DIS-2859. While MIL-STD-105E was developed for government procurement, it has become the standard for attribute inspection for industry. It is the most widely used acceptance sampling plan in the world.

Modifications to MIL-STD-105E were made by the American Society for Quality (ASQ) under the designation ANSI/ASQ Z1.4. All tables and procedures remain unchanged. There are, however, three basic changes:

1. *Nonconformity* and *nonconforming unit* are substituted for the words *defect* and *defective*.
2. The switching rule that used a limit number for one of the reduced inspection criteria is an option.
3. Additional tables for AOQL, LQ, ASN, and OC curves are added. These tables reflect scheme performance, which is the combination of switching among normal, tightened, and reduced sampling plans.

The first two changes are included in the material that follows, but the third is not.

The standard is applicable, but not limited, to attribute inspection of the following: (1) end items, (2) components and raw materials, (3) operations, (4) materials in process, (5) supplies in storage, (6) maintenance operations, (7) data or records, and (8) administrative procedures. Sampling plans of this standard are intended to be used for a continuing series of lots, but plans may be designed for isolated lots by consulting the OC curve to determine the plan with the desired protection.

The standard provides for three types of sampling: single, double, and multiple. For each type of sampling plan, provision is made for normal, tightened, or reduced inspection. Tightened inspection is used when the producer's recent quality history has deteriorated. Acceptance requirements under tightened inspection are more stringent than under normal inspection. Reduced inspection is used when the producer's recent quality history has been exceptionally good. Figure 10-1

---

[1]This section is extracted from ANSI/ASQ Z1.4—1993 by permission of the American Society for Quality.

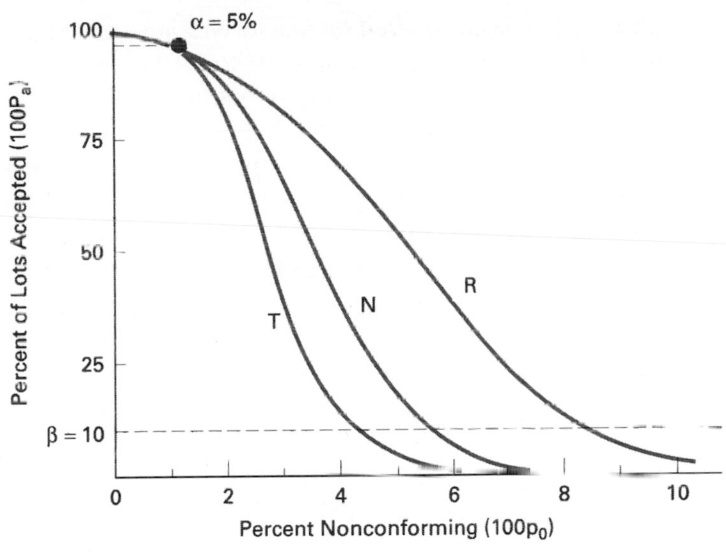

FIGURE 10-1 Comparison of normal (*N*), tightened (*T*), and reduced (*R*) inspection.

illustrates the differences among the OC curves for normal (*N*), tightened (*T*), and reduced (*R*) inspection.

The number inspected under reduced inspection is less than under normal inspection. The decision as to the type of plan to use (single, double, or multiple) is left to the responsible authority (consumer) but should be based on information given in the previous chapter. Normal inspection is used at the start of inspection with changes to tightened or reduced inspection being a function of recent quality performance.

Nonconformities and nonconforming units are classified by groups such as A, B, and C or critical, major, and minor.

Product is submitted in homogeneous lots with the manner of presentation and identification designated or approved by the responsible authority (consumer). Samples are selected at random without regard to their quality. Non-accepted lots are resubmitted after all nonconforming units are removed or nonconformities corrected. The responsible authority will determine whether reinspection should include all types or classes of nonconformities or the particular types or classes of nonconformities that caused initial non-acceptance.

Acceptable Quality Level. The Acceptable Quality Level (AQL) is the most important part of the standard because the AQL and the sample-size code letter index the sampling plan. AQL is defined as the maximum percent nonconforming (or the maximum number of nonconformities per 100 units) that, for purposes of sampling inspection, can be considered satisfactory as a process average. The phrase "can be considered satisfactory" is interpreted as a producer's risk, α, equal to 0.05; actually, α varies from 0.01 to 0.10 in the standard.

When the standard is used for percent nonconforming plans, the AQLs range from 0.010% to as high as 10%. For nonconformity-per-unit plans, AQLs are possible from 0.010 nonconformities per 100 units to 1000 nonconformities per 100 units. The AQLs are in a geometric progression, each being approximately 1.585 times the preceding one.

The AQL is designated in the contract or by the responsible authority. Different AQLs may be designated for groups of nonconformities considered collectively or for individual nonconformities. Groups of nonconformities or nonconforming units can have different AQLs, with lower values for critical ones and higher values for minor ones. AQLs are determined from (1) historical data; (2) empirical judgment; (3) engineering information, such as function, safety, interchangeable manufacturing, life testing, etc.; (4) experimentation by testing lots with various percent nonconforming or nonconformities per 100 units; (5) producer's capability; and, (6) in some situations, the consumer's requirements. AQL determination is a best-judgment decision. The standard helps to determine the AQL since only a finite number are available in the standard. It is a frequent practice to use AQL values of 0.10% or less for critical, 1.00% for major, and 2.5% for minor. The acceptance number for critical should be zero.

The AQL is a reference point on the OC curve. It does not imply that any percent nonconforming or nonconformities per 100 units is tolerable. The only way the producer can be guaranteed that a lot will be accepted is to have 0% nonconforming or to have the number of nonconforming units less than or equal to the sampling plan acceptance number.

Sample Size. The sample size is determined by the lot size and the inspection level. The inspection level to be used for a particular requirement will be prescribed by the responsible authority. Three general inspection levels (I, II, and III) are given in Table 10-1. The different levels of inspection provide approximately the same protection to the producer but different protections to the consumer. Inspection level II is the norm, with level I providing about one-half the amount of inspection and level III providing about twice the amount of inspection. Thus, level III gives a steeper OC curve and consequently more discrimination and increased inspection costs. Figure 10-2 illustrates the differences among the OC curves for inspection levels I, II, and III.

The decision on the inspection level is also a function of the type of product. For inexpensive items, for destructive testing, or for harmful testing, inspection level II should be considered. When subsequent production costs are high or when the items are complex and expensive, inspection level III may be applicable. The consumer should change the inspection level as conditions warrant.

TABLE 10-1 Sample-size Code Letters (Table I of ANSI/ASQ Z1.4—1993).

LOT OR BATCH SIZE	SPECIAL INSPECTION LEVELS S-1	S-2	S-3	S-4	GENERAL INSPECTION LEVELS I	II	III
2–8	A	A	A	A	A	A	B
9–15	A	A	A	A	A	B	C
16–25	A	A	B	B	B	C	D
26–50	A	B	B	C	C	D	E
51–90	B	B	C	C	C	E	F
91–150	B	B	C	D	D	F	G
151–280	B	C	D	E	E	G	H
281–500	B	C	D	E	F	H	J
501–1200	C	C	E	F	G	J	K
1201–3200	C	D	E	G	H	K	L
3201–10,000	C	D	F	G	J	L	M
10,001–35,000	C	D	F	H	K	M	N
35,001–150,000	D	E	G	J	L	N	P
150,001–500,000	D	E	G	J	M	P	Q
500,001 and over	D	E	H	K	N	Q	R

Source: Reprinted from ANSI/ASQ Z1.4—1993 Sampling Procedures and Tables for Inspection by Attributes by permission.

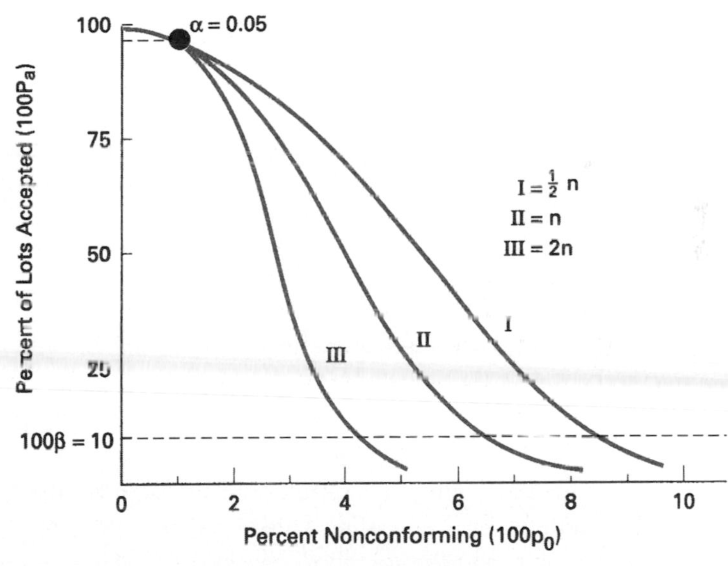

FIGURE 10-2 Comparison of inspection levels, I, II, and III.

Four additional special levels (S-1, S-2, S-3, and S-4) are given in Table 10-1 and may be used where relatively small sample sizes are necessary and large sampling risks can or must be tolerated.

Table 10-1 does not immediately provide the sample size based on the lot size and inspection level but does give a sample-size code letter. The AQL and the sample-size code letter index the desired sampling plan.

Implementation

The steps required to use the plan are as follows:

1. Determine the lot size (usually the responsibility of materials management).
2. Determine the inspection level (usually level II—it can be changed if conditions warrant).
3. Enter table and find sample-size code letter.
4. Determine the AQL.
5. Determine the type of sampling plan (single, double, or multiple).
6. Enter the appropriate table to find the sampling plan.
7. Start with normal inspection and change to tightened or reduced based on switching rules.

Example problems for single sampling plans are given in subsequent sections.

Single Sampling Plans

The single sampling plans of the standard are given in Tables 10-2, 10-3, and 10-4 for normal, tightened, and reduced inspection, respectively. In order to use the tables, the AQL, lot size, inspection level, and type of sampling plan are needed. An example problem will illustrate the technique.

EXAMPLE PROBLEM

For a lot size of 2000, an AQL of 0.65%, and an inspection level of III, determine the single sampling plans for normal, tightened, and reduced inspection.

Normal. Using the lot size $N = 2000$ and inspection level III, the sample size code letter L is obtained from Table 10-1. From Table 10-2 (Single Sampling Plans for Normal Inspection), the desired plan is obtained for code letter L and AQL 0.65%. It is $n = 200$, Ac = 3, and Re = 4. Thus, from a lot of 2000, a random sample of 200 is inspected. If 3 or fewer nonconforming units are found, the lot is accepted, if 4 or more nonconforming units are found, the lot is not accepted.

Tightened. The sample-size code letter, L, is the same as the one for normal inspection. From Table 10-3 (Single Sampling Plans for Tightened Inspection), the desired plan is obtained for code letter L and AQL 0.65%. It is $n = 200$, Ac = 2, and Re = 3. Thus, from a lot of 2000, a random sample of 200 is inspected. If 2 or less nonconforming units are found, the lot is accepted; if 3 or more nonconforming units are found, the lot is not accepted.

Reduced. The sample-size code letter, L, is the same as the one for normal inspection. From Table 10-4 (Single Sampling Plans for Reduced Inspection), the desired plan is obtained for code letter L and AQL 0.65%. It is $n = 80$, Ac = 1, and Re = 4. Thus, from a lot of 2000, a random sample of 80 is inspected. If 1 or fewer nonconforming units are found, the lot is accepted, if 4 or more nonconforming units are found, the lot is not accepted. If 2 or 3 nonconforming units are found, the lot is accepted, but the type of inspection changes from reduced to normal. A change to normal inspection is also required when a lot is not accepted.

In comparing the three plans, notice that the acceptance requirements are more stringent for tightened than for normal inspection. In fact, a sample with 3 nonconforming units is accepted under normal inspection but not accepted under tightened inspection. The sample size for reduced inspection is approximately 40% of the sample size of normal or tightened inspection, which represents a considerable saving in sampling costs

If a vertical arrow is encountered, the first sampling plan above or below the arrow is used. When this occurs, the sample-size code letter and the sample size change. For example, if a single sample tightened plan (Table 10-3) is indexed by an AQL of 4.0% and a code letter D, the code letter changes to F and the sample size changes from 8 to 20. If the vertical arrow points down, it means that the sample size is too small to make a decision; if the vertical arrow points up, it means that a decision can be made with a smaller sample size. In some cases, the sample size will exceed the lot size and, in those cases, 100% inspection is required.

Double and Multiple Sampling Plans

The standard provides for double and multiple sampling (7 samples) plans. Use of the tables is similar to the technique described for single sampling; therefore, it will not be described in this chapter.

Normal, Tightened, and Reduced Inspection

Unless otherwise directed by the responsible authority, inspection starts with the normal inspection condition. Normal, tightened, or reduced inspection will continue unchanged for each class of nonconformities or nonconforming units, or until the switching procedures given below require a change.

TABLE 10-2 Single Sampling Plans for Normal Inspection (Table II-A of ANSI/ASQ Z1.4—1993).

SAMPLE SIZE CODE LETTER	SAMPLE SIZE	ACCEPTABLE QUALITY LEVELS (NORMAL INSPECTION) 0.010 Ac Re	0.015 Ac Re	0.025 Ac Re	0.040 Ac Re	0.065 Ac Re	0.10 Ac Re	0.15 Ac Re	0.25 Ac Re	0.40 Ac Re	0.65 Ac Re	1.0 Ac Re	1.5 Ac Re	2.5 Ac Re	4.0 Ac Re	6.5 Ac Re	10 Ac Re	15 Ac Re	25 Ac Re	40 Ac Re	65 Ac Re	100 Ac Re	150 Ac Re	250 Ac Re	400 Ac Re	650 Ac Re	1000 Ac Re
A	2	↓	↓	↓	↓	↓	↓	↓	↓	↓	↓	↓	↓	↓	↓	0 1	↓	↓	1 2	2 3	3 4	5 6	7 8	10 11	14 15	21 22	30 31
B	3	↓	↓	↓	↓	↓	↓	↓	↓	↓	↓	↓	↓	↓	0 1	↑	↓	1 2	2 3	3 4	5 6	7 8	10 11	14 15	21 22	30 31	44 45
C	5	↓	↓	↓	↓	↓	↓	↓	↓	↓	↓	↓	↓	0 1	↑	↓	1 2	2 3	3 4	5 6	7 8	10 11	14 15	21 22	30 31	44 45	↑
D	8	↓	↓	↓	↓	↓	↓	↓	↓	↓	↓	↓	0 1	↑	↓	1 2	2 3	3 4	5 6	7 8	10 11	14 15	21 22	30 31	44 45	↑	↑
E	13	↓	↓	↓	↓	↓	↓	↓	↓	↓	↓	0 1	↑	↓	1 2	2 3	3 4	5 6	7 8	10 11	14 15	21 22	30 31	44 45	↑	↑	↑
F	20	↓	↓	↓	↓	↓	↓	↓	↓	↓	0 1	↑	↓	1 2	2 3	3 4	5 6	7 8	10 11	14 15	21 22	↑	↑	↑	↑	↑	↑
G	32	↓	↓	↓	↓	↓	↓	↓	↓	0 1	↑	↓	1 2	2 3	3 4	5 6	7 8	10 11	14 15	21 22	↑	↑	↑	↑	↑	↑	↑
H	50	↓	↓	↓	↓	↓	↓	↓	0 1	↑	↓	1 2	2 3	3 4	5 6	7 8	10 11	14 15	21 22	↑	↑	↑	↑	↑	↑	↑	↑
J	80	↓	↓	↓	↓	↓	↓	0 1	↑	↓	1 2	2 3	3 4	5 6	7 8	10 11	14 15	21 22	↑	↑	↑	↑	↑	↑	↑	↑	↑
K	125	↓	↓	↓	↓	↓	0 1	↑	↓	1 2	2 3	3 4	5 6	7 8	10 11	14 15	21 22	↑	↑	↑	↑	↑	↑	↑	↑	↑	↑
L	200	↓	↓	↓	↓	0 1	↑	↓	1 2	2 3	3 4	5 6	7 8	10 11	14 15	21 22	↑	↑	↑	↑	↑	↑	↑	↑	↑	↑	↑
M	315	↓	↓	↓	0 1	↑	↓	1 2	2 3	3 4	5 6	7 8	10 11	14 15	21 22	↑	↑	↑	↑	↑	↑	↑	↑	↑	↑	↑	↑
N	500	↓	↓	0 1	↑	↓	1 2	2 3	3 4	5 6	7 8	10 11	14 15	21 22	↑	↑	↑	↑	↑	↑	↑	↑	↑	↑	↑	↑	↑
P	800	↓	0 1	↑	↓	1 2	2 3	3 4	5 6	7 8	10 11	14 15	21 22	↑	↑	↑	↑	↑	↑	↑	↑	↑	↑	↑	↑	↑	↑
Q	1250	0 1	↑	↓	1 2	2 3	3 4	5 6	7 8	10 11	14 15	21 22	↑	↑	↑	↑	↑	↑	↑	↑	↑	↑	↑	↑	↑	↑	↑
R	2000	↑	↑	1 2	2 3	3 4	5 6	7 8	10 11	14 15	21 22	↑	↑	↑	↑	↑	↑	↑	↑	↑	↑	↑	↑	↑	↑	↑	↑

1 = Use first sampling plan below arrow. If sample size equals or exceeds, lot or batch size, do 100 percent inspection.

1 = Use first sampling plan above arrow.

Source: Reprinted from ANSI/ASQ standard Q3—1988 by permission.

Ac = Acceptance number.

Rc = Non-acceptance

TABLE 10-3 Single Sampling Plans for Tightened Inspection (Table II-B of ANSI/ASQ Z1.4—1993).

SAMPLE SIZE CODE LETTER	SAMPLE SIZE	ACCEPTABLE QUALITY LEVELS (TIGHTENED INSPECTION)																									
		0.010	0.015	0.025	0.040	0.065	0.10	0.15	0.25	0.40	0.65	1.0	1.5	2.5	4.0	6.5	10	15	25	40	65	100	150	250	400	650	1000
		Ac Re	Ac Re	Ac Re	Ac Re	Ac Re	Ac Re	Ac Re	Ac Re	Ac Re	Ac Re	Ac Re	Ac Re	Ac Re	Ac Re	Ac Re	Ac Re	Ac Re	Ac Re	Ac Re	Ac Re	Ac Re	Ac Re	Ac Re	Ac Re	Ac Re	Ac Re
A	2															↓			↓	1 2	2 3	3 4	5 6	8 9	12 13	18 19	27 28
B	3														↓	0 1		↓	1 2	2 3	3 4	5 6	8 9	12 13	18 19	27 28	41 42
C	5													↓	0 1		↓	1 2	2 3	3 4	5 6	8 9	12 13	18 19	27 28	41 42	↑
D	8												↓	0 1		↓	1 2	2 3	3 4	5 6	8 9	12 13	18 19	27 28	41 42	↑	
E	13											↓	0 1		↓	1 2	2 3	3 4	5 6	8 9	12 13	18 19	27 28	41 42	↑		
F	20										↓	0 1		↓	1 2	2 3	3 4	5 6	8 9	12 13	18 19	↑					
G	32									↓	0 1		↓	1 2	2 3	3 4	5 6	8 9	12 13	18 19	↑						
H	50								↓	0 1		↓	1 2	2 3	3 4	5 6	8 9	12 13	18 19	↑							
J	80							↓	0 1		↓	1 2	2 3	3 4	5 6	8 9	12 13	18 19	↑								
K	125						↓	0 1		↓	1 2	2 3	3 4	5 6	8 9	12 13	18 19	↑									
L	200					↓	0 1		↓	1 2	2 3	3 4	5 6	8 9	12 13	18 19	↑										
M	315				↓	0 1		↓	1 2	2 3	3 4	5 6	8 9	12 13	18 19	↑											
N	500			↓	0 1		↓	1 2	2 3	3 4	5 6	8 9	12 13	18 19	↑												
P	800		↓	0 1	↓	↓	1 2	2 3	3 4	5 6	8 9	12 13	18 19	↑													
Q	1250	↓	0 1			1 2	2 3	3 4	5 6	8 9	12 13	18 19	↑														
R	2000	0 1	↑	↓	1 2	2 3	3 4	5 6	8 9	12 13	18 19	↑															
S	3150			1 2																							

1 = Use first sampling plan below arrow. If sample size equals or exceeds, lot or batch size, do 100 percent inspection.

1 = Use first sampling plan above arrow.

Ac = Acceptance number.

Rc = Non-acceptance number.

Source. Reprinted from ANSI/ASQ standard Q3—1988 by permission.

TABLE 10-4 Single Sampling Plans for Reduced Inspection (Table II-C of ANSI/ASQ Z1.4—1993).

SAMPLE SIZE CODE LETTER	SAMPLE SIZE	ACCEPTABLE QUALITY LEVELS (NORMAL INSPECTION)																									
		0.010	0.015	0.025	0.040	0.065	0.10	0.15	0.25	0.40	0.65	1.0	1.5	2.5	4.0	6.5	10	15	25	40	65	100	150	250	400	650	1000
		Ac Re	Ac Re	Ac Re	Ac Re	Ac Re	Ac Re	Ac Re	Ac Re	Ac Re	Ac Re	Ac Re	Ac Re	Ac Re	Ac Re	Ac Re	Ac Re	Ac Re	Ac Re	Ac Re	Ac Re	Ac Re	Ac Re	Ac Re	Ac Re	Ac Re	Ac Re
A	2														↓	0 1		↓	1 2	2 3	3 4	5 6	7 8	10 11	14 15	21 22	30 31
B	2													↓	0 1	↑	↓	0 2	1 3	2 4	3 5	5 6	7 8	10 11	14 15	21 22	30 31
C	2												↓	0 1	↑	↓	0 2	1 3	1 4	2 5	3 6	5 8	7 10	10 13	14 17	21 24	↑
D	3											↓	0 1	↑	↓	0 2	1 3	1 4	2 5	3 6	5 8	7 10	10 13	14 17	21 24	↑	
E	5										↓	0 1	↑	↓	0 2	1 3	1 4	2 5	3 6	5 8	7 10	10 13	14 17	21 24	↑		
F	8									↓	0 1	↑	↓	0 2	1 3	1 4	2 5	3 6	5 8	7 10	10 13	↑	↑	↑			
G	13								↓	0 1	↑	↓	0 2	1 3	1 4	2 5	3 6	5 8	7 10	10 13	↑						
H	20							↓	0 1	↑	↓	0 2	1 3	1 4	2 5	3 6	5 8	7 10	10 13	↑							
J	32						↓	0 1	↑	↓	0 2	1 3	1 4	2 5	3 6	5 8	7 10	10 13	↑								
K	50					↓	0 1	↑	↓	0 2	1 3	1 4	2 5	3 6	5 8	7 10	10 13	↑									
L	80				↓	0 1	↑	↓	0 2	1 3	1 4	2 5	3 6	5 8	7 10	10 13	↑										
M	125			↓	0 1	↑	↓	0 2	1 3	1 4	2 5	3 6	5 8	7 10	10 13	↑											
N	200		↓	0 1	↑	↓	0 2	1 3	1 4	2 5	3 6	5 8	7 10	10 13	↑												
P	315	↓	0 1	↑	↓	0 2	1 3	1 4	2 5	3 6	5 8	7 10	10 13	↑													
Q	500	0 1	↑	↓	0 2	1 3	1 4	2 5	3 6	5 8	7 10	10 13	↑														
R	800	↑		0 2	1 3	1 4	2 5	3 6	5 8	7 10	10 13	↑															

1 = Use first sampling plan below arrow. If sample size equals or exceeds, lot or batch size, do 100 percent inspection.

1 = Use first sampling plan above arrow.

Ac = Acceptance number.

Rc = Non-acceptance number.

† If the acceptance number has been exceeded, but the rejection number has not been reached, accept the lot, but reinstate normal inspection.

Source: Reprinted from ANSI/ASQ standard Q3—1988 by permission.

Normal to tightened. When normal inspection is in effect, tightened inspection shall be instituted when 2 out of 5 consecutive lots or batches have not been accepted on original inspection (i.e., ignoring resubmitted lots).

Tightened to normal. When tightened inspection is in effect, normal inspection shall be instituted when 5 consecutive lots or batches are accepted on original inspection.

Normal to reduced. When normal inspection is in effect, reduced inspection shall be instituted provided all of the following conditions are satisfied.

1. The preceding 10 lots or batches have been on normal inspection and all of the lots have been accepted on original inspection.
2. The total number of nonconforming units (nonconformities) in the samples from the preceding 10 lots or batches is equal to or less than the applicable number given in Table 10-5. For example, if the total number inspected for the past 10 lots or batches is 600 and the AQL is 2.5%, the limit number is 7. Therefore, to qualify for reduced inspection, the number nonconforming in the 600 inspected must be equal to or less than 7. In some cases, more than 10 lots or batches are necessary to obtain a sufficient number of sample units for a particular AQL, as indicated by the note of Table 10-5. This condition is optional.
3. Production is at a steady rate. In other words, no difficulties, such as machine breakdowns, material shortages, or labor problems, have occurred recently.
4. Reduced inspection is considered desirable by the responsible authority (consumer). The consumer must decide if the savings from fewer inspections warrant the additional record-keeping and inspector training expenses.

Reduced to normal. When reduced inspection is in effect, normal inspection shall be instituted provided any of the four conditions below are satisfied on original inspection.

1. A lot or batch is not accepted.
2. When the sampling procedure terminates with neither acceptance nor rejection criteria having been met, the lot or batch is accepted, but normal inspection is reinstated starting with the next lot.
3. Production is irregular or delayed.
4. Other conditions, such as customer desire, warrant that normal inspection will be instituted.

If 5 consecutive lots or batches remain on tightened inspection, then inspection should be discontinued pending action to improve the quality of submitted material.

TABLE 10-5 Limit Numbers for Reduced Inspection (Table VIII of ANSI/ASQ Z1.4—1993)

NUMBER OF SAMPLE UNITS FROM LAST 10 LOTS OF BATCHES	ACCEPTABLE QUALITY LEVEL																									
	0.010	0.015	0.025	0.040	0.065	0.10	0.15	0.25	0.40	0.65	1.0	1.5	2.5	4.0	6.5	10	15	25	40	65	100	150	250	400	650	1000
20-29	*	*	*	*	*	*	*	*	*	*	*	*	*	*	*	0	0	2	4	8	14	22	40	68	115	181
30-49	*	*	*	*	*	*	*	*	*	*	*	*	*	*	0	0	1	3	7	13	22	36	63	105	178	277
50-79	*	*	*	*	*	*	*	*	*	*	*	*	*	0	0	2	3	7	14	25	40	63	110	181	301	
80-129	*	*	*	*	*	*	*	*	*	*	*	*	0	0	2	4	7	14	24	42	68	105	181	297		
130-199	*	*	*	*	*	*	*	*	*	*	*	0	0	2	4	7	13	25	42	72	115	177	301	490		
200-319	*	*	*	*	*	*	*	*	*	*	0	0	2	4	8	14	22	40	68	115	181	277	471			
320-499	*	*	*	*	*	*	*	*	*	0	0	1	4	8	14	24	39	68	113	189						
500-799	*	*	*	*	*	*	*	*	0	0	2	3	7	14	25	40	63	110	181							
800-1249	*	*	*	*	*	*	*	0	0	2	4	7	14	24	42	68	105	181								
1250-1999	*	*	*	*	*	*	0	0	2	4	7	13	24	40	69	110	169									
2000-3149	*	*	*	*	*	0	0	2	4	8	14	22	40	68	115	181										
3150-4999	*	*	*	*	0	0	1	4	8	14	24	38	67	111	186											
5000-7999	*	*	*	0	0	2	3	7	14	25	40	63	110	181												
8000-12499	*	*	0	0	2	4	7	14	24	42	68	105	181													
12500-19999	*	0	0	2	4	7	13	24	40	69	110	169														
20000-31499	0	0	2	4	8	14	22	40	68	115	181															
31500-49999	0	1	4	8	14	24	38	67	111	186																
50000 & Over	2	3	7	14	25	40	63	110	181	301																

* Denotes that the number of sample units from the last ten lots or batches is not sufficient for reduced inspection for this AQL. In this instance more than ten lots or batches may be used for the calculation, provide that the lots or batches used are the most recent ones. in sequence, that they have all been on normal inspection, and that none has been rejected while on original inspection.

Source: Reprinted from ANSI/ASQ standard Q3—1988 by permission.

Supplementary Information

The standard includes operating characteristic curves for single sampling plans with normal inspection that indicate the percentage of lots or batches that may be expected to be accepted under the various sampling plans for a given process quality. OC curves for double and multiple sampling plans are not given in the standard but are matched as closely as practical.

Table V (not reproduced in this text) of the standard gives the average outgoing quality limit for single sampling plans with normal and tightened inspection.

Average sample-size curves for double and multiple sampling as a function of the equivalent single sample size are shown in Table IX and are partially reproduced in Figure 9-17. These show the average sample sizes that may be expected to occur under the various sampling plans for a given process quality.

ANSI/ASQ Z1.4—1993 is designed for use where the units of product are produced in a continuing series of lots or batches. However, if a sampling plan is desirable for a lot or batch of an isolated nature, it should be chosen based on the Limiting Quality (LQ) and consumer's risk. Tables (not reproduced in this text) for consumers' risks of 0.05 and 0.10 are included in the standard. Therefore, a sampling plan for isolated lots can be obtained that will come close to both the producers' and consumers' criteria. However, it is much easier to use ANSI/ASQ standard Q3—1988, which follows.

ANSI/ASQ Standard Q3—1988[2]

This standard is to be used for inspection of isolated lots by attributes. It complements ANSI/ASQ Z1.4—1993, which is appropriate for a continuous stream of lots. This standard indexes tables by Limiting Quality (LQ) values and is applicable to type A or type B lots or batches. These concepts were discussed in the last chapter. The LQ values are determined by the same techniques used to determine AQL values

There are two schemes. One scheme, given in Table 10-6, is designed to be used for lots that are isolated or mixed or that have an unknown history as far as both vendor and vendee know. To use the table the lot size and LQ value must be known.

EXAMPLE PROBLEM

Given a lot size of 295 and an LQ value of 3.15%, determine the sampling plan.

From Table 10-6 the solution is

$$n = 80$$

$$Ac = 0$$

[2]This section is extracted from ANSI/ASQ standard Q3—1988 by permission of the American Society for Quality Control.

TABLE 10-6 Single Sampling Plans Indexed by Nominal Limiting Quality (LQ).

LOT SIZE		NOMINAL LIMITING QUALITY IN PERCENT (LQ)									
		0.5	**0.8**	**1.25**	**2.0**	**3.15**	**5.0**	**8.0**	**12.5**	**20**	**32**
16 to 25	*n*	→	→	→	→	→	100%	17[a]	13	9	6
	Ac					0	0	0	0	0	0
25 to 50	*n*	→	→	→	100%	100%	28[a]	22	15	10	6
	Ac				0	0	0	0	0	0	0
51 to 90	*n*	→	→	100%	50	44	34	24	16	10	8
	Ac			0	0	0	0	0	0	0	0
91 to 150	*n*	→	100%	90	80	55	38	26	18	13	13
	Ac		0	0	0	0	0	0	0	0	1
151 to 280	*n*	100%	170	130	95	65	42	28	20	20	13
	Ac	0	0	0	0	0	0	0	0	1	1
281 to 500	*n*	280	220	155	105	80	50	32	32	20	20
	Ac	0	0	0	0	0	0	0	1	1	3
501 to 1,200	*n*	380	255	170	125	125	80	50	32	32	32
	Ac	0	0	0	0	1	1	1	1	3	5
1,201 to 3,200	*n*	430	280	200	200	125	125	80	50	50	50
	Ac	0	0	0	1	1	3	3	3	5	10
3,201 to 10,000	*n*	450	315	315	200	200	200	125	80	80	80
	Ac	0	0	1	1	3	5	5	5	10	18
10,001 to 35,000	*n*	500	500	315	315	315	315	200	125	125	80
	Ac	0	1	1	3	5	10	10	10	18	18
35,001 to 150,000	*n*	800	500	500	500	500	500	315	200	125	80
	Ac	1	1	3	5	10	18	18	18	18	18
150,001 to 500,000	*n*	800	800	800	800	800	500	315	200	125	80
	Ac	1	3	5	10	18	18	18	18	18	18
> 500,000	*n*	1250	1250	1250	1250	800	500	315	200	125	80
	Ac	3	5	10	18	18	18	18	18	18	18

[a]When *n* exceeds the lot size, use 100% inspection with zero acceptance number.

→ Nominal LQ implies less than one nonconforming item in the lot. Use first available plan for higher LQ.

The nominal values of the LQ are based on $\beta = 0.10$. Because we are working with whole numbers, the actual LQ values will vary slightly from the nominal. Note that the LQ is given as a percent.

The second scheme is used when a vendor is producing a continuous stream of lots and sends one or a few to a customer who will consider them as isolated lots. This situation would frequently occur in the purchase of small quantities of a raw material. Tables are given for LQ values of 0.5, 0.8, 1.25, 2.0, 3.15, 5.0, 8.0, 12.5, 20.0, and 32.0%. Only the table for 3.15% is reproduced, as Table 10-7. The tables list the process quality in terms of AQL (as used in ANSI/ASQ Z1.4—1993), which are equivalent to the LQ for different lot sizes.

EXAMPLE PROBLEM

Given a lot size of 295, Inspection Level II, and an LQ value of 3.15%, where the isolated lot is from a vendor with a continuous stream of product, determine the sampling plan.

From Table 10-7 the solution is

$$n = 125$$

$$Ac = 1$$

Note that the information in the last 5 columns can be used to plot the OC curves. Also note that the information in Table 10-7 comes from ANSI/ASQ Z1.4. The only difference is the indexing by LQ values in order to make it easier to use.

Dodge-Romig Tables

In the 1920s H. F. Dodge and H. G. Romig developed a set of inspection tables for the lot-by-lot acceptance of product by sampling for attributes. These tables are based on two of the concepts discussed in Chapter 9, Limiting Quality (LQ)[3] and Average Outgoing Quality Limit (AOQL). For each of these concepts there are tables for single and double sampling. No provision is made for multiple sampling. Only single sampling is included in this text.

The principal advantage of the Dodge-Romig tables is a minimum amount of inspection for a given inspection procedure. This advantage makes the tables desirable for in-house inspection.

1. *Limiting Quality* (*LQ*). These tables are based on the probability that a particular lot, which has a percent nonconforming equal to the LQ, will be accepted.

[3]Dodge and Romig used the term *Lot Tolerance Percent Defective* (LTPD). In this text, *Limiting Quality* (LQ) has been substituted since it is the appropriate present-day term.

TABLE 10-7 Single Sampling Plans for Nominal Limiting Quality 3.15%.

LOT SIZES FOR INSPECTION LEVELS					SINGLE SAMPLING PLAN (NORMAL INSPECTION)			CODE LETTER	TABULATED VALUES OF SUBMITTED QUALITY ACCEPTED WITH DESIGNATED PROBABILITIES[a] (QUALITY AS PERCENT NONCONFORMING)				
S-1 TO S-3	S-4	I	II	III	AQL	*n*	AC		95%	90%	50%	10%	5%
>125[b]	>125[b]	126[b] to 35,000	126[b] to 3,200	126[b] to 1,200	0.40	125	1	K	0.284	0.426	1.34	3.11	3.80
		35,001 to 150,000	3,201 to 10,000	1,201 to 3,200	0.65	200	3	L	0.663	0.873	1.84	3.34	3.88
		>150,000	10,001 to 35,000	3,201 to 10,000	0.65	315	5	M	0.829	1.00	1.80	2.94	3.34
			>35,000	>10,000	1.00	500	10	N	1.231	1.40	2.13	3.08	3.39

[a]Probability calculated by the Poisson approximation.

[b]For fewer than 126 in the lot, inspect the lot 100%.

This probability is the consumer's risk, β, and is equal to 0.10. LQ plans give assurance that individual lots of poor quality will rarely be accepted.

There are two sets of LQ tables: one set for single sampling and one set for double sampling. Each set has tables for LQ values of 0.5, 1.0, 2.0, 3.0, 4.0, 5.0, 7.0, and 10.0%, making a total of 16 tables. For explanatory purposes, Table 10-8 is shown for single sampling, using LQ = 1.0%. Tables for other LQ values are not given.

To use the tables, an initial decision concerning single sampling or double sampling is required. This decision can be based on the information presented in Chapter 9. In addition, the LQ needs to be determined, which can be accomplished in a manner similar to that used for the AQL as described in Chapter 9. The type of sampling (single or double) and the LQ indicate the table to use.

Knowing the lot size and the process average, the acceptance sampling plan is easily obtained. For example, if the lot size, N, is 1500 and the process average is 0.25%, the required single sampling plan for LQ = 1.0% is found in Table 10-8. The answer is

$$N = 1500$$

$$n = 490$$

$$c = 2$$

The table also gives the AOQL for each plan, which for this example is 0.21%.

An analysis of LQ tables shows the following:

(a) As the lot size increases, the relative sample size decreases. Thus, for a process average of 0.25%, a lot size of 1000 has a sample size of 335, whereas a lot size of 4000 has a sample size of 645. The lot size increased by a factor of 4 while the sample size increased by a factor of about 2. Therefore, inspection costs are more economical with large lot sizes.

(b) The tables extend until the process average is one-half of the LQ. Provision for additional process averages is unnecessary, since 100% inspection becomes more economical than sampling inspection when the process average exceeds one-half of the LQ.

(c) As the process average increases, a corresponding increase occurs in the amount inspected. Therefore, an improvement in the process average results in fewer inspections and a lower sampling inspection cost.

2. *Average Outgoing Quality Limit* (AOQL). Sampling plans for the AOQL concept were developed as a practical need in certain manufacturing situations. When the lot quantity is specified, as is the case with customer lots (homogeneous), the LQ concept is applicable; however, when the inspected lot is a convenient subdivision of a flow of product for materials-handling purposes (nonhomogeneous), the AOQL concept is applicable. AOQL plans limit the amount of poor outgoing quality on an average basis but give no assurance on individual lots. Tables for the

TABLE 10-8 Dodge-Romig Single Sampling Lot Inspection Table, Based on Limiting Quality[a] LQ = 1.0%

	PROCESS AVERAGE (%)																	
	0–0.010			0.011–0.10			0.11–0.20			0.21–0.30			0.31–0.40			0.41–0.50		
LOT SIZE	n	c	AOQL (%)	n	c	AOQL (%)	n	c	AOQL (%)	n	c	AOQL (%)	n	c	AOQL (%)	n	c	AOQL (%)
1–120	All	0	0	All	0	0	All	0	0	All	0	0	All	0	0	All	0	0
121–150	120	0	0.06	120	0	0.06	120	0	0.06	120	0	0.06	120	0	0.06	120	0	0.06
151–200	140	0	0.08	140	0	0.08	140	0	0.08	140	0	0.08	140	0	0.08	140	0	0.08
201–300	165	0	0.10	165	0	0.10	165	0	0.10	165	0	0.10	165	0	0.10	165	0	0.10
301–400	175	0	0.12	175	0	0.12	175	0	0.12	175	0	0.12	175	0	0.12	175	0	0.12
401–500	180	0	0.13	180	0	0.13	180	0	0.13	180	0	0.13	180	0	0.13	180	0	0.13
501–600	190	0	0.13	190	0	0.13	190	0	0.13	190	0	0.13	190	0	0.13	305	1	0.14
601–800	200	0	0.14	200	0	0.14	200	0	0.14	330	1	0.15	330	1	0.15	330	1	0.15
801–1,000	205	0	0.14	205	0	0.14	205	0	0.14	335	1	0.17	335	1	0.17	335	1	0.17
1,001–2,000	220	0	0.15	220	0	0.15	360	1	0.19	490	2	0.21	490	2	0.21	610	3	0.22
2,001–3,000	220	0	0.15	375	1	0.20	505	2	0.23	630	3	0.24	745	4	0.26	870	5	0.26
3,001–4,000	225	0	0.15	380	1	0.20	510	2	0.24	645	3	0.25	880	5	0.28	1,000	6	0.29
4,001–5,000	225	0	0.16	380	1	0.20	520	2	0.24	770	4	0.28	895	5	0.29	1,120	7	0.31
5,001–7,000	230	0	0.16	385	1	0.21	655	3	0.27	780	4	0.29	1,020	6	0.32	1,260	8	0.34
7,001–10,000	230	0	0.16	520	2	0.25	660	3	0.28	910	5	0.32	1,150	7	0.34	1,500	10	0.37
10,001–20,000	390	1	0.21	525	2	0.26	785	4	0.31	1,040	6	0.35	1,400	9	0.39	1,980	14	0.43
20,001–50,000	390	1	0.21	530	2	0.26	920	5	0.34	1,300	8	0.39	1,890	13	0.44	2,570	19	0.48
50,001–100,000	390	1	0.21	670	3	0.29	1,040	6	0.36	1,420	9	0.41	2,120	15	0.47	3,150	23	0.50

[a] *n*, size of sample (entry of "All" indicates that each piece in lot is to be inspected); *c*, acceptance number; AOQL, Average Outgoing Quality Limit.

AOQL have one set for single sampling and one set for double sampling. Each set has tables for AOQL values of 0.1, 0.25, 0.5, 0.75, 1.0, 1.5, 2.0, 2.5, 3.0, 4.0, 5.0, 7.0, and 10.0%, making a total of 26 tables. For explanatory purposes a table for single sampling is shown in Table 10-9, using an AOQL = 3.0%. Tables for other AOQL values are not given.

In addition to determining whether single or double sampling is to be used, the AOQL is required. This can be accomplished using the same techniques as those used for finding the AQL, which was described in Chapter 9. The type of sampling (single or double) and the AOQL indicate the table to use.

Knowing the lot size and the process average, the acceptance sampling plan can be obtained. For example, if the lot size, N, is 1500 and the process average is 1.60%, then the required single sampling plan for an AOQL = 3.0% is found in Table 10-9. The answer is

$$N = 1500$$

$$n_1 = 65$$

$$c_1 = 3$$

The corresponding LQ for this plan is 10.2%.

An analysis of the AOQL tables shows the following:

(a) As the lot size increases, the relative sample size decreases.

(b) Plans are not given for process averages which exceed the AOQL, since sampling is uneconomical when the average incoming quality is poorer than the specified AOQL.

(c) The lower the process average, the smaller the sample size, resulting in lower inspection cost.

3. *Additional comments on Dodge-Romig tables*. The process average, $100p$, is obtained by the same techniques used for the p chart. Using the first 25 lots, the average percent nonconforming is obtained. For double sampling, only the first sample is included in the computation. Any lot percent nonconforming, which exceeds the limit of $100\bar{p} + 3\sqrt{100\bar{p}(1 - 100\bar{p})/n}$, is discarded (if it has an assignable cause) and a new process average calculated. However, until it is possible to obtain a process average by the technique above, the largest possible process average should be used. Thus, the last column in the tables is used until $100\bar{p}$, can be determined.

The Dodge—Romig tables do not make provision for the type of nonconformity, although different LQ or AOQL values can be used—lower ones for critical nonconformities and higher ones for minor nonconformities. No provision is made for tightened or reduced inspection, although different LQ or AOQL values can also be used. Nonconformities/100 units rather than percent nonconforming can be used for the process average. Thus, a process average of 2.00% nonconforming is the same as 2 nonconformities/100 units.

TABLE 10-9 **Dodge-Romig Single Sampling Lot Inspection Table, Based on Average Outgoing Quality Limit[a] AOQL = 3.0%.**

	PROCESS AVERAGE (%)																	
	0–0.06			0.07–0.60			0.61–1.20			1.21–1.80			1.81–2.40			2.41–3.00		
LOT SIZE	*n*	*c*	LQ (%)	*n*	*c*	LQ (%)	*n*	*c*	LQ (%)	*n*	*c*	LQ (%)	*n*	*c*	LQ (%)	*n*	*c*	LQ (%)
1–10	All	0	—	All	0	—	All	0	—	All	0	—	All	0	—	All	0	—
11–50	10	0	19.0	10	0	19.0	10	0	19.0	10	0	19.0	10	0	19.0	10	0	19.0
51–100	11	0	18.0	11	0	18.0	11	0	18.0	11	0	18.0	11	0	18.0	22	1	16.4
101–200	12	0	17.0	12	0	17.0	12	0	17.0	25	1	15.1	25	1	15.1	25	1	15.1
201–300	12	0	17.0	12	0	17.0	26	1	14.6	26	1	14.6	26	1	14.6	40	2	12.8
301–400	12	0	17.1	12	0	17.1	26	1	14.7	26	1	14.7	41	2	12.7	41	2	12.7
401–500	12	0	17.2	27	1	14.1	27	1	14.1	42	2	12.4	42	2	12.4	42	2	12.4
501–600	12	0	17.3	27	1	14.2	27	1	14.2	42	2	12.4	42	2	12.4	60	3	10.8
601–800	12	0	17.3	27	1	14.2	27	1	14.2	43	2	12.1	60	3	10.9	60	3	10.9
801–1,000	12	0	17.4	27	1	14.2	44	2	11.8	44	2	11.8	60	3	11.0	80	4	9.8
1,001–2,000	12	0	17.5	28	1	13.8	45	2	11.7	65	3	10.2	80	4	9.8	100	5	9.1
2,001–3,000	12	0	17.5	28	1	13.8	45	2	11.7	65	3	10.2	100	5	9.1	140	7	8.2
3,001–4,000	12	0	17.5	28	1	13.8	65	3	10.3	85	4	9.5	125	6	8.4	165	8	7.8
4,001–5,000	28	1	13.8	28	1	13.8	65	3	10.3	85	4	9.5	125	6	8.4	210	10	7.4
5,001–7,000	28	1	13.8	45	2	11.8	65	3	10.3	105	5	8.8	145	7	8.1	235	11	7.1
7,001–10,000	28	1	13.9	46	2	11.6	65	3	10.3	105	5	8.8	170	8	7.6	280	13	6.8
10,001–20,000	28	1	13.9	46	2	11.7	85	4	9.5	125	6	8.4	215	10	7.2	380	17	6.2
20,001–50,000	28	1	13.9	65	3	10.3	105	5	8.8	170	8	7.6	310	14	6.5	560	24	5.7
50,001–100,000	28	1	13.9	65	3	10.3	125	6	8.4	215	10	7.2	385	17	6.2	690	29	5.4

[a] *n*, size of sample (entry of "All" indicates that each piece in lot is to be inspected); *c*, acceptance number for sample; LQ, Limiting Quality corresponding to a consumer's risk (β) = 0.10.

Chain Sampling Inspection Plan[4]

A special type of lot-by-lot acceptance sampling plan for attributes was developed by H. F. Dodge. The plan was designated "Chain Sampling Plan ChSP-1." It is applicable to quality characteristics which involve destructive or costly tests.

When tests are destructive or costly, sampling plans with a small sample size are used as a matter of practical necessity. Plans with sample sizes of 5, 10, 15, etc., usually have acceptance numbers of zero ($c = 0$).

Single sampling plans for $c = 0$ have an undesirable feature, which is the poor shape of the OC curve at the producer's risk, α. Figure 10-3 shows the general shape of single sampling plans for $c = 0$ and $c = 1$ or more. The comparison shows the desirability (from a producer's viewpoint) of plans with acceptance numbers equal to 1 or more.

Chain sampling plans make use of the cumulative results of several preceding samples. The procedure is shown in Figure 10-4 and is as follows:

1. For each lot, select a sample of size n and test each for conformance to specifications.
2. If the sample has 0 nonconforming units, accept the lot; if the sample has 2 or more nonconforming units, do not accept the lot; and if the sample has 1 nonconforming unit, it may be accepted provided that there are 0 nonconforming units in the previous i samples of size n.

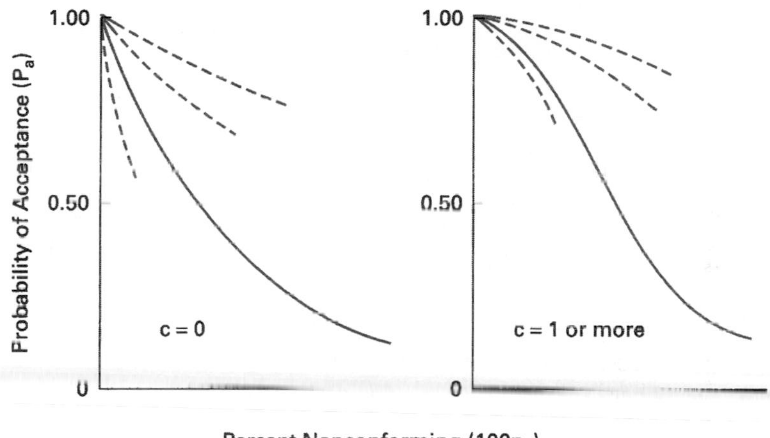

FIGURE 10-3 General shape of OC curves for single sampling plans. [Reproduced by permission from H. F. Dodge, "Chain Sampling Inspection Plan," *Industrial Quality Control*, 11, No. 4 (January 1955), 10–13.]

[4]For more information, see H. F. Dodge, "Chain Sampling Inspection Plan," *Industrial Quality Control*, 11, No. 4 (January 1955), 10–13.

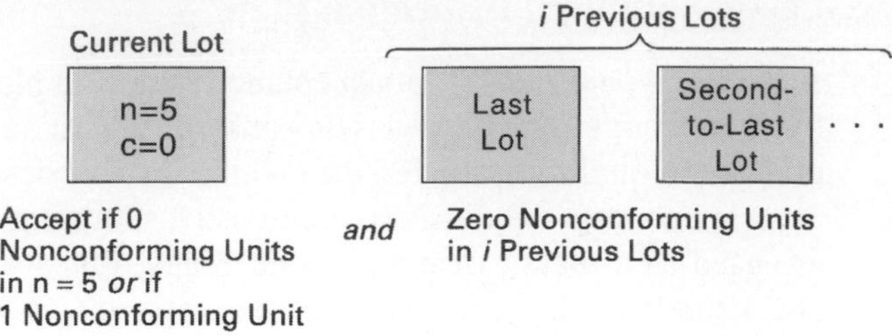

FIGURE 10-4 Chain sampling diagram.

Thus, for a chain sampling plan given by $n = 5$, $i = 3$, the lot would be accepted (1) by 0 nonconforming units in the sample of 5, or (2) by 1 nonconforming unit in the sample of 5 and 0 nonconforming units in the previous 3 (i) samples of size 5 (n).

The value of i, the number of previous samples, is determined by analysis of the Operating Characteristic (OC) curves for a given sample size, Figure 10-5 shows the OC curve for the single sample plan $n = 5$, $c = 0$, and the OC curves for ChSP-1 plans for $i = 1, 2, 3$, and 5. The OC curves for the ChSP-1 plans are obtained from the general formula

$$P_a = P_0 + P_1[P_0]^i$$

An example will illustrate the technique. For the ChSP-1 plan $n = 5$, $c = 0$, $i = 2$, and the calculations for an assumed value of $p_0 = 0.15$ are

$$P_0 = \frac{n!}{d!(n-d)!} p_0^d q_0^{n-d} = q_0^n = (0.85)^5 = 0.444$$

$$P_1 = \frac{n!}{d!(n-d)!} p_0^d q_0^{n-d} = np_0 q_0^{n-d} = 5(0.15)(0.85)^{5-1} = 0.392$$

$$P_a = P_0 + P_1[P_0]^i = 0.444 + (0.392)(0.444)^2 = 0.521$$

The point $P_a = 0.521$ is shown in Figure 10-5. The binomial is used as an approximation for the hypergeometric as long as $\frac{n}{N} \leq 0.10$.

The curve for $i = 1$ is shown dashed, since it is not a preferred choice. In practice i values of from 3 to 5 will be the most used since their OC curves approximate the single sampling plan OC curve. Where the percent nonconforming is small, the ChSP-1 plans increase the probability that a sample with 1 nonconforming unit will be accepted. This provides for the occasional nonconforming unit that is expected every now and then.

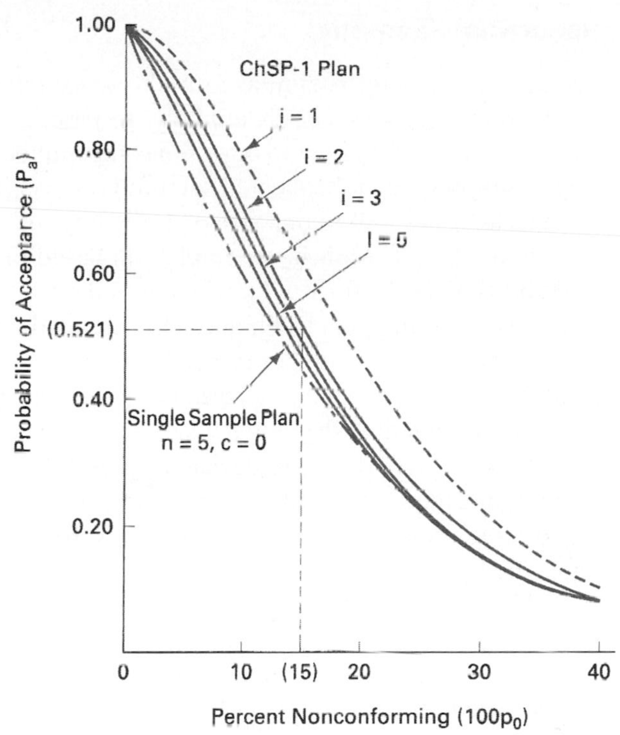

FIGURE 10-5 OC curves for ChSP-1 plans with values 1, 2, 3, 5 and for single sampling plan $n = 5, c = 0$.

For appropriate use of the chain sampling technique, the following conditions should be met:

1. The lot should be one of a continuing series of product that is sampled in substantially the order of its production.
2. The consumer can normally expect the lots to be essentially the same quality.
3. The consumer has confidence in the producer not to occasionally send an unacceptable lot that would have the optimum chance of acceptance.
4. The quality characteristic is one that involves destructive or costly tests, thereby dictating a small sample size.

Provision for an occasional nonconformity is satisfactory for major or minor classifications but not for critical.

Sequential Sampling

Sequential sampling is similar to multiple sampling except sequential sampling can theoretically continue indefinitely. In practice the plan is truncated after the number inspected is equal to three times the number inspected by a corresponding single sampling plan. Sequential sampling, which is used for costly or destructive tests, usually has a subgroup size of 1, thereby making it an item-by-item plan.

Item-by-item sequential sampling is based on the concept of the Sequential Probability Ratio Test (SPRT), which was developed by Wald.[5] Figure 10-6 illustrates the sampling plan technique. The "stepped" line shows the number nonconforming for the total number inspected and is updated with the inspection results of each item. If the cumulative results equal or are greater than the upper line, the lot is not accepted. If the cumulative results equal or are less than the lower line, the lot is accepted. If neither decision is possible, another item is inspected. Thus, if the 20th sample is found to be nonconforming, the cumulative number of nonconforming units will be 3. Since 3 exceeds the non-acceptance line for 20 inspections, the lot is not accepted.

The sequential sampling plan is defined by the producer's risk, α, and its process quality, p_α, and by the consumer's risk, β, and its process quality, p_β. Using these requirements, the equations (slope intercept form) can be determined for the acceptance line and non-acceptance line using the following formulas:

$$h_a = \log\left(\frac{1-\alpha}{\beta}\right) \Big/ \left[\log\left(\frac{p_\beta}{p_\alpha}\right) + \log\left(\frac{1-p_\alpha}{1-p_\beta}\right)\right]$$

$$h_r = \log\left(\frac{1-\beta}{\alpha}\right) \Big/ \left[\log\left(\frac{p_\beta}{p_\alpha}\right) + \log\left(\frac{1-p_\alpha}{1-p_\beta}\right)\right]$$

$$s = \log\left(\frac{1-p_\alpha}{1-p_\beta}\right) \Big/ \left[\log\left(\frac{p_\beta}{p_\alpha}\right) + \log\left(\frac{1-p_\alpha}{1-p_\beta}\right)\right]$$

$$d_a = -h_a + sn$$

$$d_r = h_r + sn$$

where s = slope of the lines
h_r = intercept for non-acceptance line
h_a = intercept for acceptance line
p_β = fraction nonconforming for consumer's risk
p_α = fraction nonconforming for producer's risk
β = consumer's risk

[5]For more information, see Abraham Wald, *Sequential Analysis* (New York: John Wiley & Sons, Inc., 1947).

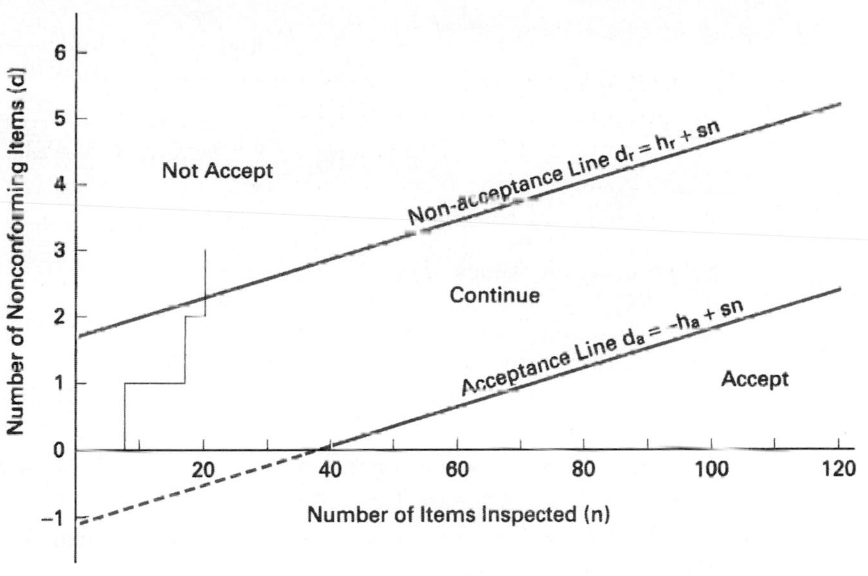

FIGURE 10-6 Graphical presentation of an item-by-item sequential plan.

α = producer's risk

d_a = number of nonconforming units for acceptance

d_r = number of nonconforming units for non-acceptance

n = number of units inspected

Thus, the equations for the sequential plan, which is defined by $\alpha = 0.05$, $p_\alpha = 0.01$, $\beta = 0.10$, and $p_\beta = 0.06$, is obtained by the following calculations:

$$h_a = \log\left(\frac{1-\alpha}{\beta}\right) \Big/ \left[\log\left(\frac{P_\beta}{P_\alpha}\right) + \log\left(\frac{1-P_\alpha}{1-P_\beta}\right)\right]$$

$$= \log\left(\frac{1-0.05}{0.10}\right) \Big/ \left[\log\left(\frac{0.06}{0.01}\right) + \log\left(\frac{0.99}{0.94}\right)\right]$$

$$= 1.22$$

$$h_r = \log\left(\frac{1-\beta}{\alpha}\right) \Big/ \left[\log\left(\frac{P_\beta}{P_\alpha}\right) + \log\left(\frac{1-P_\alpha}{1-P_\beta}\right)\right]$$

$$= \log\left(\frac{1-0.10}{0.05}\right) \Big/ \left[\log\left(\frac{0.06}{0.01}\right) + \log\left(\frac{0.99}{0.94}\right)\right]$$

$$= 1.57$$

$$s = \log\left(\frac{1 - P_\alpha}{1 - P_\beta}\right) \Big/ \left[\log\left(\frac{P_\beta}{P_\alpha}\right) + \log\left(\frac{1 - P_\alpha}{1 - P_\beta}\right)\right]$$

$$= \log\left(\frac{1 - 0.01}{1 - 0.06}\right) \Big/ \left[\log\left(\frac{0.06}{0.01}\right) + \log\left(\frac{0.99}{0.94}\right)\right]$$

$$= 0.03$$

Substituting the values of $h_a = 1.22$, $h_r = 1.57$, and $s = 0.03$ into the formulas for d_a and d_r, we obtain the following equations:

$$d_a = -1.22 + 0.03n$$

$$d_r = 1.57 + 0.03n$$

The above equations are the same as those used for the acceptance and non-acceptance lines of Figure 10-6.

While the graphical presentation of Figure 10-6 can be used as the sampling plan, it is usually more convenient to use the tabular form. This is easily accomplished by substituting values of n into the equations for the acceptance and non-acceptance lines and calculating d_a and d_r. For example, the calculations for $n = 17$ are

$$d_a = -1.22 + 0.03n \qquad d_r = 1.57 + 0.03n$$

$$= -1.22 + 0.03(17) \qquad = 1.57 + 0.03(17)$$

$$= -0.71 \qquad = 2.08$$

Since the number of nonconforming units (d_a and d_r) are whole numbers, the non-acceptance number is the next whole number above d_r, and the acceptance number is the next whole number below d_a. Thus, $n = 17$, $d_a = 0$, and $d_r = 3$. Table 10-10 illustrates the sampling plan for the first 113 samples.

It is sometimes preferable to take the sample in groups rather than singly. This is accomplished by using multiples of the desired sample size. Therefore, if the sample size is 5, the acceptance and non-acceptance numbers are determined for n values of 5, 10, 15, ...

Sequential sampling is used to reduce the number inspected for items that require costly or destructive testing. It is also applicable for any situation since the average amount inspected will be less than for single, double, and multiple sampling.

Skip-Lot Sampling

Skip-lot sampling was devised by H. F. Dodge in 1955.[6] It is a single sampling plan for minimizing inspection costs when there is a continuing supply of lots of raw

[6]H. F. Dodge, "Skip-Lot Sampling Plans," *Industrial Quality Control,* 11, No. 5 (February 1955), 3–5.

TABLE 10-10 **Unit-by-Unit Sequential Sampling Plan $\alpha = 0.05, p_\alpha = 0.01, \beta = 0.10$, and $p_\beta = 0.06$.**

NUMBER OF UNITS INSPECTED n	ACCEPTANCE NUMBER d_a	NON-ACCEPTANCE NUMBER d_r
1	[a]	[b]
2–15	[a]	2
16–40	[a]	3
41–47	0	3
48–73	0	4
74–80	1	4
81–106	1	5
107–113	2	5

[a]Acceptance not possible.

[b]Non-acceptance not possible.

material, component parts, subassemblies, and finished parts from the same source. It is particularly applicable to chemical and physical characteristics that require laboratory analyses. As companies emphasize Statistical Process Control (SPC) and Just-In-Time (JIT) procurement, this type of sampling will become more applicable.

The skip-lot sampling plan, designated SkSP–1, is based on the AOQL. However, the AOQL refers to units rather than lots, as discussed in Chapter 9. Thus, an AOQL of 1% means that on the average the plan will accept no more than 1% of the lots that are nonconforming for the characteristic under consideration.

The SkSP-1 plan begins with the inspection of every lot. When a prescribed number of lots have been accepted, a sampling of lots occurs. Figure 10-7 describes

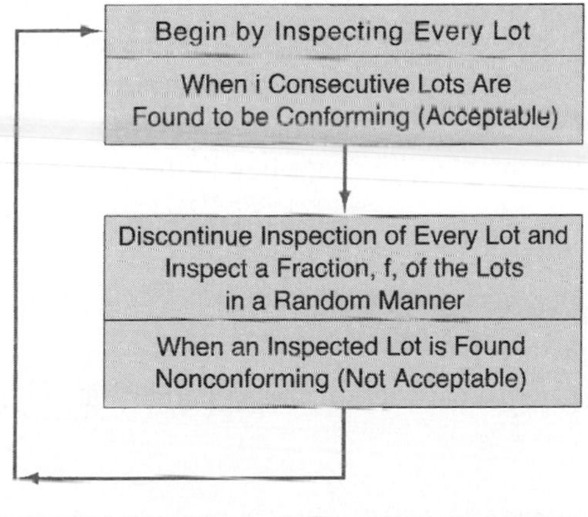

FIGURE 10-7 Procedure for SkSP-1 plans.

the SkSP-1 in a flow-chart format. When a lot is not accepted while in the sampling mode, the plan reverts to inspecting every lot.

The plan is a modification of a continuous sampling plan, CSP-1, which is described later in the chapter. The primary difference is that SkSP-1 refers to lots and CSP-1 refers to units. Table 10-11 is used for both plans to provide a family of i and f values for each AOQL value. Thus, for an AOQL of 1.22%, any of the following i and f values could be used:

f	i
$\frac{1}{2}$	23
$\frac{1}{3}$	38
$\frac{1}{4}$	49
$\frac{1}{5}$	58
⋮	⋮
$\frac{1}{200}$	255

In general, the f values used will be those at the top of the table—$\frac{1}{2}$ to $\frac{1}{5}$.

The best way to select lots to be inspected, while in the sampling mode, is to use a known probability sampling method. Thus, if $f = \frac{1}{2}$, then a head on the flip of a coin would decide if the lot is inspected; if $f = \frac{1}{3}$, a 1 or 2 on the roll of a six-sided die would determine if the lot is inspected; or, if $f = \frac{1}{4}$, a spade on a draw from a deck of cards would determine if the lot is inspected.

The plans assume that non-acceptable lots will be rectified.

ANSI/ASQ SI—1996[7]

The purpose of this standard is to provide procedures to reduce the inspection effort when the supplier's quality is superior. It is a skip-lot scheme used in conjunction with the attribute lot-by-lot plans given in ANSI/ASQZ1.4—1993; it is not to be confused with Dodge's skip-lot scheme described in the previous section. This sampling plan is an alternate to the reduced inspection of ANSI/ASQZ1.4—1993, which permits smaller sizes than normal inspection.

In order to use the plan, the supplier shall

1. Have a documented system for controlling product quality and design changes.
2. Have instituted a system that is capable of detecting and correcting changes that might adversely affect quality.
3. Not have experienced an organization change that might adversely affect quality.

[7]This section is extracted from ANSI/ASQ S1—1996 by permission of the American Society for Quality.

TABLE 10-11 Values of i for CSP-1 Plans.

f	AOQL (%)															
	0.018	**0.033**	**0.046**	**0.074**	**0.113**	**0.143**	**0.198**	**0.33**	**0.53**	**0.79**	**1.22**	**1.90**	**2.90**	**4.94**	**7.12**	**11.46**
$\frac{1}{2}$	1,540	840	600	375	245	194	140	84	53	36	23	15	10	6	5	3
$\frac{1}{3}$	2,550	1,390	1,000	620	405	321	232	140	87	59	38	25	16	10	7	5
$\frac{1}{4}$	3,340	1,820	1,310	810	530	420	303	182	113	76	49	32	21	13	9	6
$\frac{1}{5}$	3,960	2,160	1,550	965	630	498	360	217	135	91	58	38	25	15	11	7
$\frac{1}{7}$	4,950	2,700	1,940	1,205	790	623	450	270	168	113	73	47	31	18	13	8
$\frac{1}{10}$	6,050	3,300	2,370	1,470	965	762	550	335	207	138	89	57	38	22	16	10
$\frac{1}{15}$	7,390	4,030	2,890	1,800	1,180	930	672	410	255	170	108	70	46	27	19	12
$\frac{1}{25}$	9,110	4,970	3,570	2,215	1,450	1,147	828	500	315	210	134	86	57	33	23	14
$\frac{1}{50}$	11,730	6,400	4,590	2,855	1,870	1,477	1,067	640	400	270	175	110	72	42	29	18
$\frac{1}{100}$	14,320	7,810	5,600	3,485	2,305	1,820	1,302	790	500	330	215	135	89	52	36	22
$\frac{1}{200}$	17,420	9,500	6,810	4,235	2,760	2,178	1,583	950	590	400	255	165	106	62	43	26

In addition, the product shall

1. Be of stable design, which means that there have been no substantive design changes that might adversely affect the quality.
2. Have been manufactured on a continuous basis for at least 6 months unless the supplier and responsible authority agree to a longer period. The responsible authority is the purchaser or a delegated inspection agency.
3. Have been on normal and reduced inspection at the general inspection levels I, II, or III of the ANSI/ASQZ1.4—1993 during this qualification period.
4. Have maintained a quality level at or less than the AQL for at least 6 months unless the supplier and responsible authority agree to a longer period.
5. Meet the following requirements in Table 10-12 and Table 10-13.
   (a) The previous 10 or more consecutive lots have been accepted.
   (b) The minimum cumulative sample size in Table 10-12 for the last 10 or more consecutive lots have been met.
   (c) The acceptance numbers in Table 10-13 for the last 2 lots have been met.

When double or multiple sampling are used, only the results of the first sample are counted.

The example problem which follows illustrates the use of the tables.

EXAMPLE PROBLEM

A manufacturer of fireplace ornaments meets the supplier requirements and the first four product requirements. In addition, the responsible authority established an AQL of 0.25. From 12 consecutive lots, all of which were accepted, a total of 6000 units were inspected. Nine nonconforming items were found in the 12 lots, and in the last 2 lots with a sample size of 500, there were 1 and 0 nonconforming items, respectively.

Twelve lots were used because the requirements were not met on the 10th or 11th lots. The requirements of Table 10-12 are met because the minimum cumulative sample size for 9 nonconforming items is 5940, which is less than the 6000 inspected. Also, the requirements of Table 10-13 are met because for a sample size of 500, the allowable number of nonconforming items is 2, and the last 2 lots had 1 and 0, respectively. Therefore, the product qualifies for skip-lot inspection.

Percent nonconforming applies only to AQL values of 10.0 or less in tables. All AQL values are applicable to nonconformities per 100 units. Table 10-12 can be extended beyond 20 by adding the value in the last row for each additional

TABLE 10-12 Minimum Cumulative Sample Size to Initiate Skip-Lot-Inspection—Table I of ANSI/ASQ S1–1996.

NONCONFORMITIES OR NONCONFORMING ITEMS	AQL (PERCENT NONCONFORMING OR NONCONFORMITIES PER 100 UNITS)												
	0.1	0.15	0.25	0.40	0.65	1.0	1.5	2.5	4.0	6.5	10.0	15.0	25.0
0	2600	1740	1040	650	400	260	174	104	65	40	26	17	10
1	4250	2840	1700	1070	654	425	284	170	107	65	43	28	17
2	5740	3830	2300	1440	883	574	383	230	144	88	57	38	23
3	7140	4760	2860	1790	1093	714	476	286	179	110	71	48	29
4	8490	5660	3400	2120	1306	849	566	340	212	131	85	57	34
5	9800	6530	3920	2450	1508	980	653	392	245	151	98	65	39
6	11090	7390	4440	2770	1706	1109	739	444	277	171	111	74	44
7	12350	8240	4940	3090	1902	1236	824	494	309	190	124	82	49
8	13610	9070	5440	3400	2094	1361	907	544	340	209	136	91	54
9	14850	9900	5940	3710	2285	1485	990	594	371	229	149	99	59
10	16080	10720	6430	4020	2474	1608	1072	643	402	247	161	107	64
11	17290	11530	6920	4320	2660	1729	1153	692	432	266	173	115	69
12	18500	12330	7400	4630	2846	1850	1233	740	463	285	185	123	74
13	19700	13130	7880	4930	3031	1970	1313	788	493	303	197	131	79
14	20390	13930	8360	5220	3214	2089	1393	836	522	321	209	139	84
15	22080	14720	8830	5520	3397	2208	1472	883	552	340	221	147	88
16	23260	15500	9300	5820	3578	2326	1550	930	582	358	233	155	93
17	24430	16290	9770	6110	3758	2443	1629	977	611	376	244	163	98
18	25600	17070	10240	6400	3938	2560	1707	1024	640	394	256	171	102
19	26760	17840	10700	6690	4117	2676	1784	1070	669	412	268	178	107
20	27930	18620	11170	6980	4297	2793	1862	1117	698	430	279	186	112
Each additional	1170	780	470	290	180	117	78	47	29	18	12	8	5

TABLE 10-13 Acceptance Numbers to Initiate or Continue Skip-Lot Inspection (Individual Lot Criterion)—Table II of ANSI/ASQ. S1—1996.

SAMPLE	AQL (PERCENT NONCONFORMING OR NONCONFORMITIES PER 100 UNITS)												
SIZE	0.1	0.15	0.25	0.4	0.65	1.0	1.5	2.5	4.0	6.5	10.0	15.0	25.0
2	–	–	–		–	–	–	–	–	0	–	0	1
3	–	–	–	–	–	–	–	–	0	–	0	1	1
5	–	–	–	–	–	–	–	0	–	0	1	1	2
8	–	–	–	–	–	–	0	–	0	1	1	2	3
13	–	–	–	–	–	0	–	0	1	1	2	3	5
20	–	–	–	–	0	–	0	1	1	2	3	5	7
32	–	–	–	0	–	0	1	1	2	3	5	7	11
50	–	–	0	–	0	1	1	2	3	5	7	11	17
80	–	0	–	0	1	1	2	3	5	7	11	17	–
125	0	–	0	1	1	2	3	5	7	11	17	–	–
200	–	0	1	1	2	3	5	7	11	17	–	–	–
315	0	1	1	2	3	5	7	11	17	–	–	–	–
500	1	1	2	3	5	7	11	17	–	–	–	–	–
800	1	2	3	5	7	11	17	–	–	–	–	–	–
1250	2	3	5	7	11	17	–	–	–	–	–	–	–
2000	3	5	7	11	17	–	–	–	–	–	–	–	–

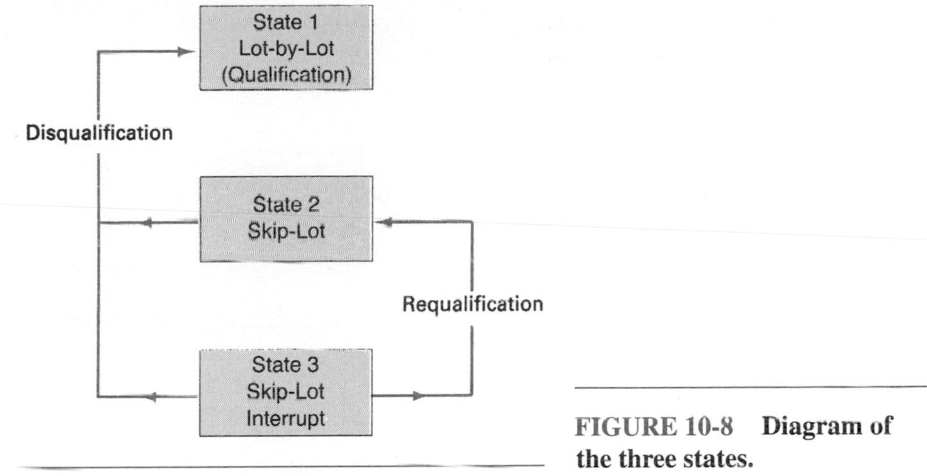

FIGURE 10-8 Diagram of the three states.

nonconforming item. Thus, for an AQL of 1.5 and 24 nonconforming items, the minimum cumulative sample size is 2174 [1862 + 4(78)].

There are three basic states to the standard. State 1 is the lot-by-lot inspection activity. When the supplier and product qualify for skip-lot inspection as described before, the scheme switches to state 2, which is the skip-lot state. State 3 is a temporary state in which skip-lot inspection can be interrupted while requalification occurs under less stringent procedures. While in state 2 or 3, disqualification can occur, in which case the program switches to state 1. Figure 10-8 shows the three states.

Multiple sampling is not allowed during states 2 and 3, and it is strongly recommended that an acceptance number of zero ($c = 0$) not be used during states 2 and 3.

Skip lot inspection (state 2) provides for four possible frequencies: 1 lot inspected in 2 submitted, 1 lot inspected in 3 submitted, 1 lot inspected in 4 submitted, and 1 lot inspected in 5 submitted. The first three frequencies are applicable for the initial skip-lot inspection frequency. Figure 10-9 shows the decision diagram for the initial frequency. It is based on the inspection results during state 1. If more than 20 lots are needed to qualify, then the frequency is 1 out of 2, which is the worst-case situation. If 20 or fewer lots are needed to qualify but some of the lots do not satisfy Table 10-13, the frequency is 1 out of 3. However, if all of the 20 or fewer lots satisfy Table 10-13, the frequency is 1 out of 4.

EXAMPLE PROBLEM

Determine the initial frequency of the previous example problem, the qualifying state (state 1). The initial frequency will be 1 in 4 because Table 10-12 was met in 20 or fewer lots and all lots met Table 10-13.

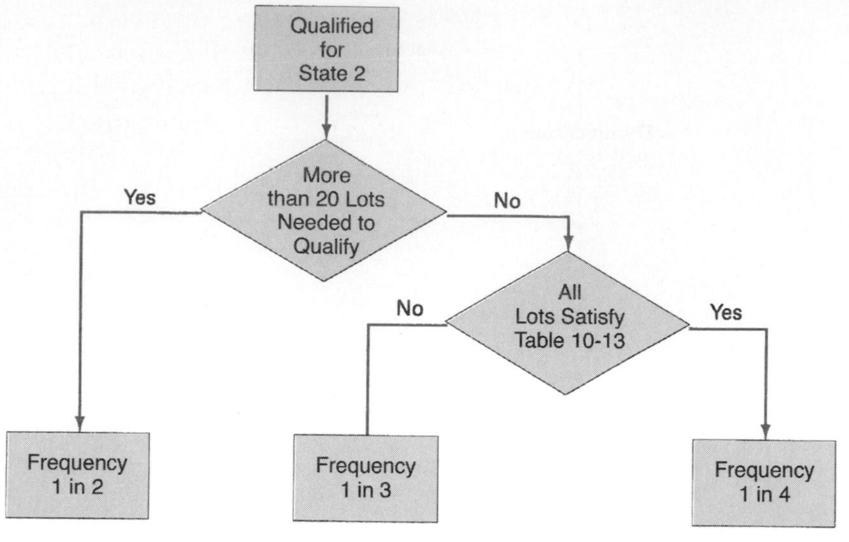

FIGURE 10-9 Diagram of the initial frequency.

The frequency of inspection can be shifted to the next lower frequency by meeting the following conditions:

1. Preceding 10 or more inspected lots accepted
2. Cumulation results satisfy Table 10-12
3. Each of the last 2 lots satisfy Table 10-13
4. Approval of the responsible authority

The first three conditions are identical to item 5 of the product qualification requirements. If a supplier's initial frequency is 1 out of every 3, it could be reduced to 1 out of every 4, which would be a substantial savings. If double sampling is used, only the first sample is counted.

EXAMPLE PROBLEM

After an initial frequency of 1 in 4, under a scheme with an AQL of 0.65, the next 10 lots inspected are accepted with a cumulative sample size of 1625 units and a total of 5 nonconforming units. If the inspection results of each of the last 2 lots are 1 nonconforming unit for sample sizes of 125 and 200, is a shift to the next lower frequency possible?

Since the requirements of Table 10-12 and Table 10-13 are met, a shift to a frequency of 1 in 5 is authorized, provided the responsible authority approves.

State 3, the skip-lot interrupt state, occurs whenever the last inspected lot does not meet the requirements of Table 10-13. When this situation occurs, the inspection is on a lot-by-lot basis under normal, level I, II, or III. If 4 consecutive lots are accepted and the last 2 meet the requirements of Table 10-13, skip-lot inspection is reinstated. However, the frequency is increased to the next higher level unless the previous level was 1 out of 2. Thus, if the previous frequency was 1 out of 4, the next higher is 1 out of 3.

The product shall be disqualified for skip-lot inspection and lot-by-lot resumed when any of the following criteria are met:

1. A lot is not accepted during state 3.
2. Requalification is not achieved within 10 lots.
3. There is no production activity during a period specified by the supplier and responsible authority (if no period is agreed to, it is 2 months).
4. The supplier significantly deviates from the supplier qualifications or product qualifications.
5. The responsible authority decides to return to the lot-by-lot inspection of ANSI/ASCZ1.4—1993.

Skip-lot inspection should be used when it is more cost-effective than reduced inspection under ANSI/ASQ Z1.4—1993. Just-in-time procurement activities increase inspection costs because of smaller lot sizes; therefore, reduced inspection and skip-lot inspection are attractive alternates to normal inspection. One feature of skip-lot is the fact that its OC curves closely approximate the corresponding normal plans.

ACCEPTANCE SAMPLING PLANS FOR CONTINUOUS PRODUCTION

Introduction

Acceptance sampling plans that have been discussed in this chapter are lot-by-lot plans. Many manufacturing operations do not create lots as a normal part of the production process, since they are produced by a continuous process on a conveyor or other straight-line system. In such cases, acceptance sampling plans for continuous production are required.

Plans for continuous production consist of alternating sequences of sampling inspection and screening (100%) inspection. These plans usually begin with 100% inspection, and if a stated number of units (clearance number, i) is free of nonconformities, sampling inspection is instituted. Sampling continues until a specific number of nonconforming units are found, at which time 100% inspection is reinstated.

Sampling plans for continuous production are applicable to attribute, nondestructive inspection of moving product. The inspection must be of such a nature that it is relatively easy and rapid so that no "bottlenecks" occur because of the inspection activity. In addition, the process must be capable of manufacturing homogeneous product. Production personnel usually handle the 100% inspection and quality personnel the sampling. Critical, major, and minor classifications in a unit will have different AOQL and i values, but usually the same f value.

The concept of sampling for continuous production was first devised by H. F. Dodge in 1943, with a sampling plan that has been commonly referred to as CSP-1. This plan and two additional plans, CSP-2 and CSP-3, are categorized as single-level plans. In 1955, the theory of multilevel continuous plans was presented by G. Licherman and H. Soloman. Multilevel plans provide for reduced levels of sampling inspection when the quality continues to be superior.[8] Much of this early work was incorporated into MIL-STD-1235 (ORD), which was superseded by MIL-STD-1235A (MU) on June 28. 1974. The designation for the standard was changed to MIL-STD-1235B when the U.S. Navy adopted the plan on December 10, 1981.

CSP-1[9] Plans

This plan begins by 100% inspection (screening) of the product in the order of production until a certain number of successive units are free of nonconformities. When that number is obtained, 100% inspection is discontinued and sampling inspection begins. The sample is a fraction of the flow of the product and is selected in such a manner as to minimize any bias. If a nonconformity occurs, sampling inspection is discontinued and 100% inspection begins. Figure 10-10 shows the procedure for CSP-1 plans. The clearance number i is the number of conforming units in 100% inspection, and the sampling frequency f is the ratio of units inspected to the total units passing an inspection station during periods of sampling inspection. Thus, an f value of 1/20 means that 1 sampling inspection is made for every 20 units of product.

CSP-1 plans are indexed by AOQL. For a particular AOQL, there are different combinations of i and f, which are given in Table 10-11. Thus, one plan for an AOQL of 0.79 is $i = 59$ and $f = \frac{1}{3}$. This plan specifies that sampling inspection of 1 out of every 3 products is instigated after 59 consecutive products are free of nonconformities. Sampling continues until a nonconformity is found, at which time screening inspection is reinstated. Some other plans for an AOQL of 0.79 are

$$i = 113 \qquad f = \tfrac{1}{7}$$
$$= 270 \qquad = \tfrac{1}{50}$$

Analysis of the table shows that as the f value decreases, the i value increases.

---

[8] G. Licherman and H. Soloman, "Multi-level Continuous Sampling Plans," *Annals of Mathematical Statistics*, 26 (December 1955), 686–704.

[9] H. F. Dodge, "A Sampling Inspection Plan for Continuous Production," *Annals of Mathematical Statistics*, 14 (September 1943), 264–279.

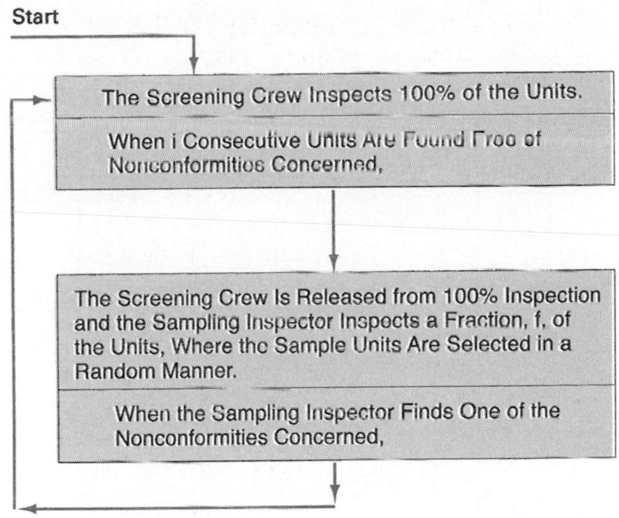

FIGURE 10-10 Procedure for CSP-1 and CSP-F plans.

The choice of *i* and *f* values for a particular AOQL are based on practical considerations. As *f* gets smaller, the protection from spotty quality decreases, especially for values less than $\frac{1}{50}$. Another practical consideration is the amount of production per shift; as the amount increases, the value of *f* can decrease. Also, the value of *f* can be influenced by the sampling inspector's workload.

CSP-2 Plans

Continuous sampling inspection plan, designated CSP-2, is a modification of CSP-1. Plan CSP-1 requires a return to 100% inspection wherever a nonconformity is found during the sampling inspection. CSP-2, however, does not require a return to 100% inspection unless a second nonconformity is found in the next *i* or fewer sample units.[10] Figure 10-11 gives the procedure for CSP-2 plans.

The purpose of CSP-2 plans is to provide protection against the occurrence of an isolated nonconformity that would initiate a return to 100% inspection.

Plans are indexed by a specific AOQL, which provides for different combinations of *i* and *f*, as shown in Table 10-14. Thus, $i = 35, f = \frac{1}{5}$ and $i = 59, f = \frac{1}{15}$ are two of many plans for an AOQL of 2.90.

For the latter plan, $i = 59$ and $f = \frac{1}{15}$, sampling inspection of 1 out of every 15 continues after one nonconformity is found. If a second nonconformity is found

[10] H. F. Dodge and M. N. Torrey, "Additional Continuous Sampling Inspection Plans," *Industrial Quality Control*, 7, No. 5 (March 1951), 7–12.

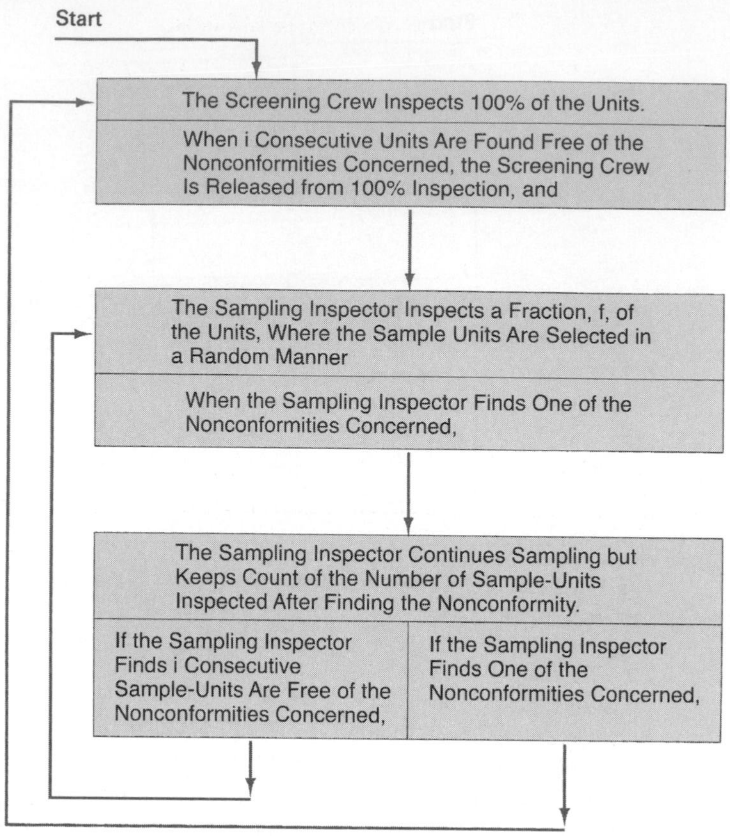

FIGURE 10-11 Procedure for CSP-2 plans.

in the next 59 sample units, 100% inspection is invoked. If a second nonconformity does not occur, sampling continues without the conditional stipulation.

MIL-STD-1235B

The standard is composed of five different continuous sampling plans. Inspection is by attributes for nonconformities or nonconforming units using three classes of severity: critical, major, and minor.

Continuous sampling plans are designed based on the AOQL. In order to be comparable with ANSI/ASQ Z1.4—1993 and other standards, the plans are also indexed by the AQL. The AQL is merely an index to the plans and has no other meaning.

The standard has a special provision for critical nonconformities. Only two plans, CSP-1 and CSP-F, can be used for critical nonconformities. Even in these cases, the responsible authority (consumer) can require 100% inspection at all times.

TABLE 10-14 **Values of *i* for CSP-2 Plans.**

	AOQL (%)							
f	0.53	0.79	1.22	1.90	2.90	4.94	7.12	11.46
$\frac{1}{2}$	80	54	35	23	15	9	7	4
$\frac{1}{3}$	128	86	55	36	24	14	10	7
$\frac{1}{4}$	162	109	70	45	30	18	12	8
$\frac{1}{5}$	190	127	81	52	35	20	14	9
$\frac{1}{7}$	230	155	99	64	42	25	17	11
$\frac{1}{10}$	275	185	118	76	50	29	20	13
$\frac{1}{15}$	330	220	140	90	59	35	24	15
$\frac{1}{25}$	395	265	170	109	71	42	29	18
$\frac{1}{50}$	490	330	210	134	88	52	36	22

In each of the five sampling plans, provision is made for the discontinuation of inspection. The consumer can suspend product acceptance when the product quality is such that 100% inspection continues beyond a prescribed number of units, s. In other words, if sampling inspection does not occur within s units, the product quality is below standard and product acceptance can be suspended. The table for s values is not reproduced in the text.

Sampling plans are designated by code letters. Table 10-15 provides a range of permissible code letters based on the number of units in the production interval (usually an 8-h shift). Factors that influence the selection of the code letter are inspection time per unit of product, production rate, and proximity to other inspection stations. When idle inspection time is a significant consideration, a plan with a higher sampling frequency and lower clearance number is usually preferred.

TABLE 10-15 **Sampling-Frequency Code Letters**

NUMBER OF UNITS IN PRODUCTION INTERVAL	PERMISSIBLE CODE LETTERS
2–8	A, B
9–25	A–C
26–90	A–D
91–500	A–E
501–1,200	A–F
1,201–3,200	A–G
3,201–10,000	A–H
10,001–35,000	A–I
35,001–150,000	A–J
150,001–up	A–K

CSP-1 and CSP-2 Plans Both of Dodge's plans, CSP-1 and CSP-2, are incorporated into the standard, except the form of the plan is different. It includes sample-size code letters and AQLs as shown in Table 10-17 for the CSP-T plan.

CSP-F Plans CSP-F is a single-level continuous sampling procedure that provides for alternating sequences of 100% inspection and sampling inspection. The procedure is the same as the CSP-1 plan, which is shown in Figure 10-10. CSP-F plans are indexed by the AOQL and also by the amount of product manufactured in a production interval. This allows smaller clearance numbers to be used, which permits CSP-F plans to be applied for short production-run situations or to be applied where the inspection operation is time-consuming.

There are 12 tables for the CSP-F plans; each table represents a different AOQL value. Table 10-16 is an example of the table for AOQL = 0.33%. Tables for other AOQL values are not included in the book. The *i* values in the last row of the table are the same as those given for the CSP-1 plan with an AOQL = 0.33%.

An example problem will illustrate the procedure. An AOQL value of 0.33%, an *f* value of $\frac{1}{4}$, and a lot size of 7500 gives an answer of $i = 177$ from Table 10-16.

TABLE 10-16 Values of *i* for CSP-F Plans (AQL,[a] 0.25%; AOQL, 0.33%) [Table 3-A-8 of MIL-STD-1235B]

SAMPLE-FREQUENCY CODE LETTER	A	B	C	D	E	F	G
f	$\frac{1}{2}$	$\frac{1}{3}$	$\frac{1}{4}$	$\frac{1}{5}$	$\frac{1}{7}$	$\frac{1}{10}$	$\frac{1}{15}$
N							
1–500	70	99	114	123	133	140	146
501–1,000	77	116	140	155	174	188	200
1,001–2,000	81	127	158	181	211	236	258
2,001–3,000	82	132	166	192	228	261	291
3,001–4,000	83	134	170	198	237	276	312
4,001–5,000	83	135	173	201	244	286	327
5,001–6,000	84	136	174	204	248	293	338
6,001–7,000	84	137	176	206	251	298	346
7,001–8,000	84	137	177	207	254	302	353
8,001–9,000	84	138	177	209	256	305	358
9,001–10,000	84	138	178	209	257	308	362
10,001–11,000	84	138	178	210	259	310	366
11,001–12,000	84	139	179	211	260	312	369
12,001–15,000	84	139	180	212	262	316	376
15,001–20,000	84	140	181	214	265	320	384
20,001 and over	84	140	182	217	270	335	410

[a]AQLs are provided as indices to simplify use of this table but have no other meaning relative to the plans.

CSP-T Plans. CSP-T is a multilevel continuous sampling procedure that provides for alternate sequences of 100% inspection and sampling inspection. It differs from the previous inspection plans in that it provides for a reduced sampling frequency upon demonstration of superior product quality. Figure 10-12 shows the CSP-T procedure. Table 10-17 gives the values of i and f for a specified AOQL. Note that the AOQL values are at the bottom of the table.

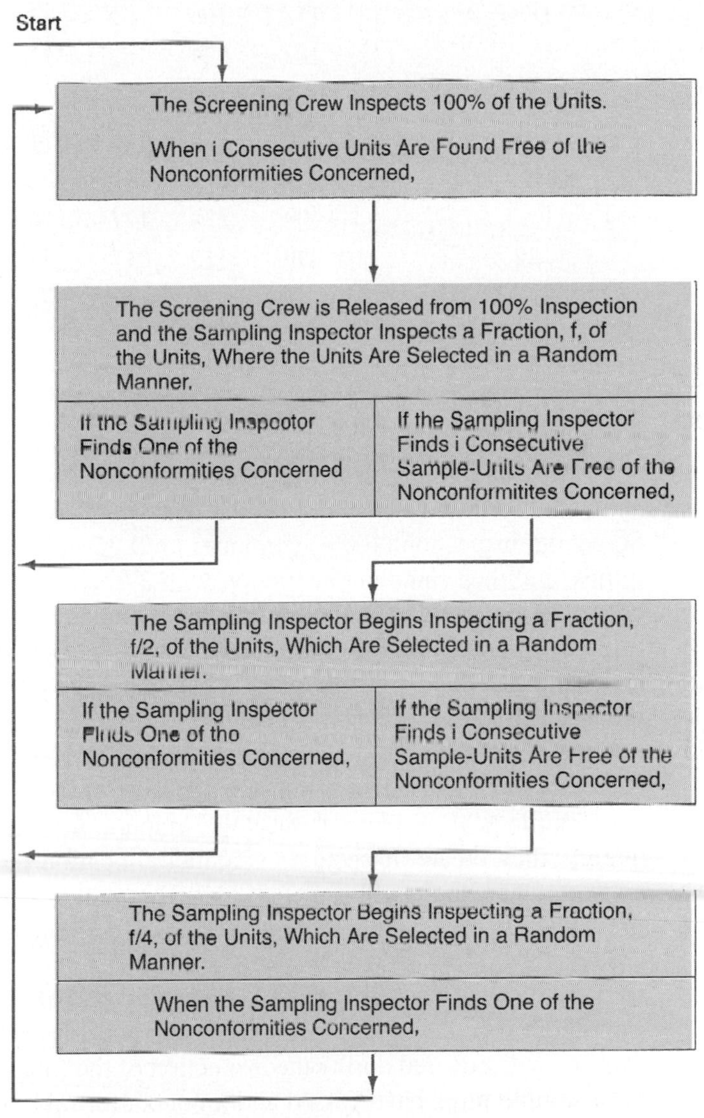

FIGURE 10-12 Procedure for CSP-T plans.

TABLE 10-17 Values of *i* for CSP-T Plans [Table 5-A of MIL-STD-1235B (MU)].

SAMPLING-FREQUENCY CODE LETTER	f	AQL[a] (%) 0.40	0.65	1.0	1.5	2.5	4.0	6.5	10.0
A	$\frac{1}{2}$	87	58	38	25	16	10	7	5
B	$\frac{1}{3}$	116	78	51	33	22	13	9	6
C	$\frac{1}{4}$	139	93	61	39	26	15	11	7
D	$\frac{1}{5}$	158	106	69	44	29	17	12	8
E	$\frac{1}{7}$	189	127	82	53	35	21	14	9
F	$\frac{1}{10}$	224	150	97	63	41	24	17	11
G	$\frac{1}{15}$	226	179	116	74	49	29	20	13
H	$\frac{1}{25}$	324	217	141	90	59	35	24	15
I	$\frac{1}{50}$	409	274	177	114	75	44	30	19
J,K	$\frac{1}{100}$	499	335	217	139	91	53	37	23
AOQL (%)		0.53	0.79	1.22	1.90	2.90	4.94	7.12	11.46

[a]AQLs are provided as indices to simplify use of this table but have no other meaning relative to the plans.

An example problem will illustrate the use of the procedure. For an AOQL value of 2.90% and an f value of $\frac{1}{7}$, the corresponding i value from Table 10-17 is 35. Screening inspection (100%) continues until 35 units are found free of nonconformities, and then sampling inspection with a frequency of $\frac{1}{7}$ commences. If no nonconformities are found in the next 35 sample units, the sample frequency is changed to $f/2$ or $\frac{1}{14}$. Sampling continues with this new frequency of $\frac{1}{14}$ until 35 sample units are found free of nonconformities, at which point the sampling frequency is further reduced. This last reduction is changed to $f/4$ or $\frac{1}{28}$, and sampling continues at this rate until production of the item is completed. Of course, any time a nonconformity is found, 100% inspection is reinstated and the procedure starts over again.

While CSP-T plans reduce the amount inspected as a result of superior quality, they create inspection personnel allocation problems. For example, with an f value of $\frac{1}{4}$, there will need to be 16 people for 100% inspection, 4 people at the first level, 2 people at the second level, and 1 person at the last level.

CSP-V Plans. The fifth plan in MIL-STD-1235B is a single-level continuous sampling procedure. A return to 100% inspection is required whenever a nonconformity is discovered during the inspection of the first i sample units. Once the initial i sample units have passed and a nonconformity occurs, a return to 100% inspection is required; however, the clearance number, i, is reduced by 2/3. Thus, if the original i value is 39 from the CSP-T plan, the clearance number, i, is reduced to 13. This type of plan can be beneficially applied in those situations where there is no advantage to reducing the sampling frequency, f. The situation occurs when inspection personnel cannot be assigned to other duties.

Figure 10-13 shows the CSP-V procedure. This plan simplifies the inspection personnel allocation problem. In addition, it will minimize the amount inspected when, and if, a nonconformity occurs.

ACCEPTANCE SAMPLING PLANS FOR VARIABLES

Introduction

While attribute sampling plans are the most common type of acceptance sampling, there are situations where variable sampling is required. Variable sampling plans are based on the sample statistics of average and standard deviation and the type of frequency distribution.

Advantages and Disadvantages. Variable sampling has the principal advantage that the sample size is considerably less than with attribute sampling. In addition, variable sampling provides a better basis for improving quality and gives more information for decision making.

One of the disadvantages of variable sampling is that only one characteristic can be evaluated; a separate plan is required for each quality characteristic. Variable sampling usually involves greater administrative, clerical, and equipment costs. Furthermore, the distribution of the population has to be known or estimated.

Types of Sampling Plans. There are two types of variable plans—percent nonconforming and process parameter. Variable plans for percent nonconforming are designed to determine the proportion of product that is outside specifications. Of the variable plans for percent nonconforming, two will be discussed in this section. These are the Shainin lot plot and ANSI/ASQ Z1.9—1993.

Variable plans for process parameter are designed to control the average and standard deviation of the distribution of the product to specified levels. Plans of this type are acceptance control chart, sequential sampling for variables, and hypothesis testing. Because of the limited application of these plans, they are briefly discussed at the end of this chapter.

Shainin Lot Plot Plan

The Shainin lot plot plan is a variable sampling plan used in some industries. It was developed by Dorian Shainin while he was Chief Inspector at Hamilton Standard

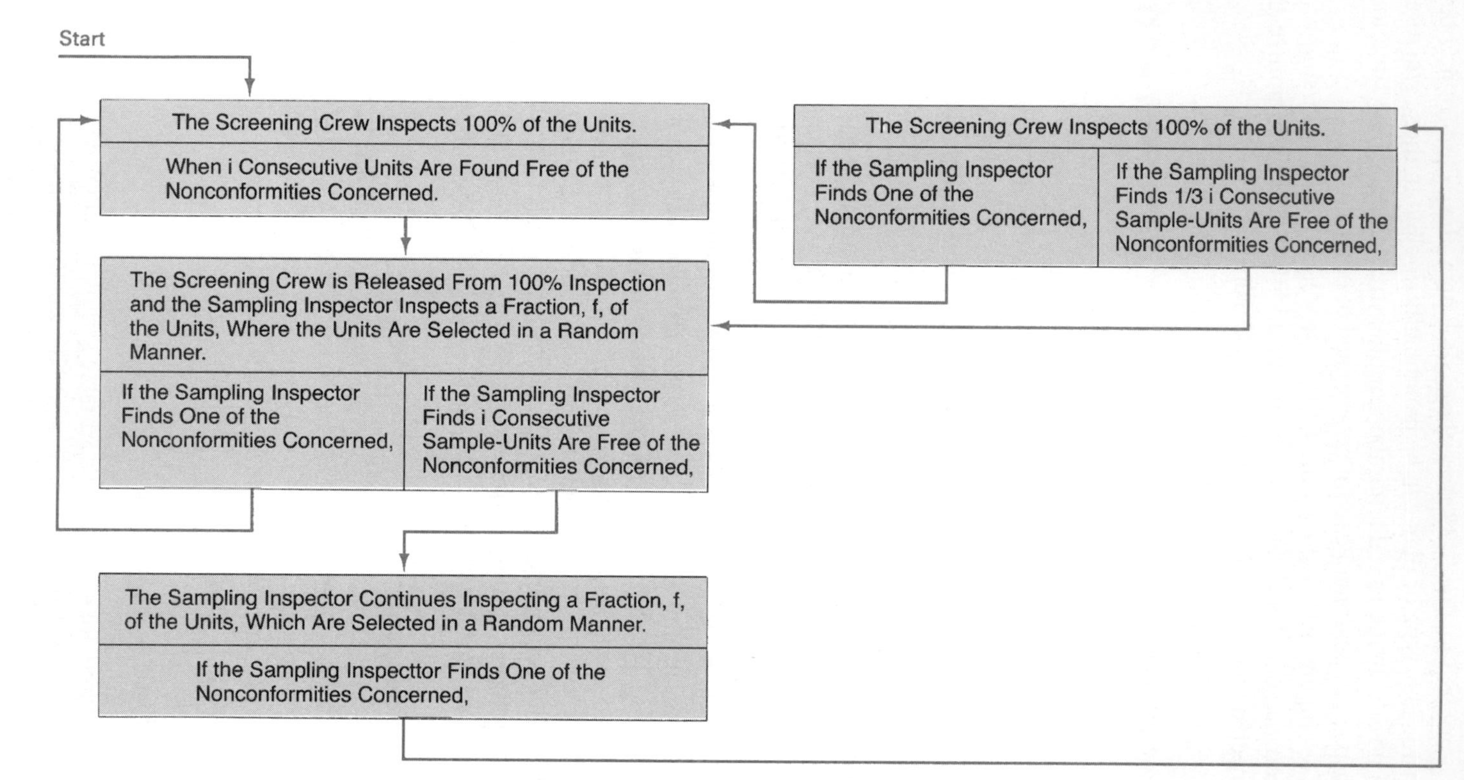

FIGURE 10-13 Procedure for CSP-V plans.

Division of United Aircraft Corporation.[11] The plan uses a plotted frequency distribution (histogram) to evaluate a sample for decisions concerning acceptance or non-acceptance of a lot. The most significant feature of the plan is the fact that it is applicable to both normal and nonnormal frequency distributions. Another feature is its simplicity. It is a practical plan for in-house inspection as well as receiving inspection.

Lot Plot Method.

The method[12] for obtaining the lot plots is as follows:

1. A random sample of 10 subgroups of 5 each for a total of 50 items is obtained from the lot. Table 10-18 shows the inspection results.

2. The average, $\overline{X}$, and range, R, are calculated for each subgroup and are shown in Table 10-18.

3. A histogram is constructed using the techniques described in Chapter 4. The Shainin plan states that the number of cells should be between 7 and 16, which is somewhat larger than the guidelines given previously. The histogram with an interval of 0.3 and 9 cells is shown in Figure 10-14.

4. The average of the averages, $\overline{\overline{X}}$ and the average of the ranges, $\overline{R}$, are

$$\overline{\overline{X}} = \frac{\Sigma \overline{X}}{g} = \frac{976.8}{10} = 97.7 \qquad \overline{R} = \frac{\Sigma R}{g} = \frac{13.7}{10} = 1.37$$

TABLE 10-18 Random Sample of 10 Subgroups of 5 Each for a Total of 50 (Data for the Width of a Brass Plate, in Millimeters).

	1	2	3	4	5	6	7	8	9	10
	96.7	97.0	98.0	97.8	97.5	98.5	98.3	98.2	97.9	97.4
	97.7	98.3	99.0	97.2	96.7	97.1	97.7	97.9	97.7	96.5
	98.4	97.2	98.3	97.6	98.1	96.8	97.6	97.8	97.8	96.9
	97.4	97.2	97.5	98.0	97.1	97.6	98.8	98.1	97.1	97.3
	97.0	97.8	97.7	97.4	96.9	98.2	98.0	98.8	98.3	98.4
Avg.	97.4	97.5	98.1	97.6	97.3	97.6	98.1	98.2	97.8	97.3
Range	1.7	1.3	1.5	0.8	1.4	1.7	1.2	1.0	1.2	1.9

[11]Dorian Shainin, "The Hamilton Standard Lot Plot Method of Acceptance Sampling by Variables," *Industrial Quality Control*, 7, No. 1 (July 1950), 15–34.

[12]The method has been modified to utilize modern calculation techniques and prior information given in the book.

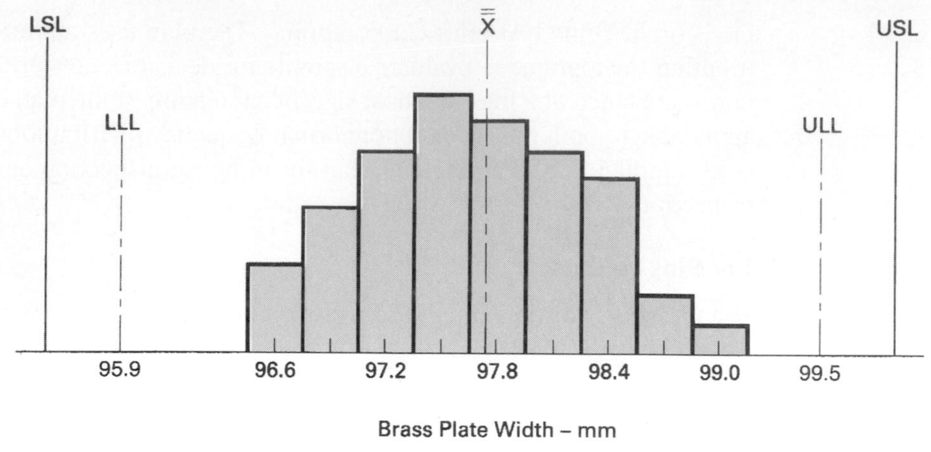

FIGURE 10-14 Lot plot histogram showing lot limits and specifications.

5. Using these values, the upper lot limit and lower lot limit are calculated as follows:

$$\text{ULL} = \overline{\overline{X}} + \frac{3\overline{R}}{d_2} \qquad\qquad \text{LLL} = \overline{\overline{X}} - \frac{3\overline{R}}{d_2}$$

$$= 97.7 + \frac{(3)(1.37)}{2.326} \qquad\qquad = 97.7 - \frac{(3)(1.37)}{2.326}$$

$$= 99.5 \qquad\qquad = 95.9$$

These values are shown in Figure 10-14.

Lot Plot Evaluation. Once the lot plot and the lot limits are obtained, the decision concerning acceptance or non-acceptance is made. This decision is based on a comparison of the lot plot with 11 different types of lot plots, which are shown in Figure 10-15.

The first four types are applicable to lot plots which are approximately normally distributed. In the type 1 situation, the lot plot is well within specification limits, and the lot is accepted without the need to calculate the lot limits. If the lot limits are within the specifications, as illustrated by type 2, the lot is accepted. When the lot limits are outside the specifications, as shown by types 3 and 4, the percentage of product beyond specifications is obtained and a review board determines the final disposition of the stock. In some cases an attribute plan is employed to determine the lot acceptability when one or two values are beyond the lot limits.

The other types of lot plots are used for nonnormal distributions. For example, type 5 is skewed; types 6 and 9 indicate that the lot was screened or sorted;

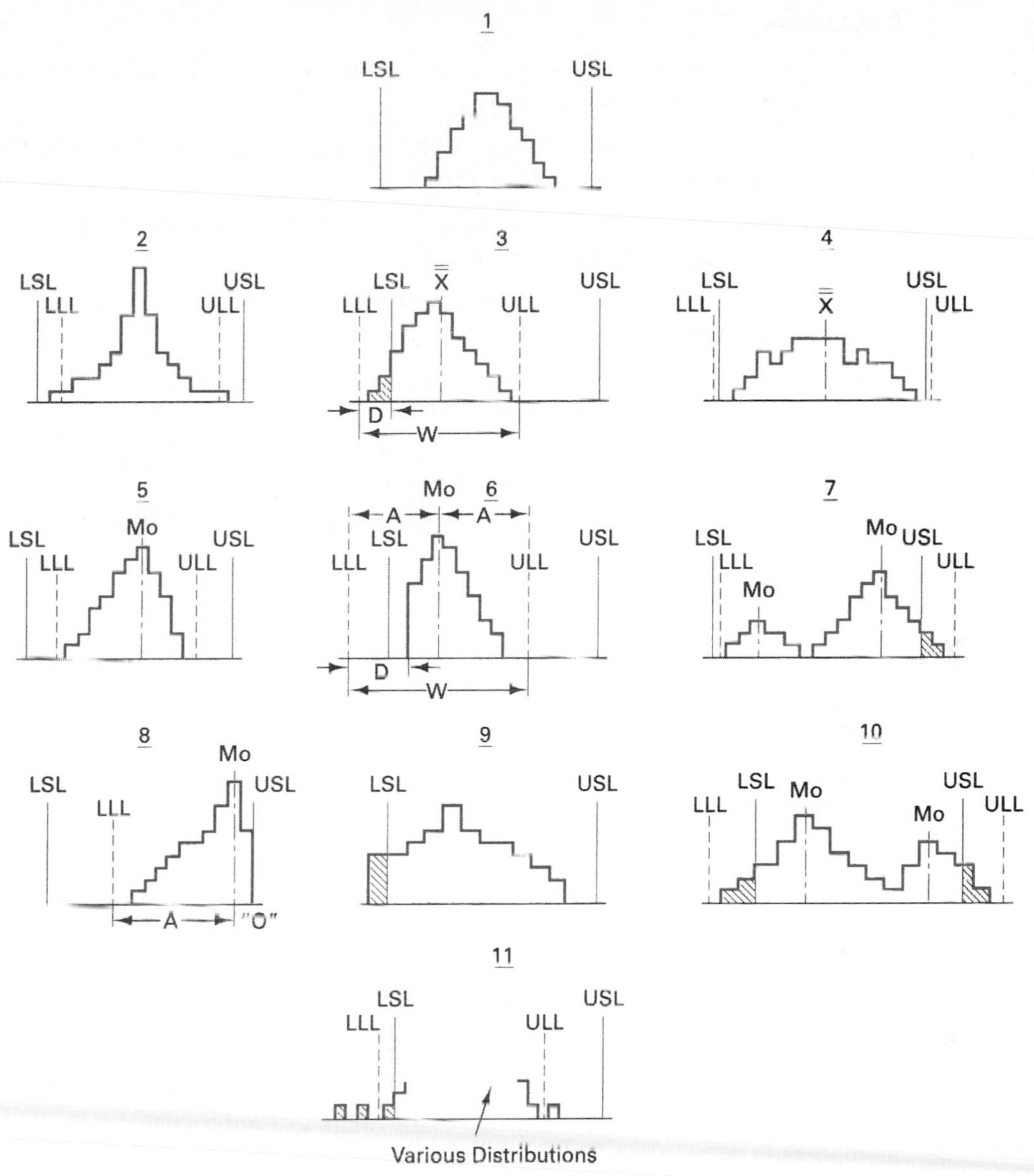

FIGURE 10-15 Eleven typical types of lot plots. [Reproduced by permission from Dorian Shainin. "The Hamilton Standard Lot Plot Method." *Industrial Quality Control*, 7, No. 1 (July 1950), 17.]

types 7 and 10 illustrate the bimodal condition; and type 11 is for stray values. The example problem, as shown by Figure 10-15, illustrates the type 5 lot plot and the lot would be accepted. Special techniques are specified for analyzing the non-normal lot plots.

Comments.

1. Once learned, the lot plot procedure is relatively simple and has resulted in improved quality and lower inspection costs.
2. Unacceptable lots are returned to the producer, and this action will cause a subsequent improvement in quality.
3. Inspectors can accept lots; however, disposition of unsatisfactory lots is left to a material-review board.
4. Many users of the lot plot method have modified the Shainin method for their own situation.
5. The major criticism of the plan is that the shape of the lot plot does not always give an accurate indication of the true distribution. Shainin states that the lot plot is close enough to have no practical effect on the final decision, or if there are any errors, they are in a safe direction.
6. For additional information, the reader is referred to the published articles.[13]

ANSI/ASQ Z1.9—1993[14]

ANSI/ASQ Z1.9—1993 is a lot-by-lot acceptance sampling plan by variables. Modifications to MIL-STD-414 were made by the American Society for Quality so that it would more closely match ANSI/ASQ Z1.4 and ISO/DIS 3951. These modifications have been included in this book.

The standard is indexed by numerical values of the AQL that range from 0.10 to 10.0%. Provision is made for normal, tightened, and reduced inspection. Sample sizes are a function of the lot size and the inspection level. The standard assumes a normally distributed random variable. Since the standard is 101 pages long, only a portion of the tables and the procedures will be given.

The standard makes provision for nine different procedures that can be used to evaluate a lot for acceptance or non-acceptance. Figure 10-16 shows the composition of the standard. If the variability (σ) of the process is known and stable, the variability known plan is the most economical. When the variability is unknown, the standard deviation method or the range method is used. Since the range method requires a larger sample size, the standard deviation method is recommended. There are two types of specifications—single and double. Two alternative procedures, Forms 1 and 2, are available and will lead to the same decision. While Form 1 is somewhat easier, it is only applicable to single specification situations. Therefore, Form 2 is the preferred procedure.

---

[13]Dorian Shainin, "Recent Lot Plot Experiences Around the Country," *Industrial Quality Control,* 8, No. 5 (March 1952), 22.

[14]This section is extracted from ANSI/ASQ Z1.9—1993 by permission of the American Society for Quality.

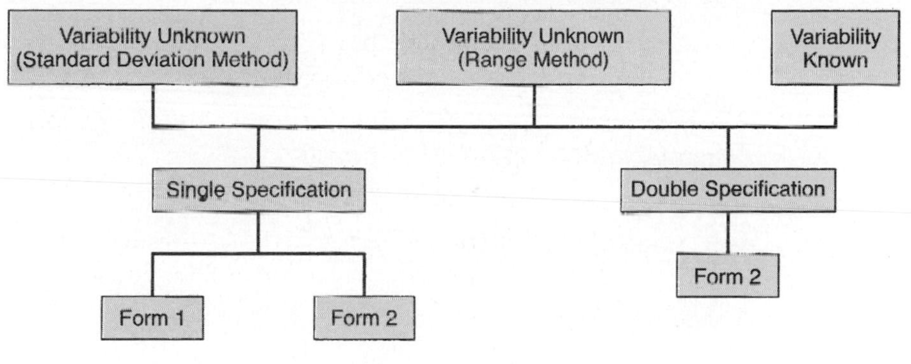

FIGURE 10-16 Composition of ANSI/ASQC Z1.9—1993.

The standard is divided into four sections. Section A contains a general description, sample-size code letters, and OC curves for the sampling plans. Procedure and examples for the unknown variability—standard deviation method are given in Section B; procedures and examples for unknown variability—range method are given in Section C; and procedures and examples for known variability are given in Section D.

The sample size for all the methods is designated by code letters. These code letters are based on the lot size and the inspection level as shown in Table 10-19. There are five inspection levels: Special Levels S3 and S4 and General Levels I, II, and III. The special levels are used when small sample sizes are necessary and large risks can and must be tolerated. An analysis of the general inspection levels is similar to ANSI/ASQ Z1.4. Unless otherwise specified, inspection level II will be used. Inspection level III gives a steeper OC curve and therefore reduces consumer's risk. When greater consumer's risks can be tolerated, inspection level I can be used.

An example problem for unknown variability—standard deviation method, single specification, and Form 2 are used to demonstrate the procedure.

EXAMPLE PROBLEM

The minimum temperature of operation for a certain device is specified as 180°C. A lot of 40 items is submitted for inspection where inspection level II, normal inspection, and AQL = 1.0% are the criteria.

From Table 10-19 the code letter is *D*, which gives a sample $n = 5$ (from Table 10-20). The temperatures for the five samples are 197°, 188°, 184°, 205°, and 201°C.

$$\overline{X} = \frac{\Sigma X}{n} = \frac{197 + 188 + 184 + 205 + 201}{5} = 195°\text{C}$$

$$s = \sqrt{\frac{\Sigma X^2 - \frac{(\Sigma X)^2}{n}}{n-1}} = \sqrt{\frac{190{,}435 - 190{,}125}{5-1}} = 8.80$$

Lower-quality index:

$$Q_L = \frac{\overline{X} - L}{s} = \frac{195 - 180}{8.80} = 1.70$$

Estimate a lot percent nonconforming below L: p_L

From Table 10-21, $p_L = 0.66\%$

Maximum allowable percent nonconforming: M

From Table 10-20, $M = 3.33\%$

The lot meets acceptance criterion if $p_L \leqq M$:

Since $0.66\% < 3.33\%$, accept lot.

TABLE 10-19 Sample-Size Code Letters (Table A-2 of ANSI/ASQ Z1.9).

	INSPECTION LEVELS				
	SPECIAL		GENERAL		
LOT SIZE	S3	S4	I	II	III
2–8	B	B	B	B	C
9–15	B	B	B	B	D
16–25	B	B	B	C	E
26–50	B	B	C	D	F
51–90	B	B	D	E	G
91–150	B	C	E	F	H
151–280	B	D	F	G	I
281–400	C	E	G	H	J
401–500	C	E	G	I	J
501–1,200	D	F	H	J	K
1,201–3,200	E	G	I	K	L
3,201–10,000	F	H	J	L	M
10,001–35,000	G	I	K	M	N
35,001–150,000	H	J	L	N	P
150,001–500,000	H	K	M	P	P
500,001 and over	H	K	N	P	P

TABLE 10-20 **Master Table for Normal and Tightened Inspection for Plans Based on Variability Unknown, Standard Deviation Method (Double Specification Limit and Form 2-Single Specification Limit)—Table B-3 of ANSI/ASQ Z1.9.**

SAMPLE SIZE CODE LETTER	SAMPLE SIZE	ACCEPTABLE QUALITY LEVELS (NORMAL INSPECTION)											
		T	.10	.15	.25	.40	.65	1.00	1.50	2.50	4.00	6.50	10.00
		M	M	M	M	M	M	M	M	M	M	M	M
B	3							↓	↓	7.59	18.86	26.94	33.69
C	4					↓	↓	1.49	5.46	10.88	16.41	22.84	29.43
D	5		↓	↓	↓	0.041	1.34	3.33	5.82	9.80	14.37	20.19	26.55
E	7	↓	0.005	0.087	0.421	1.05	2.13	3.54	5.34	8.40	12.19	17.34	23.30
F	10	0.077	0.179	0.349	0.714	1.27	2.14	3.27	4.72	7.26	10.53	15.17	20.73
G	15	0.186	0.311	0.491	0.839	1.33	2.09	3.06	4.32	6.55	9.48	13.74	18.97
H	20	0.228	0.356	0.531	0.864	1.33	2.03	2.93	4.10	6.18	8.95	13.01	18.07
I	25	0.250	0.378	0.551	0.874	1.32	2.00	2.86	3.97	5.98	8.65	12.60	17.55
J	35	0.253	0.373	0.534	0.833	1.24	1.87	2.66	3.70	5.58	8.11	11.89	16.67
K	50	0.243	0.355	0.503	0.778	1.16	1.73	2.47	3.44	5.21	7.61	11.23	15.87
L	75	0.225	0.326	0.461	0.711	1.06	1.59	2.27	3.17	4.83	7.10	10.58	15.07
M	100	0.218	0.315	0.444	0.684	1.02	1.52	2.18	3.06	4.67	6.88	10.29	14.71
N	150	0.202	0.292	0.412	0.636	0.946	1.42	2.05	2.88	4.42	6.56	9.86	14.18
P	200	0.204	0.294	0.414	0.637	0.945	1.42	2.04	2.86	4.39	6.52	9.80	14.11
		.10	.15	.25	.40	.65	1.00	1.50	2.50	4.00	6.50	10.00	
		Acceptable Quality Levels (tightened inspection)											

All AQL values are in percent nonconforming. T denotes plan used exclusively on tightened inspection and provides symbol for identification of appropriate OC curve. Use first sampling plan below arrow; that is, both sample size as well as M value. When sample size equals or exceeds lot size, every item in the lot must be inspected.

TABLE 10-21 Table for Estimating the Lot Percent Nonconforming (p_L or p_U) Using Standard Deviation Method (Values in percent) (Table B-5 of ANSI/ASQ Z1.9[a]).

	SAMPLE SIZE							
Q_U OR Q_L	5	10	20	30	40	50	100	200
0	50.00	50.00	50.00	50.00	50.00	50.00	50.00	50.00
0.10	46.44	46.16	46.08	46.05	46.04	46.04	46.03	46.02
0.20	42.90	42.35	42.19	42.15	42.13	42.11	42.09	42.08
0.30	39.37	38.60	38.37	38.31	38.28	38.27	38.24	38.22
0.40	35.88	34.93	34.65	34.58	34.54	34.53	34.49	34.47
0.50	32.44	31.37	31.06	30.98	30.95	30.93	30.89	30.87
0.60	29.05	27.94	27.63	27.55	27.52	27.50	27.46	27.44
0.70	25.74	24.67	24.38	24.31	24.28	24.26	24.23	24.21
0.80	22.51	21.57	21.33	21.27	21.25	21.23	21.21	21.20
0.90	19.38	18.67	18.50	18.46	18.44	18.43	18.42	18.41
1.00	16.36	15.97	15.89	15.88	15.87	15.87	15.87	15.87
1.10	13.48	13.50	13.52	13.53	13.54	13.54	13.55	13.56
1.20	10.76	11.24	11.38	11.42	11.44	11.46	11.48	11.49
1.30	8.21	9.22	9.48	9.55	9.58	9.60	9.64	9.66
1.40	5.88	7.44	7.80	7.90	7.94	7.97	8.02	8.05
1.50	3.80	5.87	6.34	6.46	6.52	6.55	6.62	6.65
1.60	2.03	4.54	5.09	5.23	5.30	5.33	5.41	5.44
1.70	0.66	3.41	4.02	4.18	4.25	4.30	4.38	4.42
1.80	0.00	2.49	3.13	3.30	3.38	3.43	3.51	3.55
1.90	0.00	1.75	2.40	2.57	2.65	2.70	2.79	2.83
2.00	0.00	1.17	1.81	1.98	2.06	2.10	2.19	2.23
2.10	0.00	0.74	1.34	1.50	1.58	1.62	1.71	1.75
2.20	0.00	0.437	0.968	1.120	1.192	1.233	1.314	1.352
2.30	0.00	0.233	0.685	0.823	0.888	0.927	1.001	1.037
2.40	0.00	0.109	0.473	0.594	0.653	0.687	0.755	0.787
2.50	0.00	0.041	0.317	0.421	0.473	0.503	0.563	0.592
2.60	0.00	0.011	0.207	0.293	0.337	0.363	0.415	0.441
2.70	0.00	0.001	0.130	0.200	0.236	0.258	0.302	0.325
2.80	0.00	0.000	0.079	0.133	0.162	0.181	0.218	0.237
2.90	0.00	0.000	0.046	0.087	0.110	0.125	0.155	0.171
3.00	0.00	0.000	0.025	0.055	0.073	0.084	0.109	0.122

[a]The actual Table B-5 of ANSI/ASQ Z1.9 contains more sample sizes and about 10 times as many values for Q_u or Q_L.

The example problem pertained to a lower specification. If the single specification had pertained to an upper specification, U, the method would have been the same, except Q_U would have been calculated using the formula

$$Q_U = \frac{U - \overline{X}}{s}$$

The estimate of the percent nonconforming above U, p_U, is obtained from Table 10-21 and compared to M for the acceptance-non-acceptance decision.

If the problem involves an upper and lower specification, then both p_U and p_L are calculated and compared to M.

EXAMPLE PROBLEM

Assuming that there is also an upper specification of 209°C for the previous example problem, determine the status of the lot.

Upper quality index:

$$Q_U = \frac{U - \overline{X}}{s} = \frac{209 - 195}{8.80} = 1.59 \text{ (say, 1.60)}$$

Estimate of lot percent nonconforming above $U = p_U$:

From Table 10-21, $p_U = 2.03\%$

The lot meets acceptance criteria if $p_L + p_U \leq M$

Since $(0.66 + 2.03)\% \leq 3.32\%$, accept lot.

The formula for the quantity index, Q, is very similar to the formula for the Z value, which is given in Chapter 4. Table 10-21 is based on Q and the sample size, whereas Table A in the Appendix is based on the Z value and the infinite situation. The value of p is the estimate of the percent nonconforming, which is above or below the specification limit as shown in Figure 10-17. As long as p_L, p_U, or $p_L + p_U$ is less than the maximum allowable percent nonconforming M (for a particular AQL and n), the lot is accepted.

Normal and tightened inspection use the same table. The AQL values for normal inspection are indexed from the top of the table, and for tightened inspection, they are indexed from the bottom of the table. Switching rules are the same as ANSI/ASQ Z1.4.

The standard contains a special procedure for application of mixed variable—attribute sampling plans. If the lot does not meet the acceptability criterion of the variable plan, an attribute single sampling plan, with tightened inspection

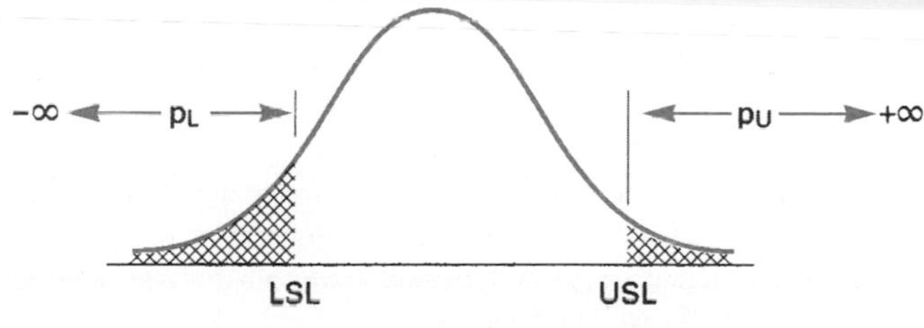

FIGURE 10-17 Percent nonconforming below and above specifications.

and the same AQL, is obtained from ANSI/ASQ Z1.4. A lot can be accepted by either of the plans. Non-acceptance of a lot requires both plans to not accept it.

Other Acceptance Sampling Plans for Variables

There arc three other types of acceptance sampling plans by variables that are occasionally used. These types of variable plans are concerned with the average quality or the variability in the quality of the product and not with the percent nonconforming. They may be used for sampling bulk material that is shipped in bags, drums, tank cars, and so on. A brief discussion of each is given in this section.

Acceptance control charts provide a technique for not accepting or accepting a lot using the sample average. Acceptance control limits and the sample size are established from the known standard deviation, specification limits, AQL, and values of the consumer and producer risks. The use of a control chart allows personnel to observe quality trends.[15]

Sequential sampling by variables can be used when the quality characteristic is normally distributed and when the standard deviation is known. The technique for this sampling plan is similar to the sequential plan by attributes that was discussed previously. However, the variable plan plots the cumulative sum, ΣX, and the attribute plan plots the number of nonconforming units, d. Sequential sampling can result in reduced sampling inspection.[16]

A third type of sampling by variables is referred to as *hypothesis testing*. There are a number of different tests to evaluate the sample average or sample deviation for acceptance or non-acceptance decisions.[17]

PROBLEMS

1. Using ANSI/ASQ Z1.4, a General Services Administration inspector needs to determine the single sampling plans for the following information:

	INSPECTION LEVEL	INSPECTION	AQL	LOT SIZE
(a)	II	Tightened	1.5%	1,400
(b)	I	Normal	65	115
(c)	III	Reduced	0.40%	160,000
(d)	III	Normal	2.5%	27

[15]For more information, see R. A. Freund, "Acceptance Control Charts," *Industrial Quality Control,* October 1957, 13–23.

[16]For more information, see A. J. Duncan, *Quality Control and Industrial Statistics* (Homewood, III.: Richard D. Irwin, Inc., 1987), pp. 346–360.

[17]For more information, see J. M. Juran, ed., *Quality Control Handbook,* 4th ed. (New York: McGraw-Hill Book Company, 1988), Sec. 23, pp. 60–81.

2. Explain the meaning of the sampling plan determined in Problem 1(c) if (a) 6 nonconforming units are found in the sample, (b) 8 nonconforming units are found, and (c) 4 nonconforming units are found.

3. Inspection results for the last 8 lots using ANSI/ASQ Z1.4 and the single sampling plan of $n = 225$, $c = 3$ are

I.	1 nonconforming unit	V.	3 nonconforming units
II.	4 nonconforming units	VI.	0 nonconforming units
III.	5 nonconforming units	VII.	2 nonconforming units
IV.	1 nonconforming unit	VIII.	2 nonconforming units

If normal inspection was used for lot I, what inspection should be used for lot IV?

4. Using the information from Problem 3, what was the status after lot V? after lot VIII?

5. For a single sampling plan using ANSI/ASQ Z1.4, code letter C, normal inspection, and AQL = 25 nonconformities/100 units, the number inspected and the count of nonconformities for the last 10 lots are

	n	c		n	c
I.	5	0	VI.	5	3
II.	5	1	VII.	5	1
III.	5	2	VIII.	5	2
IV.	5	2	IX.	5	3
V.	5	1	X.	5	1

If production is at a steady rate and reduced inspection is authorized, can a change from normal to reduced inspection be initiated?

6. Using ANSI/ASQ Standard Q3 for a lot size of 3500 and an LQ of 5.0%, determine the sampling plan for an isolated lot.

7. A few lots of raw material are purchased from a vendor whose process is a continuous stream of product. Using ANSI/ASQ Standard Q3, determine the sampling plan for Inspection Level II, LQ = 3.15%, and lot size = 4000.

8. Using the Dodge-Romig tables, the quality manager of a telephone manufacturer wants to determine the single sampling plan for minor nonconformities for an AOQL = 3.0% when the process average is 0.80% and the lot size is 2500. What is the LQ?

9. What would be the sampling plan of Problem 8 if the lot is a new product and the process average is unknown?

10. An insurance company is using the Dodge Romig tables for LQ to determine a single sampling plan for an LQ = 1.0% when the process average is 0.35% and $N = 600$. What is the AOQL?

11. If the insurance company cited in Problem 10 has initiated a new form and the process average is not available, what plan is recommended?

12. If the process average is 0.19% nonconforming, what single sampling plan is recommended using the Dodge-Romig LQ tables? LQ is 1.0% and the lot size is 8000. What is the AOQL?

13. Determine the probability of acceptance of product that 0.15% nonconforming using the sampling plan for Problem 10.

14. Determine the OC curve for a ChSP-1 where $n = 4$, $c = 0$, and $i = 3$. Use five points to determine the curve.

15. A chain sampling plan, ChSP-1, is being used for the inspection of lots of 250 pieces. Six samples are inspected. If none are nonconforming, the lot is accepted; if 1 nonconforming unit is found, the lot is accepted if the 3 previous lot samples were free of nonconforming units. Determine the probability of acceptance of a lot that is 3% nonconforming.

16. A unit sequential sampling plan is defined by $p_\alpha = 0.08$, $\alpha = 0.05$, $p_\beta = 0.18$, and $\beta = 0.10$. Determine the equations for the acceptance/non-acceptance line and draw the graphical plan.

17. For a unit sequential sampling plan that is defined by $\alpha = 0.08$, $p_\alpha = 0.05$, $\beta = 0.15$, and $p_\beta = 0.12$, determine the equations for the acceptance and non-acceptance lines. Using these equations, establish a table of the non-acceptance number, acceptance number, and number of units inspected. The table can be stopped when the non-acceptance number equals 6.

18. A food distribution warehouse is evaluating Dodge's SkSP-1 using an AOQL of 1.90%. Determine the i values for $f = \frac{1}{2}, \frac{1}{3}$ and $\frac{1}{4}$.

19. For SkSP-1 determine the i values for $f = \frac{1}{2}, \frac{1}{3}$, and $\frac{1}{4}$ using an AOQL of 0.79%.

20. A hospital supplier of disposable thermometers meets the supplier's first four product requirements. Do they meet Tables 10-12 and 10-13 of ANSI/ASQ S1 for an AQL of 0.25? The sample size and nonconforming items for the first 10 consecutive lots are as follows:

LOT	SAMPLE SIZE	NUMBER NONCONFORMING	LOT	SAMPLE SIZE	NUMBER NONCONFORMING
1	315	0	6	315	0
2	315	2	7	315	0
3	315	0	8	315	0
4	315	0	9	315	1
5	315	1	10	315	0

21. The next 14 consecutive lots of Problem 20 are as follows:

LOT	SAMPLE SIZE	NUMBER NONCONFORMING	LOT	SAMPLE SIZE	NUMBER NONCONFORMING
11	315	1	18	315	1
12	315	0	19	315	2
13	315	0	20	315	1
14	315	0	21	315	0
15	315	0	22	315	0
16	315	0	23	315	0
17	315	0	24	315	0

Describe what happens in terms of states 1, 2, and 3. Be sure to specify the initial frequency and any changes. Note that the ANSI/ASQ Z1.4 sampling plan for single sample normal is $n = 315$ and $c = 2$.

22. Does a manufacturer of capacitors meet Table 10-12 and Table 10-13 requirements and why? Data are as follows: AQL of 0.65%; 20 consecutive lots accepted with a total sample size of 2650; 11 nonconforming units; and the last lots have 1 nonconforming unit each with a sample size of 200.

23. If the product meets Table 10-12 and Table 10-13 requirements for the conditions in Problem 22, describe the initial sampling frequency if (a) all 20 lots meet the individual lot criteria, and if (b) 1 of the 20 lots does not meet the individual lot criteria.

24. What state occurs if lot 25 of Problem 21 has 3 nonconforming units?

25. A microwave manufacturer wants to evaluate three sampling plans for an AOQL = 0.143% using CSP-1. Determine the i values for $f = \frac{1}{2}, \frac{1}{4}$ and $\frac{1}{10}$.

26. For Dodge's CSP-2 plan, determine the value of i for an AOQL value of 4.94% and a frequency of 20%.

27. A computer-paper manufacturer is using MIL-STD-1235B with an AOQL of 1.22%. Determine the value of i for CSP-T with a sampling frequency of $\frac{1}{15}$. What is the sampling frequency for the second and third levels?

28. For CSP-1, determine the value of i for an AOQL = 0.198% and a frequency of $\frac{1}{4}$.

29. Determine the i value for a CSP-F plan with an AOQL of 0.33%, a lot size of 3000, and a frequency of $\frac{1}{5}$. What is the code letter?

30. If the original i value for a CSP-V plan is 150, what is the value once the initial 150 units have passed and a nonconformity occurs?

31. Using the Shainin lot plot, compute the lot limits and draw the lot plot. The hardness inspection results of 50 sample units using Rockwell-C are as follows:

SUBGROUP	DATA	AVERAGE
1	50, 49, 53, 49, 56	51.4
2	52, 50, 47, 50, 51	50.0
3	49, 49, 53, 51, 48	50.0
4	49, 52, 50, 52, 51	50.8
5	51, 53, 51, 52, 53	52.0
6	54, 50, 54, 53, 52	52.6
7	53, 51, 52, 47, 50	50.6
8	46, 55, 54, 52, 52	51.8
9	49, 53, 51, 51, 50	50.8
10	51, 48, 55, 51, 52	51.4

What type of lot plot does the distribution above represent? If the specifications are from 41 to 60, is the lot accepted?

32. The diameter of a 3/8–in. thread has specifications of 9.78 mm and 9.65 mm. Sample results from 50 random inspections are given below. Determine the lot limits and draw the lot plot. What type of lot plot does the distribution below represent?

SUBGROUP	DATA	AVERAGE
1	9.77, 9.76, 9.75, 9.76, 9.76	9.760
2	9.73, 9.74, 9.77, 9.74, 9.77	9.750
3	9.73, 9.77, 9.76, 9.77, 9.75	9.756
4	9.78, 9.77, 9.77, 9.76, 9.78	9.772
5	9.72, 9.78, 9.77, 9.78, 9.74	9.758
6	9.75, 9.77, 9.76, 9.77, 9.77	9.764
7	9.78, 9.76, 9.77, 9.76, 9.78	9.770
8	9.77, 9.77, 9.77, 9.78, 9.78	9.774
9	9.78, 9.77, 9.76, 9.76, 9.77	9.768
10	9.75, 9.78, 9.77, 9.78, 9.76	9.768

33. A lot of 480 items is submitted for inspection with an inspection level of II. Determine the sample-size code letter and the sample size for inspection by variables using ANSI/ASQ Zl.9.

34. Assuming normal inspection, ANSI/ASQ Zl.9, variability unknown-standard deviation method, code letter D, AQL = 2.50%, and a single lower specification of 200 g, determine the acceptance decision using Form 2. The inspection results of the 5 samples are: 204, 211, 199, 209, and 208 g.

35. If the lower specification of Problem 34 is 200.5 g, what is the acceptance decision?

36. For tightened inspection, ANSI/ASQ Z1.9, variability unknown—standard deviation method, code letter F, AQL = 0.65%, and an upper single specification of 4.15 mm, determine whether the lot is accepted. Use Form 2. The results of the 10 sample inspections are 3.90, 3.70, 3.40, 4.20, 3.60, 3.50, 3.70, 3.60, 3.80, and 3.80 mm.

37. If Problem 36 has normal inspection, what is the decision?

38. If Problem 34 has tightened inspection, what is the decision?

39. If Problem 34 also has an upper specification of 212 g, what is the decision?

40. If Problem 36 also has a lower specification of 3.25 mm, what is the decision?

41. Using EXCEL write a computer program for chain sampling.

11 RELIABILITY

FUNDAMENTAL ASPECTS

Definition

Simply stated, reliability is quality over the long run. It is the ability of the product to perform its intended function over a period of time. A product that "works" for a long period of time is a reliable one. Since all units of a product will fail at different times, reliability is a probability.

A more precise definition is: *Reliability is the probability that a product will perform its intended function satisfactorily for a prescribed life under certain stated environmental conditions*. From the definition, there are four factors associated with reliability: (1) numerical value, (2) intended function, (3) life, and (4) environmental conditions.

The numerical value is the probability that the product will function satisfactorily during a particular time. Thus, a value of 0.93 would represent the probability that 93 of 100 products would function after a prescribed period of time and 7 products would not function after the prescribed period of time. Particular probability distributions can be used to describe the failure[1] rate of units of product.

The second factor concerns the intended function of the product. Products are designed for particular applications and are expected to be able to perform those applications. For example, an electric hoist is expected to lift a certain design load; it is not expected to lift a load that exceeds the design specification. The screwdriver is designed to turn screws, not open paint cans.

The third factor in the definition of reliability is the intended life of the product; in other words, how long the product is expected to last. Thus, the life of automobile tires is specified by different values, such as 36 months or 70,000 km, depending on the construction of the tire. Product life is specified as a function of usage, time, or both.

The last factor in the definition involves environmental conditions. A product that is designed to function indoors, such as an upholstered chair, cannot be expected to function reliably outdoors in the sun, wind, and precipitation. Environmental conditions also include the storage and transportation aspects of the product. These aspects may be more severe than actual use.

Achieving Reliability

Emphasis. Increased emphasis is being given to product reliability. One of the reasons for this emphasis is due to the Consumer Protection Act of 1972. Another reason is the fact that products are more complicated. At one time the washing machine was a simple device that agitated the clothes in a hot, soapy solution. Today, a washing machine has different agitating speeds, different rinse speeds, different cycle times, different water temperatures, different water levels, and provisions to dispense a number of washing ingredients at precise times in the cycle. An additional reason for the increased emphasis on reliability is due to automation; people are, in many cases, not able to manually operate the product if an automated component does not function.

System Reliability. As products become more complex (have more components), the chance that they will not function increases. The method of arranging the components affects the reliability of the entire system. Components can be arranged in series, parallel, or a combination. Figure 11-1 illustrates the various arrangements.

---

[1]The word *failure* is used in this chapter in its limited technical sense and refers to the testing activity rather than usage.

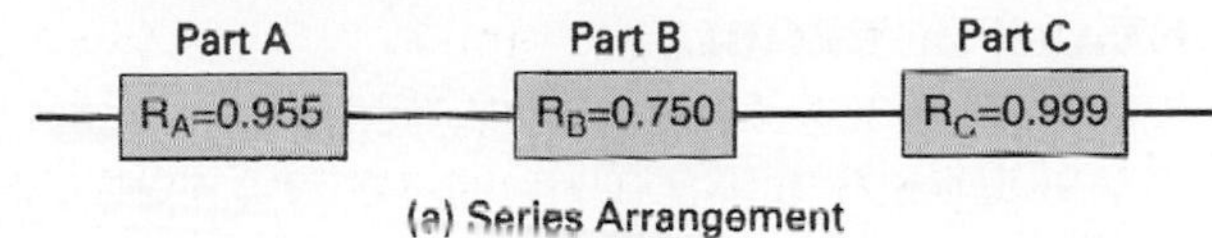

(a) Series Arrangement

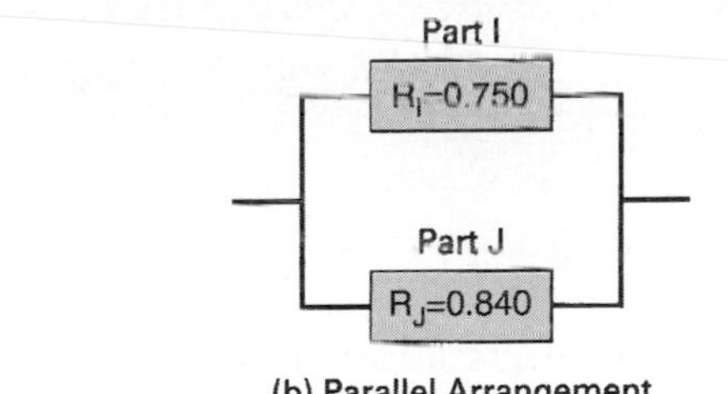

(b) Parallel Arrangement

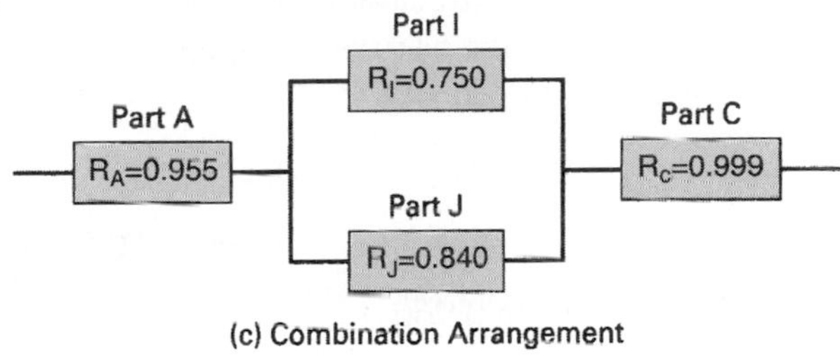

(c) Combination Arrangement

FIGURE 11-1 Methods of arranging components.

When components are arranged in series, the reliability of the system is the product of the individual components. Thus, for the series arrangement of Figure 11-1 (a), the multiplicative theorem is applicable and the series reliability, R_S, is calculated as follows:

$$
\begin{aligned}
R_S &= (R_A)(R_B)(R_C) \\
&= (0.955)(0.750)(0.999) \\
&= 0.716
\end{aligned}
$$

Note that R_A, R_B, and R_C are the probability (P_A, P_B, and P_C) that components A, B, and C will work. As components are added to the series, the system reliability decreases. Also, the system reliability is always less than its lowest value. The cliché that a chain is only as strong as its weakest link is a mathematical fact.

EXAMPLE PROBLEM

A system has 5 components, A, B, C, D, and E, with reliability values of 0.985, 0.890, 0.985, 0.999, and 0.999, respectively. If the components are in series, what is the system reliability?

$$\begin{aligned} R_S &= (R_A)(R_B)(R_C)(R_D)(R_E) \\ &= (0.985)(0.890)(0.985)(0.999)(0.999) \\ &= 0.862 \end{aligned}$$

When components are arranged in series and a component does not function, then the entire system does not function. This is not the case when the components are arranged in parallel. When a component does not function, the product continues to function using another component until all parallel components do not function. Thus, for the parallel arrangement in Figure 11-1(b), the system parallel, R_S, is calculated as follows:

$$\begin{aligned} R_S &= 1 - (1 - R_I)(1 - R_J) \\ &= 1 - (1 - 0.750)(1 - 0.840) \\ &= 0.960 \end{aligned}$$

Note that $(1 - R_I)$ and $(1 - R_J)$ are the probabilities that components I and J will not function. As the number of components in parallel increases, the reliability increases. The reliability for a parallel arrangement of components is greater than the reliability of the individual components.

EXAMPLE PROBLEM

Determine the system reliability of 3 components—A,B and C—with individual reliabilities of 0.989, 0.996, and 0.994 when they are arranged in parallel.

$$\begin{aligned} R_S &= 1 - (1 - R_A)(1 - R_B)(1 - R_C) \\ &= 1 - (1 - 0.989)(1 - 0.996)(1 - 0.994) \\ &= 0.999999736 \end{aligned}$$

Note that 9 significant figures are used in the answer to emphasize the parallel component principle.

Most complex products are a combination of series and parallel arrangements of components. This is illustrated in Figure 11-1(c), wherein part B is replaced by the parallel components, part I and part J. The reliability of the system, R_S, is calculated as follows:

$$
\begin{aligned}
R_S &= (R_A)(R_{I,J})(R_C) \\
&= (0.95)(0.96)(0.99) \\
&= 0.90
\end{aligned}
$$

EXAMPLE PROBLEM

Find the reliability of the system below where components 1, 2, 3, 4, 5, and 6 have reliabilities of 0.900, 0.956, 0.982, 0.999, 0.953, and 0.953.

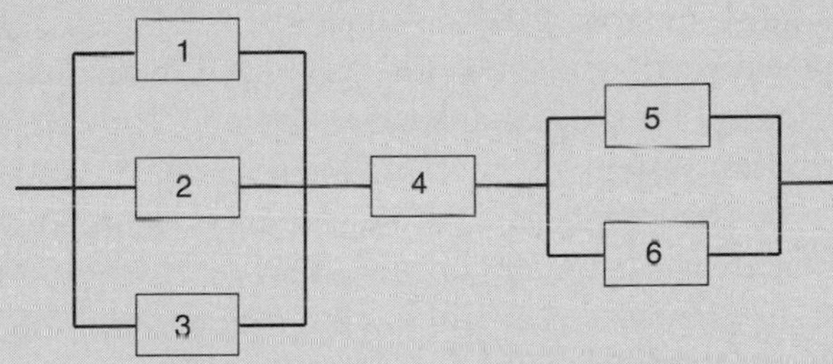

$$
\begin{aligned}
R_S &= (R_{1,2,3})(R_4)(R_{5,6}) \\
&= [1 - (1 - R_1)(1 - R_2)(1 - R_3)][R_4] \\
&\quad [1 - (1 - R_5)(1 - R_6)] \\
&= [1 - (1 - 0.900)(1 - 0.956)(1 - 0.982)] \\
&\quad [0.999][1 - (1 - 0.953)(1 - 0.953)] \\
&= .997
\end{aligned}
$$

While most products are comprised of series and parallel systems, there are complex systems, such as a wheatstone bridge or standby redundancy, which are more difficult to analyze.

Design. The most important aspect of reliability is the design. It should be as simple as possible. As previously pointed out, the fewer the number of components, the greater the reliability. If a system has 50 components in series and each component has a reliability of 0.990, the system reliability is

$$R_S = R^n = 0.990^{50} = 0.605$$

If the system has 20 components in series, the system reliability is

$$R_S = R^n = 0.990^{20} = 0.818$$

Although this example may not be realistic, it does support the fact that the fewer the components, the greater the reliability.

Another way of achieving reliability is to have a backup or redundant component. When the primary component does not function, another component is activated. This concept was illustrated by the parallel arrangement of components. It is frequently cheaper to have inexpensive redundant components to achieve a particular reliability than to have a single expensive component.

Reliability can also be achieved by overdesign. The use of large factors of safety can increase the reliability of a product. For example, a 1–in. rope may be substituted for a $\frac{1}{2}$-in. rope even though the $\frac{1}{2}$-in. rope would have been sufficient.

When an unreliable product can lead to a fatality or substantial financial loss, a fail-safe type of device should be used. Thus, disabling extremity injuries from power-press operations are minimized by the use of a clutch. The clutch must be engaged for the ram and die to descend. If there is a malfunction of the clutch-activation system, the press will not operate.

The maintenance of the system is an important factor in reliability. Products that are easy to maintain will likely receive better maintenance. In some situations it may be more practical to eliminate the need for maintenance. For example, oil-impregnated bearings do not need lubrication for the life of the product.

Environmental conditions such as dust, temperature, moisture, and vibration can be the cause of an unreliable product. The designer must protect the product from these conditions. Heat shields, rubber vibration mounts, and filters are used to increase the reliability under adverse environmental conditions.

There is a definite relationship between investment in reliability (cost) and reliability. After a certain point, there is only a slight improvement in reliability for a large increase in product cost. For example, assume that a $50 component has a reliability of 0.750. If the cost is increased to $100, the reliability becomes 0.900; if the cost is increased to $150, the reliability becomes 0.940; and if the cost is increased to $200, the reliability becomes 0.960. As can be seen by this hypothetical example, there is a diminishing reliability return for the investment dollar.

Production. The production process is the second most important aspect of reliability. Basic quality techniques that have been described in earlier chapters will minimize the risk of product unreliability. Emphasis should be placed on those components which are least reliable.

Production personnel can take action to ensure that the equipment used is right for the job and investigate new equipment as it becomes available. In addition, they can experiment with process conditions to determine which conditions produce the most reliable product.

Transportation. The third aspect of reliability is the transportation of the product to the customer. No matter how well conceived the design or how carefully produced, the actual performance of the product by the customer is the final evaluation. The reliability of the product at the point of use can be greatly affected by the type of handling the product receives in transit. Good packaging techniques and shipment evaluation are essential.

Maintenance. While designers try to eliminate the need for customer maintenance, there are many situations where it is not practical or possible. In such cases, the customer should be given ample warning: for example, a warning light or buzzer when a component needs a lubricant. Maintenance should be simple and easy to perform.

ADDITIONAL STATISTICAL ASPECTS

Distributions Applicable to Reliability

Types of continuous probability distributions used in reliability studies are exponential, normal, and Weibull.[2] Their frequency distributions as a function of time are given in Figure 11-2(a).

Reliability Curves

Reliability curves for the exponential, normal, and Weibull distributions as a function of time are given in Figure 11-2(b). The formulas for these distributions are also given in the figure. For the exponential and Weibull curves the formulas are $R_t = e^{-t/\theta}$ and $R_t = e^{-\alpha t^{\beta}}$, respectively. The formula for the normal distribution is

$$R_t = 1.0 - \int_0^t f(t)dt$$

which requires integration. However, Table A in the Appendix can be used to find the area under the curve, which is the $\int_0^t f(t)dt$.

Failure-Rate Curve

Failure-rate is important in describing the life-history curve of a product. The failure-rate curves and formulas for the exponential, normal, and Weibull as a function of time are shown in Figure 11-2(c).

---

[2]A fourth type, the gamma distribution, is not given because of its limited application. Also, the discrete probability distributions, geometric and negative binomial, are not given for the same reason.

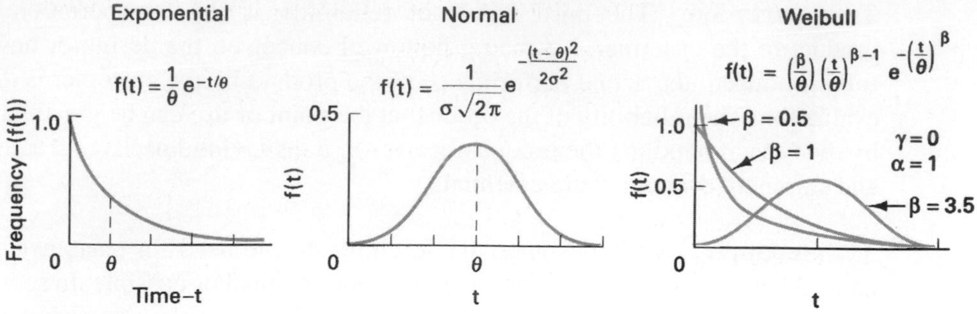

a) Frequency Distribution as a Function of Time

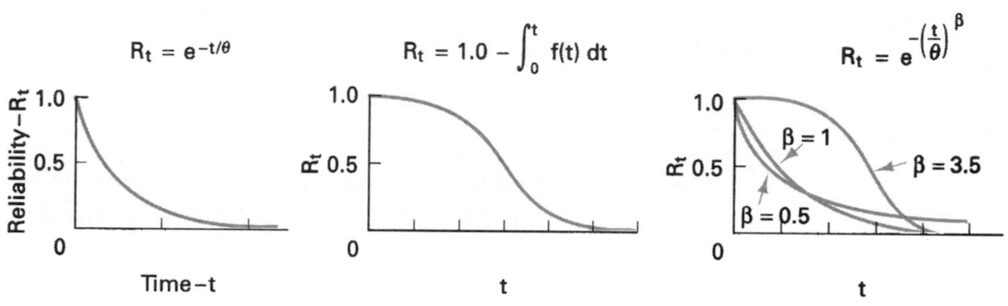

b) Reliability as a Function of Time

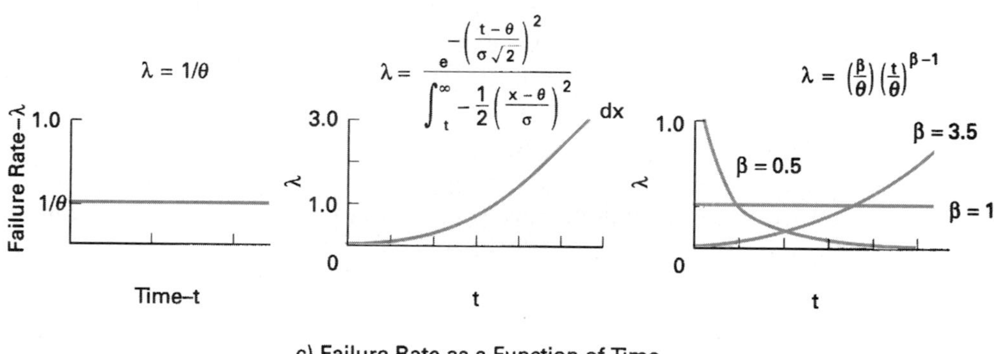

c) Failure Rate as a Function of Time

FIGURE 11-2 Probability distributions, failure-rate curves, and reliability curves as a function of time.

Failure rate can be estimated from test data by use of the formula

$$\lambda_{est} = \frac{\text{number of test failures}}{\text{sum of test times or cycles}} = \frac{r}{\Sigma t + (n - r)T}$$

where λ = failure rate, which is the probability that a unit will fail in a stated unit of time or cycles

r = number of test failures

t = test time for a failed item

n = number of items tested

T = termination time

The formula is applicable for the time terminated without a replacement situation. It is modified for the time terminated with replacement and failure terminated situations. The following examples will illustrate the difference.

EXAMPLE PROBLEM (TIME TERMINATED WITHOUT REPLACEMENT)

Determine the failure rate for an item that has the test of 9 items terminated at the end of 22 hours. Four of the items failed after 4, 12, 15, and 21 h, respectively. Five items were still operating at the end of 22h.

$$\begin{aligned}\lambda_{est} &= \frac{r}{\Sigma t + (n - r)T} \\ &= \frac{4}{(4 + 12 + 15 + 21) + (9 - 4)22} \\ &= 0.025\end{aligned}$$

EXAMPLE PROBLEM (TIME TERMINATED WITH REPLACEMENT)

Determine the failure rate for 50 items that are tested for 15h. When a failure occurs, the item is replaced with another unit. At the end of 15 h, 6 of the items had failed.

$$\begin{aligned}\lambda_{est} &= \frac{r}{\Sigma t} \\ &= \frac{6}{50(15)} \\ &= 0.008\end{aligned}$$

Note that the formula was simplified because the total test time is equal to Σt.

EXAMPLE PROBLEM (FAILURE TERMINATED)

Determine the failure rate for 6 items that are tested to failure. Test cycles are 1025, 1550, 2232, 3786, 5608, and 7918.

$$\lambda = \frac{r}{\Sigma t}$$

$$= \frac{6}{1025 + 1550 + 2232 + 3786 + 5608 + 7918}$$

$$= 0.00027$$

Note that the formula was simplified because the total test time is equal to Σt.

For the exponential distribution and for the Weibull distribution when β, the shape parameter, equals 1, there is a constant failure rate. When the failure rate is constant, the relationship between mean life and failure rate is as follows:[3]

$$\theta = \frac{1}{\lambda} \quad \text{(for constant failure rate)}$$

where θ = mean life or Mean Time Between Failures (MTBF)

EXAMPLE PROBLEM

Determine the mean life for the three previous example problems. Assume that there is a constant failure rate.

$$\theta = \frac{1}{\lambda} = \frac{1}{0.025} = 40 \text{ h}$$

$$\theta = \frac{1}{\lambda} = \frac{1}{0.008} = 125 \text{ h}$$

$$\theta = \frac{1}{\lambda} = \frac{1}{0.00027} = 3704 \text{ cycles}$$

Life-History Curve

Figure 11-3 shows a typical life-history curve of a complex product for an infinite number of items. The curve, sometimes referred to as the "bathtub" curve, is a comparison of failure rate with time. It has three distinct phases: the debugging phase, chance failure phase, and wear-out phase. The probability distributions shown in Figure 11-2(c) are used to describe these phases.

---

[3]Failure rate is also equal to $f(t)/R_t$.

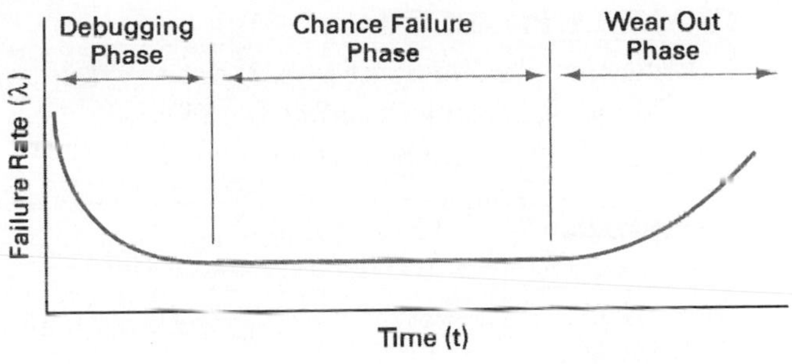

FIGURE 11-3 Typical life history of a complex product for an infinite number of items.

The *debugging phase,* which is also called the burn-in or infant-mortality phase, is characterized by marginal and short-life parts that cause a rapid decrease in the failure rate. While the shape of the curve varies somewhat due to the type of product, the Weibull distribution with shaping parameters less than 1, $\beta < 1$, is used to describe the occurrence of failures. The debugging phase may be part of the testing activity prior to shipment for some products. For other products, this phase is usually covered by the warranty period. In either case, it is a significant quality cost.

The *chance failure phase* is shown in the figure as a horizontal line, thereby making the failure rate constant. Failures occur in a random manner due to the constant failure rate. The assumption of a constant failure rate is valid for most products; however, some products may have a failure rate that increases with time. In fact, a few products show a slight decrease, which means that the product is actually improving over time. The exponential distribution and the Weibull distribution with shape parameter equal to 1 are used to describe this phase of the life history. When the curve increases or decreases, a Weibull shape parameter greater or less than 1 can be used. Reliability studies and sampling plans are, for the most part, concerned with the chance failure phase. The lower the failure rate, the better the product.

The third phase is the *wear-out phase,* which is depicted by a sharp rise in the failure rate. Usually the normal distribution is the one that best describes the wear-out phase. However, the Weibull distribution with shape parameters greater than 1, $\beta > 1$, can be used depending on the type of wear-out distribution.

The curve shown in Figure 11-3 is the type of failure pattern exhibited by most products; however, there will be some products that deviate from this curve. It is important to know the type of failure pattern so that known probability distributions can be used for analysis and prediction of product reliability. By changing the shape parameter, β, all three phases can be modeled using the Weibull. Test results from samples are used to determine the appropriate probability distribution. An example will illustrate the construction of the life-history curve.

EXAMPLE PROBLEM

Determine the life-history curve for the test data in cycles for 1000 items. Assume that failure occurred at $\frac{1}{2}$ the cycle range and survivors went to the end of the cycle range. Data are

NUMBER OF CYCLES	NUMBER OF FAILURES	NUMBER OF SURVIVORS	CALCULATIONS $\Lambda = r/\Sigma t$
0–10	347	653	347/((5)(347) + (10)(653)) = 0.0420
11–20	70	583	70/((15)(70) + (20)(583)) = 0.0055
21–30	59	524	59/((25)(59) + (30)(524)) = 0.0034
31–40	53	471	53/((35)(53) + (40)(471)) = 0.0026
41–50	51	420	51/((45)(51) + (50)(420)) = 0.0022
51–60	60	360	60/((55)(60) + (60)(360)) = 0.0024
61–70	79	281	79/((65)(79) + (70)(281)) = 0.0032
71–80	92	189	92/((75)(92) + (80)(189)) = 0.0043
81–90	189	0	189/((85)(189) + (90)(0)) = 0.0118

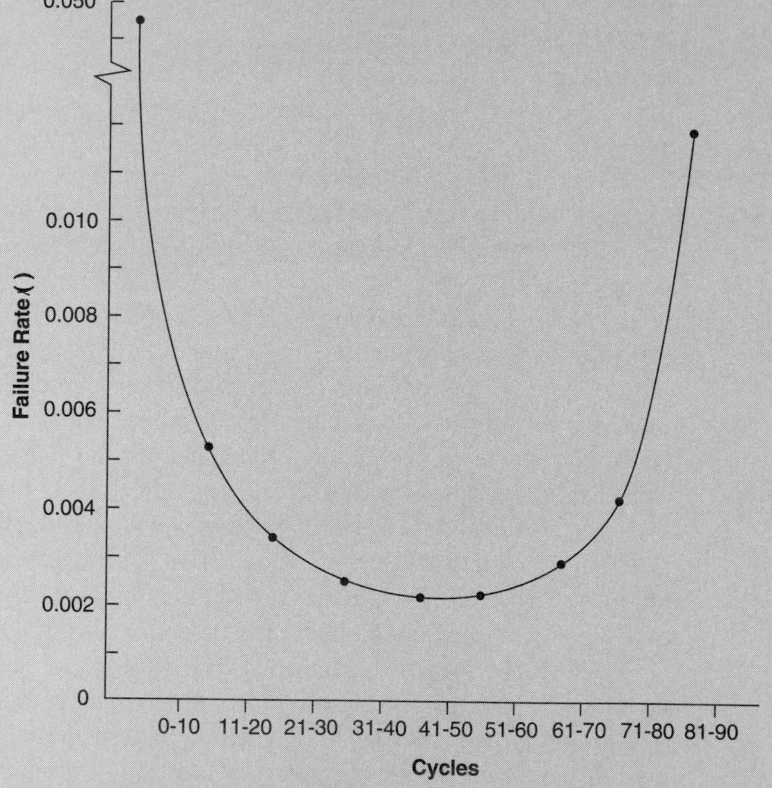

Note that the points are plotted between the cycle values because the failure are is for the entire cell.

Normal Failure Analysis

While the normal curve is applicable to the wear-out phase, the Weibull is usually used. The normal curve is introduced first because the reader is familiar with its use. From Figure 11-2(b), the formula for reliability is

$$R_t = 1.0 - \int_0^t f(t)dt$$

However, the integral, $\int_0^t f(t)dt$ (see Figure 11-2[a]), is the area under the curve to the left of time, t, and is obtained from Appendix Table A. Thus, our equation becomes

$$R_t = 1.0 - P(t)$$

where R_t = reliability at time t

$P(t)$ = probability of failure or area of the normal curve to the left of time t

The process is the same as that learned in Chapter 4. An example problem will illustrate the technique.

EXAMPLE PROBLEM

A 75 W light bulb has a mean life of 750 h with a standard deviation of 50 h. What is the reliablity at 850 h?

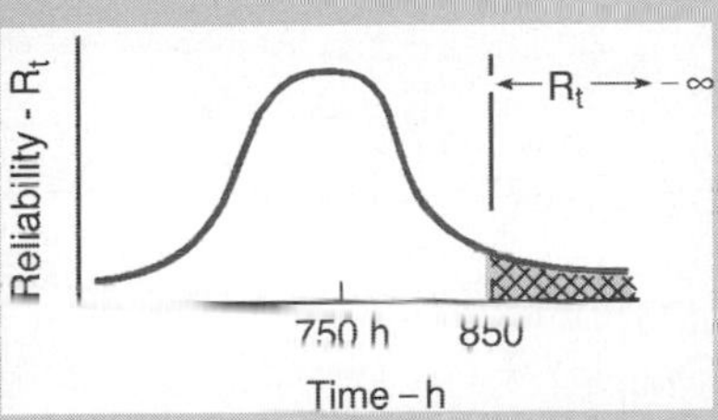

$$Z = \frac{X - \theta}{\sigma} = \frac{850 - 750}{50} = 2.0$$

From Table A, $P(t) = 0.9773$

$$R_{t=850} = 1.0 - P(t)$$
$$= 1.0 - 0.9773$$
$$= 0.0127 \text{ or } 1.27\%$$

On the average a light bulb will last 850 h. 1.27% of the time. Stated another way, 127 light bulbs our of 10,000 will last 850 h or more.

[a]Note that θ is substituted for μ in the equation for z.

Exponential Failure Analysis

As previously stated, the exponential distribution and the Weibull distribution with shape parameter 1 are used to describe the constant failure rate. Knowing the failure rate and its reciprocal, the mean life, we can calculate the reliability using the formula

$$R_t = e^{-t/\theta}$$

where t = time or cycles

θ = mean life

EXAMPLE PROBLEM

Determine the reliability at $t = 30$ for the example problem where the mean life for a constant failure rate was 40 h.

$$R_t = e^{-t/\theta}$$
$$= e^{-30/40}$$
$$= .472$$

What is the reliability at 10 h?

$$R_t = e^{-t/\theta}$$
$$= e^{-10/40}$$
$$= .453$$

What is the reliability at 50 h?

$$R_t = e^{-t/\theta}$$
$$= e^{-50/40}$$
$$= .287$$

The example problem shows that the reliability of the item is less as the time increases. This fact is graphically illustrated by Figure 11-2(b).

Weibull Failure Analysis

The Weibull distribution can be used for the debugging phase ($\beta < 1$), the chance failure phase ($\beta = 1$), and the wear-out phase ($\beta > 1$). By setting $= 1$, the Weibull equals the exponential; by setting $\beta = 3.4$, the Weibull approximates the normal.

From Figure 11-2(b) the formula for the reliability is

$$R_t = e^{-\left(\frac{t}{\theta}\right)^\beta}$$

where β = the Weibull slope

The estimation of the parameters θ and β can be accomplished either graphically, analytically, or by an electronic spreadsheet such as EXCEL. Graphical analysis uses special Weibull probability paper. The data is plotted on the paper and a best fit line is "eye balled." From this information θ and β are determined. The computer has made this technique obsolete. An electronic spreadsheet can accomplish the same goal as probability paper, and it is more accurate.[4]

EXAMPLE PROBLEM

The failure pattern of a new type of battery fits the Weibull distribution with slope 4.2 and mean life 103 h. Determine its reliability at 120 h.

$$R_t = e^{-\left(\frac{t}{\theta}\right)^{\beta}}$$

$$= e^{-\left(\frac{120}{103}\right)^{4.2}}$$

$$= 0.150$$

OC Curve Construction

The Operating Characteristic (OC) curve is constructed in a manner similar to that given in Chapter 9. However, the fraction nonconforming, p_0, is replaced by the mean life θ. The shape of the OC curve as shown in Figure 11-4 is different than those of

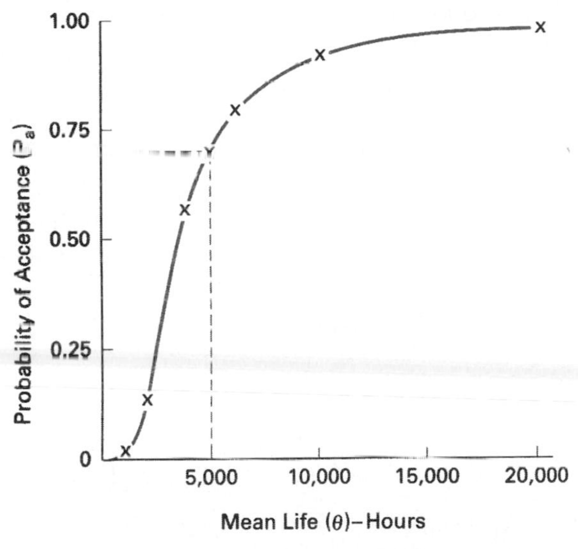

FIGURE 11-4 OC curve for the sampling plan $n = 16$, $T = 600\ h$, $c = 2$, and $r = 3$.

[4]For more information, see D. L. Grosh, *A Primer of Reliability Theory* (New York: John Wiley & Sons, 1989), pp. 67–69.

TABLE 11-1 **Calculations for the OC Curve for the Sampling Plan $n = 16, T = 600$ h, $c = 2, r = 3$.**

MEAN LIFE θ	FAILURE RATE $\lambda = 1/\theta$	EXPECTED AVERAGE NUMBER OF FAILURES $nT\lambda$	P_a $c = 2$
20,000	0.00005	0.48	0.983[a]
10,000	0.0001	0.96	0.927[a]
5,000	0.0002	1.92	0.698[a]
2,000	0.0005	4.80	0.142
1,000	0.0010	9.60	0.004
4,000	0.00025	2.40	0.570
6,000	0.00017	1.60	0.783

[a]By interpolation.

Chapter 9. If lots are submitted with a mean life of 5000 h, the probability of acceptance is 0.697 using the sampling plan described by the OC curve of Figure 11-4.

An example problem for a constant failure rate will be used to illustrate the construction. A lot-by-lot acceptance sampling plan with replacement is as follows:

Select a sample of 16 units from a lot and test each item for 600 h. If 2 or less items fail, accept the lot; if 3 or more items fail, do not accept the lot. In symbols, the plan is $n = 16$, $T = 600$ h, $c = 2$, and $r = 3$. When an item fails, it is replaced by another one from the lot. The first step in the construction of the curve is to assume values for the mean life θ. These values are converted to the failure rate, λ, as shown in the second column of Table 11–1. The expected average number of failures for this sampling plan is obtained by multiplying nT [$nT = (16)(600)$] by the failure rate as shown in the third column of the table.

The value $nT\lambda$ performs the same function as the value of np_0, which was previously used for the construction of an OC curve. Values for the probability of acceptance of the lot are found in Table C of the Appendix for $c = 2$. Typical calculations are as follows (assume that $\theta = 2000$):

$$\lambda = \frac{1}{\theta} = \frac{1}{2000} = 0.0005$$

$$nT\lambda = (16)(600)(0.0005) = 4.80$$

From Table C of the Appendix for $nT\lambda = 4.80$ and $c = 2$,

$$P_a = 0.142$$

Additional calculations for other assumed values of θ are shown in Table 11-1.

Since this OC curve assumes a constant failure rate, the exponential distribution is applicable. The Poisson[5] distribution is used to construct the OC curve since it approximates the exponential.

Because of the constant failure rate, there are other sampling plans that will have the same OC curve. Some of these are

$$n = 4, \qquad T = 2400 \text{ h}, \qquad c = 2$$

$$n = 8, \qquad T = 1200 \text{ h}, \qquad c = 2$$

$$n = 24, \qquad T = 450 \text{ h}, \qquad c = 2$$

Any combination of n and T values that give 9600 with $c = 2$ will have the same OC curve.

OC curves for reliability sampling plans are also plotted as a function of θ/θ_0, which is the actual mean life/acceptable mean life. When the OC curve is constructed in this manner, all OC curves for life tests with or without replacement have one point in common. This point is the producer's risk α and $\theta/\theta_0 = 1.0$.

LIFE AND RELIABILITY TESTING PLANS

Types of Tests

Since reliability testing requires the use of the product and sometimes its destruction, the type of test and the amount of testing is usually an economic decision. Testing is normally done on the end product; however, components and parts can be tested if they are presenting problems. Since testing is usually done in the laboratory, every effort should be made to simulate the real environment under controlled conditions.

Life tests are of the following three types:

Failure-Terminated. These life-test sample plans are terminated when a preassigned number of failures occurs to the sample. Acceptance criteria for the lot are based on the accumulated item test times when the test is terminated.

Time-Terminated. This type of life-test sampling plan is terminated when the sample obtains a predetermined test time. Acceptance criteria for the lot are based on the number of failures in the sample during the test time.

Sequential. A third type of life-testing plan is a sequential life-test sampling plan whereby neither the number of failures nor the time required to reach a decision are

[5] $P(c) = \frac{(np_0)^c}{c!} e^{-np_0}$ (Poisson formula) Substituting $\lambda T = np_0$ and $c = 0$, then $R_t = P(0) = \frac{\lambda T^0}{0!} e^{-\lambda T} = e^{-\lambda T}$

fixed in advance. Instead, decisions depend on the accumulated results of the life test. The sequential life-test plans have the advantage that the expected test time and the expected number of failures required to reach a decision as to lot acceptability are less than the failure-terminated or the time-terminated types.

Testing may be conducted with replacement of a failed unit or without replacement. *With replacement* occurs when a failure is replaced with another unit. Test time continues to be accumulated with the new sample unit. This situation is possible when there is a constant failure rate and the replaced unit has an equal chance of failure. The *without-replacement* situation occurs when the failure is not replaced.

Tests are based on one or more of the following characteristics:

1. *Mean life*—the average life of the product.
2. *Failure rate*—the percentage of failures per unit time or number of cycles.
3. *Hazard rate*—the instantaneous failure rate at a specified time. This varies with age except in the special case of a constant failure rate wherein the failure rate and hazard rate are the same. The Weibull distribution is applicable and the hazard rate increases with age if the shape parameter β is greater than 1 and decreases with age if the shape parameter is less than 1.
4. *Reliable life*—the life beyond which some specified portion of the items in the lot will survive. The Weibull distribution and the normal distribution as they pertain to the wear-out phase are applicable.

Table 11-2 gives a summary of some of the life-testing and reliability plans. The time-terminated tests in terms of mean-life criteria are the most common plans.

Handbook H108

Quality Control and Reliability Handbook H108 gives sampling procedures and tables for life and reliability testing. Sampling plans in the handbook are based on the exponential distribution. The handbook provides for the three different types of tests: failure-terminated, time-terminated, and sequential. For each of these types of tests, provision is made for the two situations: with replacement of failed units during the test or without replacement. Essentially, the plans are based on the mean-life criterion, although failure rate is used in one part of the handbook.

Since the handbook is over 70 pages long, only one of the plans will be illustrated. This plan is a time-terminated, with-replacement, mean-life plan, which is a common plan. There are three methods of obtaining this plan. Example problems will be used to illustrate the methods.

1. *Stipulated producer's risk, consumer's risk, and sample size*. Determine the time-terminated, with-replacement, mean-life sampling plan, where the producer's

TABLE 11-2 **Summary of Some Life-Testing and Reliability Plans.**

	PLANS IN TERMS OF:					TYPE OF TEST		
DOCUMENT	BASIC DISTRIBUTION AND TYPE OF PLAN	MEAN LIFE	HAZARD RATE	RELIABLE LIFE	FAILURE RATE (FR)	FAILURE-TERMINATED	TIME-TERMINATED	SEQUENTIAL
H 103[a]	Exponential lot by lot	X			X	X	X	X
MIL-STD-690B[b]	Exponential, lot by lot				X		X	
MIL-STD-781C[c]	Exponential sampling scheme	X					X	X
TR-3[d]	Weibull, lot by lot	X					X	
TR-4[e]	Weibull, lot by lot		X				X	
TR-6[f]	Weibull, lot by lot			X			X	
TR-7[g]	Weibull, lot by lot, converts MIL-STD-105D	X	X	X			X	

[a]"H108, Sampling Procedures and Tables for Life and Reliability Testing (Based on Exponential Distribution)," U.S. Department of Defense, Quality Control and Reliability Handbook, Government Printing Office, Washington, D.C., 1960.

[b]"MIL-STD-690B, Failure Rate Sampling Plans and Procedures," U.S. Department of Defense, Military Standard, Government Printing Office, Washington, D.C., 1974.

[c]"MIL-STD-781C, Reliability Testing for Engineering Development, Qualification, and Production," U.S. Department of Defense, Military Standard, Government Printing Office, Washington, D.C., 1986.

[d]"TR-3, Sampling Procedures and Tables for Life and Reliability Testing Based on the Weibull Distribution (Mean Life Criterion)," U.S. Department of Defense," Quality Control and Reliability Technical Report, Government Printing Office, Washington, D.C., 1961.

[e]"TR-4, Sampling Procedures and Tables for Life and Reliability Testing Based on the Weibull Distribution (Hazard Rate Criterion)," U.S. Department of Defense, Quality Control and Reliability Technical Report, Government Printing Office, Washington, D.C., 1962.

[f]"TR-6, Sampling Procedures and Tables for Life and Reliability Testing Based on the Weibull Distribution (Reliable Life Criterion)," U.S. Department of Defense, Quality Control and Reliability Technical Report, Government Printing Office, Washington, D.C., 1963.

[g]"TR-7, Factors and Procedures for Applying MIL-STD-105D Sampling Plans to Life and Reliability Testing," U.S. Department of Defense, Quality Control and Reliability Technical Report, Government Printing Office, Washington, D.C., 1965.

Source: Reproduced by permission from J. M. Juran ed., Quality Control Handbook (New York: McGraw-Hill Book Company, 1988), Sec. 25, p. 80.

risk, α, of rejecting lots with mean life $\theta_0 = 900$ h is 0.05 and the customer's risk, β, of accepting lots with mean life $\theta_1 = 300$ h is 0.10. The ratio θ_1/θ_0 is

$$\frac{\theta_1}{\theta_0} = \frac{300}{900} = 0.333$$

From Table 11–3, for $\alpha = 0.05$, $\beta = 0.10$, and $\theta_1/\theta_0 = 0.333$, the code letter B-8 is obtained. Since the calculated ratio will rarely equal the one in the table, the next larger one is used.

For each code letter, A, B, C, D, and E, there is a table to determine the non-acceptance number and the value of the ratio T/θ_0, where T is the test time. Table 11–4 gives the value for code letter B. Thus, for code B-8, the non-acceptance number r is 8. The value of T/θ_0 is a function of the sample size.

The sample size is selected from one of the multiples of the non-acceptance number: $2r$, $3r$, $4r$, $5r$, $6r$, $7r$, $8r$, $9r$, $10r$, and $20r$. For the life-test plans, the sample size depends on the relative cost of placing large numbers of units of products on test and on the expected length of time the life tests must continue

TABLE 11-3 Life-Test Sampling Plan Code Designation[a] (Table 2A-1 of H108)

$\alpha = 0.01$, $\beta = 0.01$		$\alpha = 0.05$, $\beta = 0.10$		$\alpha = 0.10$, $\beta = 0.10$		$\alpha = 0.25$, $\beta = 0.10$		$\alpha = 0.50$, $\beta = 0.10$	
CODE	θ_1/θ_0	CODE	θ_1/θ_0	CODE	θ_1/θ_0	CODE	θ_1/θ_0	CODE	θ_1/θ_0
A-1	0.004	B-1	0.022	C-1	0.046	D-1	0.125	E-1	0.301
A-2	0.038	B-2	0.091	C-2	0.137	D-2	0.247	E-2	0.432
A-3	0.082	B-3	0.154	C-3	0.207	D-3	0.325	E-3	0.502
A-4	0.123	B-4	0.205	C-4	0.261	D-4	0.379	E-4	0.550
A-5	0.160	B-5	0.246	C-5	0.304	D-5	0.421	E-5	0.584
A-6	0.193	B-6	0.282	C-6	0.340	D-6	0.455	E-6	0.611
A-7	0.221	B-7	0.312	C-7	0.370	D-7	0.483	E-7	0.633
A-8	0.247	B-8	0.338	C-8	0.396	D-8	0.506	E-8	0.652
A-9	0.270	B-9	0.361	C-9	0.418	D-9	0.526	E-9	0.667
A-10	0.291	B-10	0.382	C-10	0.438	D-10	0.544	E-10	0.681
A-11	0.371	B-11	0.459	C-11	0.512	D-11	0.608	E-11	0.729
A-12	0.428	B-12	0.512	C-12	0.561	D-12	0.650	E-12	0.759
A-13	0.470	B-13	0.550	C-13	0.597	D-13	0.680	E-13	0.781
A-14	0.504	B-14	0.581	C-14	0.624	D-14	0.703	E-14	0.798
A-15	0.554	B-15	0.625	C-15	0.666	D-15	0.737	E-15	0.821
A-16	0.591	B-16	0.658	C-16	0.695	D-16	0.761	E-16	0.838
A-17	0.653	B-17	0.711	C-17	0.743	D-17	0.800	E-17	0.865
A-18	0.692	B-18	0.745	C-18	0.774	D-18	0.824	E-18	0.882

[a]Producer's risk, α is the probability of not accepting lots with mean life θ_2; consumer's risk, β, is the probability of accepting lots with mean life θ_1.

in order to determine acceptability of the lots. Increasing the sample size will, on one hand, cut the average time required to determine acceptability but, on the other hand, will increase the cost due to placing more units of product on test. For this example problem, the multiple $3r$ is selected, which gives a sample size $n = 3(8) = 24$. The corresponding value of $T/\theta_0 = 0.166$, which gives a test time T of

$$\begin{aligned} T &= 0.166(\theta_0) \\ &= 0.166(900) \\ &= 149.4 \text{ or } 149 \text{ h} \end{aligned}$$

A sample of 24 items is selected from a lot and all are tested simultaneously. If the eighth failure occurs before the termination time of 149 h, the lot is not accepted; if the eighth failure still has not occurred after 149 test hours, the lot is accepted.

2. *Stipulated producer's risk, rejection number, and sample size.* Determine the time-terminated, with-replacement, mean-life sampling plan where the producer's risk of not accepting lots with mean life $\theta_0 = 1200$ h is 0.05, the non-acceptance number is 5, and the sample size is 10, or $2r$. The same set of tables is used for this method as for the previous one. Table 11-4 is the table for the code letter B designation as well as for $\alpha = 0.05$. Thus, using Table 11-4, the value for $T/\theta_0 = 0.197$ and the value for T is

$$\begin{aligned} T &= 0.197(\theta_0) \\ &= 0.197(1200) \\ &= 236.4 \text{ or } 236 \text{ h} \end{aligned}$$

A sample of 10 items is selected from a lot and all are tested simultaneously. If the fifth failure occurs before the termination time of 236 h, the lot is not accepted; if the fifth failure still has not occurred after 236 h, the lot is accepted.

3. *Stipulated producer's risk, consumer's risk, and test time.* Determine the time-terminated, with-replacement, mean-life sampling plan which is not to exceed 500 h and which will accept a lot with mean life of 10,000 h (θ_0) at least 90% of the time ($\beta = 0.10$) but will not accept a lot with mean life of 2000 h (θ_1) about 95% of the time ($\alpha = 0.05$). The first step is to calculate the two ratios, θ_1/θ_0 and T/θ_0.

$$\frac{\theta_1}{\theta_0} = \frac{2000}{10{,}000} = \frac{1}{5}$$

$$\frac{T}{\theta_0} = \frac{500}{10{,}000} = \frac{1}{20}$$

TABLE 11-4 Values of T/θ_0 for $\alpha = 0.05$—Time-Terminated, with Replacement Code Letter B [Table 2C-2(b) of H108].

		SAMPLE SIZE									
CODE	r	$2r$	$3r$	$4r$	$5r$	$6r$	$7r$	$8r$	$9r$	$10r$	$20r$
B-1	1	0.026	0.017	0.013	0.010	0.009	0.007	0.006	0.006	0.005	0.003
B-2	2	0.089	0.059	0.044	0.036	0.030	0.025	0.022	0.020	0.018	0.009
B-3	3	0.136	0.091	0.068	0.055	0.045	0.039	0.034	0.030	0.027	0.014
B-4	4	0.171	0.114	0.085	0.068	0.057	0.049	0.043	0.038	0.034	0.017
B-5	5	0.197	0.131	0.099	0.079	0.066	0.056	0.049	0.044	0.039	0.020
B-6	6	0.218	0.145	0.109	0.087	0.073	0.062	0.054	0.048	0.044	0.022
B-7	7	0.235	0.156	0.117	0.094	0.078	0.067	0.059	0.052	0.047	0.023
B-8	8	0.249	0.166	0.124	0.100	0.083	0.071	0.062	0.055	0.050	0.025
B-9	9	0.261	0.174	0.130	0.104	0.087	0.075	0.065	0.058	0.052	0.026
B-10	10	0.271	0.181	0.136	0.109	0.090	0.078	0.068	0.060	0.054	0.027
B-11	15	0.308	0.205	0.154	0.123	0.103	0.088	0.077	0.068	0.062	0.031
B-12	20	0.331	0.221	0.166	0.133	0.110	0.095	0.083	0.074	0.066	0.033
B-13	25	0.348	0.232	0.174	0.139	0.116	0.099	0.087	0.077	0.070	0.035
B-14	30	0.360	0.240	0.180	0.144	0.120	0.103	0.090	0.080	0.072	0.036
B-15	40	0.377	0.252	0.189	0.151	0.126	0.108	0.094	0.084	0.075	0.038
B-16	50	0.390	0.260	0.195	0.156	0.130	0.111	0.097	0.087	0.078	0.039
B-17	75	0.409	0.273	0.204	0.164	0.136	0.117	0.102	0.091	0.082	0.041
B-18	100	0.421	0.280	0.210	0.168	0.140	0.120	0.105	0.093	0.084	0.042

Using the values of θ_1/θ_0, T/θ_0, α, and β, the values of r and n are obtained from Table 11-5 and are $n = 27$ and $r = 4$.

The sampling plan is to select a sample of 27 items from a lot. If the fourth failure occurs before the termination time of 500 h, the lot is not accepted; if the fourth failure still has not occurred after 500 h, the lot is accepted.

When using this technique the tables provided for values of $\alpha = 0.01, 0.05, 0.10$, and 0.25; $\beta = 0.01, 0.05, 0.10$, and 0.25; $\theta_1/\theta_0 = \frac{2}{3}, \frac{1}{2}, \frac{1}{3}, \frac{1}{5}, \frac{1}{10}$; and $T\theta_0 = \frac{1}{3}, \frac{1}{5}, \frac{1}{10}$, and $\frac{1}{20}$.

TABLE 11-5 Sampling Plans for Specified α, β, θ_1/θ_0, and T/θ_0 (Table 2C-4 of H108).

		T/θ_0					T/θ_0			
		1/3	1/5	1/10	1/20		1/3	1/5	1/10	1/20
θ_1/θ_0	r	n	n	n	n	r	n	n	n	n
		$\alpha = 0.01$		$\beta = 0.01$			$\alpha = 0.05$		$\beta = 0.01$	
2/3	136	331	551	1103	2207	95	238	397	795	1591
1/2	46	95	158	317	634	33	72	120	241	483
1/3	19	31	51	103	206	13	25	38	76	153
1/5	9	10	17	35	70	7	9	16	32	65
1/10	5	4	6	12	25	4	4	6	13	27
		$\alpha = 0.01$		$\beta = 0.05$			$\alpha = 0.05$		$\beta - 0.05$	
2/3	101	237	395	790	1581	67	162	270	541	1082
1/2	35	68	113	227	454	23	47	78	157	314
1/3	15	22	37	74	149	10	16	27	54	108
1/5	8	8	14	29	58	5	6	10	19	39
1/10	4	3	4	8	16	3	3	4	8	16
		$\alpha - 0.01$		$\beta = 0.10$			$\alpha = 0.05$		$\beta = 0.10$	
2/3	83	189	316	632	1265	55	130	216	433	867
1/2	30	56	93	187	374	19	37	62	124	248
1/3	13	18	30	60	121	8	11	19	39	79
1/5	7	7	11	23	46	4	4	7	13	27
1/10	4	2	4	8	16	3	3	4	8	16
		$\alpha = 0.01$		$\beta = 0.25$			$\alpha = 0.05$		$\beta = 0.25$	
2/3	60	130	217	434	869	35	77	129	258	517
1/2	22	37	62	125	251	13	23	38	76	153
1/3	10	12	20	41	82	6	7	13	26	52
1/5	5	4	7	13	25	3	3	4	8	16
1/10	3	2	2	4	8	2	1	2	3	7

(continued on the next page)

TABLE 11-5 **Sampling Plans for Specified α, β, θ_1/θ_0, and T/θ_0 (Table 2C-4 of H108) (continued).**

		T/θ_0					T/θ_0			
		1/3	**1/5**	**1/10**	**1/20**		**1/3**	**1/5**	**1/10**	**1/20**
θ_1/θ_0	r	n	n	n	n	r	n	n	n	n
		$\alpha = 0.10$		$\beta = 0.01$			$\alpha = 0.25$		$\beta = 0.01$	
2/3	77	197	329	659	1319	52	140	234	469	939
1/2	26	59	98	197	394	17	42	70	140	281
1/3	11	21	35	70	140	7	15	25	50	101
1/5	5	7	12	24	48	3	5	8	17	34
1/10	3	3	5	11	22	2	2	4	9	19
		$\alpha = 0.10$		$\beta = 0.05$			$\alpha = 0.25$		$\beta = 0.05$	
2/3	52	128	214	429	859	32	84	140	280	560
1/2	18	38	64	128	256	11	25	43	86	172
1/3	8	13	23	46	93	5	10	16	33	67
1/5	4	5	8	17	34	2	3	5	10	19
1/10	2	2	3	5	10	2	2	4	9	19
		$\alpha = 0.10$		$\beta = 0.10$			$\alpha = 0.25$		$\beta = 0.10$	
2/3	41	99	165	330	660	23	58	98	196	392
1/2	15	30	51	102	205	8	17	29	59	119
1/3	6	9	15	31	63	4	7	12	25	50
1/5	3	4	6	11	22	2	3	4	9	19
1/10	2	2	2	5	10	1	1	2	3	5
		$\alpha = 0.10$		$\beta = 0.25$			$\alpha = 0.25$		$\beta = 0.25$	
2/3	25	56	94	188	376	12	28	47	95	190
1/2	9	16	27	54	108	5	10	16	33	67
1/3	4	5	8	17	34	2	2	4	9	19
1/5	3	3	5	11	22	1	1	2	3	6
1/10	2	1	2	5	10	1	1	1	2	5

The method to use for obtaining the desired life-test sampling plan is determined by the available information.

AVAILABILITY AND MAINTAINABILITY

For long-lasting products and services such as refrigerators, electric power lines, and front-line service, the time-related factors of availability, reliability, and maintainability are interrelated. For example, when a water line breaks (reliability) it is no longer available to provide water to customers and must be repaired or maintained.

Availability is a time-related factor that measures the ability of a product or service to perform its designated function. The product or service is available when it is in the operational state, which includes active and standby use. A calculator is operational (uptime) when it is being used to perform calculations and when it is being carried in a book bag. Availability can be quantified by the ratio:

$$A = \frac{\text{Uptime}}{\text{Uptime} + \text{Downtime}} = \frac{\text{MTBF}}{\text{MTBT} + \text{MDT}}$$

For a repairable item, mean downtime (MDT) is the mean time to repair (MTTR); for items that are not repairable, mean downtime is the time to obtain a replacement. For the calculator example, the downtime may just require the time to change the batteries, the time to send the calculator back to the manufacturer for repair (not cost effective), or the time to purchase a new calculator. For a manufacturing process such as a steel rolling mill, mean time to repair is critical and may even require an overnight shipment of repair parts. Downtime by a person waiting on a telephone for a service representative to become available to respond to an inquiry may require the organization to increase the number of service representatives or the number of telephone lines. Downtime takes on many different aspects depending on the product or service.

Maintainability is the ease with which preventative and corrective maintenance on a product or service can be achieved. One of the best times to improve maintainability is in the design phase of a product or service. Improvements in design have resulted in 100,000 miles between automotive tune-ups, self-lubricating bearings, expert systems for service activities, and so forth. Production processes rely on total productivity maintenance to improve maintainability. Maintainability uses a number of different figures of merit such as mean time to repair, mean time to service, repair hours per 1000 number of operating hours, preventative maintenance cost, and downtime probability.

According to Professor Mulder[6], keeping maintainability low may be a more cost effective method of keeping availability high than concentrating on reliability. For example, a Rolls Royce automobile is an extremely reliable automobile; however, when it does breakdown, the wait time to find a dealer, obtain parts, and make repairs may be many days.

COMPUTER PROGRAM

Using EXCEL the software in the diskette inside the back cover will solve for β and θ of the Weibull distribution. Its file name is *Weibull*.

[6]Statement and example from David C. Mulder, "Comparing the Quality Measurements." Unpublished Manuscript

PROBLEMS

1. A system has 4 components, A, B, C, and D, with reliability values of 0.98, 0.89, 0.94, and 0.95, respectively. If the components are in series, what is the system reliability?

2. A flashlight has 4 components: 2 batteries with reliability of 0.998, a light bulb with reliability of 0.999, and a switch with reliability of 0.997. Determine the reliability of this series system.

3. Christmas tree light bulbs used to be manufactured in series—if one bulb went out, they all did. What would be the reliability of this system if each bulb had a reliability of 0.999 and there were 20 bulbs in the system?

4. What is the reliability of the system below?

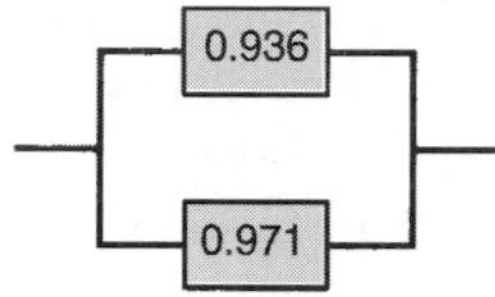

5. If component B of Problem 1 is changed to 3 parallel components and each has the same reliability, what is the system reliability now?

6. What is the reliability of the system below where the reliabilities of components A, B, C, and D are 0.975, 0.985, 0.988, and 0.993, respectively?

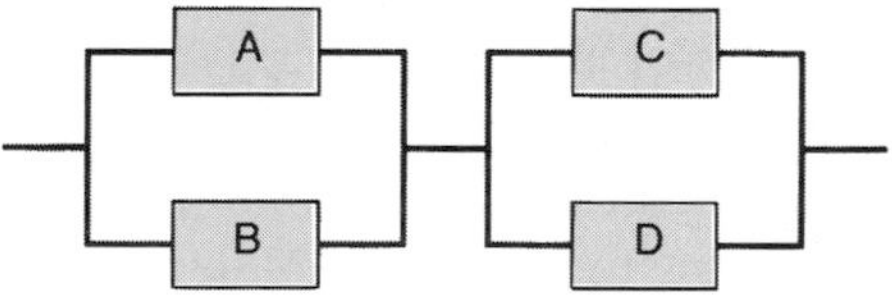

7. Using the same reliabilities of Problem 6, what is the reliability of the system below?

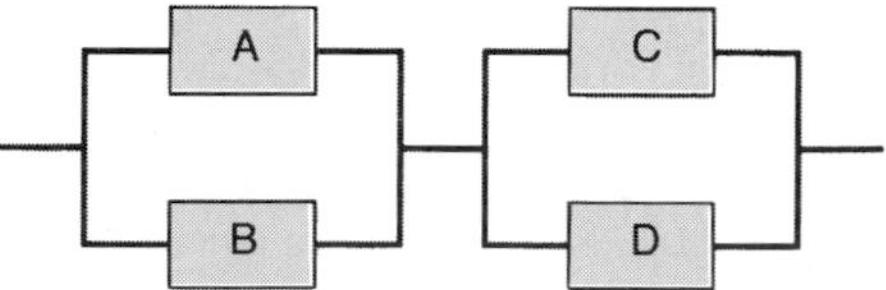

8. A system is composed of 5 components in series, and each has a reliability of 0.96. If the system can be changed to 3 components in series, what is the change in the reliability?

9. Determine the failure rate for 5 items that are tested to failure. Test data in hours are 184, 96, 105, 181, and 203.

10. Twenty-five parts are tested for 15 hours. At the end of the test, 3 parts had failed at 2, 5, and 6 hours. What is the failure rate?

11. Fifty parts are tested for 500 cycles each. When a part fails it is replaced by another one. At the end of the test 5 parts had failed. What is the failure rate?

12. Assume a constant failure rate and determine the mean life of problems 9, 10, and 11.

13. Determine the failure rate for a 150-h test of 9 items where 3 items failed without replacement at 5, 76, and 135 h. What is the mean life for a constant failure rate?

14. If the mean life for a constant failure rate is 52 h, what is the failure rate?

15. Construct the life-history curve for the following test data.

TEST HOURS	FAILURES	SURVIVORS
0–69	150	350
70–139	75	275
140–209	30	245
210–279	27	218
280–349	23	194
350–419	32	163
420–489	53	110
490–559	62	48
560–629	32	16
630–699	16	0
	500	500

16. Using normal distribution, determine the reliability at 6000 cycles of a switch with a mean life of 5500 cycles and a standard deviation of 165 cycles.

17. Determine the reliability at $t = 80$ h for the example problem where $\theta = 125$ and there is a constant failure rate. What is the reliability at $t = 125$ h? At $t = 160$ h?

18. Determine the reliability at $t = 3500$ cycles for the example problem where the mean life of a constant failure rate is 3704 cycles. What is the reliability at $t = 3650$ cycles? At 3900 cycles?

19. Using the Weibull distribution for Problem 16 rather than the normal distribution, determine the reliability when $\beta = 3.5$.

20. The failure pattern of an automotive engine water pump fits the Weibull distribution with $\beta = 0.7$. If the mean life during the debugging phase is 150 h, what is the reliability at 50 h?

21. Construct the OC curve for a sampling plan specified as $n = 24$, $T = 149$, $c = 7$, and $r = 8$.

22. Construct the OC curve for a sampling plan specified as $n = 10$, $T = 236$, $c = 4$, and $r = 5$.

23. Determine the time-terminated, with-replacement, mean-life sampling plan where the producer's risk of rejecting lots with mean life of 800 h is 0.05 and the consumer's risk of accepting lots with mean life $\theta_1 = 220$ is 0.10. The sample size is 30.

24. Determine the time-terminated, with-replacement sampling plan that has the following specifications: $T = 160$, $\theta_1 = 400$, $\beta = 0.10$, $\theta_0 = 800$, and $\alpha = 0.05$.

25. Determine the time-terminated, with-replacement sampling plan where the producer's risk of rejecting lots with mean life $\theta_0 = 900$ h is 0.05, the rejection number is 3, and the sample size is 9.

26. Find a replacement life-test sampling plan of 300 h that will accept a lot with mean life of 3000 h 95% of the time but will reject a lot with mean life of 1000 h 90% of the time.

27. If the probability of accepting a lot with a mean life of 1100 cycles is 0.95 and the probability of not accepting a lot with mean life of 625 cycles is 0.90, what is the sampling plan for a sample size of 60?

28. Find a life-test, time-terminated sampling plan with replacement that will accept a lot with a mean life of 900 h with probability of 0.95 ($\alpha = 0.05$). The test is to be stopped after the occurrence of the second failure, and 12 units of product are to be placed on test.

29. Using EXCEL design a template for the construction of an OC curve and test it by solving Problem 21.

30. Using the Weibull software provided determine β and θ for the ordered data set below: 20, 32, 40, 46, 54, 62, 73, 85, 89, 99, 102, 118, 140, 151.

12 MANAGEMENT AND PLANNING TOOLS[1]

INTRODUCTION

While the statistical process control (SPC) tools are excellent problem solving tools, there are many situations where they are not appropriate. This chapter discusses some additional tools that can be very effective for teams and, in some cases, for individuals. They do not use hard data but rely on subjective information. Application of these tools has been proven useful in process improvement, cost reduction, policy deployment, and new-product development.

[1]Reproduced, with permission, from Besterfield et. al., *Total Quality Management*, 3e (Upper Saddle River, NJ: Prentice-Hall, 2003).

WHY, WHY

Although this tool is very simple, it is effective. It can be a key to finding the root cause of a problem by focusing on the process rather than on people. The procedure is to describe the problem in specific terms and then ask why. You may have to ask why three or more times to obtain the root cause. An example will help illustrate the concept.

Why did we miss the delivery date?
  It wasn't scheduled in time.
Why?
  There were a lot of engineering changes.
Why?
  Customer requested them.
The team suggested changing the delivery date whenever engineering changes occured.

This tool is very beneficial in developing critical thinking. It is frequently a quick method of solving a problem.

FORCED FIELD ANALYSIS

This analysis is used to identify the forces and factors that may influence the problem or goal. It helps an organization to better understand promoting or driving and restraining or inhibiting forces so that the positives can be reinforced and the negatives reduced or eliminated. The procedure is to define the objective, determine criteria for evaluating the effectiveness of the improvement action, brainstorm the forces that promote and inhibit achieving the goal, prioritize the forces from greatest to least, and take action to strengthen the promoting forces and weaken the inhibiting forces. An example will illustrate the tool.

OBJECTIVE: STOP SMOKING	
Promoting Forces ⟶	⟵ ***Inhibiting Forces***
Poor Health ⟶	⟵ Habit
Smelly Clothing ⟶	⟵ Addiction
Poor Example ⟶	⟵ Taste
Cost ⟶	⟵ Stress
Impact on Others ⟶	⟵ Advertisement

The benefits are the determination of the positives and negatives of a situation, encouraging people to agree and prioritize the competing forces, and identify the root causes.

NOMINAL GROUP TECHNIQUE

This technique provides for issue/idea input from everyone on the team and for effective decisions. An example will illustrate the technique. Let's assume that the team wants to decide which problem to work on. Everyone writes on a piece of paper the problem they think is most important. The papers are collected, and all problems are listed on a flip chart. Then each member of the team uses another piece of paper to rank the problems from least important to most important. The rankings are given a numerical value starting at 1 for least important and continuing to the most important. Points for each problem are totaled, and the item with the highest number of points is considered to be the most important.

AFFINITY DIAGRAM

This diagram allows the team to creatively generate a large number of issues/ideas and then logically group them for problem understanding and possible breakthrough solution. The procedure is to state the issue in a full sentence, brainstorm using short sentences on self-adhesive notes, post them for the team to see, sort ideas into logical groups, and create concise descriptive headings for each group Figure 12-1 illustrates the technique.

Large groups should be divided into smaller groups with appropriate headings. Notes that stand alone could become headers or placed in a miscellaneous category. Affinity diagrams encourage team creativity, break down barriers, facilitate breakthroughs, and stimulate ownership of the process.

INTERRELATIONSHIP DIAGRAM[2]

The interrelationship diagram (ID) clarifies the interrelationship of many factors of a complex situation. It allows the team to classify the cause-and-effect relationships among all the factors so that the key drivers and outcomes can be used to solve the problem. The procedure is somewhat more complicated than the previous tools, thus, it will be itemized.

1. The team should agree on the issue or problem statement.

2. All of the ideas or issues from other techniques or from brainstorming should be laid out, preferably in a circle as shown in Figure 12-2(a).

---

[2]This section adapted, with permission, from Michael Brassard, *The Memory Jogger Plus* + (Methuen, Mass.: GOAL/QPC, 1989).

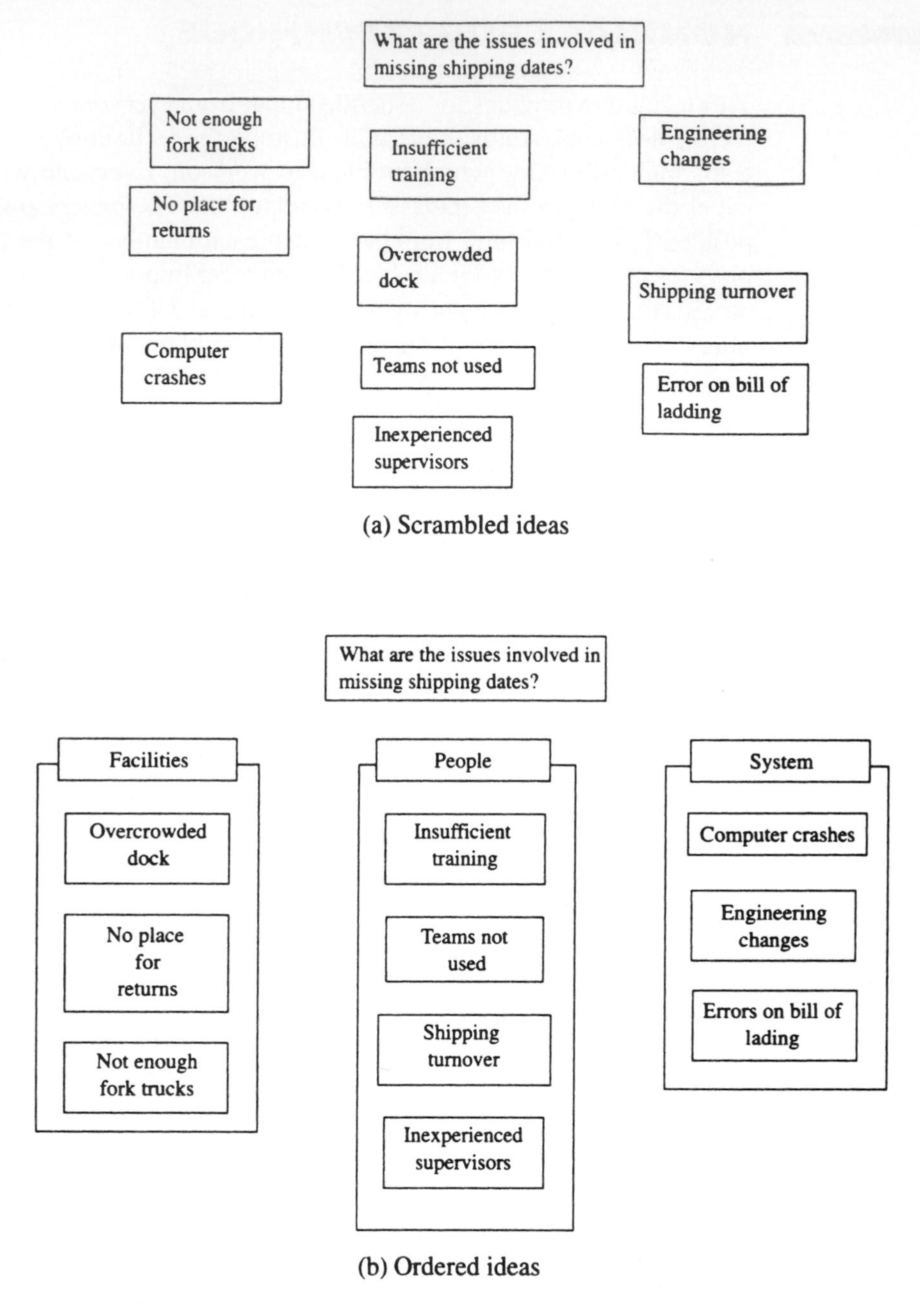

FIGURE 12-1 Affinity Diagram.

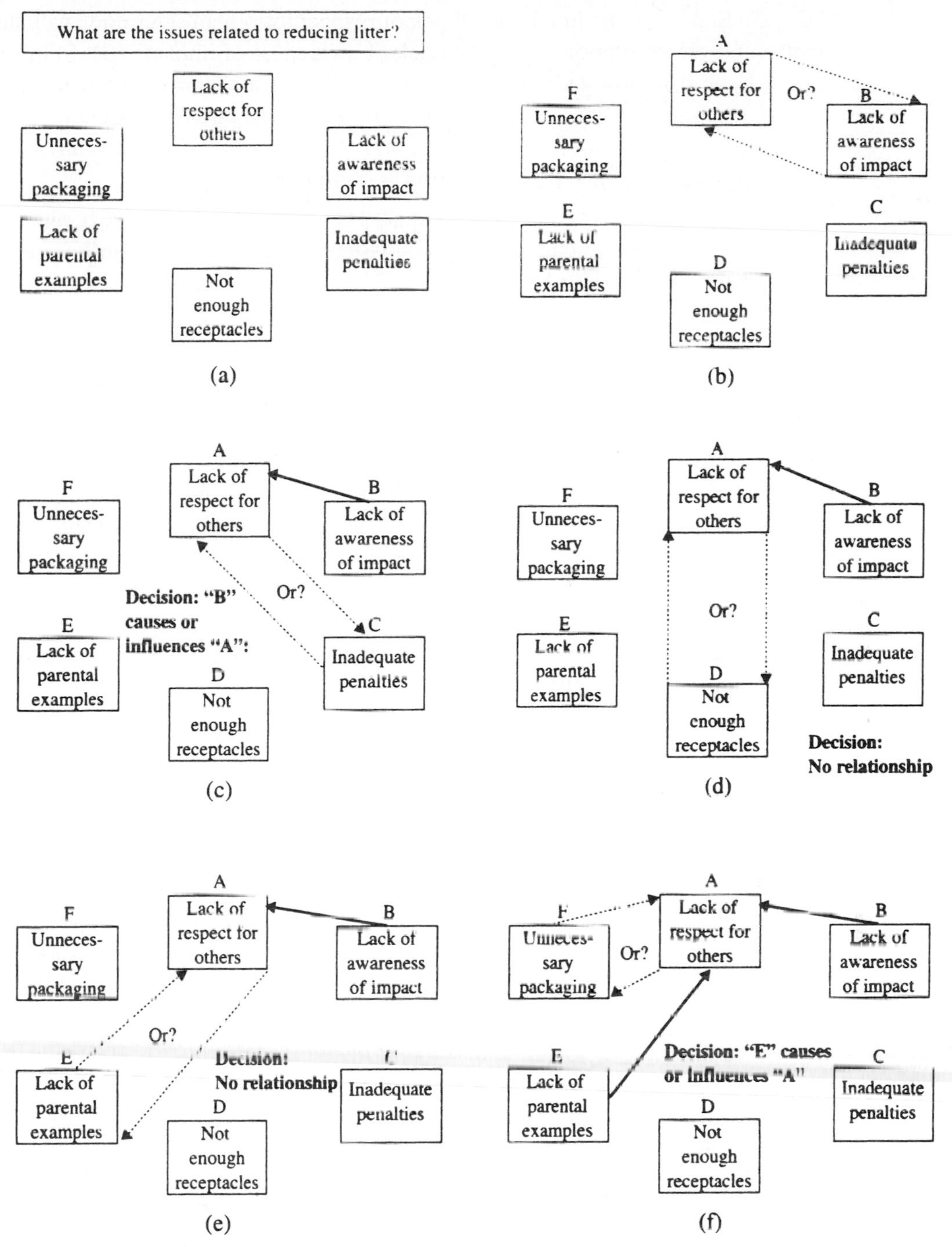

FIGURE 12-2 Interrelationship Diagram for First Iteration.

3. Start with the first issue, "Lack of respect for others" (A), and evaluate the cause-and-effect relationship with "Lack of awareness of impact" (B). In this situation, Issue B is stronger than Issue A; therefore, the arrow is drawn from Issue B to Issue A as shown in Figure 12-2(c). Each issue in the circle is compared to Issue A as shown in Figure 12-2(c), (d), (e), and (f). Only Issues B and E have a relationship with Issue A. The first iteration is complete.

4. The second iteration is to compare Issue B with Issues C, D, E, and F. The third iteration is to compare Issue C with Issues D, E, and F. The fourth iteration is to compare Issue D with Issues E and F. The fifth iteration is to compare Issue E with Issue F.

5. The entire diagram should be reviewed and revised where necessary. It is a good idea to obtain information from other people on upstream and downstream processes.

6. The diagram is completed by tallying the incoming and outgoing arrows and placing this information below the box. Figure 12-3(d) shows a completed diagram. Issue B is the "driver" because it has zero incoming arrows and five outgoing arrows. It is usually the root cause. The issue with the highest incoming arrows is Issue E. It is a meaningful measure of success.

A relationship diagram allows a team to identify root causes from subjective data, systematically explores cause-and-effect relationships, encourages members to think multidirectionally, and develops team harmony and effectiveness.

TREE DIAGRAM

This tool is used to reduce any broad objective into increasing levels of detail in order to achieve the objective. The procedure is to first choose an action-oriented objective statement from the interrelationship diagram, affinity diagram, brainstorming, team mission statement, and so forth. Second, using brainstorming, choose the major headings as shown in Figure 12-4 under Means.

The third step is to generate the next level by analyzing the major headings. Ask, "What needs to be addressed to achieve the objective?" Repeat this question at each level. Three levels below the objective are usually sufficient to complete the diagram and make appropriate assignments. The diagram should be reviewed to determine if these actions will give the results anticipated or if something has been missed.

The tree diagram encourages team members to think creatively, makes large projects manageable, and generates a problem-solving atmosphere.

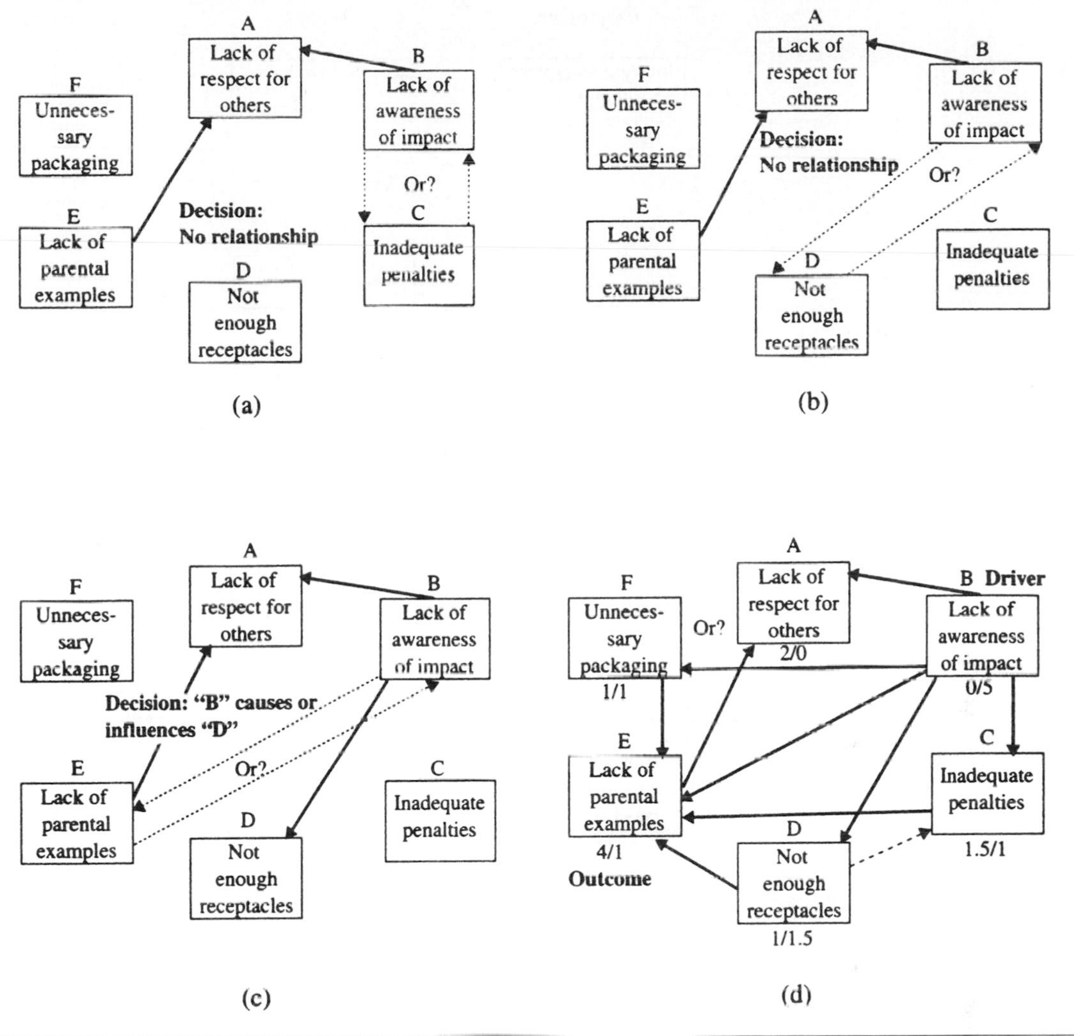

FIGURE 12-3 **Completed Interrelationship Diagram.**

MATRIX DIAGRAM

The matrix diagram allows individuals or teams to identify, analyze, and rate the relationship among two or more variables. Data are presented in table form and can be objective or subjective, which can be given symbols with or without numerical values. Quality function deployment (QFD), which was briefly discussed in Chapter 3, is an outstanding example of the use of the matrix diagram. There are at least five standard formats: L-shaped (2 variables), T-shaped (3 variables),

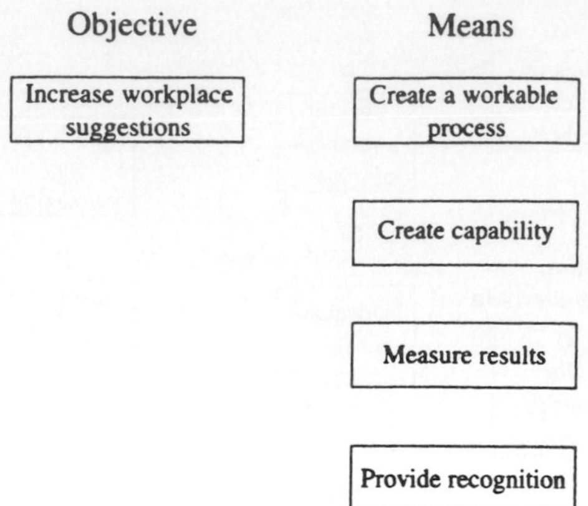

(a) Objective and means

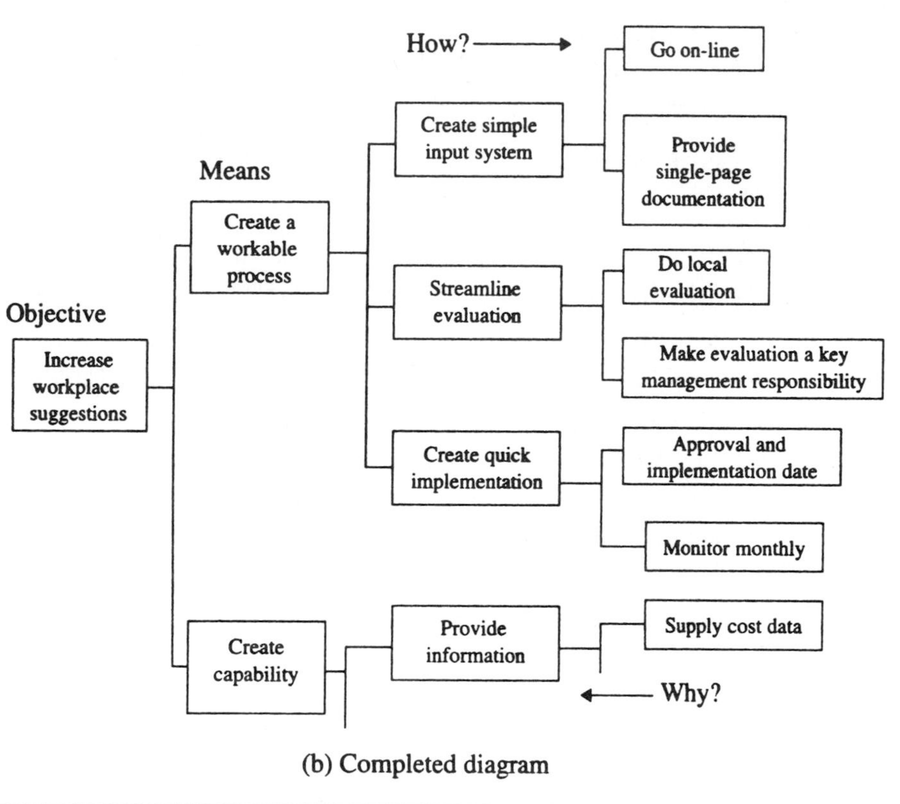

(b) Completed diagram

FIGURE 12-4 **Tree Diagram.**

Tool \ Use	Creativity	Analysis	Consensus	Action
Affinity diagram	○		○	△
Interrelationship digraph		○	◎	
Tree diagram		◎		◎
Prioritization matrix			○	
Matrix diagram		○	◎	○
PDPC	◎	◎	◎	○
Activity network diagram			◎	○

Legend: Always ○ Frequently ◎ Occasionally △

FIGURE 12-5 Matrix Diagram for Uses of the Seven Management Tools. Reproduced, with permission, from Ellen R. Domb, "7 New Tools: The Ingredients for Successful Problem Solving," *Quality Digest* (December 1994).

Y-shaped (3 variables), C-shaped (3 variables), and X-shaped (4 variables). Our discussion will be limited to the L-shaped format, which is the most common.[3]

Figure 12-5 illustrates a matrix diagram for using the seven management and planning tools. The procedure for the diagram is for the team to first select the factors affecting a successful plan. Next select the appropriate format, which in this case is the L-shaped diagram. That step is followed by determining the relationship symbols. Any symbols can be adopted, provided the diagram contains a legend as shown in the bottom of the figure. Numerical values are sometimes associated with the symbol as we previously did with QFD. The last step is to complete the matrix by analyzing each cell and inserting the appropriate symbol.

The matrix diagram clearly shows the relationship of the two variables. It encourages the team to think in terms of relationships, their strength, and any patterns.

PRIORITIZATION MATRICES

These tools prioritize issues, tasks, characteristics, and so forth, based on weighted criteria using a combination of tree and matrix diagram techniques. Once prioritized, effective decisions can be made. Prioritization matrices are designed to reduce the team's options rationally before detailed implementation planning occurs. It utilizes

[3]Detailed information on the other formats is available from Michael Brassard. *The Memory Jogger Plus+* (Methuen, Mass.: GOAL/QPC, 1996).

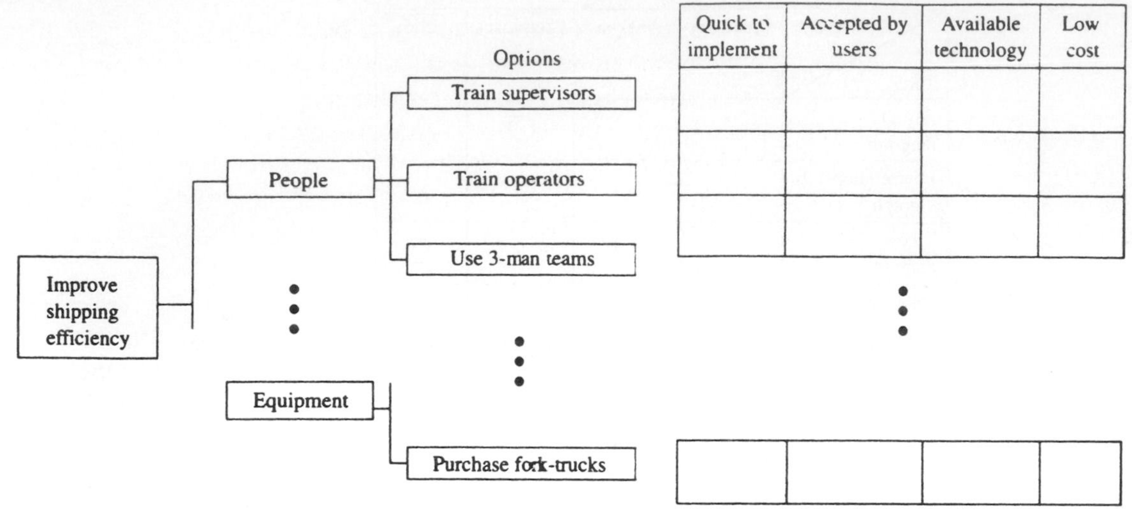

FIGURE 12-6 Prioritization Matrix for Improving Shipping Efficiency.

a combination of tree and matrix diagrams as shown in Figure 12-6. There are 15 implementation options; however, only the first three, beginning at "train supervisors," and the last one "purchase fork-trucks," are shown in the tree diagram. There are four implementation criteria, however, as shown at the top of the matrix. Prioritization matrices are the most difficult of the tools in this chapter, therefore, we will list the steps for creating one.

1. Construct an L-shaped matrix combining the options, which are the lowest-level of detail of the tree diagram with the criteria. This information is given in Table 12-1.

TABLE 12-1 Improve Shipping Efficiency Using the Consensus Criteria Method

	CRITERIA				
OPTIONS	QUICK TO IMPLEMENT	ACCEPTED BY USERS	AVAILABLE TECHNOLOGY	LOW COST	TOTAL
Train Operators	13(2.10) = 27.3	15(1.50) = 22.5	11(0.45) = 5.0	13(0.35) = 4.6	59.4
Train Supervisors	12(2.10) = 25.2	11(1.50) = 16.5	12(0.45) = 5.4	8(0.35) = 2.8	49.9
Use 3-person Teams	8(2.10) = 16.8	3(1.50) = 4.5	13(0.45) = 5.9	14(0.35) = 4.9	32.1
⋮	⋮	⋮	⋮	⋮	⋮
Purchase Fort-trucks	6(2.10) = 12.5	12(1.50) = 18	10(0.45) = 4.5	1(0.35) = 0.4	35.5

2. Determine the implementation criteria using the nominal group technique (NGT) or any other technique that will satisfactorily weight the criteria. Using NGT, each team member submits the most important criteria on a piece of paper. They are listed on a flip chart, and the team members submit another piece of paper rank ordering those listed on the flip chart. Those criteria with the greatest value are the most important. The team decides how many of the criteria to use. In this situation, the team decides to use the four criteria shown at the top of the matrix.

3. Prioritize the criteria using the NGT. Each team member weights the criteria so the total weight equals 1.00, and the results are totaled for the entire team as shown below:

CRITERIA	MEMBER #1	MEMBER #2		TOTAL
Accepted by users	.30	.25		1.50
Low cost	.15	.20	. . .	0.35
Quick to implement	.40	.30		2.10
Available technology	.15	.25		0.45
	1.00	1.00		

4. Using NDT, rank order the options in terms of importance by each criterion, average the results, and round to the nearest whole number. Thus, this ranking should be from 1 to the number of options for each criterion. For example, train operators is ranked 13 for quick to implement.

5. Compute the option importance score under each criterion by multiplying the rank by the criteria weight as shown in Table 12-1. The options with the highest total are those that should be implemented first.

There are two other techniques that are more complicated, and the reader is referred to *The Memory Jogger Plus+* for information.

PROCESS DECISION PROGRAM CHART

Programs to achieve particular objectives do not always go according to plan, and unexpected developments may have serious consequences. The process decision program chart (PDPC) avoids surprises and identifies possible countermeasures. Figure 12-7 illustrates the PDPC.

The procedure starts with the team stating the objective, which is to plan a successful conference. That activity is followed by the first level, which is the conference activities of registration, presentations, and facilities. Only the presentation activity is illustrated. In some cases a second level of detailed activities may be used. Next, the team brainstorms to determine what could go wrong with the conference, and these are shown as the "what-if" level. Countermeasures are brainstormed and placed in a

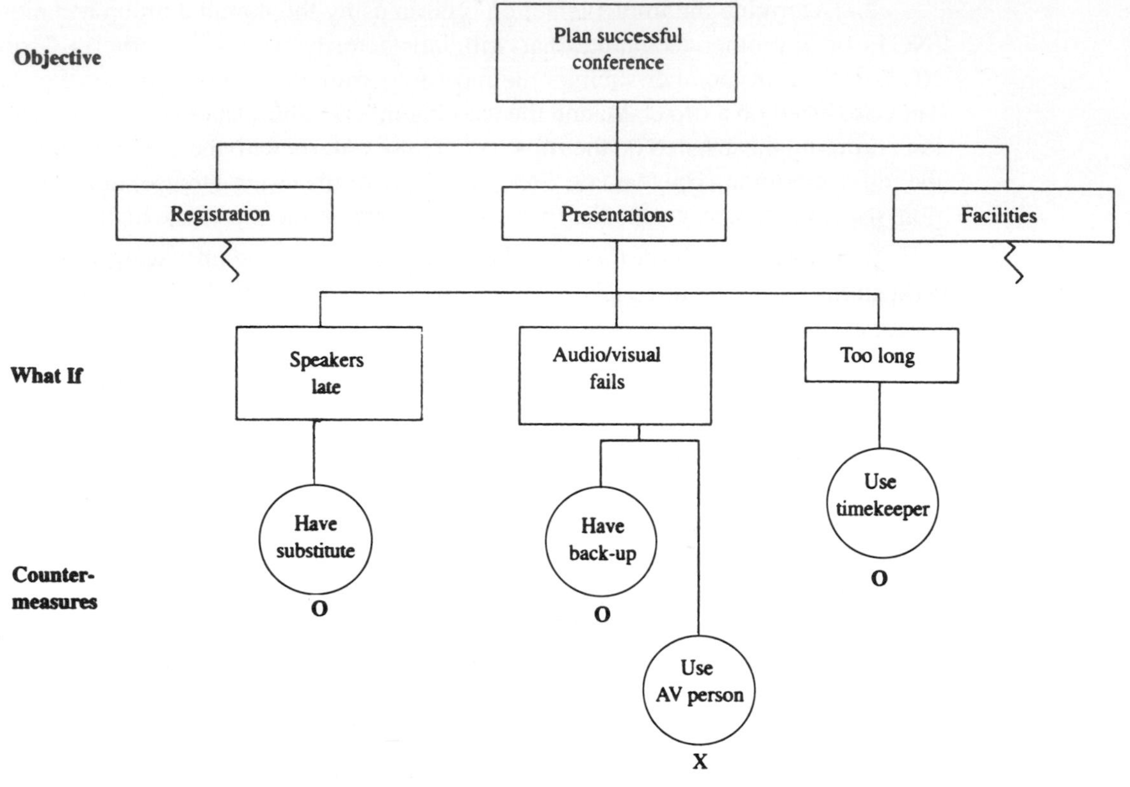

FIGURE 12-7 **PDPC for Conference Presentation.**

balloon in the last level. The last step is to evaluate the countermeasures and select the optimal ones by placing an *O* underneath. Place an *X* under those that are rejected.

The example has used a graphical format. PDPC can also use an outline format with the activities listed. The probability, in percent, that a "what-if" will occur can be included in the box. Countermeasures should be plausible. PDPC should be used when the task is new or unique, complex, or potential failure has great risks. This tool encourages team members to think about what can happen to a process and how countermeasures can be taken. It provides the mechanism to effectively minimize uncertainty in an implementation plan.

ACTIVITY NETWORK DIAGRAM

This tool goes by a number of different names and deviations, such as program evaluation and review technique (PERT), critical path method (CPM), arrow diagram, and activity on node (AON). It allows the team to schedule a project efficiently.

The diagram shows completion times, simultaneous tasks, and critical activity path. Given below is the procedure to follow:

1. The team brainstorms or documents all the tasks to complete a project. These tasks are recorded on self-adhesive notes so all members can see them.

2. The first task is located and placed on the extreme left of a large view work surface, as shown in Figure 12-8(a).

3. Any tasks that can be done simultaneously are placed below, as shown in Figure 12-8(b).

4. Repeat Steps 2 and 3 until all tasks are placed in their correct sequence, as illustrated in Figure 12-8(c). Note: Because of space limitations, not all of the tasks are shown.

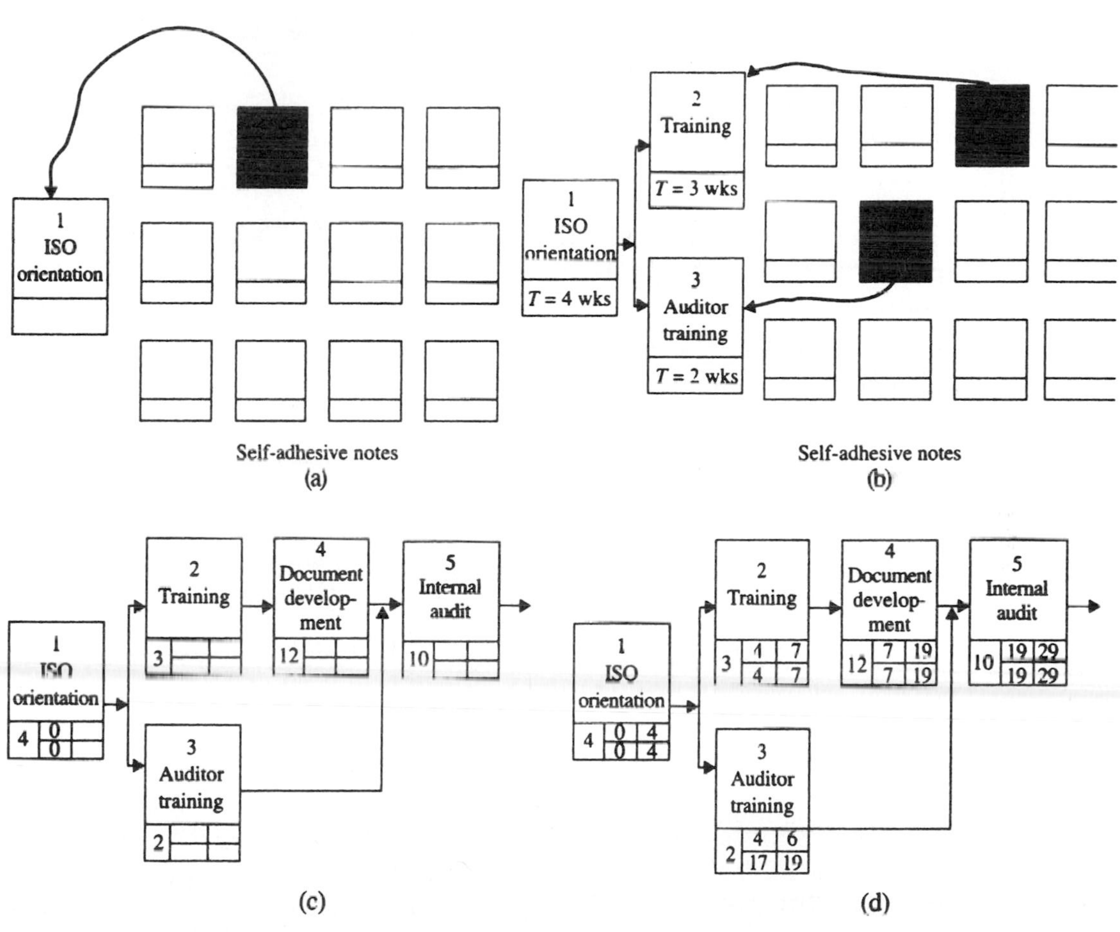

FIGURE 12-8 Activity Network Diagram.

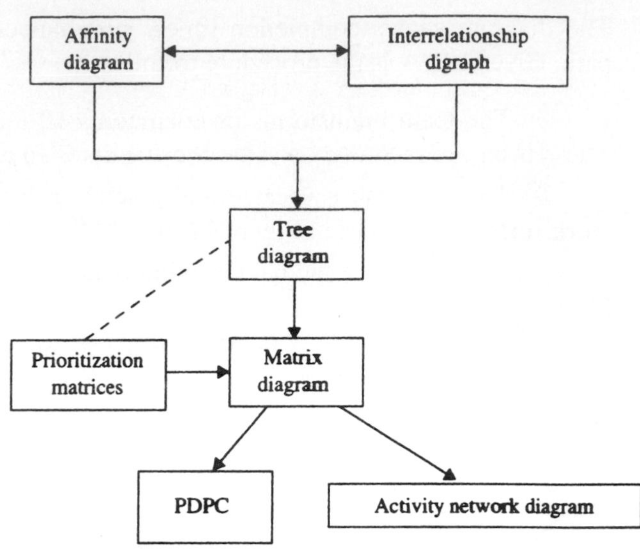

FIGURE 12-9 System Flow Diagram.

5. Number each task and draw connecting arrows. Determine the task completion time and post it in the lower left box. Completion times are recorded in hours, days, or weeks.

6. Determine the critical path by completing the four remaining boxes in each task. As shown below, these boxes are used for the earliest start time (ES), earliest finish (EF), latest start (LS), and latest finish (LF).

Activity time [T]	Earliest Start [ES]	Earliest Finish [EF]
	Latest Start [LS]	Latest Finish [LF]

The ES for Task 1 is 0, and the EF is 4 weeks later using the equation $EF = ES \div T$; the ES for Task 2 is 4 weeks, which is the same as the EF of Task 1, and the EF of Task 2 is $4 + 3 = 7$. This process is repeated for Tasks 4 and 5, which gives a total time of 29 weeks through the completion of the internal audit. If the project is to stay on schedule, the LS and LF for each of these tasks must equal the ES and EF, respectively. These values can be calculated by working backwards—subtracting the task time. They are shown in Figure 12-8(d).

Task 3, auditor training, does not have to be in sequence with the other tasks. It does have to be completed during the 19th week, because the ES for Task 5 is 19. Therefore, the LF for Task 3 is also 19 and the LS is 17. Auditor training could start after Task 1, which would give an ES of 4 and an EF of 6. The slack for Task 3 equals LS − ES [17 − 4 = 13]. The critical path is the longest cumulative time of connecting activities and occurs when the slack of each task is zero; thus, it is 1, 2, 4, and 5.

The benefits of an activity network diagram are (1) a realistic timetable determined by the users, (2) team members understand their role in the overall plan, (3) bottlenecks can be discovered and corrective action taken, and (4) members focus on the critical tasks. For this tool to work, the task times must be correct or reasonably close.

SUMMARY'

The first three tools can be used in a wide variety of situations. They are simple to use by individuals and/or teams.

The last seven tools in the chapter are called the seven management and planning tools. Although these tools can be used individually, they are most effective when used as a system to implement an improvement plan. Figure 12-9 shows a suggested flow diagram for this integration.

The team may wish to follow this sequence or modify it to meet their needs.

PROBLEMS

1. Determine why you did poorly on a recent examination by using the why, why tool.

2. Use the forced field analysis to
   (a) Lose weight.
   (b) Improve your GPA.
   (c) Increase your athletic ability in some sport.

3. Prepare an affinity diagram, using a team of three or more people, to plan
   (a) An improvement in the cafeteria.
   (b) A spring-break vacation.
   (c) A field trip to a local organization.

4. Using a team of three or more people, prepare an interrelationship digraph for the
   (a) Computer networking of nine locations in the organization's facility.
   (b) Implementation of a recognition and reward system.
   (c) Performance improvement of the accounting department or any other work group.
5. Develop a tree diagram, using a team of three or more people, for
   (a) The customer requirements for a product or service.
   (b) Planning a charity walk-a-thon.
6. The church council is planning the activities for a successful carnival. Using a team of three or more people, design a tree diagram to determine detailed assignments.
7. Develop a matrix diagram to design an organization-wide training or employee involvement program. Use a team of three or more people.
8. Using a team of three or more people, construct a matrix diagram to
   (a) Determine customer requirements for a new product or service.
   (b) Allocate team assignments to implement a project such as new student week.
   (c) Compare teacher characteristics with potential student performance.
9. Develop a prioritization matrix, using the tree diagram developed in Problem 6.
10. Construct a PDPC for
    (a) A charity walk-a-thon (see Problem 5).
    (b) The church carnival of Problem 6.
    (c) The matrix diagram developed in Problem 7.
11. Using a team of three or more people, construct an activity network diagram for
    (a) Constructing a cardboard boat.
    (b) An implementation schedule for a university event such as a graduation.
    (c) Developing a new instructional laboratory.
12. Select a problem or situation and, with a team of three or more people, use the seven management and planning tools to implement an action plan. If one of the tools doesn't fit, justify its exclusion.

APPENDIX

TABLE A Areas Under the Normal Curve.[a]

$\frac{X_i - \mu}{\sigma}$	0.09	0.08	0.07	0.06	0.05	0.04	0.03	0.02	0.01	0.00
−3.5	0.00017	0.00017	0.00018	0.00019	0.00019	0.00020	0.00021	0.00022	0.00022	0.00023
−3.4	0.00024	0.00025	0.00026	0.00027	0.00028	0.00029	0.00030	0.00031	0.00033	0.00034
−3.3	0.00035	0.00036	0.00038	0.00039	0.00040	0.00042	0.00043	0.00045	0.00047	0.00048
−3.2	0.00050	0.00052	0.00054	0.00056	0.00058	0.00060	0.00062	0.00064	0.00066	0.00069
−3.1	0.00071	0.00074	0.00076	0.00079	0.00082	0.00085	0.00087	0.00090	0.00094	0.00097
−3.0	0.00100	0.00104	0.00107	0.00111	0.00114	0.00118	0.00122	0.00126	0.00131	0.00135
−2.9	0.0014	0.0014	0.0015	0.0015	0.0016	0.0016	0.0017	0.0017	0.0018	0.0019
−2.8	0.0019	0.0020	0.0021	0.0021	0.0022	0.0023	0.0023	0.0024	0.0025	0.0026
−2.7	0.0026	0.0027	0.0028	0.0029	0.0030	0.0031	0.0032	0.0033	0.0034	0.0035
−2.6	0.0036	0.0037	0.0038	0.0039	0.0040	0.0041	0.0043	0.0044	0.0045	0.0047
−2.5	0.0048	0.0049	0.0051	0.0052	0.0054	0.0055	0.0057	0.0059	0.0060	0.0062
−2.4	0.0064	0.0066	0.0068	0.0069	0.0071	0.0073	0.0075	0.0078	0.0080	0.0082
−2.3	0.0084	0.0087	0.0089	0.0091	0.0094	0.0096	0.0099	0.0102	0.0104	0.0107
−2.2	0.0110	0.0113	0.0116	0.0119	0.0122	0.0125	0.0129	0.0132	0.0136	0.0139
−2.1	0.0143	0.0146	0.0150	0.0154	0.0158	0.0162	0.0166	0.0170	0.0174	0.0179
−2.0	0.0183	0.0188	0.0192	0.0197	0.0202	0.0207	0.0212	0.0217	0.0222	0.0228
−1.9	0.0233	0.0239	0.0244	0.0250	0.0256	0.0262	0.0268	0.0274	0.0281	0.0287
−1.8	0.0294	0.0301	0.0307	0.0314	0.0322	0.0329	0.0336	0.0344	0.0351	0.0359
−1.7	0.0367	0.0375	0.0384	0.0392	0.0401	0.0409	0.0418	0.0427	0.0436	0.0446
−1.6	0.0455	0.0465	0.0475	0.0485	0.0495	0.0505	0.0516	0.0526	0.0537	0.0548
−1.5	0.0559	0.0571	0.0582	0.0594	0.0606	0.0618	0.0630	0.0643	0.0655	0.0668
−1.4	0.0681	0.0694	0.0708	0.0721	0.0735	0.0749	0.0764	0.0778	0.0793	0.0808
−1.3	0.0823	0.0838	0.0853	0.0869	0.0885	0.0901	0.0918	0.0934	0.0951	0.0968
−1.2	0.0895	0.1003	0.1020	0.1038	0.1057	0.1075	0.1093	0.1112	0.1131	0.1151
−1.1	0.1170	0.1190	0.1210	0.1230	0.1251	0.1271	0.1292	0.1314	0.1335	0.1357
−1.0	0.1379	0.1401	0.1423	0.1446	0.1469	0.1492	0.1515	0.1539	0.1562	0.1587
−0.9	0.1611	0.1635	0.1660	0.1685	0.1711	0.1736	0.1762	0.1788	0.1814	0.1841
−0.8	0.1867	0.1894	0.1922	0.1949	0.1977	0.2005	0.2033	0.2061	0.2090	0.2119
−0.7	0.2148	0.2177	0.2207	0.2236	0.2266	0.2297	0.2327	0.2358	0.2389	0.2420
−0.6	0.2451	0.2483	0.2514	0.2546	0.2578	0.2611	0.2643	0.2676	0.2709	0.2743
−0.5	0.2776	0.2810	0.2843	0.2877	0.2912	0.2946	0.2981	0.3015	0.3050	0.3085
−0.4	0.3121	0.3156	0.3192	0.3228	0.3264	0.3300	0.3336	0.3372	0.3409	0.3446
−0.3	0.3483	0.3520	0.3557	0.3594	0.3632	0.3669	0.3707	0.3745	0.3783	0.3821
−0.2	0.3859	0.3897	0.3936	0.3974	0.4013	0.4052	0.4090	0.4129	0.4168	0.4207
−0.1	0.4247	0.4286	0.4325	0.4364	0.4404	0.4443	0.4483	0.4522	0.4562	0.4602
−0.0	0.4641	0.4681	0.4721	0.4761	0.4801	0.4840	0.4880	0.4920	0.4960	0.5000

TABLE A (Continued)

$\frac{X_i - \mu}{\sigma}$	0.00	0.01	0.02	0.03	0.04	0.05	0.06	0.07	0.08	0.09
+0.0	0.5000	0.5040	0.5080	0.5120	0.5160	0.5199	0.5239	0.5279	0.5319	0.5359
+0.1	0.5398	0.5438	0.5478	0.5517	0.5557	0.5596	0.5636	0.5675	0.5714	0.5753
+0.2	0.5793	0.5832	0.5871	0.5910	0.5948	0.5987	0.6026	0.6064	0.6103	0.6141
+0.3	0.6179	0.6217	0.6255	0.6293	0.6331	0.6368	0.6406	0.6443	0.6480	0.6517
+0.4	0.6554	0.6591	0.6628	0.6664	0.6700	0.6736	0.6772	0.6808	0.6844	0.6879
+0.5	0.6915	0.6950	0.6985	0.7019	0.7054	0.7088	0.7123	0.7157	0.7190	0.7224
+0.6	0.7257	0.7291	0.7324	0.7357	0.7389	0.7422	0.7454	0.7486	0.7517	0.7549
+0.7	0.7580	0.7611	0.7642	0.7673	0.7704	0.7734	0.7764	0.7794	0.7823	0.7852
+0.8	0.7881	0.7910	0.7939	0.7967	0.7995	0.8023	0.8051	0.8079	0.8106	0.8133
+0.9	0.8159	0.8186	0.8212	0.8238	0.8264	0.8289	0.8315	0.8340	0.8365	0.8389
+1.0	0.8413	0.8438	0.8461	0.8485	0.8508	0.8531	0.8554	0.8577	0.8599	0.8621
+1.1	0.8643	0.8665	0.8686	0.8708	0.8729	0.8749	0.8770	0.8790	0.8810	0.8830
+1.2	0.8849	0.8869	0.8888	0.8907	0.8925	0.8944	0.8962	0.8980	0.8997	0.9015
+1.3	0.9032	0.9049	0.9066	0.9082	0.9099	0.9115	0.9131	0.9147	0.9162	0.9177
+1.4	0.9192	0.9207	0.9222	0.9236	0.9251	0.9265	0.9279	0.9292	0.9306	0.9319
+1.5	0.9332	0.9345	0.9357	0.9370	0.9382	0.9394	0.9406	0.9418	0.9429	0.9441
+1.6	0.9452	0.9463	0.9474	0.9484	0.9495	0.9505	0.9515	0.9525	0.9535	0.9545
+1.7	0.9554	0.9564	0.9573	0.9582	0.9591	0.9599	0.9608	0.9616	0.9625	0.9633
+1.8	0.9641	0.9649	0.9656	0.9664	0.9671	0.9678	0.9686	0.9693	0.9699	0.9706
+1.9	0.9713	0.9719	0.9726	0.9732	0.9738	0.9744	0.9750	0.9756	0.9761	0.9767
+2.0	0.9773	0.9778	0.9783	0.9788	0.9793	0.9798	0.9803	0.9808	0.9812	0.9817
+2.1	0.9821	0.9826	0.9830	0.9834	0.9838	0.9842	0.9846	0.9850	0.9854	0.9857
+2.2	0.9861	0.9864	0.9868	0.9871	0.9875	0.9878	0.9881	0.9884	0.9887	0.9890
+2.3	0.9893	0.9896	0.9898	0.9901	0.9904	0.9906	0.9909	0.9911	0.9913	0.9916
+2.4	0.9918	0.9920	0.9922	0.9925	0.9927	0.9929	0.9931	0.9932	0.9934	0.9936
+2.5	0.9938	0.9940	0.9941	0.9943	0.9945	0.9946	0.9948	0.9949	0.9951	0.9952
+2.6	0.9953	0.9955	0.9956	0.9957	0.9959	0.9960	0.9961	0.9962	0.9963	0.9964
+2.7	0.9965	0.9966	0.9967	0.9968	0.9969	0.9970	0.9971	0.9972	0.9973	0.9974
+2.8	0.9974	0.9975	0.9976	0.9977	0.9977	0.9978	0.9979	0.9979	0.9980	0.9981
+2.9	0.9981	0.9982	0.9983	0.9983	0.9984	0.9984	0.9985	0.9985	0.9986	0.9986
+3.0	0.99865	0.99869	0.99874	0.99878	0.99882	0.99886	0.99889	0.99893	0.99896	0.99900
+3.1	0.99903	0.99906	0.99910	0.99913	0.99915	0.99918	0.99921	0.99924	0.99926	0.99929
+3.2	0.99931	0.99934	0.99936	0.99938	0.99940	0.99942	0.99944	0.99946	0.99948	0.99950
+3.3	0.99952	0.99953	0.99955	0.99957	0.99958	0.99960	0.99961	0.99962	0.99964	0.99965
+3.4	0.99966	0.99967	0.99969	0.99970	0.99971	0.99972	0.99973	0.99974	0.99975	0.99976
+3.5	0.99977	0.99978	0.99978	0.99979	0.99980	0.99981	0.99981	0.99982	0.99983	0.99983

[a]Proportion of total area under the curve that is under the proportion of the curve from $-\infty$ to $(X_i - \mu)/\sigma$ (X_i represents any desired value of the variable X).

TABLE B Factors for Computing Central Lines and 3σ Control Limits for $\overline{X}$, s, and R Charts.

OBSERVATIONS IN SAMPLE, n	CHART FOR AVERAGES: FACTORS FOR CONTROL LIMITS			CHART FOR STANDARD DEVIATIONS: FACTOR FOR CENTRAL LINE	FACTORS FOR CONTROL LIMITS				CHART FOR RANGES: FACTOR FOR CENTRAL LINE	FACTORS FOR CONTROL LIMITS				
	A	A_2	A_3	c_4	B_3	B_4	B_5	B_6	d_2	d_3	D_1	D_2	D_3	D_4
2	2.121	1.880	2.659	0.7979	0	3.267	0	2.606	1.128	0.853	0	3.686	0	3.267
3	1.732	1.023	1.954	0.8862	0	2.568	0	2.276	1.693	0.888	0	4.358	0	2.574
4	1.500	0.729	1.628	0.9213	0	2.266	0	2.088	2.059	0.880	0	4.698	0	2.282
5	1.342	0.577	1.427	0.9400	0	2.089	0	1.964	2.326	0.864	0	4.918	0	2.114
6	1.225	0.483	1.287	0.9515	0.030	1.970	0.029	1.874	2.534	0.848	0	5.078	0	2.004
7	1.134	0.419	1.182	0.9594	0.118	1.882	0.113	1.806	2.704	0.833	0.204	5.204	0.076	1.924
8	1.061	0.373	1.099	0.9650	0.185	1.815	0.179	1.751	2.847	0.820	0.388	5.306	0.136	1.864
9	1.000	0.337	1.032	0.9693	0.239	1.761	0.232	1.707	2.970	0.808	0.547	5.393	0.184	1.816
10	0.949	0.308	0.975	0.9727	0.284	1.716	0.276	1.669	3.078	0.797	0.687	5.469	0.223	1.777
11	0.905	0.285	0.927	0.9754	0.321	1.679	0.313	1.637	3.173	0.787	0.811	5.535	0.256	1.744
12	0.866	0.266	0.886	0.9776	0.354	1.646	0.346	1.610	3.258	0.778	0.922	5.594	0.283	1.717
13	0.832	0.249	0.850	0.9794	0.382	1.618	0.374	1.585	3.336	0.770	1.025	5.647	0.307	1.693
14	0.802	0.235	0.817	0.9810	0.406	1.594	0.399	1.563	3.407	0.763	1.118	5.696	0.328	1.672
15	0.775	0.223	0.789	0.9823	0.428	1.572	0.421	1.544	3.472	0.756	1.203	5.741	0.347	1.653
16	0.750	0.212	0.763	0.9835	0.448	1.552	0.440	1.526	3.532	0.750	1.282	5.782	0.363	1.637
17	0.728	0.203	0.739	0.9845	0.466	1.534	0.458	1.511	3.588	0.744	1.356	5.820	0.378	1.622
18	0.707	0.194	0.718	0.9854	0.482	1.518	0.475	1.496	3.640	0.739	1.424	5.856	0.391	1.608
19	0.688	0.187	0.698	0.9862	0.497	1.503	0.490	1.483	3.689	0.734	1.487	5.891	0.403	1.597
20	0.671	0.180	0.680	0.9869	0.510	1.490	0.504	1.470	3.735	0.729	1.549	5.921	0.415	1.585

TABLE C **The Poisson Distribution** $P(c) = \frac{(np_0)^c}{c!} e^{-np_0}$ **(Cumulative Values Are in Parentheses).**

c \ np_0	0.1	0.2	0.3	0.4	0.5
0	0.905 (0.905)	0.819 (0.819)	0.741 (0.741)	0.670 (0.670)	0.607 (0.607)
1	0.091 (0.996)	0.164 (0.983)	0.222 (0.963)	0.268 (0.938)	0.303 (0.910)
2	0.004 (1.000)	0.016 (0.999)	0.033 (0.996)	0.054 (0.992)	0.076 (0.986)
3		0.010 (1.000)	0.004 (1.000)	0.007 (0.999)	0.013 (0.999)
4				0.001 (1.000)	0.001 (1.000)

c \ np_0	0.6	0.7	0.8	0.9	1.0
0	0.549 (0.549)	0.497 (0.497)	0.449 (0.449)	0.406 (0.406)	0.368 (0.368)
1	0.329 (0.878)	0.349 (0.845)	0.359 (0.808)	0.366 (0.772)	0.368 (0.736)
2	0.099 (0.977)	0.122 (0.967)	0.144 (0.952)	0.166 (0.938)	0.184 (0.920)
3	0.020 (0.997)	0.028 (0.995)	0.039 (0.991)	0.049 (0.987)	0.061 (0.981)
4	0.003 (1.000)	0.005 (1.000)	0.008 (0.999)	0.011 (0.998)	0.016 (0.997)
5			0.001 (1.000)	0.002 (1.000)	0.003 (1.000)

c \ np_0	1.1	1.2	1.3	1.4	1.5
0	0.333 (0.333)	0.301 (0.301)	0.273 (0.273)	0.247 (0.247)	0.223 (0.223)
1	0.366 (0.699)	0.361 (0.662)	0.354 (0.627)	0.345 (0.592)	0.335 (0.558)
2	0.201 (0.900)	0.217 (0.879)	0.230 (0.857)	0.242 (0.834)	0.251 (0.809)
3	0.074 (0.974)	0.087 (0.966)	0.100 (0.957)	0.113 (0.947)	0.126 (0.935)
4	0.021 (0.995)	0.026 (0.992)	0.032 (0.989)	0.039 (0.986)	0.047 (0.982)
5	0.004 (0.999)	0.007 (0.999)	0.009 (0.998)	0.011 (0.997)	0.014 (0.996)
6	0.001 (1.000)	0.001 (1.000)	0.002 (1.000)	0.003 (1.000)	0.004 (1.000)

c \ np_0	1.6	1.7	1.8	1.9	2.0
0	0.202 (0.202)	0.183 (0.183)	0.165 (0.165)	0.150 (0.150)	0.135 (0.135)
1	0.323 (0.525)	0.311 (0.494)	0.298 (0.463)	0.284 (0.434)	0.271 (0.406)
2	0.258 (0.783)	0.264 (0.758)	0.268 (0.731)	0.270 (0.704)	0.271 (0.677)
3	0.138 (0.921)	0.149 (0.907)	0.161 (0.892)	0.171 (0.875)	0.180 (0.857)
4	0.055 (0.976)	0.064 (0.971)	0.072 (0.964)	0.081 (0.956)	0.090 (0.947)
5	0.018 (0.994)	0.022 (0.993)	0.026 (0.990)	0.031 (0.987)	0.036 (0.983)
6	0.005 (0.999)	0.006 (0.999)	0.008 (0.998)	0.010 (0.997)	0.012 (0.995)
7	0.001 (1.000)	0.001 (1.000)	0.002 (1.000)	0.003 (1.000)	0.004 (0.999)
8					0.001 (1.000)

(continued on the next page)

TABLE C (Continued)

c \ np_0	2.1	2.2	2.3	2.4	2.5
0	0.123 (0.123)	0.111 (0.111)	0.100 (0.100)	0.091 (0.091)	0.082 (0.082)
1	0.257 (0.380)	0.244 (0.355)	0.231 (0.331)	0.218 (0.309)	0.205 (0.287)
2	0.270 (0.650)	0.268 (0.623)	0.265 (0.596)	0.261 (0.570)	0.256 (0.543)
3	0.189 (0.839)	0.197 (0.820)	0.203 (0.799)	0.209 (0.779)	0.214 (0.757)
4	0.099 (0.938)	0.108 (0.928)	0.117 (0.916)	0.125 (0.904)	0.134 (0.891)
5	0.042 (0.980)	0.048 (0.976)	0.054 (0.970)	0.060 (0.964)	0.067 (0.958)
6	0.015 (0.995)	0.017 (0.993)	0.021 (0.991)	0.024 (0.988)	0.028 (0.986)
7	0.004 (0.999)	0.005 (0.998)	0.007 (0.998)	0.008 (0.996)	0.010 (0.996)
8	0.001 (1.000)	0.002 (1.000)	0.002 (1.000)	0.003 (0.999)	0.003 (0.999)
9				0.001 (1.000)	0.001 (1.000)

c \ np_0	2.6	2.7	2.8	2.9	3.0
0	0.074 (0.074)	0.067 (0.067)	0.061 (0.061)	0.055 (0.055)	0.050 (0.050)
1	0.193 (0.267)	0.182 (0.249)	0.170 (0.231)	0.160 (0.215)	0.149 (0.199)
2	0.251 (0.518)	0.245 (0.494)	0.238 (0.469)	0.231 (0.446)	0.224 (0.423)
3	0.218 (0.736)	0.221 (0.715)	0.223 (0.692)	0.224 (0.670)	0.224 (0.647)
4	0.141 (0.877)	0.149 (0.864)	0.156 (0.848)	0.162 (0.832)	0.168 (0.815)
5	0.074 (0.951)	0.080 (0.944)	0.087 (0.935)	0.094 (0.926)	0.101 (0.916)
6	0.032 (0.983)	0.036 (0.980)	0.041 (0.976)	0.045 (0.971)	0.050 (0.966)
7	0.012 (0.995)	0.014 (0.994)	0.016 (0.992)	0.019 (0.990)	0.022 (0.988)
8	0.004 (0.999)	0.005 (0.999)	0.006 (0.998)	0.007 (0.997)	0.008 (0.996)
9	0.001 (1.000)	0.001 (1.000)	0.002 (1.000)	0.002 (0.999)	0.003 (0.999)
10				0.001 (1.000)	0.001 (1.000)

c \ np_0	3.1	3.2	3.3	3.4	3.5
0	0.045 (0.045)	0.041 (0.041)	0.037 (0.037)	0.033 (0.033)	0.030 (0.030)
1	0.140 (0.185)	0.130 (0.171)	0.122 (0.159)	0.113 (0.146)	0.106 (0.136)
2	0.216 (0.401)	0.209 (0.380)	0.201 (0.360)	0.193 (0.339)	0.185 (0.321)
3	0.224 (0.625)	0.223 (0.603)	0.222 (0.582)	0.219 (0.558)	0.216 (0.537)
4	0.173 (0.798)	0.178 (0.781)	0.182 (0.764)	0.186 (0.744)	0.189 (0.726)
5	0.107 (0.905)	0.114 (0.895)	0.120 (0.884)	0.126 (0.870)	0.132 (0.858)
6	0.056 (0.961)	0.061 (0.956)	0.066 (0.950)	0.071 (0.941)	0.077 (0.935)
7	0.025 (0.986)	0.028 (0.984)	0.031 (0.981)	0.035 (0.976)	0.038 (0.973)
8	0.010 (0.996)	0.011 (0.995)	0.012 (0.993)	0.015 (0.991)	0.017 (0.990)
9	0.003 (0.999)	0.004 (0.999)	0.005 (0.998)	0.006 (0.997)	0.007 (0.997)
10	0.001 (1.000)	0.001 (1.000)	0.002 (1.000)	0.002 (0.999)	0.002 (0.999)
11				0.001 (1.000)	0.001 (1.000)

TABLE C (Continued)

c \ np_0	3.6	3.7	3.8	3.9	4.0
0	0.027 (0.027)	0.025 (0.025)	0.022 (0.022)	0.020 (0.020)	0.018 (0.018)
1	0.098 (0.125)	0.091 (0.116)	0.085 (0.107)	0.079 (0.099)	0.073 (0.091)
2	0.177 (0.302)	0.169 (0.285)	0.161 (0.268)	0.154 (0.253)	0.147 (0.238)
3	0.213 (0.515)	0.209 (0.494)	0.205 (0.473)	0.200 (0.453)	0.195 (0.433)
4	0.191 (0.706)	0.193 (0.687)	0.194 (0.667)	0.195 (0.648)	0.195 (0.628)
5	0.138 (0.844)	0.143 (0.830)	0.148 (0.815)	0.152 (0.800)	0.157 (0.785)
6	0.083 (0.927)	0.088 (0.918)	0.094 (0.909)	0.099 (0.899)	0.104 (0.889)
7	0.042 (0.969)	0.047 (0.965)	0.051 (0.960)	0.055 (0.954)	0.060 (0.949)
8	0.019 (0.988)	0.022 (0.987)	0.024 (0.984)	0.027 (0.981)	0.030 (0.979)
9	0.008 (0.996)	0.009 (0.996)	0.010 (0.994)	0.012 (0.993)	0.013 (0.992)
10	0.003 (0.999)	0.003 (0.999)	0.004 (0.998)	0.004 (0.997)	0.005 (0.997)
11	0.001 (1.000)	0.001 (1.000)	0.001 (0.999)	0.002 (0.999)	0.002 (0.999)
12			0.001 (1.000)	0.001 (1.000)	0.001 (1.000)

c \ np_0	4.1	4.2	4.3	4.4	4.5
0	0.017 (0.017)	0.015 (0.015)	0.014 (0.014)	0.012 (0.012)	0.011 (0.011)
1	0.068 (0.085)	0.063 (0.078)	0.058 (0.072)	0.054 (0.066)	0.050 (0.061)
2	0.139 (0.224)	0.132 (0.210)	0.126 (0.198)	0.119 (0.185)	0.113 (0.174)
3	0.190 (0.414)	0.185 (0.395)	0.180 (0.378)	0.174 (0.359)	0.169 (0.343)
4	0.195 (0.609)	0.195 (0.590)	0.193 (0.571)	0.192 (0.551)	0.190 (0.533)
5	0.160 (0.769)	0.163 (0.753)	0.166 (0.737)	0.169 (0.720)	0.171 (0.704)
6	0.110 (0.879)	0.114 (0.867)	0.119 (0.856)	0.124 (0.844)	0.128 (0.832)
7	0.064 (0.943)	0.069 (0.936)	0.073 (0.929)	0.078 (0.922)	0.082 (0.914)
8	0.033 (0.976)	0.036 (0.972)	0.040 (0.969)	0.043 (0.965)	0.046 (0.960)
9	0.015 (0.991)	0.017 (0.989)	0.019 (0.988)	0.021 (0.986)	0.023 (0.983)
10	0.006 (0.997)	0.007 (0.996)	0.008 (0.996)	0.009 (0.995)	0.011 (0.994)
11	0.002 (0.999)	0.003 (0.999)	0.003 (0.999)	0.004 (0.999)	0.004 (0.998)
12	0.001 (1.000)	0.001 (1.000)	0.001 (1.000)	0.001 (1.000)	0.001 (0.999)
13					0.001 (1.000)

c \ np_0	4.6	4.7	4.8	4.9	5.0
0	0.010 (0.010)	0.009 (0.009)	0.008 (0.008)	0.008 (0.008)	0.007 (0.007)
1	0.046 (0.056)	0.043 (0.052)	0.039 (0.047)	0.037 (0.045)	0.034 (0.041)
2	0.106 (0.162)	0.101 (0.153)	0.095 (0.142)	0.090 (0.135)	0.084 (0.125)
3	0.163 (0.325)	0.157 (0.310)	0.152 (0.294)	0.146 (0.281)	0.140 (0.265)
4	0.188 (0.513)	0.185 (0.495)	0.182 (0.476)	0.179 (0.460)	0.176 (0.441)

(continued on the next page)

TABLE C (Continued)

c \ np_0	4.6	4.7	4.8	4.9	5.0
5	0.172 (0.685)	0.174 (0.669)	0.175 (0.651)	0.175 (0.635)	0.176 (0.617)
6	0.132 (0.817)	0.136 (0.805)	0.140 (0.791)	0.143 (0.778)	0.146 (0.763)
7	0.087 (0.904)	0.091 (0.896)	0.096 (0.887)	0.100 (0.878)	0.105 (0.868)
8	0.050 (0.954)	0.054 (0.950)	0.058 (0.945)	0.061 (0.939)	0.065 (0.933)
9	0.026 (0.980)	0.028 (0.978)	0.031 (0.976)	0.034 (0.973)	0.036 (0.969)
10	0.012 (0.992)	0.013 (0.991)	0.015 (0.991)	0.016 (0.989)	0.018 (0.987)
11	0.005 (0.997)	0.006 (0.997)	0.006 (0.997)	0.007 (0.996)	0.008 (0.995)
12	0.002 (0.999)	0.002 (0.999)	0.002 (0.999)	0.003 (0.999)	0.003 (0.998)
13	0.001 (1.000)	0.001 (1.000)	0.001 (1.000)	0.001 (1.000)	0.001 (0.999)
14					0.001 (1.000)

c \ np_0	6.0	7.0	8.0	9.0	10.0
0	0.002 (0.002)	0.001 (0.001)	0.000 (0.000)	0.000 (0.000)	0.000 (0.000)
1	0.015 (0.017)	0.006 (0.007)	0.003 (0.003)	0.001 (0.001)	0.000 (0.000)
2	0.045 (0.062)	0.022 (0.029)	0.011 (0.014)	0.005 (0.006)	0.002 (0.002)
3	0.089 (0.151)	0.052 (0.081)	0.029 (0.043)	0.015 (0.021)	0.007 (0.009)
4	0.134 (0.285)	0.091 (0.172)	0.057 (0.100)	0.034 (0.055)	0.019 (0.028)
5	0.161 (0.446)	0.128 (0.300)	0.092 (0.192)	0.061 (0.116)	0.038 (0.066)
6	0.161 (0.607)	0.149 (0.449)	0.122 (0.314)	0.091 (0.091)	0.063 (0.129)
7	0.138 (0.745)	0.149 (0.598)	0.140 (0.454)	0.117 (0.324)	0.090 (0.219)
8	0.103 (0.848)	0.131 (0.729)	0.140 (0.594)	0.132 (0.456)	0.113 (0.332)
9	0.069 (0.917)	0.102 (0.831)	0.124 (0.718)	0.132 (0.588)	0.124 (0.457)
10	0.041 (0.958)	0.071 (0.902)	0.099 (0.817)	0.119 (0.707)	0.125 (0.582)
11	0.023 (0.981)	0.045 (0.947)	0.072 (0.889)	0.097 (0.804)	0.114 (0.696)
12	0.011 (0.992)	0.026 (0.973)	0.048 (0.937)	0.073 (0.877)	0.095 (0.791)
13	0.005 (0.997)	0.014 (0.987)	0.030 (0.967)	0.050 (0.927)	0.073 (0.864)
14	0.002 (0.999)	0.007 (0.994)	0.017 (0.984)	0.032 (0.959)	0.052 (0.916)
15	0.001 (1.000)	0.003 (0.997)	0.009 (0.993)	0.019 (0.978)	0.035 (0.951)
16		0.002 (0.999)	0.004 (0.997)	0.011 (0.989)	0.022 (0.973)
17		0.001 (1.000)	0.002 (0.999)	0.006 (0.995)	0.013 (0.986)
18			0.001 (1.000)	0.003 (0.998)	0.007 (0.993)
19				0.001 (0.999)	0.004 (0.997)
20				0.001 (1.000)	0.002 (0.999)
21					0.001 (1.000)

TABLE C (Continued)

c \ np_0	11.0	12.0	13.0	14.0	15.0
0	0.000 (0.000)	0.000 (0.000)	0.000 (0.000)	0.000 (0.000)	0.000 (0.000)
1	0.000 (0.000)	0.000 (0.000)	0.000 (0.000)	0.000 (0.000)	0.000 (0.000)
2	0.001 (0.001)	0.000 (0.000)	0.000 (0.000)	0.000 (0.000)	0.000 (0.000)
3	0.004 (0.005)	0.002 (0.002)	0.001 (0.001)	0.000 (0.000)	0.000 (0.000)
4	0.010 (0.015)	0.005 (0.007)	0.003 (0.004)	0.001 (0.001)	0.001 (0.001)
5	0.022 (0.037)	0.013 (0.020)	0.007 (0.011)	0.004 (0.005)	0.002 (0.003)
6	0.041 (0.078)	0.025 (0.045)	0.015 (0.026)	0.009 (0.014)	0.005 (0.008)
7	0.065 (0.143)	0.044 (0.089)	0.028 (0.054)	0.017 (0.031)	0.010 (0.018)
8	0.089 (0.232)	0.066 (0.155)	0.046 (0.100)	0.031 (0.062)	0.019 (0.037)
9	0.109 (0.341)	0.087 (0.242)	0.066 (0.166)	0.047 (0.109)	0.032 (0.069)
10	0.119 (0.460)	0.105 (0.347)	0.086 (0.252)	0.066 (0.175)	0.049 (0.118)
11	0.119 (0.579)	0.114 (0.461)	0.101 (0.353)	0.084 (0.259)	0.066 (0.184)
12	0.109 (0.688)	0.114 (0.575)	0.110 (0.463)	0.099 (0.358)	0.083 (0.267)
13	0.093 (0.781)	0.106 (0.681)	0.110 (0.573)	0.106 (0.464)	0.096 (0.363)
14	0.073 (0.854)	0.091 (0.772)	0.102 (0.675)	0.106 (0.570)	0.102 (0.465)
15	0.053 (0.907)	0.072 (0.844)	0.088 (0.763)	0.099 (0.669)	0.102 (0.567)
16	0.037 (0.944)	0.054 (0.898)	0.072 (0.835)	0.087 (0.756)	0.096 (0.663)
17	0.024 (0.968)	0.038 (0.936)	0.055 (0.890)	0.071 (0.827)	0.085 (0.748)
18	0.015 (0.983)	0.026 (0.962)	0.040 (0.930)	0.056 (0.883)	0.071 (0.819)
19	0.008 (0.991)	0.016 (0.978)	0.027 (0.957)	0.041 (0.924)	0.056 (0.875)
20	0.005 (0.996)	0.010 (0.988)	0.018 (0.975)	0.029 (0.953)	0.042 (0.917)
21	0.002 (0.998)	0.006 (0.994)	0.011 (0.986)	0.019 (0.972)	0.030 (0.947)
22	0.001 (0.999)	0.003 (0.997)	0.006 (0.992)	0.012 (0.984)	0.020 (0.967)
23	0.001 (1.000)	0.002 (0.999)	0.004 (0.996)	0.007 (0.991)	0.013 (0.980)
24		0.001 (1.000)	0.002 (0.998)	0.004 (0.995)	0.008 (0.988)
25			0.001 (0.999)	0.003 (0.998)	0.005 (0.993)
26			0.001 (1.000)	0.001 (0.999)	0.003 (0.996)
27				0.001 (1.000)	0.002 (0.998)
28					0.001 (0.999)
29					0.001 (1.000)

TABLE D Random Numbers.

63271	59986	71744	51102	15141	80714	58683	93108
88547	09896	95436	79115	08303	01041	20030	63754
55957	57243	83865	09911	19761	66535	40102	26646
46276	87453	44790	67122	45573	84358	21625	16999
55363	07449	34835	15290	76616	67191	12777	21861
69393	92785	49902	58447	42048	30378	87618	26933
13186	29431	88190	04588	38733	81290	89541	70290
17726	28652	56836	78351	47327	18518	92222	55201
36520	64465	05550	30157	82242	29520	69753	72602
81628	36100	39254	56835	37636	02421	98063	89641
84649	48968	75215	75498	49539	74240	03466	49292
63291	11618	12613	75055	43915	26488	41116	64531
70502	53225	03655	05915	37140	57051	48393	91322
06426	24771	59935	49801	11081	66762	94477	02494
20711	55609	29430	70165	45406	78484	31699	52009
41990	70538	77191	25860	55204	73417	83920	69468
72452	36618	76298	26678	89334	33938	95567	29380
37042	40318	57099	10528	09925	89773	41335	96244
53766	52875	15987	46962	67342	77592	57651	95508
90585	58955	53122	16025	84299	53310	67380	84249
32001	96293	37203	64516	51530	37069	40261	61374
62606	64324	46354	72157	67248	20135	49804	09226
10078	28073	85389	50324	14500	15562	64165	06125
91561	46145	24177	15294	10061	98124	75732	08815
13091	98112	53959	79607	52244	63303	10413	63839
73864	83014	72457	26682	03033	61714	88173	90835
66668	25467	48894	51043	02365	91726	09365	63167
84745	41042	29493	01836	09044	51926	43630	63470
48068	26805	94595	47907	13357	38412	33318	26098
54310	96175	97594	88616	42035	38093	36745	56702
14877	33095	10924	58013	61439	21882	42059	24177
78295	23179	02771	43464	59061	71411	05697	67194
67524	02865	39593	54278	04237	92441	26602	63835
58268	57219	68124	73455	83236	08710	04284	55005
97158	28672	50685	01181	24262	19427	52106	34308
04230	16831	69085	30802	65559	09205	71829	06489
94879	56606	30401	02602	57658	70091	54986	41394
71446	15232	66715	26385	91518	70566	02888	79941
32886	05644	79316	09819	00813	88407	17461	73925
62048	33711	25290	21526	02223	75947	66466	06232

TABLE E Commonly Used Conversion Factors.

QUANTITY	CONVERSION	MULTIPLY BY
Length	in. to m	2.54[a] E − 02
Area	in.^{2} to m^2	6.451 600 E − 04
Volume	in.^{3} to m^3	1.638 706 E − 05
	U.S. gallon to m^3	3.785 412 E − 03
Mass	oz (avoir) to kg	2.834 952 E − 02
Acceleration	ft/s^2 to m/s^2	3.048[a] E − 01
Force	poundal to N	1.382 550 E − 01
Pressure, stress	poundal/ft^2 to Pa	1.488 164 E + 00
	lb$_f$/in.^{2} to Pa	6.894 757 E + 03
Energy, work	(ft) (lb$_f$) to J	1.355 818 E + 00
Power	hp (550 ft) (lb$_f$/s) to W	7.456 999 E + 02

[a]Relationship is exact and needs no additional decimal points.

SELECTED BIBLIOGRAPHY

ANSI/ASQ B1–B3—1996, Quality Control Chart Methodologies. Milwaukee, Wis.: American Society for Quality, 1996.

ANSI/ASQ SI—1996, *An Attribute Skip-Lot Sampling Program*. Milwaukee, Wis.: American Society for Quality, 1996.

ANSI/ASQ S2—1995, *Introduction to Attribute Sampling*. Milwaukee, Wis.: American Society for Quality, 1995.

ANSI/ASQ Standard Q3—1988, *Sampling Procedures and Tables for Inspection by Isolated Lots by Attributes*. Milwaukee, Wis.: American Society for Quality, 1988.

ANSI/ASQ Z1.4—1993, *Sampling Procedures and Tables for Inspection by Attributes*. Milwaukee, Wis.: American Society for Quality, 1993.

ANSI/ASQ Z1.9—1993, *Sampling Procedures and Tables for Inspection by Variables for Percent Nonconforming*. Milwaukee, Wis.: American Society for Quality, 1993.

ANSI/ISO/ASQ A3534-1—1993, *Statistics—Vocabulary and Symbols—Probability and General Statistical Terms*. Milwaukee, Wis.: American Society for Quality, 1993.

ANSI/ISO/ASQ A3534-2—1993, *Statistics—Vocabulary and Symbols—Statistical Quality*. Milwaukee, Wis.: American Society for Quality, 1993.

ASQ/AIAG Task Force, *Fundamental Statistical Process Control*. Troy, Mich.: Automobile Industry Action Group, 1991.

ASQ Quality Cost Committee, *Guide for Reducing Quality Costs, 2d ed.* Milwaukee, Wis.: American Society for Quality, 1987.

ASQ Quality Cost Committee, *Principles of Quality Costs*. Milwaukee, Wis.: American Society for Quality, 1986.

ASQ Statistics Division, *Glossary and Tables for Statistical Quality Control*. Milwaukee, Wis.: American Society for Quality, 1983.

Bossert, James L., *Quality Function Deployment: A Practitioner's Approach*. Milwaukee, Wis.: ASQ Quality Press, 1991.

Brassard, Michael, *The Memory Jogger Plus*, Methuen, Mass.: GOAL/QPC, 1989.

Camp, Robert C., *Benchmarking: The Search for Industry Best Practices That Lead to Superior Practice*. Milwaukee, Wis.: ASQ Quality Press, 1989.

Crosby, Phillip B., *Quality Is Free*. New York: McGraw-Hill Book Company, 1979.

Crosby, Phillip B., *Quality Without Tears*. New York: McGraw-Hill Book Company, 1984.

Deming, W. Edwards, *Quality, Productivity, and Competitive Position*. Cambridge, Mass.: Massachusetts Institute of Technology, 1982.

Duncan, Acheson J., *Quality Control and Industrial Statistics, 5th ed.* Homewood, Ill.: Irwin, Inc., 1986.

Fellers, Gary, *SPC for Practitioners: Special Cases and Continuous Processes*. Milwaukee, Wis.: ASQ Quality Press, 1991.

Gitlow, H. S., and S. J. Gitlow, *The Deming Guide to Quality and Competitive Position*. Englewood Cliffs, N.J.: Prentice Hall, Inc., 1987.

Grosh, Doris L., *A Primer of Reliability Theory*. New York: John Wiley & Sons, 1989.

Henley, Ernest J., and Hiromitsu Kumamoto, *Reliability Engineering and Risk Assessment*. Englewood Cliffs, N.J.: Prentice Hall, Inc., 1981.

Ishikawa, K., *What Is Total Quality Control*? Englewood Cliffs, N.J.: Prentice Hall, Inc., 1985.

JURAN, JOSEPH M. (ed.), *Quality Control Handbook, 4th ed*. New York: McGraw-Hill Book Company, 1988.

JURAN, JOSEPH M., AND FRANK M. GRYNA, JR., *Quality Planning and Analysis, 2nd ed.* New York: McGraw-Hill Book Company, 1980.

PEACH, ROBERT W., Editor, *The ISO 9000 Handbook*. Fairfax, Va: CEEM Information Services, 1992.

SHAPIRO, SAMUEL S., The ASQ Basic References in Quality Control: Statistical Techniques, Edward J. Dudewicz, Ph.D., Editor, *Volume 3: How to Test Normality and Other Distributional Assumptions*. Milwaukee, Wis.: American Society for Quality,1980.

TAGUCHI, G., *Introduction to Quality Engineering*. Tokyo: Asian Productivity Organization, 1986.

WHEELER, DONALD J., *Short Run SPC*. Knoxville, Tenn.: SPC Press, Inc., 1991.

WINCHELL, WILLIAM, *TQM: Getting Started and Achieving Results with Total Quality Management*. Dearborn, Mich.: Society of Manufacturing Engineers, 1992.

GLOSSARY

Acceptable Quality Level (AQL) The maximum percent nonconforming that can be considered satisfactory for the purpose of acceptance sampling.

Acceptance Sampling A system by which a lot is accepted or rejected based on the results of inspecting samples in that lot.

Activity Network Diagram A group of diagrams that facilitates the efficient scheduling of a project.

Affinity Diagram A diagram that allows the team to creatively generate a large number of issues/ideas and logically group them.

Assignable Cause A cause of variation that is large in magnitude and easily identified; also called special cause.

Attribute A quality characteristic classified as either conforming or not conforming to specifications.

Availability A time-related factor that measures the ability of a product or service to perform its designated function.

Average The sum of all observations divided by the total number of observations.

Average Outgoing Quality (AOQ) Curve A curve that shows the average quality level for lots leaving the acceptance sampling system for different percent nonconforming values.

Average Sample Number (ASN) Curve A curve that shows the average amount inspected per lot by the consumer for different percent nonconforming values.

Average Total Inspection (ATI) Curve A curve that shows the amount inspected by both the consumer and the producer for different percent nonconforming values.

Cause and Effect Diagram A picture composed of lines and symbols designed to represent a meaningful relationship between an effect and its causes.

Cell A grouping of observed values within specified upper and lower boundaries.

Chance Cause A cause of variation that is small in magnitude and difficult to identify; also called random or common cause.

Check Sheets A device used to carefully and accurately record data.

Combination A counting technique that requires the unordered arrangement of a set of objects.

Consumer's Risk The probability that an unacceptable lot will be accepted.

Control Chart A graphical record of the variation in quality of a particular characteristic(s) during a specified time period.

Control Limits The limits on a control chart used to evaluate the variations in quality from subgroup to subgroup—not to be confused with specification limits.

Failure Rate The probability that a unit of product will fail in a specific unit of time or cycles.

Forced Field Analysis A technique to identify the forces and factors that may influence the problem or goal.

Frequency Distribution The arrangement of data to show the repetition of values in a category.

Histogram A graphical display, in rectangle form, of a frequency distribution.

Interrelationship Diagram A diagram that clarifies the inter relationship of many factors of a complex situation.

Kurtosis A value used to describe the peakedness of a distribution.

Limiting Quality (LQ) The percent nonconforming in a lot or batch that, for acceptance sampling purposes, the consumer wishes the probability of acceptance to be low.

Maintainability The ease with which preventative and corrective maintenance on a product or service can be achieved.

Matrix A diagram that provides for the identity, analysis, and evaluation of the interrelationship among variables.

Mean The average of a population.

Median The value that divides a series of ordered observations so that the number of items above it is equal to the number of items below it.

Mode The value that occurs with the greatest frequency in a set of numbers.

Multi-Vari Chart A graphical record of variation due to within parts, between parts, and time-to-time.

Nominal Group Technique A technique that process provides issues/ideas from everyone on the team and for effective decisions.

Nonconforming Unit A product or service which contains at least one nonconformity.

Nonconformity The departure of a quality characteristic from its intended level with a severity sufficient to cause a product or service not to meet specifications.

Operating Characteristic (OC) Curve A curve that shows the probability that a lot with a certain percent nonconforming will be accepted.

Pareto Diagram A method used to identify and communicate the vital few causes and the useful many of a situation.

Percent Tolerance Percent Chart A chart that evaluates the percent deviation from target for precontrol data.

Permutation A counting technique that requires the ordered arrangement of a set of objects.

Population The total set of observations being considered in a statistical procedure.

Precontrol A technique to compare samples of two with specifications.

Prioritization Matrices ThIS technique priorities issues, tasks, characteristics, and so forth based on weighted criterion.

Probability The mathematical calculation of the likelihood that an event will occur.

Process Capability The spread of the process. It is equal to six standard deviations when the process is in a state of statistical control.

Process Decision Program Chart This chart avoids surprises and identifies possible countermeasures.

Process Flow Diagram A diagram that shows the movement of a product or service as it travels through various processing stations.

Producer's Risk The probability that an acceptable lot is not accepted.

Quality Meeting or exceeding customer's expectations; customer satisfaction.

Range The difference between the largest and smallest observed values.

Reliability The probability that a product will perform its intended function for a prescribed life under certain stated conditions.

Sample A small portion of a population used to represent the entire population.

Skewness The departure of data from symmetrical.

Specification Limits The limits that define the boundaries of an acceptable product or service.

Standard Deviation A measure of the dispersion about the mean of a population or average of a sample.

Target The intended value of a quality characteristic; also called the nominal value.

Tolerance The permissible variation in the size of a quality characteristic.

Tree Diagram A tool used to reduce a broad objective into increasing levels of detail.

Variable A quality characteristic that is measurable, such as weight, length, etc.

ANSWERS TO SELECTED PROBLEMS

CHAPTER 3

1. 30.1%, 19.0%, 17.4%, 12.7%, 8.8%, 3.5%, 3.0%, 5.5%

3. 30.1%, 28.1%, 8.1%, 3.6%, 3.3%, 3.2%, 23.6%

5. 30.9%, 23.1%, 12.1%, 11.3%, 6.7%, 5.8%, 2.7%

7. 39.1%, 21.7%, 13.1%, 10.9%, 8.7%, 4.3%, 2.2%

9. Total nonconformities are decreasing.

15. As hours of machine use increases, millimeters off target increases; 1.61

17. No relationship

CHAPTER 4

1. (a) 0.86, (b) 0.63, (c) 0.15, (d) 0.48

3. (a) 0.0006, (b) 0.001, (c) 0.002, (d) 0.3

5. (a) 66.4, (b) 379.1, (c) 5, (d) 4.652, (e) 6.2×10^2

7. Frequencies starting at 5.94 are: 1, 2, 4, 8, 16, 24, 20, 17, 13, 3, 1, 1

9. Frequencies starting at 0.3 are: 3, 15, 34, 29, 30, 22, 15, 2

11. (a) Relative frequencies starting at 5.94 (in %) are: 0.9, 1.8, 3.6, 7.3, 14.5, 21.8, 18.2, 15.4, 11.8, 2.7, 0.9, 0.9
    (b) Cumulative frequencies starting at 5.945 are: 1, 3, 7, 15, 31, 55, 75, 92, 105, 108, 109, 110
    (c) Relative cumulative frequencies starting at 5.945 (in %) are: 0.9, 2.7, 6.4, 13.6, 28.2, 50.0, 68.2, 83.6, 95.4, 98.2, 99.1, 100.0

13. (a) Relative frequencies starting at 0.3 are: 0.020, 0.100, 0.227, 0.193, 0.200, 0.147, 0.100, 0.013
    (b) Cumulative frequencies starting at 0.3 are: 3, 18, 52, 81, 111, 133, 148, 150
    (c) Relative cumulative frequencies starting at 0.3 are: 0.020, 0.120, 0.347, 0.540, 0.740, 0.888, 0.987, 1.000

17. 116

19. 95

21. 3264

23. (a) 15; (b) 35.5

25. (a) 55, (b) none, (c) 14, 17

27. (a) 11, (b) 6, (c) 14, (d) 0.11

29. 0.004

31. (a) 0.8 (b) 20

35. (b) Frequencies beginning at 0.5 are: 1, 17, 29, 39, 51, 69, 85, 88

39. (a) Relative frequencies beginning at 0.5 (in %) are: 1.1, 18.2, 13.6, 11.4, 13.6, 20.5, 18.2, 3.4
    (b) Cumulative relative frequencies beginning at 0.5 (in %) are: 1.1, 19.3, 33.0, 44.3, 58.0, 78.4, 96.6, 100.0

41. (b) −0.14, 3.11

43. Process is not capable—5 out of 65 above specification and 6 out of 65 below specification.

45. (a) 0.0268, (b) 0.0099, (c) 0.9914 (based on rounded Z values)

47. 0.606

49. (a) Normal
    (b) Not normal, but symmetrical

CHAPTER 5

1. $\overline{\overline{X}} = 0.72$; CLs $= 0.83, 0.61$; $\overline{R} = 0.148$; CLs $= 0.34, 0$

3. $\overline{X}_0 = 482$; CLs $= 500, 464$; $R_0 = 25$; CLs $= 57, 0$

5. $\overline{X}_0 = 2.08$; CLs $= 2.42, 1.74$; $R_0 = 0.47$; CLs $= 1.08, 0$
7. $\overline{X}_0 = 81.9$; CLs $= 82.8, 81.0$; $s_0 = 0.7$; CLs $= 1.4, 0.0$
11. 0.47% scrap, 2.27% rework, $\overline{X}_0 = 305.32$ mm, 6.43% rework
13. 0.13
15. $6\sigma = 160$
17. (a) $6\sigma = 0.80$, (b) 1.38
19. 0.82, change specifications or reduce σ
21. $C_{pk} = 0.82; 0.41; 0; -0.41$
23. $\overline{\overline{X}} = 4.56$; CLs $= 4.76, 4.36$; $\overline{R} = 0.20$; CLs $= 0.52, 0$
25. $Md_{Md} = 6.3$; CLs $= 7.9, 4.7$; $R_{Md} = 1.25$; CLs $= 3.4, 0$
27. $\overline{X} = 7.59$; CLs $= 8.47, 6.71$; MR $= 0.33$; CLs $= 1.08, 0$
29. $\overline{X}_0 = 20.40$; RLs $= 20.46, 20.34$
31. Histogram is symmetrical, while run chart slopes downward.

CHAPTER 6

1. 400, r = 4
3. At 1400 h, time-to-time variation occurred and within-piece variation increased at 2100 h.
5. $\overline{X}_0 = 25.0$; CLs $= 25.15, 24.85$; $R_0 = 0.11$, CLs $=$ 47, 0
7. Yes, ratio is 1.17
9. $\overline{Z}_o = 0$; CLs $= +1.023, -1.023$; $\overline{W}_o = 1.00$; CLs $= 2.574, 0$; $\overline{Z}$ plotted points $= -0.2, 1.6, 0.4$; W plotted points $= 0.8, 0.6, 1.2$
11. Z points $= 1.67, -3.00$ (out of control); $MW = 1.33, 2.67$
13. PC $= 31.5, 32.5$
15. 73.5%, 24.5%
17. (a) −10, 80, (b) −80, 0, (c) −10, 64, (d) −20, −30
19. % R&R = 21.14%; gage may be acceptable based on practical conditions.

CHAPTER 7

1. 1.000, 0
3. 0.833
5. 0.50, 0.81
7. 0.40
9. 0.57

11. 0.018
13. 0.989
15. 520
17. 3.13×10^{15}
19. 161,700
21. 25,827,165
23. $1.50696145 \times 10^{-16}$
25. $C_r^n = C_{n-r}^n$
27. If $n = r$, then $C = 1$
29. 0.255, 0.509, 0.218, 0.018, P(4) is impossible.
31. 0.087, 0.997
33. 0.0317
35. 0.246
37. Binomial 0.075; Poisson 0.076
39. 0.475
41. Hypergeometric 0.435; Poisson 0.329; Poisson is poor estimator.
43. 0.084; Poisson is a good approximation.

CHAPTER 8

1. (a) Plotted points = 0.010, 0.018, 0.023, 0.015, 0.025
   (b) $\overline{p} = 0.0297$; CLs = 0.055, 0.004
   (c) $p_0 = 0.0242$; CLs = 0.047, 0.001
3. $100p_0 = 1.54$; CLs = 0.0367, 0; $q_0 = 0.9846$; CLs = 1.0000, 0.9633; $100q_0 = 98.46$; CLs = 100.00, 96.33
5. CLs = 0.188, 0; in control
7. $p_0 = 0.011$
9. $p_0 = 0.144$, CLs = 0.227; 0.061; Nov. 15
11. (a) $100p_0 = 1.54$; CLs = 3.67, 0
    (b) $100p_0 = 2.62$; CLs = 3.76, 1.48
13. $np_0 = 45.85$; CLs = 66.26; np
15. (a) $q_0 = 0.9846$; CLs = 1.00, 0.9633
    $100q_0 = 98.46$; CLs = 100; 96.33
    $nq_0 = 295$; CLs = 300, 289
    (b) $q_0 = 0.9738$; CLs = 0.9850, 0.9624
    $100q_0 = 97.38$; CLs = 98.50, 96.24
    $nq_0 = 1704$; CLs = 1724, 1684

17. $c_0 = 6.86$; CLs = 14.72, 0
19. Process in control; $\bar{c} = 10.5$; CLs = 20.2, 0.78
21. $u_0 = 3.34$; CLs = 5.08, 161
23. $u_0 - 1.092$, CLs = 1.554, 0.630

CHAPTER 9

1. (p, P_a) pairs are (0.01, 0.974), (0.02, 0.820), (0.04, 0.359), (0.05, 0.208), (0.06, 0.109), (0.08, 0.025), (0.10, 0.005)
3. $(P_a)_{\mathrm{I}} = P$ (2 or less)
   $(P_a)_{\mathrm{II}} = P(3)_{\mathrm{I}}$ or less$)_{\mathrm{II}} + P(4)_{\mathrm{I}} P$ (2 or less$)_{\mathrm{II}} + P(5)_{\mathrm{I}} P$ (1 or less$)_{\mathrm{II}}$
   $(P_a)_{\mathrm{both}} = (P_a)_{\mathrm{I}} + (P_a)_{\mathrm{II}}$
5. (100p, AOQ) pairs are (1, 0.974), (2, 1.640), (4, 1.436), (5, 1.040), (6, 0.654), (8, 0.200); AOQL ≅ 1.7%
7. AQL = 0.025%; AOQL = 0.19%
9. (p, ASN) pairs are (0, 125), (0.01, 140), (0.02, 169), (0.03, 174), (0.04, 165), (0.05, 150), (0.06, 139)
11. (p, ATI) pairs are (0, 80), (0.00125, 120), (0.01, 311), (0.02, 415), (0.03, 462), (0.04, 483)
13. (100p, AOQ) pairs are (0.5, 0.493), (1.0, 0.848), (1.5, 0.885), (2.0, 0.694), (2.5, 0.430)
15. 3,55; 6, 155; 12, 407
17. 2, 82; 6, 162; 14, 310
19. 1, 332; 3, 502; 5, 655
21. 3, 195
23. 4, 266
25. 5, 175
27. 0.69

CHAPTER 10

1. (a) $n = 125$, $Ac = 3$, $Re = 4$; (b) $n = 8$, $Ac = 10$, $Re = 11$; (c) $n = 500$, $Ac = 5$, $Re = 8$; (d) $n = 20$, $Ac = 1$, $Re = 2$
3. Tightened, the last two lots were rejected so criteria are met.
5. No, limit number is exceeded.
7. $n = 200$, $Ac = 3$
9. $n = 140$, $c = 7$, LQ = 8.2%

11. $n = 305, c = 1$, AOQL = 0.14%

13. 0.753

15. 0.923

17. $d_a = -1.91 + 0.080n$; $d_r = 2.48 + 0.080n$

19. 36, 59, 76

21. Start in state 1 ($f = 1/3$); lot 11 go to state 2; lot 19 go to state 3; lot 23 back to state 2.

23. (a) $f = 1/4$; (b) $f = 1/3$

25. 194, 420, 762

27. $i = 116$, 1/30, 1/60

29. $i = 192$, D

31. 57.85, 44.43, Type 1

33. I, 25

35. $10.76 \geq 9.80$, *Re* lot

37. $Q_u = 1.91$, *Ac* lot

39. *Re* lot

CHAPTER 11

1. 0.78

3. 0.980

5. 0.87

7. 0.99920

9. 6.5×10^{-3}

11. 0.0002

13. 0.0027; 370 h

15. Plotted points are (35, 0.0051), (105, 0.0016), (175, 0.0005), (245, 0.0004), (315, 0.0003), (385, 0.0004), (455, 0.0007), (525, 0.0010), (595, 0.0011), (665, 0.0015)

17. 0.527; 0.368; 0.278

19. 0.257

21. Plotted points are (1300, 0.993), (1000, 0.972), (500, 0.576), (400, 0.332), (300, 0.093), (250, 0.027), (750, 0.891), (600, 0.750)

23. $n = 30, r = 5, T = 57$ h

25. $n = 9, r = 3, T = 160$ h

27. $n = 60, r = 30, T = 542$ cycles

INDEX